This book presents a comprehensive survey of the climatology and meteorology of Antarctica. As well as describing the climate which prevails in the Antarctic, the book also considers the processes by which this climate is maintained and explores links between the Antarctic and the global climate system.

The first section of the book reviews the methods by which we can observe the Antarctic atmosphere and presents a synthesis of climatological measurements. In the second section, the processes which maintain the observed climate are considered, starting with large-scale atmospheric circulation and moving through synoptic-scale weather systems to mesoscale and small-scale processes. The final section reviews our current knowledge of the variability of the Antarctic climate and considers changes that may occur in Antarctica as a result of 'greenhouse' warming. Throughout the book, the links between the Antarctic atmosphere and other elements of the Antarctic climate system (oceans, sea ice and ice sheets) are stressed and the processes which couple the Antarctic with the global climate system are examined. The instruments and platforms used in Antarctic climate studies are discussed (including automatic stations and international data centres), with special emphasis on the role of remote sensing from satellites and numerical modelling techniques.

This volume will be of greatest interest to meteorologists and climatologists with a specialised interest in Antarctica and the Southern Ocean, but it will also appeal to researchers in Antarctic glaciology, oceanography and biology. Graduates and undergraduates studying physical geography or the earth, atmospheric and environmental sciences will find much useful background material in the book.

Antarctic Meteorology and Climatology

Cambridge Atmospheric and Space Science Series

TITLES IN PRINT IN THIS SERIES

M. H. Rees
Physics and chemistry of the upper atmosphere

Roger Daley
Atmosphere data analysis

Ya. L. Al'pert
Space plasma, Volumes 1 and 2

J. R. Garratt
The atmospheric boundary layer

J. K. Hargreaves
The solar–terrestrial environment

Sergei Sazhin
Whistler-mode waves in a hot plasma

S. Peter Gary
Theory of space plasma microinstabilities

Martin Walt
Introduction to geomagnetically trapped radiation

Tamas I. Gombosi
Gas kinetic theory

Boris A. Kagan
Ocean–atmosphere interaction and climate modelling

Ian N. James
Introduction to circulating atmospheres

J. C. King and J. Turner
Antarctic meteorology and climatology

Antarctic Meteorology and Climatology

J. C. King
Ice and Climate Division, British Antarctic Survey

J. Turner
Ice and Climate Division, British Antarctic Survey

CAMBRIDGE UNIVERSITY PRESS
Cambridge, New York, Melbourne, Madrid, Cape Town, Singapore, São Paulo

Cambridge University Press
The Edinburgh Building, Cambridge CB2 8RU, UK

Published in the United States of America by Cambridge University Press, New York

www.cambridge.org
Information on this title: www.cambridge.org/9780521465601

First published 1997
This digitally printed version 2007

A catalogue record for this publication is available from the British Library

Library of Congress Cataloguing in Publication data

King, J. C. (John Christopher), 1955–
Antarctic meteorology and climatology / J. C. King and J. Turner.
p. cm. – (Cambridge atmospheric and space science series)
Includes bibliographical references and index.
ISBN 0 521 46560 5
I. Antarctica – Climate. 2. Meteorology – Antarctica. I. Turner, J. (John) II. Title. III. Series.

ISBN 978-0-521-46560-1 hardback
ISBN 978-0-521-03984-0 paperback

Contents

Preface

Over the last decade there have been great advances in our understanding of the meteorology and climatology of the Antarctic and a consequential increased awareness of the role of the continent in the global climate system. Improved technologies in electronics and communications have facilitated the development of unmanned observing systems, such as automatic weather systems and drifting buoys. These systems have been deployed in previously inaccessible areas, such as the remote interior of the continent and the winter sea ice zone, and have vastly increased the data available for meteorological and climatological studies. In parallel with this development, remote sensing techniques and technologies have been improved and measurements from satellite sensors have contributed greatly to our understanding of the structure and dynamics of Antarctic weather systems. Finally, numerical models have been used increasingly to study the behaviour of the Antarctic atmosphere on scales ranging from the mesoscale to the hemispheric.

In 1984 the late Werner Schwerdtfeger's survey of the meteorology and climatology of the Antarctic regions (Schwerdtfeger, 1984) was published. This important work covered the advances in our understanding of the Antarctic atmosphere that came about as a result of the International Geophysical Year (IGY) of 1957–8 and the First GARP Global Experiment (FGGE) in 1979. Although there have been no co-ordinated efforts comparable in scope to IGY and FGGE over the last decade, the technological advances described above and developments in our theoretical understanding of atmospheric processes have allowed great progress to be made in the study of the major role that the polar regions play in the global climate system, but there is also growing concern about the contribution that the Antarctic ice sheets may make to changes in global sea level. We therefore feel that it is appropriate to review once again our current understanding of the meteorology and climatology of the Antarctic and to assess the research priorities for the coming years.

When planning this book we had to make a number of decisions on the range of topics to be covered. Firstly, we decided that the geographical focus should be on the region south of 60° S since this is the usual geographical and political definition of Antarctica. However, this notional limit is somewhat arbitrary in a climatological context and we have not allowed it to prevent us discussing interactions between the Antarctic atmosphere and the circulation of lower latitudes. It was also necessary to set an upper limit to our region of interest. Although there has been a great deal of interest in the Antarctic stratosphere since the discovery of the so-called 'ozone hole' in the early 1980s, we decided that this region probably merits a book in its own right and our discussion generally stops at the tropopause. We also decided that a discussion of the long palaeoclimate records obtained from Antarctic ice cores was outside the scope of the book and we have restricted ourselves to considering the Antarctic climate of the present and recent past (with a speculative look to the future in Chapter 7). Finally, although our main concern is with the Antarctic atmosphere, we have touched on the other components of the climate system – the oceans, sea ice and ice sheets – where we felt it necessary, although our coverage of Antarctic oceanography and glaciology is far from complete.

Within the limits described above, we have attempted to produce a comprehensive survey of Antarctic meteorology and climatology, with the emphasis on developments over the last ten years. The book divides naturally into three parts. The first part, comprising Chapters 1–3, introduces the reader to the physical environment of Antarctica, discusses the instruments and platforms that have been used to study the Antarctic climate and concludes with a description of the physical climatology of the continent and surrounding oceans. In the second part we consider the processes that maintain the observed climate, starting, in Chapter 4, with the large-scale atmospheric circulation, then moving down in scale to synoptic scale weather systems (Chapter 5) and small and mesoscale phenomena (Chapter 6). Finally, Chapter 7 surveys our current knowledge of the variability of the Antarctic climate and the interactions between the Antarctic atmosphere and the circulation at mid- and low latitudes. We have included a detailed bibliography, with the emphasis on recent research.

Many people have helped in the preparation of this book. Dr I. Allison, Dr A. Carleton, Mr S. R. Colwell, Dr W. M. Connolley, Mr S. A. Harangozo, Dr T. A. Lachlan-Cope, Mr R. S. Ladkin, Mr D. W. S. Limbert, Mr S. Pendlebury, Dr M. Pook, Dr G. Marshall, Professor S. D. Mobbs, Professor E.M. Morris, Dr K. W. Nicholls, Dr D. A. Peel, Dr I. Simmonds, Mr J. P. Thomas and Mr M. Thorley read draft versions of sections of the book and provided may useful suggestions on how it could be improved. Figures were contributed by Dr P. S. Anderson, Dr D. H. Bromwich, Dr W. M. Connolley, Mr S. R. Colwell, Dr T. Lachlan-Cope, Dr S. Leonard, Dr T. R. Parish, Dr I. Simmonds and Dr D. G. Vaughan. Mr S. Wattam, who had spent one summer season forecasting at Rothera Station, provided much useful material on the forecasting techniques applied in the Antarctic and concerning the quality of the numerical products. Dr I. Allison, Dr M. Colacino and Professor C. R. Stearns provided information on automatic weather stations operating in the Antarctic. Ms J. Rae advised on the periods during which meteorological data were collected at the British Antarctic stations. Mr R. S. Ladkin processed most of the satellite images used throughout the book and Mr R. Missing dealt with the image optimization and printing. Mr A. Sylvester and Mr R. Missing co-ordinated the not inconsiderable task of preparing the figures. Mrs K. Salisbury provided secretarial support in the preparation of the final manuscript. Finally, we thank the Director of the British Antarctic Survey for allowing us to embark on this project.

Chapter 1

Introduction

1.1 Physical characteristics of the Antarctic

Antarctica comprises the area of the Earth south of 60° S and includes the ice-covered continent, isolated islands and a large part of the Southern Ocean. The continent itself makes up about 10% of the land surface of the Earth with the combined area of the ice sheets and ice shelves being about 14×10^6 km^2. It lies entirely within the Antarctic Circle, except for the northern part of the Antarctic Peninsula and the region south of the Indian Ocean (see the map on the end pages). However, the Pole of Relative Inaccessibility, which is the location furthest away on average from the coast in all directions, is close to 81° S, 58° W, some 900 km from the South Pole. This displacement of the highest plateau away from the geographic pole undoubtedly has an effect on the general circulation of the atmosphere and will be discussed in later chapters of the book.

In the eastern hemisphere, where the coastline follows the 62° S line of latitude for a considerable distance, the continent has a circular, symmetric form. The only major indentation in this part of the coast is Prydz Bay containing the Lambert Glacier–Amery Ice Shelf system which is located near 70° E. However, in the western hemisphere, the coastline is much more sinuous and is dominated by the northward extension of the Antarctic Peninsula and the two great embayments containing the Ross and Weddell Seas.

The Antarctic ice sheet consists of three distinct morphological zones – East (or Greater) Antarctica, West (or Lesser) Antarctic and the Antarctic Peninsula (see the map on the end pages). The largest is East Antarctica which covers 10.35×10^6 km^2 and is dominated by the high Antarctic plateau, which rises quickly inland of the coast. Here, except for the narrow coastal strip, the huge mass of ice is all above 2 km in elevation and small areas of the plateau even extend to just above 4 km elevation at Dome Argus. Inland of the coast, the gradients on the plateau are very small and to an observer the surface appears completely flat. Nevertheless, the surface is not smooth but often covered by sastrugi, which are ridges several centimetres high formed by wind erosion and snow deposition. The orientation of the sastrugi ridges is parallel to the prevailing wind and so provides a very valuable indicator of the surface wind flow in remote areas.

West Antarctica has an area of 1.97×10^6 km^2 and an average elevation of 850 m (see the topographic map in Figure 1.1). Although generally lower than East Antarctica, some areas reach more than 2000 m elevation on the plateau with nunataks (the exposed peaks of buried mountains) reaching to more than 4000 m. The Transantarctic Mountains, which stretch from Victoria Land to the Ronne Ice Shelf, separate East and West Antarctica and rise to a maximum elevation of 4528 m at Mt Kirkpatrick.

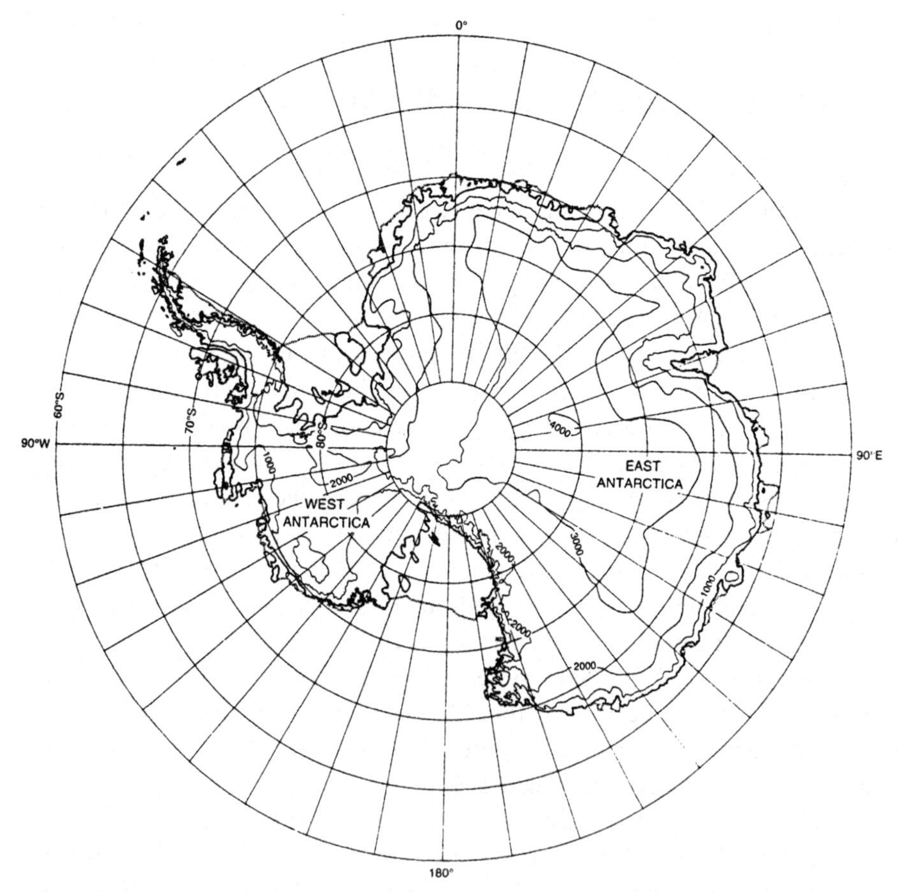

Figure 1.1 A topographic map of the Antarctic. Contours are drawn every 1000 m.

The third area is the Antarctic Peninsula which has an area of 0.52×10^6 km^2 and extends northwards from the main mass of the Antarctic continent. The Peninsula is a narrow mountainous barrier with an average width of 70 km and a mean height of 1500 m with the highest point being Mt Jackson at just over 3000 m. Such a high north–south barrier extending northwards to 63° S has a profound effect on the atmospheric circulation and results in markedly different climatic conditions over the Bellingshausen and Weddell Seas to the west and east respectively.

At 5440 m, Vinson Massif is the highest point in Antarctica. This peak is located in the Ellsworth Mountains between the Antarctic Peninsula and the Transantarctic Mountains, close to the western edge of the Ronne Ice Shelf.

The vast majority of the surface of Antarctica is covered with ice, with the continent as a whole containing around 30×10^6 km^3 or 90% of the world's fresh water ice. 86.5% of this is in East Antarctica with the remainder being held in West Antarctica (11.5%) and in the floating ice shelves (2%). Although Antarctica is best known for its uniform regions of ice and snow, there are many local topographic and morphological features on the surface, such as exposed rock outcrops (nunataks), which are the only visible evidence of the large mountain ranges that exist below the ice surface. This exposed rock makes up about 2.4% of the area of the continent. Figure 1.2 shows the Traverse Mountains extending through the ice sheet of the Antarctic Peninsula.

The Antarctic ice sheet varies in thickness across the continent and is of greatest depth in East Antarctica, some 400 km inland of Dumont d'Urville, where it has a thickness of 4776 m (Drewry, 1983). The mass of ice conceals the details of the land below, which is made up of sub-glacial mountains and lower elevation topographic features. In recent years these features have been investigated extensively by remote sensing techniques, such as radio echo sounding from both airborne and ground-based instrumentation.

The ice that builds up in the interior of the Antarctic gradually flows down to the edge of the continent in ice streams that move at speeds of up to 500 m per year. Some of the most rapidly moving ice streams are in West Antarctica and feed into the Ross and Ronne Ice Shelves. When the ice streams reach the edge of the continent either they form floating ice shelves or the ice breaks into blocks, which become the tabular icebergs that are a common feature of the Southern Ocean. Ice shelves are a permanent feature of certain parts of the Antarctic and make up 11% of the area of the Antarctic. The largest of these are the Ross Ice Shelf (0.54×10^6 km^2) and the Filchner–Ronne Ice Shelf (0.53×10^6 km^2). The Brunt Ice Shelf, on the eastern side of the Weddell Sea, is shown in Figure 1.3. It can be seen that this, like many of the ice shelves, is extremely flat and provides an excellent natural laboratory for studies concerning the very stable atmospheric boundary layer that develops during the Antarctic winter.

The edges of ice shelves make up 44% of the coastline of the continent, with

Figure 1.2 An example of a nunatak on the Antarctic Peninsula. Here the Traverse Mountains are largely buried and only the peaks are visible above the ice sheet. (Photograph courtesy Dr J. Paren.)

the remainder consisting of grounded ice walls (38%), ice streams and outlet glaciers (13%) and rock (5%) (Drewry, 1983). Most ice shelves are 100–500 m thick, although thicker areas do occur, such as near the seawards edge of the Filchner Ice Shelf. It has been shown that the thickness of the Filchner–Ronne Ice Shelf varies from about 300 m at the seawards edge to up to 1600 m at the grounding line (Jenkins and Doake, 1991).

The continent is ringed by a belt of sea ice that, at its maximum extent during the late summer, reaches 60° S around most of the continent and close to 55° S to the north of the Weddell Sea. Unlike in the Arctic, most of the sea ice melts by the later summer and the edge retreats back almost to the coast around much of East Antarctica. Most Antarctic sea ice is therefore first-year ice and is relatively thin. However, some multi-year ice is found in the western part of the Weddell Sea and along the coast of West Antarctica. Over the year the areal extent of the sea ice varies from a minimum of about 3×10^6 km^2 in March to a maximum of 20×10^6 km^2 in September–October (Budd, 1991).

1.2 A brief history of Antarctic meteorology

Belief in an isolated, high-latitude southern continent extends back to before the sixteenth century when such a landmass was shown on contemporary maps, such

Figure 1.3 The edge of the Brunt Ice Shelf on the eastern side of the Weddell Sea. The very flat conditions across the shelf are apparent, as is the iceberg calving that takes place at the edge.

as that of Orontius Finaeus (Fogg, 1992). However, this depiction of a Terra Australis was based on the theory that the two hemispheres were symmetrical rather than on any observational data, since there is no evidence that seafarers journeyed any great distance to the south in the Southern Hemisphere. Discovering the true nature of the Antarctic and surrounding ocean areas therefore had to await the first voyages into the high-latitude areas of the Southern Hemisphere by the great explorers of the seventeenth and eighteenth centuries. Although many early explorations were organised with the goal of making territorial claims, scientific discovery was regarded as important and Edmund Halley led the first scientific expedition into the Southern Hemisphere in 1698–1700. Here the primary goal was to study geomagnetism, but he reported floating ice at his most southerly location of 52° S, 43° W suggesting that significantly colder regions were to be found further south. Roggeveen confirmed this picture of cold conditions during his 1721–22 voyage which reached 62° 30' S in the western Pacific Ocean and encountered extensive, impenetrable sea ice. However, it was the voyages of Cook (1772–75) and Bellingshausen (1819–21) circumnavigating the continent at high southern latitudes that gradually accumulated evidence for the existence of the southern continent. As they crossed the oceanic Antarctic Polar Front (see Section 3.5.2) and ventured further south, encountering extensive pack ice, the dream of a temperate paradise, that had been put forward in certain quarters, receded and our current image of a frigid environment became accepted.

Meteorological observations were made on many of the early expeditions, with these becoming increasingly sophisticated and more instrumented with time. During the voyage by Captain Cook, observations of air temperature were made that showed the position of the atmospheric Polar Front. Surface air pressure was recorded by Bellingshausen and his southerly circumnavigation of the continent was the first to register the increasingly high pressure found at the more southerly latitudes. In 1829 Henry Foster left a minimum thermometer on Deception Island, which was recovered during the 1841–42 voyage by Smyley on his ship the *Ohio*. It registered a minimum temperature of −5°F (−20.3°C), which, although probably too high from our current knowledge of the climate of the island, still indicated the extremes that occurred at these southerly latitudes.

During the latter part of the nineteenth century there were large numbers of sealers in the Southern Ocean frequenting the sub-Antarctic islands and these were the first people to land on the Antarctic continent and the first to winter in the Antarctic region. However, scientific activity in the area also increased in association with the transit of Venus which took place in 1874 and the first International Polar Year of 1882–83, which involved the establishment of a number of observing stations on the peri-Antarctic islands. These stations made meteorological observations, although for only a relatively short period. Most of the stations established as part of the International Polar Year were in the Arctic but observations were collected at sites on Tierra del Fuego and South Georgia, providing some high latitude data in the Southern Hemisphere.

The first set of meteorological observations collected throughout an Antarctic winter were made when de Gerlache's ship the *Belgica* was trapped in sea ice at 72° S, 90° W during the winter of 1898. This allowed the scientists Arctowski and Dobrowolski to make a range of meteorological observations of cloud, frost and optical properties. Just a year later in 1899, the first land station on the Antarctic continent was established at Cape Adare by the British Antarctic Expedition lead by Borchgrevink. Although concerned primarily with exploration, they made a range of meteorological observations. In the years up to the First World War there were many important expeditions to the Antarctic, including those of Shackleton, Scott, Amundsen, Filchner and Mawson. Many of these expeditions established stations that made meteorological observations, allowing the climatology of the fringes of the continent to be established. In fact the number of stations on the continent had increased to the point at which simultaneous observations were available for all coastal sectors of the continent for the years 1902–04 and 1910–12. Although during these years the emphasis was on the continent, in 1903 the Scottish National Antarctic Expedition established a meteorological observatory on Laurie Island, South Orkney Islands. This station was handed over to the Argentinian Government in 1904 and has been operated since that time as Orcadas Station, providing the longest climatological record of any of the sub-Antarctic Islands. During this time, one of the few stations operated on the eastern side of the Antarctic Peninsula was established in 1902 by Nordenskjöld and the Swedish South Polar Expedition. This was at Snow Hill Island and was operated longer than intended, until November 1903, since the relief ship *Antarctic* was crushed by sea ice. The meteorologist in this party was Bodman who ran a comprehensive observing programme over the three-year period of operation.

This flurry of activity at the start of the twentieth century marked the start of scientific investigation of the Antarctic continent itself as stations were first operated year-round in the coastal region. The period is best remembered for the competition in the quest for the Pole between the teams led by Scott and Amundsen. However, like most expeditions, Scott's party consisted both of explorers and of scientists and his 1910–13 expedition in particular had a strong meteorological observing programme, which was organised by G. C. Simpson, who was later to become the Director of the British Meteorological Office. This ensured that reliable meteorological data were collected and made available to the larger meteorological community upon the completion of the expedition through publication in the literature. The expeditions of this period also discovered completely new phenomena in the Antarctic, such as the strong katabatic winds which occur in many of the coastal areas of the continent. These katabatic winds were experienced by the Australasian Antarctic Expedition of 1911–14, led by Sir Douglas Mawson, at their permanent base at Cape Denison, Adélie Land, where they encountered remarkably strong, persistent winds from the interior that, during the months of March and April 1912, blew nearly continuously at 60–80 miles per hour with gusts frequently over 100 miles per hour. Although the early expedition meteorologists made only surface measurements, data on conditions above the surface could be inferred by indirect means, such as through observing the smoke plume from Mt Erebus, Ross Island which gave an indication of the wind direction at the level of the top of the mountain. The first *in situ* upper air measurements made in the Antarctic were carried out by scientists on Scott's first expedition and from Drygalski's ship the *Gauss*, which was beset in ice throughout the winter of 1902. This latter work was carried out using a tethered hydrogen balloon with the observations relayed to the ship via a telephone allowing data to be collected up to a height of 480 m. Unmanned upper-air measurements were soon to follow with the release by Simpson of 14 meteorographs in 1911.

Data from the early expeditions were also of great value in the early attempts to understand the general circulation of the Southern Hemisphere and the part played by the Antarctic. Even though the data in the high-latitude areas were very sparse, observations from the many ships which plied the waters of the mid-latitude areas of the Southern Hemisphere and observations from over the continent were used to construct synoptic charts for the period October 1901 to March 1904. These charts were used by Meinardus to determine the basic components of the atmospheric circulation across the hemisphere (Meinardus and Mecking, 1911).

During the inter-war years the most important expeditions, insofar as meteorological research and operations were concerned, were led by the American Admiral Byrd in the Ross Sea sector of the continent. Beginning in 1928, a series of stations called Little America were established on the Ross Ice Shelf at the Bay of Whales from where meteorological observations were made. Here, in 1940, the first Antarctic radiosonde ascent was made using a hydrogen-filled balloon so that for the first time the vertical structure of the Antarctic atmosphere could be investigated (Court, 1949). Admiral Byrd also spent a five-month period alone at a hut 160 km south of Little America II, making the first long series of observations from a station in the interior of the continent.

After the Second World War, the establishment of further permanent scientific research stations on the sub-antarctic islands and on the Antarctic Peninsula gave a great boost to understanding the meteorology of the continent. These stations maintained full synoptic observing programmes that allowed the production of surface charts on a regular basis. The chart series produced have been reviewed by Schmitt (1957).

Possibly the most influential scientific expedition of that time was the Norwegian–British–Swedish Expedition of 1949–52 to 'Maudheim', Dronning Maud Land, on the eastern shore of the Weddell Sea. The quality and quantity of meteorological research, particularly the studies of radiation and energy balance, the boundary layer and aerology had a great influence on meteorologists planning and involved in the International Geophysical Year (IGY) of 1957–58. Throughout the early 1950s, however, there were a number of disputes regarding territorial claims, especially on the Antarctic Peninsula. However, under the impetus of the IGY, local intergovernmental rivalry was supplanted by international co-operation and paved the way for the current state of affairs in which the Antarctic is freely available to all nations for scientific investigations. The IGY brought about a significant advance in the collection of *in situ* Antarctic observations with the establishment of over 60 research stations by 12 nations. In the years prior to the IGY, the bases established included Mawson (1954), Horseshoe Island (1955), McMurdo (1956), Halley Bay (1956), Dumont d'Urville (1956), Mirny (1956), Davis (1957) and Wilkes (1957) and the first in the interior of the continent at Vostok (1958) and the South Pole (1957). This resulted in additional surface observations and also an increase in the radiosonde data collected. Radiosondes had increasingly been flown on an experimental basis in the Antarctic following the pioneering work at Little America during the Second World War. In the post-war years further ascents had been made at Maudheim (1950–52) and Port Martin (1951–52), but the IGY prompted the establishment of routine upper air programmes at many stations so that, for the first time, a synoptic picture of upper air conditions could be obtained across the continent.

Until the middle of the twentieth century it was almost impossible to forecast the weather for the Antarctic because of the lack of *in situ* data, the wide spacing of the stations making synoptic observations and the poor communications throughout the Antarctic. However, by the mid-1940s there were enough observations being received from the Antarctic Peninsula for reasonable surface analyses to be drawn, at least for the northern part of the area, so that forecasts could be issued for the region. These data were used by the British, initially through the wartime Operation Tabarin and later the Falkland Islands and Dependencies Meteorological Service, which was established in 1950, to provide forecasts for the peri-Antarctic islands and the area of operation of the whaling fleet which extended to 65° S.

In 1957 the International Weather Central Forecast Office was established at Little America V (Bay of Whales) where 12-hourly surface synoptic charts for the Antarctic and southern oceans were regularly prepared and, whenever possible, 500 hPa and 300 hPa charts also (Astapenko, 1964). Little America V was closed at the end of 1958 and the work transferred to McMurdo and thence to the Australian Weather Bureau in Melbourne.

The major advance in analysis and short-period forecasting for the Antarctic came with the introduction of operational polar orbiting satellites in 1967. These satellites provided a synoptic view of the whole continent and greatly aided the detection of major weather systems and fronts, especially over the Southern Ocean and the continental interior. Accurate medium term (1–3 day) prediction of atmospheric conditions in the Southern Hemisphere high latitudes became possible with the introduction of global numerical models as powerful computers became available in the 1970s. Since then, surface and 500 hPa Southern Hemisphere analyses and forecasts have been issued at 12-hourly intervals over the Global Telecommunications System (GTS) from Melbourne. Other meteorological centres with global analysis capability, mainly in the Northern Hemisphere, have since the 1980s also provided Southern Hemisphere analyses and 1–3 day period forecasts that have proved beneficial to logistic operations around the continent.

The IGY surface and upper-air data were published in standard form by the World Meteorological Organisation, while the South African Weather Bureau produced a series of surface pressure and 500 hPa charts based on these data. The IGY data were used extensively in studies of the atmospheric circulation over the high southern latitudes and are discussed in the relevant sections of the book. Many of the IGY stations remained in operation after the end of the project and some are still in operation today, forming a valuable network of stations for monitoring climate variation in the Antarctic. The success of the IGY was also a major factor in establishing a firm foundation for international co-operation in Antarctic science and was important in paving the way for the Antarctic Treaty, which came into force in 1961. It also laid the foundations for the Scientific Committee for Antarctic Research (SCAR), which continues to provide an important framework for co-ordination and co-operation in Antarctic research and logistical operations.

The last few years have been characterised by a high degree of international co-operation between research scientists and operational meteorologists, resulting in the establishment of complex operational communications systems for the transfer of data and the organisation of major research projects. Today, many nations operate forecast offices on the research stations which support the logistical operations that are carried out by ships, aircraft and surface-based field parties. The forecasters here use satellite imagery, *in situ* data and the output of numerical models in a similar way to a forecaster in the extra-polar regions and face many of the forecasting problems experienced in mid-latitudes. However, a number of problems specific to the polar regions, such as forecasting clear sky precipitation, katabatic winds and ice fog, are found in certain areas. The World Meteorological Organisation has been responsible for the standardisation of meteorological reporting codes used worldwide and their Working Group on Antarctic Meteorology, established in 1964, played a major role in improving the transfer of observational data within the Antarctic and in enabling this to reach the major analysis and forecast centres outside the continent in time to be of value in operational meteorology.

The largest international meteorological research project organised to date was the intensive data gathering and analysis exercise known as the First GARP Global Experiment (FGGE). This took place from 1 December 1978 to 30 November 1979, when many additional observing systems, such as drifting buoys, automatic weather stations and constant level balloons were deployed in the high southern latitudes. The FGGE allowed the numerical analysis and assimilation schemes to be improved and also provided guidance on how the observing network should be developed. Since the FGGE the experiments conducted in the Antarctic have been of more limited extent and have consisted of regional investigations into katabatic winds, the surface boundary layer, mesocyclones and the interaction between weather systems and pack ice movement and distribution, such as in the 1986 Winter Weddell Sea Project (Limbert *et al.*, 1989).

1.3 The role of the Antarctic atmosphere in the global climate system

The global atmosphere may be thought of as a heat engine, driven by excess heating in the tropics and cooling in high latitudes, with the global circulation transporting heat between the tropics and the poles. Antarctica is the heat sink for the Southern Hemisphere and thus exerts considerable control over the circulation of the atmosphere at high and mid-latitudes. The net cooling at high latitudes results partly from simple geometry – these regions receive less solar radiation at the top of the atmosphere averaged over the year than do the tropics – but the presence of snow and ice in the polar regions gives rise to a strong positive feedback. Snow and ice surfaces have a high albedo and thus enhance the cooling at high latitudes by reflecting more solar radiation than would bare rock or ocean surfaces. A decrease (increase) in temperature in the polar regions is likely to lead to an increase (decrease) in ice coverage, which will then tend to amplify the original temperature change. This ice–albedo feedback mechanism can be an important process for generating variations and changes in the polar climate that will have an impact on the global climate system. The high-latitude continental areas of the Northern Hemisphere have large areas of seasonal snow cover that can respond rapidly to temperature changes and thus contribute strongly to the ice–albedo feedback. The Antarctic continent, in contrast, is covered by permanent (on a timescale of centuries, at least) ice sheets and in the South Polar regions it is the sea ice around Antarctica which has the greatest potential to contribute to climate variability and change. Unlike the pack ice of the Arctic ocean, it is unconstrained by surrounding land masses and exhibits considerable year-to-year variations in extent that can affect the atmospheric circulation of the Southern Hemisphere.

The Antarctic continent is an important feature in a hemisphere otherwise largely devoid of large-scale orography. The Antarctic Peninsula provides an effective barrier to zonal flow in both atmosphere and ocean and thus exerts an important influence on the Southern Hemisphere's circulation. Of even greater significance are the high ice sheets of the continental interior. Strong cooling over the surface of the dome-shaped ice sheets generates a persistent katabatic circulation, with cold surface air flowing out from the interior towards the coasts. In Chapter 4 we shall see that this flow plays a central role in shaping the high-latitude circulation of the Southern Hemisphere atmosphere and that its influence may extend into mid-latitudes. Since the Antarctic orography is not symmetric about the South Pole, the Antarctic acts as an efficient generator of long Rossby waves. These may propagate northwards and affect the mid-latitude circulation (see Chapter 7).

Antarctica also exerts an important influence on the global ocean circulation. The seas surrounding the Antarctic continent are known to be one of the most important areas for the production of dense, saline ocean bottom water. Cold air flowing outwards from the continent over the Southern Ocean rapidly cools the surface waters, promoting downwelling. This tendency is enhanced by the rejection of dense, brine-rich water as sea ice forms on the cooled ocean surface. Melting at the base of ice shelves contributes further to the formation of dense bottom water. Atmospheric processes in Antarctica thus play an important role in the formation of ocean bottom water, which moves northwards and affects the ocean circulation on a global scale (see Section 3.5).

The Antarctic atmosphere can also have an impact on the global oceans in another way. The Antarctic ice sheets contain about 30×10^6 km^3 of ice. If all this ice were to melt, global sea levels would rise by about 65 m. The mass of the

Antarctic ice sheets is maintained by a close balance between snowfall over the continent and the discharge of ice across the coast. Both of these processes may be affected by changes in atmospheric conditions. Precipitation over Antarctica is sustained by the transport of moist air southwards from mid-latitudes (Section 4.4) whereas ice discharge rates may be affected by changes both in atmospheric and in oceanic temperatures. Changes in the local climate of Antarctica will thus have global implications through their effect on the Antarctic ice mass budget and consequent effects on global sea level.

Chapter 2

Observations and instrumentation

2.1 Observing in the Antarctic

The Antarctic environment presents a number of challenges to those involved in making meteorological observations. Extreme cold, strong winds, icing and blowing snow can all adversely affect instruments, making the observer's task a difficult one. In the following section we shall look at problems associated with making measurements using standard meteorological equipment. However, before doing this, we should remember that some important observations cannot be made instrumentally and we still rely on the skill of the human observer for accurate recording of these elements.

Foremost amongst these elements is the observation of cloud cover, height and type. Although satellites can now provide useful cloud climatologies (see Section 3.4) the detail provided by the human observer is still essential for many studies. Accurate estimation of cloud cover at Antarctic stations, as elsewhere, depends on having well-trained and experienced observing staff. Observers at Antarctic stations are usually employed on short-term contracts, spending only one or two years in Antarctica, and are often less experienced than observers working for national weather services. Without guidance from more experienced colleagues, it can be difficult for them to recognise any subjective bias that may be creeping into their observations. These problems can be overcome to some extent by ensuring that incoming observers have a reasonable 'hand over' period with outgoing staff. The British system of two-winter contracts ensures that this happens, but it is becoming increasingly difficult to find observers prepared to commit themselves to this length of time in Antarctica. Cloud heights are particularly difficult to estimate over the flat, featureless ice sheets and ice shelves and, during the long polar night, estimation of cloud cover can be problematical. However, in compensation, cloud searchlights can be used to make objective measurements of cloudbase height. 'White-out' conditions caused by blowing snow can make cloud observation impossible, leaving the observer no option but to record 'sky obscured'.

Estimation of visibility can be difficult over ice sheets, where there are few natural reference points. It is often necessary to construct artificial visibility markers at various ranges. Even with the addition of such markers, many stations only report visibility on the coarse scale normally used for observations at sea.

The third important subjective observation is the reporting of present and past weather types. Such observations are useful for case studies of weather systems and the frequency of occurrence of certain weather types can be a useful proxy for precipitation amount in climatological studies (see Section 3.4). Once again, observer training and experience are paramount. Observers must learn to distinguish between blowing snow and true precipitation and must be able to categorise precipitation as light, moderate or heavy. They must also learn to recognise phenomena such as clear-sky precipitation ('diamond dust') that rarely occur outside the polar regions.

2.2 Instruments for meteorological measurements

2.2.1 Pressure

There are no particular problems associated with the measurement of atmospheric pressure at Antarctic stations. In recent years conventional mercury and aneroid barometers have been increasingly replaced by transducers connected to automatic logging systems in the general move towards automation. At stations in very windy areas, static pressure anomalies associated with windflow around the station buildings can become significant and some thought may need to be given to siting the barometer.

Atmospheric pressure observations are conventionally reduced to sea level values. This presents no problem at coastal stations but, over the high interior of the continent, the correction becomes very large. It is standard practice to reduce pressure observations to mean sea level (MSL) by assuming that the (hypothetical) atmosphere below the station is isothermal at the station surface temperature. The presence of a strong surface inversion over the Antarctic interior means that surface temperatures are anomalously cold compared with the bulk of the atmosphere; hence, when pressures are reduced to MSL, an anomalously strong anticyclone is seen over the continent. Juckes *et al.* (1994) suggest using the atmospheric temperature just above the inversion to avoid this problem. However, the presentation of MSL pressure fields over the continent should be avoided if possible; analyses of the 500 hPa height field are to be preferred since this is the lowest standard level which is everywhere above the surface of Antarctica. When pressure observations from interior Antarctic stations are required for research purposes, efforts should be made to obtain the original station level pressure data since the procedures for reduction of pressure to MSL vary from nation to nation.

2.2.2 Temperature

Thermometers at Antarctic stations are usually housed in screens of conventional design, which are not entirely satisfactory. Louvred screens can become choked with blowing snow and, during the Antarctic summer, conventional screens are subjected to high levels of solar radiation, both from above and reflected from the snow surface, and may not provide sufficient shielding, particularly when the wind speed is low. Reports of high maximum temperatures during the summer months need to be treated cautiously for this reason. At some stations the thermometers are placed in artificially aspirated radiation shields to overcome this problem.

Traditionally, mercury- and alcohol-in-glass thermometers were used for temperature measurements, the latter for temperatures below about −30°C. In automated measurement systems these have been replaced by platinum resistance thermometers or semiconductor temperature transducers.

2.2.3 Humidity

Humidity is one of the more problematic elements for the Antarctic meteorologist to measure. Because the saturation vapour pressure of water is a strong function of temperature, the absolute humidity of Antarctic air is generally very low, making the application of traditional psychrometric techniques difficult. At a temperature of −10°C and 1000 hPa pressure, air with a relative humidity of 65% generates an ice-bulb temperature depression of 0.92°C. At −30°C this is reduced to 0.06°C. Clearly, wet-and-dry bulb psychrometry is of little use at these low temperatures. Maintaining an ice bulb on a thermometer requires considerable skill and patience and cannot be automated. A number of humidity transducers are available for use in automated systems. These include carbon film hygristors and polymer capacitative sensors. Such sensors are generally supplied with calibrations appropriate to temperate conditions and additional calibration may be required to extend their range to low temperatures (Anderson, 1994). For very accurate measurement of humidity at low temperatures it may be necessary to use a cooled-mirror frost point hygrometer. However, such instruments are expensive and require careful maintenance so they are rarely used for synoptic observations.

When using humidity observations from Antarctic stations it should be noted that the standard meteorological definition of relative humidity is

$$RH = \frac{e}{e_s} \times 100\% \tag{2.1}$$

where e is the water vapour pressure and e_s is the saturation vapour pressure over *liquid* water. In other words, relative humidity is expressed relative to saturation with respect to liquid water, even when the temperature is below freezing.

2.2.4 Wind velocity

Wind speed and direction are usually measured using cup anemometers and wind vanes at the WMO standard height of 10 m. Alternatively a propellor anemometer–vane combination may be used. All mechanical wind speed and direction sensors are prone to icing in the Antarctic environment, necessitating frequent manual attention if records are to be kept accurate. Gates and Thompson (1985) studied the icing of anemometers in a wind tunnel and concluded that large anemometers were less affected than smaller instruments. Ultrasonic anemometers and pressure anemometers offer an attractive alternative to traditional sensors since they have no moving parts and can be kept clear of ice by the application of electrical heating. However, ultrasonic anemometers are badly affected by blowing snow (King and Anderson, 1988).

2.2.5 Precipitation

No truly satisfactory technique exists for measuring precipitation at Antarctic stations. Conventional snowgauges are almost useless since the catch is greatly affected by wind speed and it is impossible to discriminate between precipitation and blowing snow. On the high plateau most precipitation occurs as a near-continuous fall of minute ice crystals, totalling less than 50 mm water equivalent per year.

Daily measurements of snow accumulation using an array of snow stakes can provide some indication of precipitation providing that compaction, evaporation and removal of blowing snow are taken into account (Limbert, 1963). As mentioned earlier, observations of the frequency, type and intensity of precipitation can be used as an indicator of precipitation variability (Turner and Colwell, 1995) but it is difficult to convert them to absolute totals without some means of calibration. Recently developed optical precipitation sensors offer some potential for improving Antarctic precipitation measurements.

2.2.6 Radiation

The only radiation measurement carried out routinely at most Antarctic stations is the recording of the duration of bright sunshine for climatological purposes. Traditionally, these measurements were made using Campbell–Stokes recorders, which burn a track onto a recording card. For stations south of the Antarctic Circle, two recorders are required during the period of 24-hour daylight. Nowadays, these instruments are increasingly being replaced by photoelectric devices and automatic recorders.

Long-term measurements of solar and long-wave radiation have been made at a few stations using pyranometers, pyrgeometers and net radiometers. The problems involved in making these measurements in Antarctica are similar to those faced elsewhere. However, special attention needs to be given to keeping the instrument domes free of frost and rime and care needs to be taken to avoid the instruments or their supports shading themselves during the period of 24-hour daylight.

Interest in potential increases in ground-level ultraviolet radiation associated with the springtime depletion of stratospheric ozone over Antarctica has prompted the installation of ultraviolet radiation sensors at several stations. These vary in complexity from simple broad-band pyranometers to sophisticated spectral measurement systems.

2.2.7 Upper-air measurements

Since the International Geophysical Year, long-term aerological soundings have been carried out at about 17 Antarctic stations. At the time of writing, 13 of these are still in operation. Apart from Amundsen–Scott (South Pole) Station, all are situated on the coast of the continent and there are no stations in West Antarctica between 60° W and 170° W.

Stations that have run radiosonde programmes are listed in Appendix A. Since Antarctic upper-air stations have been operated by a number of nations, a variety of radiosondes and ground equipment has been used in the region. Connolley and King (1993) listed the equipment in use at Antarctic radiosonde stations during the 1980s. Users of Antarctic upper air data need to be aware that systematic differences between stations may result, at least in part, from the use of different equipment. Additionally, variations in operating and data processing procedures adopted by the various nations can contribute further to inter-station differences. Examples of such procedural differences include the schemes used to correct temperature measurements for radiation errors and the acceptance criteria applied to humidity measurements. Current American practice is not to report humidities when the temperature is below −40°C or the indicated relative humidity is below 20% (Elliot and Gaffen, 1991). The effects of these differing practices may need to be assessed in any study using data from a

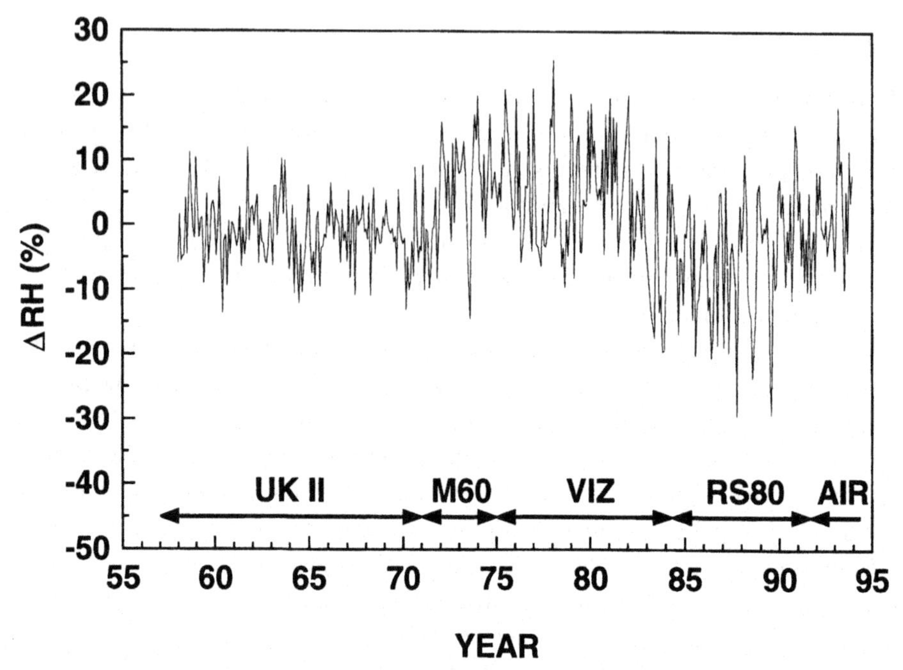

Figure 2.1 Monthly mean 850 hPa humidity anomalies relative to the 1957–1994 mean at Halley Station. Radiosonde types used during the period of the record are shown along the time axis. UK II, UK Meteorological Office Mk IIb sonde; M60, Graw M60 sonde; VIZ, VIZ sonde; RS80, Vaisala RS80 sonde; and AIR, AIR Intellisonde.

number of upper-air stations. Even in single-station climatological studies, changes in sonde type or in procedures can introduce spurious variations into the climatic record, as demonstrated by Figure 2.1, which shows monthly mean 850 hPa relative humidity anomalies at Halley Station, where five different radiosonde types were used between 1957 and 1993. Although there is considerable natural variability in humidity on all timescales, it is fairly clear that there are step changes both in the mean humidity and in its month-to-month variability coincident with the times that the sonde type changed.

Several techniques have been used for tracking radiosonde balloons for windfinding purposes. Radar tracking is the most reliable wind-finding technique and has the advantage that it provides an independent estimate of the balloon's altitude, which may be used to check the accuracy of the radiosonde pressure and temperature readings if these are used to compute altitude using the hydrostatic equation. However, wind-finding radars are expensive to purchase and difficult to maintain with the limited resources available at an Antarctic station. The radiosondes must carry a radar reflector, which increases the cost of each sounding, both by virtue of the cost of the reflector itself and because of the cost of the additional gas required to provide increased lift. In recent years there has been a move to alternative forms of windfinding. Radiotheodolites (highly directional steerable antennae) may be used to track the radiosonde. Since they only give the sonde azimuth and elevation, the radiosonde height must be deduced from pressure and temperature readings in order to calculate the sonde position and hence the wind velocity. At elevation angles of less than 15° or so, errors in the computed winds become very large because of the unfavourable geometry and possible multipathing of signals. Nevertheless, radiotheodolite systems are considerably cheaper and simpler than wind-finding radars and have been used increasingly in recent years. A development of this system which has been used for some time at stations operated by the USSR and subsequently by Russia is the addition of a transponder unit to the radiosonde. This gives a direct reading of slant range, making the radiotheodolite system equivalent to radar.

Manual tracking of balloons using optical theodolites has been used at a number of Antarctic stations. The disadvantages of the technique are obvious – it is impossible to obtain winds under cloudy conditions or during the long polar night. Nowadays, use of this technique is mainly restricted to the study of boundary-layer winds at temporary field stations where more sophisticated equipment is not available.

At some Antarctic stations, radiosonde balloons are tracked using the Omega very low-frequency (VLF) navigation aid. Omega is a positioning system that uses phase delays from a global network of VLF transmitters to compute position. A VLF receiver in the radiosonde receives these signals and telemeters them to the ground station, where position and winds are computed. The system is very simple to operate and does not require complex ground equipment. Unfortunately, Omega reception in the Antarctic is poor because VLF signals propagating over the ice cap are strongly absorbed. Even at coastal stations, signals from only a few Omega stations are received. In order to calculate accurate winds, signals from at least three stations are required and the great circle paths from these stations to the radiosonde station must have a reasonable angular separation. It is impossible to satisfy these criteria at some Antarctic locations. A further disadvantage of the Omega wind-finding system is that it is necessary to average the Omega signals over quite a long period to compute position, thus smoothing out any fine structure in the wind profiles. In the future, radiosondes carrying Global Positioning System (GPS) satellite navigation system receivers may offer the ease of use of Omega wind-finding with none of its disadvantages.

In addition to routine radiosonde programmes, a number of special upper-air investigations have been carried out at Antarctic stations. Ozonesondes have been used to measure stratospheric ozone profiles at Faraday, Halley and Marambio. During the winter of 1963, radiometersondes were launched at three-day intervals from Amundsen–Scott, Byrd, Hallett and Wilkes Stations to study atmospheric radiative cooling rates (Kuhn *et al.*, 1967). Similar measurements have been made from Syowa Station. At the time of writing (1995) sondes carrying frost point hygrometers are being launched from the Amundsen–Scott

Station in order to make accurate measurements of water vapour levels in the upper troposphere and stratosphere.

Strong winds, blowing snow and winter darkness all contrive to make the launching of radiosonde balloons a difficult task in Antarctica and it is a tribute to the skill and dedication of the observing staff that few soundings are missed because of adverse conditions. However, soundings are most often missed during periods of strong winds and, as discussed by Bromwich (1979), in any climatological study using upper-air data it is important to ensure that this selective sampling is not biasing the results. During the Antarctic winter, when upper-tropospheric temperatures drop below −80°C, radiosonde balloon envelopes can become brittle, causing the balloon to burst before reaching the 100 hPa level. Experience at British Antarctic Survey stations has shown that winter balloon performance can be greatly improved by dipping balloons into a mixture of oil and aviation fuel prior to inflation. Balloons treated in this way reach well into the stratosphere before bursting.

2.3 Automatic weather stations

The harsh environment of the high southern latitudes means that it is very expensive to establish and operate manned research stations in the Antarctic and especially so on the high interior plateau of the continent. Most stations are therefore located in a ring around the coastal margin with, even today, distances of several hundred kilometres between stations. Although data from these bases are able to provide a broad indication of the synoptic-scale circulation for operational analysis, there are not enough data to resolve the often complex structure of weather systems around the Antarctic or provide details of the many mesoscale systems observed on satellite imagery. For climatological and research studies the observing network is also very coarse and the many local climatic regimes and variations in the wind field resulting from the sharp topographic gradients cannot be resolved with the present data. The concept of using un-manned automatic weather stations (AWSs) as a supplement to the data from the research stations is therefore very attractive, provided that the systems are reliable and the data easily collected. Early experiments with automatic observing instruments were carried out by the Australian National Antarctic Research Expedition in 1955 when equipment was installed at Mt Henderson and on an island north of Mawson Station. However, it was not possible to receive the data in real time and the observations could only be used for research and climatological investigations.

The first Antarctic AWSs that transmitted data via a satellite link were designed and built by A. Peterson's group in the Radio Science Department at Stanford University and deployed during February 1980 in the Adélie Land region of East Antarctica. They were installed in a chain extending between

Dome C and the French Dumont d'Urville station, with the data being used in conjunction with aircraft observations and numerical model output as part of the research project IAGO (Interaction Atmosphère–Glace–Océan) to examine the katabatic winds. Since that time, the AWS activities of the USA's Antarctic Program have been taken over by the Department of Atmospheric and Oceanic Sciences at the University of Wisconsin-Madison. Since 1980 more than 60 of the Madison AWSs have been deployed on the continent and surrounding islands and the experience gained has allowed the systems to be developed into reliable and efficient platforms for the collection of data in remote regions. Appendix B lists the AWSs known to have been deployed in the Antarctic by various nations for operational and research activities.

2.3.1 Hardware and data collection

Because the Madison AWSs are installed in many locations in the Antarctic and their observations have been used in a large number of investigations concerned with Antarctic meteorology and climatology (Stearns and Wendler, 1988), their design and instrumentation will be described here. Figure 2.2, which has been

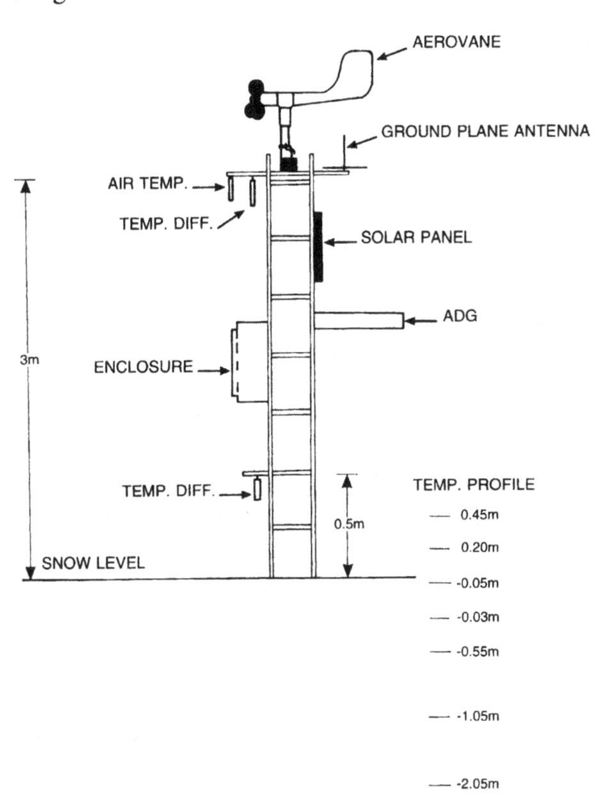

Figure 2.2 The layout of a Madison AWS. After Keller *et al.* (1990).

reproduced from Keller *et al.* (1990), shows a typical Madison system with a full set of instrumentation. The AWS consists of a 3 m high tower with an enclosure for the controlling hardware at the mid-point and most of the instruments on a horizontal boom at the top. The sealed enclosure contains a computer that controls the observing schedule and the radio transmitter which uplinks the data to a polar orbiting satellite. The whole system is powered by batteries located at the base of the tower, which are recharged during periods of sunlight by a north-facing solar panel fixed near the top of the tower. The instruments fitted to a particular AWS depend on where it is to be installed and the applications of its data. However, the following are the most commonly encountered instruments on the Madison AWSs.

(i) A Bendix or Belfort aerovane or, on more recent systems, a R. M. Young wind monitor for measuring wind speed and direction. This is located on the top of the tower and is capable of measuring wind speeds of up to 64.8 m s^{-1}.

(ii) A platinum resistance thermometer at the top of the tower to measure air temperature. This can give a resolution of 0.124°C. On some AWS the temperature difference between the top of the tower and 50 cm from the surface is measured using a pair of thermocouples with the difference being measured with a resolution of 0.05°C.

(iii) A Vaisala HMP-31UT or HMP-35A humidity sensor mounted on the top of the tower. This sensor has a resolution of about 1%.

(iv) A Paroscientific Digiquartz pressure transducer located in the enclosure close to the mid-point of the tower. The accuracy of the sensor is ±0.25 mbar.

(v) A series of eight temperature sensors extending down into the snow to a depth of about 4 m to provide a temperature profile for studies of heat conduction.

AWSs are installed at their remote locations during the austral summer by field parties using helicopters or aircraft equipped with skis. With the small, but nevertheless significant, amounts of snowfall that occur in the Antarctic, the AWSs gradually become buried if left unattended and eventually fail. However, most AWSs are visited periodically when the whole system is raised so that the sensors are once again at their nominal height and any failed or suspect components can be replaced.

The observations from the Madison AWS are collected via the ARGOS data collection system (DCS) on the USA's NOAA series of polar orbiting meteorological satellites (Schwalb, 1982). The ARGOS system receives information uplinked by the AWSs at a frequency of 401.650 MHz during overpasses of the satellites. These data are sent as 256 bit data words transmitted over a 1 s period and repeated every 200 s. The data transmitted are updated every 10 min so that this is the highest frequency of observations that can be collected. The data transmission contains the current and four previous sets of temperature, pressure and wind observations made at T−10, T−20, T−30 and T−40 min. Relative humidity and temperature difference between the two sensors are sent for the present time and T−20 and T−40 min. Obviously, data can only be collected during the time that a NOAA satellite is within the line-of-sight of the AWS and this overpass time can vary from around 2 min, with a low elevation pass, to about 20 min when the satellite passes nearly overhead of the site. The data transmission times from the various AWSs are staggered so that there are no conflicts between units and the data collection is maximised. With the AWSs at latitudes south of about 75° S, the approximately 100 min orbital period of the satellites and the maintenance of two NOAA spacecraft in orbit at all times mean that there is usually one satellite within view of each AWS about every 50 min, so that data losses are minimal.

The NOAA spacecraft download the data collected by the DCS on each orbit to receiving stations in North America and Europe. From there the data are passed to Service ARGOS in France who make tapes of the raw data collected available to the Madison group on a monthly basis.

The DCS on the NOAA satellites also immediately re-transmits any messages received as part of the high-resolution picture transmission (HRPT) broadcast so that any station with a steerable HRPT receiver can collect the nearby AWS observations in near real-time manner. This is a valuable means of checking that the AWSs are still operating correctly and also provides data more rapidly for those concerned with real-time meteorological analysis and forecasting in the Antarctic. For example, the HRPT receiver at McMurdo Station can collect observations from the group of AWSs on the Ross Ice Shelf. These data are of particular value to air operations taking place over that sector of the Antarctic by virtue of the provision of information on advancing weather systems that may affect the station.

A recent development in Antarctic ionospheric and magnetospheric research has been the move towards the implementation of automatic, un-manned stations for the collection of upper atmosphere data from remote sites in the interior of the continent. The first of these automatic geophysical observatories (AGOs) has recently been deployed in Coats Land and incorporates basic meteorological instruments (pressure, temperature, wind speed and direction sensors). This arrangement allows both the surface meteorological and the upper atmosphere instruments to be serviced during one visit by an aircraft. It is hoped that, as the AGO network expands, similar co-operation between the meteorological and geospace communities can take place, allowing expensive logistical resources to be shared.

The AWS observations are now of a high standard, with much of the data collected being as good as that from manned stations. However, problems can and do occur with un-manned equipment located in such a harsh environment. The most vulnerable part of the AWS is the wind sensor and it is difficult to design a system that can withstand the sustained, very high wind speeds encountered in certain coastal areas of the continent. A major problem during the winter months is the build-up of hoar frost on the aerovane, which can be so severe that the unit stops working completely. Until higher air temperatures melt the frost, these conditions result in a constant wind direction and zero wind speed being transmitted. Data from one or more sensors of the AWSs can be lost for a variety of reasons, including the rise in surface snow level, corrosion of components, battery failure or a transmitter drift or failure. However, considering the environment in which the units operate, the volume and quality of the data collected to date are remarkable and a testament to the skills of the designers and those who maintain the systems.

2.3.2 Processing and distribution of AWS data

The basic observations made every 10 min by the Madison AWSs consist of wind speed, wind direction, air temperature and air pressure, with relative humidity and the 3−0.5 m air temperature difference being collected every 20 min. With two satellites collecting data, this results in at least 100 sets of 10 min observations per 24 h period, although all 144 are often received. In Madison, the tapes of raw data received from Service ARGOS are processed to remove duplicate reports and convert the data to geophysical units. Quality control is also carried out to check that the data are not outside the range expected for the location and time of year and that no communications errors have occurred. Monthly data summaries and monthly means of three-hourly measurements of temperature, pressure, wind speed and direction, together with extremes measured during the month and resultant wind are then prepared for distribution. Monthly statistics are only computed when at least 25% of the three-hourly reports are available so that the figures are representative for the month. The distribution of the data takes place by a number of means including the following.

(i) Floppy discs containing the three-hourly data. This method allows the observations to be input into a personal computer.

(ii) An annual volume containing the three-hourly data and monthly and annual means. These books are widely distributed to researchers and polar libraries (see Section 2.8).

(iii) The 10 min and three-hourly data are available via an Internet anonymous FTP account.

The operational numerical weather prediction community now run global models that have very few observations over the Antarctic continent so that efforts have been made in recent years to get the AWS observations on to the

Global Telecommunications System (GTS). Since 1990 data from the Madison units have been put onto the GTS via the Naval Oceanography Center, Monterey, California, USA, who obtain the observations every 3 or 6 h. More recently, the Australian AWS observations have been injected via the Australian Bureau of Meteorology in Melbourne and the data from the German AWSs at Toulouse. In fact, by 1994 there were around 150 AWS observations on the GTS each day, which is more than all the observations from manned stations. AWSs are therefore making a great contribution to operational meteorology, as well as to Antarctic research activities.

2.3.3 The network of AWSs

Since 1980 there has been a rapid expansion in the network of AWSs in the Antarctic and there are now over 50 systems currently operational. Figure 2.3 shows the current distribution, although it should be noted that each year some changes occur as a result of units failing, new projects requiring systems in areas previously sparse in data and changes in forecasting requirements. Figure 2.3 shows that the present distribution is very uneven, with systems being clustered in certain areas as a result of particular projects and national programmes. The areas with most AWSs are the following.

(i) The Ross Ice Shelf and around Ross Island, with approximately 12 AWSs. These have a dual function in that they support air operations between Christchurch, New Zealand and McMurdo Station as well as research activities on the ice shelf concerned with barrier winds and mesoscale weather systems. All these AWSs are operated by the University of Wisconsin.

(ii) The vicinity of Terra Nova Bay and Reeves Glacier and along the coast of Victoria Land, with about 12 AWSs. These are a mixture of American and Italian systems used in research studies into the very strong katabatic winds that flow down from the high plateau.

(iii) The Antarctic Peninsula, which has several AWSs in a line down its eastern side. These have been installed by the British Antarctic Survey on behalf of the Madison group to study barrier winds.

(iv) Close to the South Pole, where a group of five AWSs have been installed by the USA as part of a boundary-layer study.

(v) The Adélie Coast with several American AWSs, which provide data for the study of katabatic winds.

(vi) Around and inland of Casey Station, where there are several Australian AWSs. Data from these systems are used within research activities (including katabatic wind studies in conjunction with the Adélie Land AWSs) and for forecasting in the vicinity of the station.

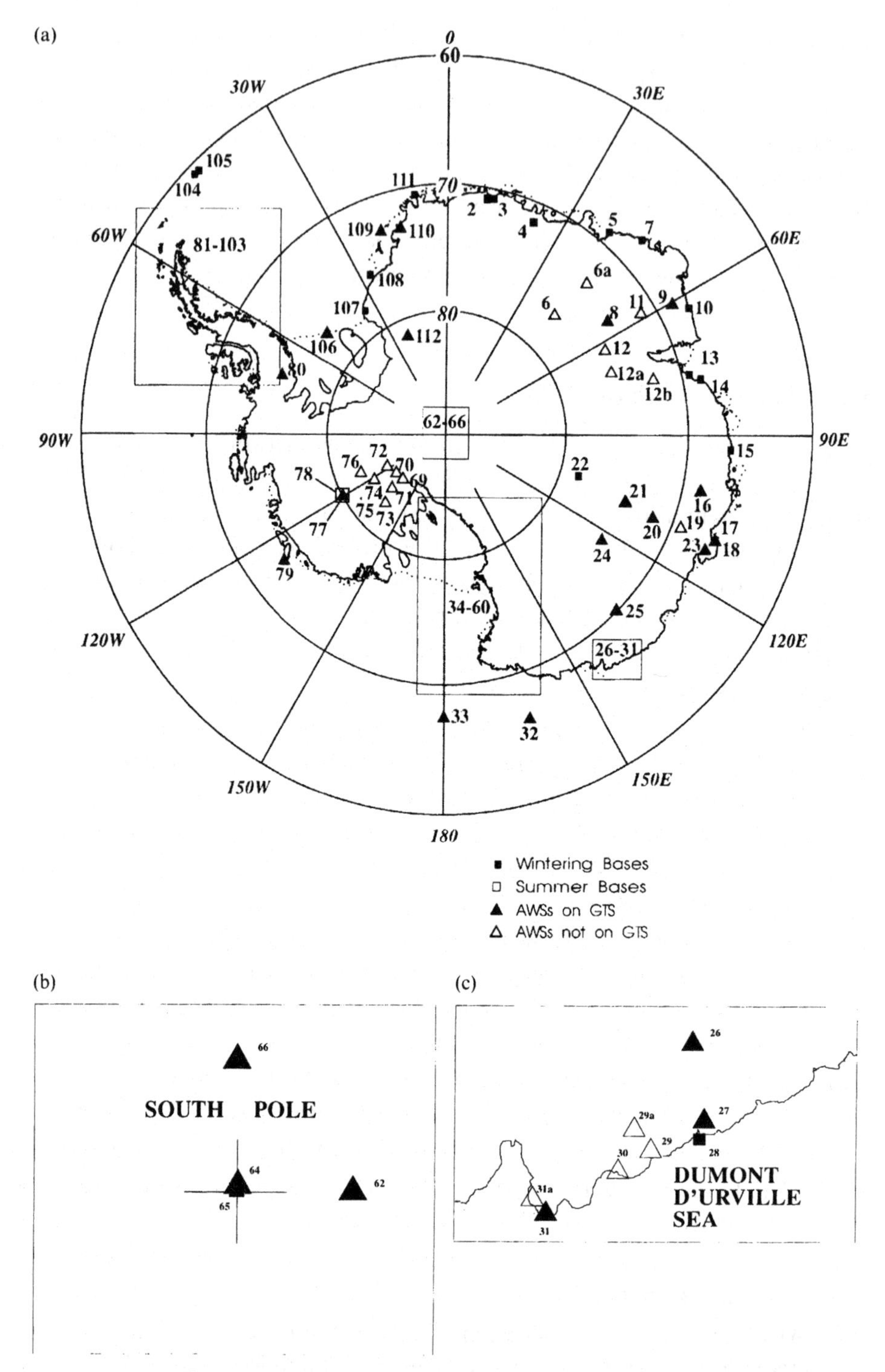

Figure 2.3 Maps showing the locations of AWSs and research stations making meteorological observations in 1995. The key to the station numbers is in Appendix A. Appendix B provides details on the AWSs.

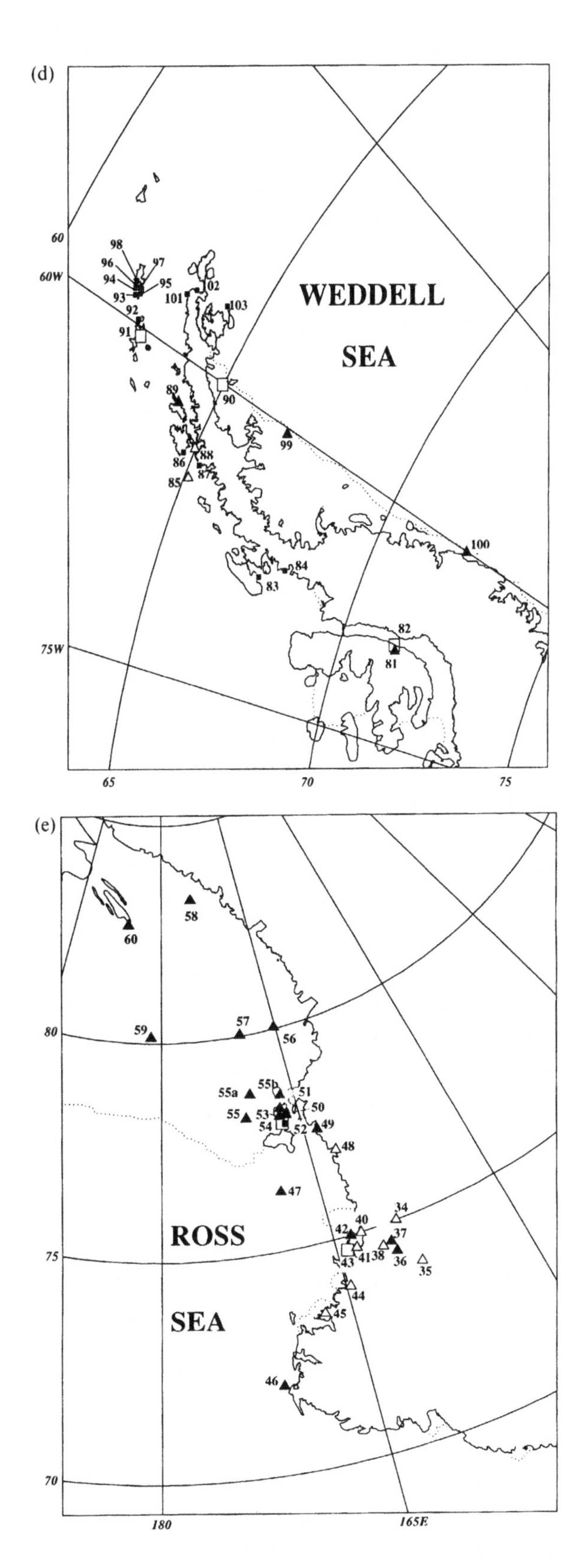

Figure 2.3 (*cont.*)

(vii) Around the Lambert Glacier Basin, at approximately 2500 m elevation, there is a network of six Australian AWSs used to support forecasting operations in the Prince Charles Mountains and for climatological and katabatic wind research in the region.

Although the early AWSs were American systems deployed by the Madison group, there are now several other nations participating in projects that involve AWSs, although many of the systems have been built by Madison in co-operative ventures. Because the technology of the AWSs is being further developed and the hardware made more reliable, it is expected that the present network will expand in coming years, with many of the large data gaps across the continent being filled. A list of all AWSs that have been used in the Antarctic, their locations and periods of operation is given in Appendix B.

2.3.4 Research applications of AWS data

The observations from AWSs have been used in many research studies since their first deployment in 1980 and the results of this work are reported throughout this book in the relevant sections. Here we will briefly review some of the main applications of AWS data in the light of the strengths and weaknesses of the observing systems and the particular data requirements.

(i) The wind field on the Antarctic continent has been of great research interest for many years and, as described above, the earliest and many of the subsequent AWS deployments have been for investigations into the katabatic and barrier winds. These studies have resulted in great advances in our knowledge of the katabatic flow as well as the mesoscale wind field in certain sectors of the continent. The only limitation on their use has been the harshness of the environment in certain area where even the robust hardware of the Madison AWS has not been able to withstand the conditions. This was the case on some islands around the continent where the icing was so severe that it destroyed the aerovanes, making it impossible to measure wind speed or direction.

(ii) Satellite imagery has shown that a large number of mesoscale weather systems occur around the coast of the Antarctic and over the ice shelves. These systems have a horizontal length scale well below the resolution of the manned stations and have had to be studied using other means. On the Ross Ice Shelf the network of AWSs has been particularly useful in showing the wind field associated with these lows and the horizontal thermal gradients that are thought to play a role in their formation. However, data on conditions in the lower and middle troposphere above such systems are urgently needed and it is hoped that one day AWSs may include some sounding capability and provide profiles of temperature.

(iii) Our knowledge of the climatology of large sectors of the continent is very poor and here AWSs will be able to play a role provided that they can be maintained and serviced on a regular basis. In particular, it is hoped that climatological data can be obtained for areas of the plateau where manned stations are especially difficult to operate (Allison *et al.*, 1993a).

(iv) A recent application of AWS data has been in the production of 500 hPa height fields over the high Antarctic plateau from the surface temperature observations (see Section 5.5.1). With, at the time of writing, only the American Amundsen–Scott base at the South Pole making radiosonde ascents in the interior of the continent, such techniques for determining the synoptic conditions over the plateau are particularly valuable. It has been shown that this method can give estimates of the 500 hPa height with acceptable accuracy at elevations of 2500 m and above.

2.4 Drifting buoys

For many years ways have been sought to fill the large data void over the Southern Ocean which is a major handicap to operational meteorology and research studies. Earlier in the twentieth century there were many commercial ships operating in the Southern Hemisphere and a large number of whaling vessels could be found in the peri-Antarctic islands. Most of these vessels maintained meteorological log books and some even provided observations in near real time for use in weather forecasting. However, with the decline in the number of ships in the Southern Ocean, satellite data have come to be used extensively instead of these observations. Since the 1970s, data from satellite radiometers have been used to derive sea surface temperatures under cloud-free conditions and infra-red and microwave sounders have allowed the production of temperature and humidity profiles (see Section 2.6.2). However, satellites could not provide information directly on the main surface meteorological variables of pressure, wind speed and direction and near-surface air temperature and humidity. Since the launch of the ERS-1 satellite in 1991, the scatterometer on that platform has been able to provide surface wind vectors over the ice-free ocean (see Section 2.6.4), but at present it is not possible to determine the other surface data with sufficient accuracy from satellite data alone. For these reasons buoys have been, and continue to be, deployed in the open ocean and in the Antarctic sea ice to provide data for operational forecasting and research into the atmosphere and sea ice.

The major advance in the development and deployment of buoys came with the First GARP Global Experiment (FGGE) when over 300 systems were deployed in the Southern Ocean as part of this very large scientific project to investigate atmospheric predictability and the requirements for an optimum observing system. These buoys were contributed by eight different countries with the goal of having no point in the ocean more than 500 km from a buoy. The success of this project can be seen from the map of FGGE drifting buoy tracks for the period 22 November 1978 to 3 December 1979 shown in Figure 2.4.

Since the FGGE, the high density of systems in operation has never been equalled but there have been many advances in buoy technology and a huge increase in the applications of the data. In this section we will review the capabilities of the current generation of buoys and examine some of the applications to atmospheric and sea ice research in the Antarctic.

2.4.1 Buoy design and data collection

Many different types of drifting buoy have been deployed in recent years and currently a number of commercial companies and research institutes manufacture buoys. These are of varying degrees of sophistication, from low-cost ocean

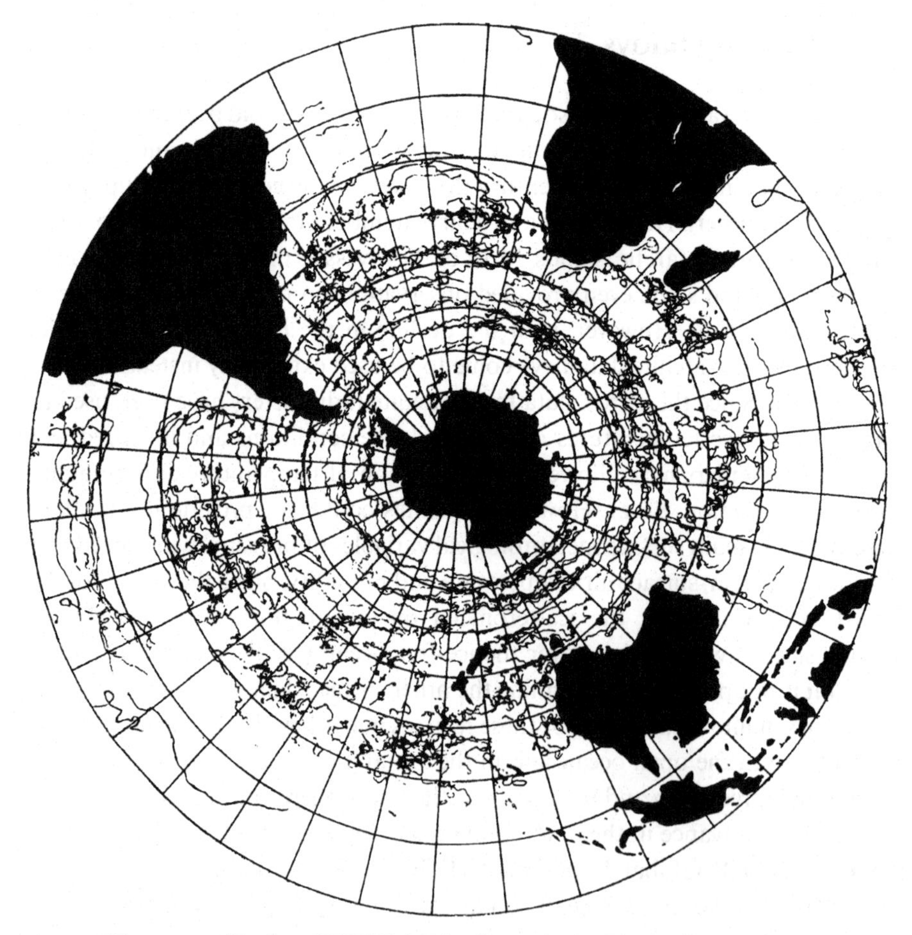

Figure 2.4 Tracks of FGGE drifting buoys in the Southern Hemisphere between 22 November 1978 and 3 December 1979. From Garrett (1980).

drifters with no meteorological sensors to very advanced systems making a wide range of atmospheric and oceanographic measurements.

Various instrumentation can be attached to the basic buoy platform, depending on the data requirements and the experiments that are to be carried out. Measurements made by buoys in recent years have included atmospheric pressure, wind speed and direction, air temperature and humidity at various levels above the surface and, in the case of buoys on ice floes, the surface temperature of the snow or ice and snow thickness. Buoys in the ice-free ocean can measure sea surface temperature and salinity and also make measurements of the temperature profile in the ocean to a depth of 200 m or more. A number of buoys have recently carried GPS receivers, which provide a more accurate alternative to the ARGOS Doppler method for determining the location of the buoy. A schematic diagram showing the dimensions and instrument locations on a typical buoy for insertion in the sea ice is shown in Figure 2.5.

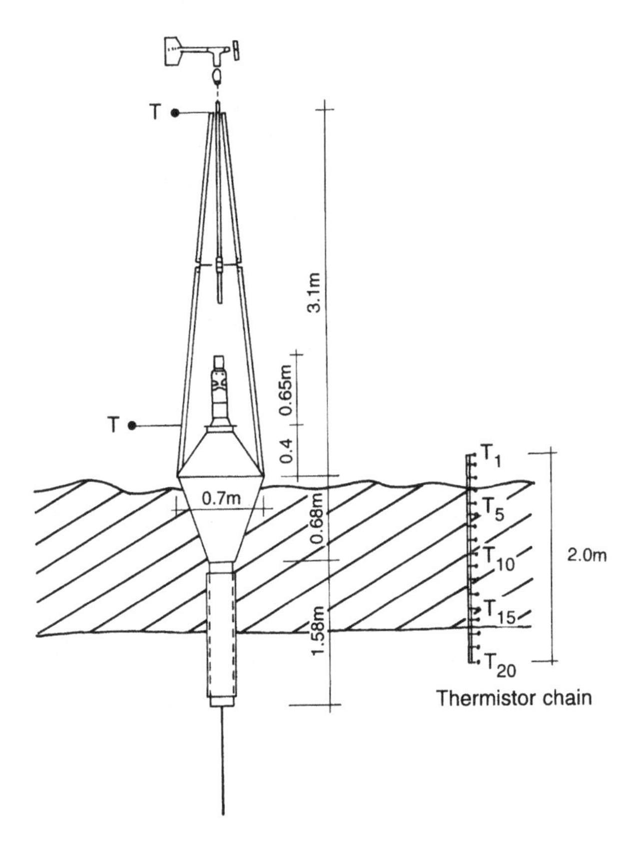

Figure 2.5 A schematic diagram of a typical ice-strengthened buoy. Courtesy Dr Ch. Kottmeier, Alfred-Wegener-Institut, Germany.

Table 2.1 *Meteorological parameters measured by a series of five buoys deployed in the Weddell Sea during 1990–93. Also included are the number of sensors mounted on the buoys and the accuracy of the measurements. From Launiainen and Vihma (1994).*

Parameter	Number of sensors on each buoy	Accuracy
Atmospheric pressure	1	1 hPa
Air temperature	4	0.05°C
Relative humidity	2	2%
Buoy hull temperature	1	0.2°C
Water temperature	10 or 20	0.1 or 0.05°C
Wind speed	2	0.3 m s^{-1}
Wind direction	1	5°
Snow depth	1	2 cm

Because of the very harsh environment experienced around the Antarctic the sensors on a buoy are often duplicated in an attempt to extend its lifetime and to ensure high data quality. Nevertheless, despite the poor conditions, many buoys deployed on sea ice do survive for long periods and eventually, when the sea ice melts, float northwards to joint the main eastwards-flowing ocean currents. Some buoys, however, can enter areas of very heavy ice and be crushed between the floes with the loss of the buoy being indicated by the cessation of the transmitted data. Other reasons for buoys being lost include exhaustion of the battery and failure of the transmitter.

The accuracy of data collected by a drifting buoy will clearly not be as good as that of data from a manned station where the instrumentation can be checked periodically and cleared of rime or other deposition. Nevertheless, improvements in systems over recent years have produced data that are acceptable for most investigations. The accuracy of data collected by the sensors on a group of five buoys deployed in the Weddell Sea during 1990–93, as reported by Launiainen and Vihma (1994) is given in Table 2.1.

Although the values in Table 2.1 appear acceptable for most applications, a number of major problems can befall sensors on buoys. In common with many humidity sensors used at manned stations, those on buoys perform poorly at low temperatures and are often very inaccurate at temperatures below about −10°C. Radiation errors are also a major problem with the temperature sensors on many buoys. Ice accretion on the anemometer can be a major problem and, under severe conditions, can stop collection of data or lead to very poor results. A further problem, especially at more northerly latitudes, is that, if the snow accumulation is particularly heavy then the anemometers may become buried and stop working. The experiences of Launiainen and Vihma (1994) suggest that cup anemometers perform less well than do systems based on propellers if ice accretion is heavy.

Data collection

The primary means of collecting data from drifting buoys is via the ARGOS DCS on the USA's NOAA polar orbiting meteorological satellites (Schwalb, 1982). Most buoys transmit data every few minutes with data being collected at every satellite pass. As NOAA aim to maintain two polar orbiting satellites in operation at all times and since the satellites have an orbital period of about 100 min, the frequent passes over the polar regions result in buoy data being collected about every hour from all platforms. In an area such as the Weddell Sea, which is at a relatively high latitude, about 20 satellite passes per day can collect data when two satellites are in orbit. This gives a mean interval between passes of 1.2 h (Launiainen and Vihma, 1994).

The transmissions from the buoys are collected at ground stations after being downloaded from the polar orbiting satellites when they pass over North America or Europe. Here the buoy data are processed to convert the raw observations into geophysical measurements and various quality control checks are carried out. A very important step is to compute the location of each buoy using a knowledge of the location of the satellite that collected the data and the characteristics of the signal from the buoy. Once this has been completed the data are usually coded according to the WMO standard drifting buoy code (FM 18-X BUOY) and input to the GTS at a suitable node.

2.4.2 The deployment of drifting buoys

Drifting buoys have been deployed both in association with limited experiments organised within national research programmes and as part of major international projects. Experiments such as the Winter Weddell Sea Project have involved the deployment of buoys by a number of nations and have provided extensive data sets for atmospheric and sea ice research.

An International Programme for Antarctic Buoys (IPAB) has recently been established within the World Climate Research Programme. The goal of this initiative is to coordinate and develop the buoy network to an acceptable density over the coming years. To get a spacing of 500 km between buoys in the sea ice zone would require around 50 buoys to be in place at any time.

At present, a number of nations are involved in the deployment of drifting buoys around the Antarctic including Australia, Brazil, Finland, Germany, Italy, Japan, South Africa, the UK and the USA. The actual deployment of the buoys on the sea ice or in the ocean usually takes place from research vessels during the austral summer or autumn when they are often engaged in combined research and

Figure 2.6 A buoy being deployed in the Antarctic sea ice. Courtesy Dr Ch. Kottmeier, Alfred-Wegener-Institut, Germany.

logistical re-supply operations. A photograph of a buoy being deployed in sea ice is shown in Figure 2.6. Although buoys to be used to study sea ice are usually deployed on suitable ice floes, they can initially be placed in open water and allowed to become frozen into the ice and be carried forward within the advancing pack. Despite the pressure exerted on the buoys during the freezing process, it is usually possible for them to operate successfully under such conditions (Allison, 1989).

Buoys can also be deployed from aircraft, although there is a much greater risk of damage to the instrumentation when this is done. However, the advantages are that many systems can be dropped within a short period of time and the work is not dependent on the tight logistical programmes of research vessels.

Because of the divergent flow of the sea ice around the Antarctic, most buoys drift northwards and emerge into the open water within several months. This means that Antarctic buoys have a much shorter useful life in the ice than do those in the Arctic, although they can provide useful ocean data.

2.4.3 Research applications of drifting buoy data

Data from drifting buoys have been used in a wide variety of research activities concerned with the Antarctic sea ice zone and the Southern Ocean. Although data from individual buoys may be used both in meteorological and in sea ice studies we will here consider these areas of research separately.

Meteorology

The main benefit of meteorological data from drifting buoys is the provision of valuable observations of near-surface conditions in areas that are otherwise completely devoid of objective data. Such data are therefore very valuable for the study of synoptic and mesoscale weather systems in the coastal region, especially when systems have passed on to the sea ice and surface wind vectors cannot be computed from scatterometer measurements. The arrays of several buoys that have been deployed around the Antarctic have recently been able to provide information on the surface pressure distribution on scales of 100–500 km, which is extremely valuable in investigations of mesoscale weather systems.

Buoy data have also proved to be of value in studies of the boundary layer and have provided data on the heat exchange between the atmosphere, ice and ocean. One such study by Launiainen and Vihma (1994) considered the turbulent surface fluxes of sensible and latent heat in the Weddell Sea area using data from three buoys deployed on the sea ice, in the open ocean and at the edge of an ice shelf.

Sea ice research

The ability to determine the position of a DCS uplink transmitter on a buoy using the ARGOS system means that a buoy embedded in the sea ice can be tracked throughout the whole of its lifetime. As a position can be produced at each pass of the satellite, the movement of the buoy, and therefore that of the sea ice, can be determined several times each day, if required. Single buoys can provide information on the broad-scale movement of ice whereas arrays of six or more buoys can provide data on mesoscale ice deformation.

The drift of pack ice around the Antarctic has been investigated by many workers using buoy data. Allison (1989) examined sea ice motion around the coast of East Antarctica using three satellite-tracked buoys and showed the highly mobile nature of the ice, even when it was located many hundreds of kilometres from the ice edge. Limbert *et al.* (1989) used data on the position of a buoy embedded in the Weddell Sea sea ice to study links between atmospheric weather systems and movement of the pack in the Weddell Sea.

2.4.4 Operational applications of drifting buoy data

From the experiences of using data from drifting buoys during the FGGE experiment it became clear that the data gave a significant improvement in the operational forecasts produced by the main meteorological centres (Fitt *et al.*, 1979). This was not only in terms of improving the numerical analyses which were able to use the objective measurements but also in the manual analysis process, in which it was found that depression centres could be identified more accurately. Today, buoy observations are an important element of the World Weather Watch system and make a major contribution to analysis and forecasting over the Southern Ocean. Unfortunately, at the present time, a number of the observations made by the buoys are still not disseminated on the GTS and a high priority for the future must be to ensure that the observations are made available to the main forecast centres within several hours so that they can be assimilated into the numerical models.

2.5 Surface-based remote sensing

Some research programmes require the measurement of vertical profiles of wind and temperature at more frequent intervals than could be easily or economically accomplished using radiosondes. In order to provide such data a number of surface-based remote sensing techniques have been used in the Antarctic.

Sodar (sonic radar) has proved a useful technique for studying the structure of the atmosphere in the lowest kilometre. A simple sodar system (Figure 2.7) operates by firing a sonic pulse vertically and recording sound backscattered from the atmosphere. Scattering takes place from temperature fluctuations which have a length scale comparable to the wavelength of the sonic pulse, hence sodar echoes indicate regions of the atmosphere that are both turbulent and thermally stratified. It is usual to display the strength of the returned echoes on a time–height chart, similar to a ship's echo sounder record, such as that shown

Figure 2.7 Installing a sodar at Halley research station.

in Figure 6.9 later. Although these records are somewhat qualitative, if used in conjunction with other data they can provide useful insight into the time evolution of the boundary layer and lower troposphere. Studies using simple sodars have been conducted at South Pole (Neff, 1980) and Halley (Culf and McIlveen, 1993) Stations.

More quantitative information can be obtained if the frequency of the return echoes is recorded as well as their amplitude. One can then determine the component of wind speed along the sonic beam by measuring the Doppler shift of the return signal. By making measurements along three different beams it is possible to produce a vertical profile of the wind vector at frequent intervals. Doppler sodars have proved particularly valuable for investigating the structure and development of katabatic winds in the Antarctic (Liu and Bromwich, 1993; Argentini *et al.*, 1992).

A further development of sodar is the radio acoustic sounding system (RASS). Here, the sonic pulse fired by a sodar is tracked by a centimetric radar. This enables one to determine the speed of sound, and hence the temperature, at a number of levels in the lower atmosphere. Bromwich and Liu (1995) used Doppler sodar and RASS to study the dynamics of the katabatic wind confluence zone along the Siple Coast.

Outside the Antarctic, wind-profiling Doppler radars have found application both in research studies and as replacements for operational radiosonde stations. As yet this technique has not been widely used in Antarctica. The extreme dryness of the cold Antarctic atmosphere is likely to limit the strength of clear-air radar returns and may well preclude useful wind profile measurements using such systems.

2.6 Satellites, space-based observing systems and ground stations

Since the first Earth-observing satellites were launched in the early 1960s, data from these platforms have played a major role in the study of the Antarctic because they were able to provide frequent data for areas with few *in situ* measurements. The early satellites carried only simple imaging instruments, which provided coarse-resolution pictures in the visible and infra-red parts of the spectrum. Nevertheless, these were able to give information on the structure of the major weather systems and were also valuable in climatological investigations, since they showed the frequency of cloud and cyclones. Since that time there have been many developments in satellite technology, the extraction of new geophysical information from the raw satellite measurements and the integration of observations with model fields and *in situ* data. In this section we will consider the various forms of satellite data that are now available and consider how they have been applied to operational and research meteorological activities in the Antarctic. The receiving stations for the collection of data are also discussed, together with the satellite systems planned for the next few years. Here we will only discuss polar orbiting satellites because the Antarctic is on the very edge of the imagery available from the geosynchronous satellites and their main use, insofar as Antarctic meteorology is concerned, is in relaying synoptic observations via their data collection systems (DCSs) for injection into the GTS (see Section 2.7). Here there is insufficient space to provide a great deal of detail on the satellite systems and all the applications of the data that have emerged in recent years, so other texts should be consulted for more information. Massom (1991) provided a full description of the satellites and observing instruments that have been used in the whole range of polar studies. Two books, although not specifically concerned with the polar regions, that provide a great deal of information on the interpretation of imagery and satellite sounding of the atmosphere are respectively by Scorer (1986) and Houghton (1984). These books should be consulted for more information on satellite meteorology and remote sensing techniques.

2.6.1 Visible and infra-red imagery

Visible and infra-red imagery are the most important forms of satellite data for the study of the meteorology and climatology of the Antarctic in that they provide information on the location of synoptic and mesoscale weather systems, frontal bands and areas of cloud over the continent and surrounding sea areas. With so few *in situ* observations, this imagery is the only means of monitoring the broad-scale synoptic environment over the continent for operational forecasting and it is also a very powerful tool in research investigations. A typical example of visible-wavelength imagery is shown in Figure 2.8(a). Here a major depression can be seen over the Bellingshausen Sea, revealed by the swirl of frontal cloud that extends from the centre of the low towards lower latitudes. Because the unfrozen ocean has a very low albedo and the cloud much higher values, the details of the cloud band are very clearly revealed and an analyst can easily determine the centre of the depression and the location of the front. Similarly, the difference between the albedo of the ocean and that of sea ice allows the accurate determination of the extent of the ice, provided that it is not covered by thick cloud. Such albedo differences can also be exploited for routine monitoring of iceberg calving, the extent of the ice shelves and the opening of ice-free areas near the coast (polynyas).

At night and during the winter months when there is no solar illumination, use must be made of the thermal infra-red (TIR) imagery which provides data on the temperature of the surface and cloud tops. An example of 11 μm TIR imagery for the situation shown in Figure 2.8(a) is illustrated in Figure 2.8(b). Here the sea surface temperatures over the Bellingshausen Sea are relatively high while the cloud top temperatures of the frontal band are much lower, providing good resolution of the structure of the cloud associated with the depression. Infra-red imagery is excellent for observing the major weather systems which have high, cold cloud associated with them but is less useful for detecting areas of cloud that have a similar temperature to the surface. Under these conditions it may be impossible to differentiate the cloud from the surface and multi-spectral techniques, such as those described below, have to be employed. Nevertheless, TIR imagery has found many applications in the Antarctic and some that have emerged recently, which are discussed elsewhere in the book, are the investigation of the katabatic drainage flow via warm signatures on the ice shelves (Section 6.1), the study of mesocyclones (Section 6.5) and the determination of the motion of sea ice (Section 3.5).

Over the last few years other wavelengths in the infra-red part of the spectrum have been used in addition to the TIR data. In particular, imagery at 3.7 μm has proved to be of great value for separating different types of cloud. The 3.7 μm data for the case already discussed above are shown in Figure 2.8(c). During the day, imagery at this wavelength contains a combination of emitted terrestrial radiation and reflected solar radiation, which complicates the interpretation of the data. At night, with no solar component, the imagery is very similar to TIR data and can be interpreted in much the same way. During the day, the reflected component is dominant and the imagery primarily contains

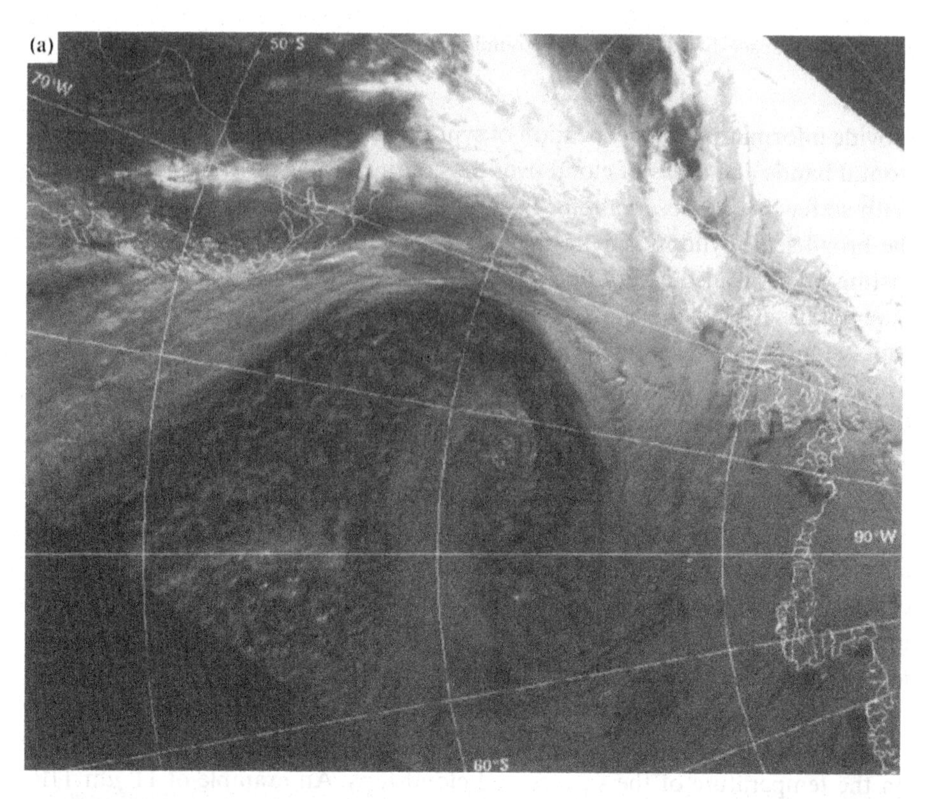

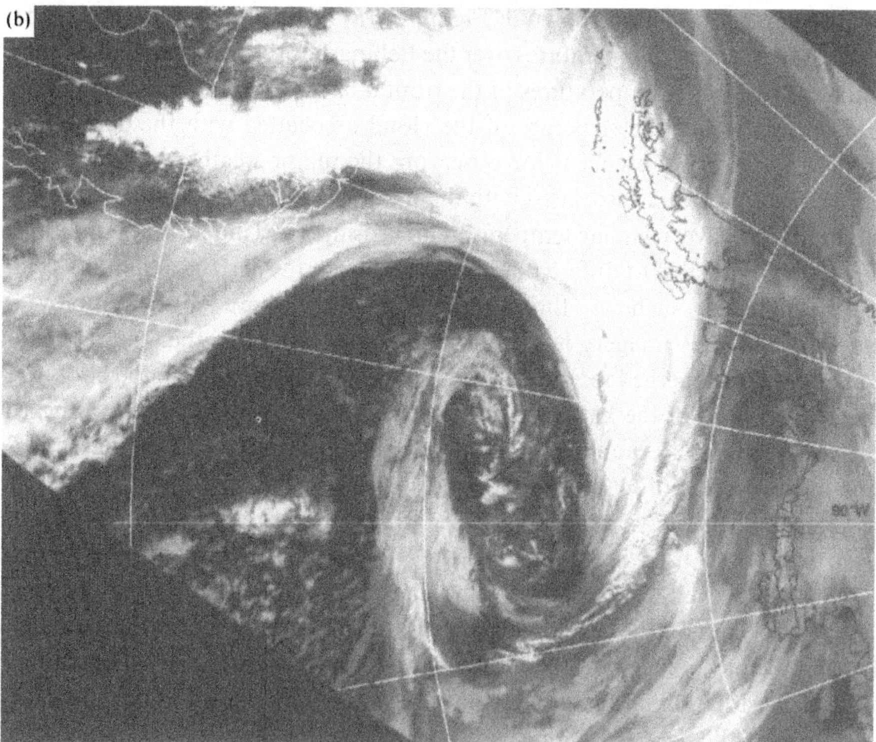

Figure 2.8 A major depression over the Bellingshausen Sea observed with (a) visible (0.6 μm), (b) thermal infra-red (11 μm) and (c) 3.7 μm AVHRR imagery.

information on the albedo of the cloud and surface at this wavelength. At first glance the image in Figure 2.8(c) appears rather strange because the reflectivity of some of the surfaces is very different from that in the visible part of the spectrum. For example, while the unfrozen ocean appears dark and clouds composed of water droplets have a high albedo, all ice, whether it is in the form of land ice, sea ice or ice crystal clouds, has a very low albedo and appears black. This results in the low cloud associated with the depression appearing white, while the higher cloud composed of ice crystals looks black. It can be seen in Figure 2.8(c) that the snow and ice on the surface of the continent appear black and resemble the ocean, which makes it difficult to detect the coastline. However, because clouds composed of water droplets appear white, even when they are supercooled, it is very easy to detect them over the continent during the day. Imagery at this wavelength is therefore very useful as an aid to forecasting cloud movement over the continent and in cloud analysis within research investigations.

Visible and TIR imagery has been a powerful tool in climatological studies of the Antarctic since at least one of these forms of data is always available, allowing Antarctic-wide, year-round investigations to be carried out. During

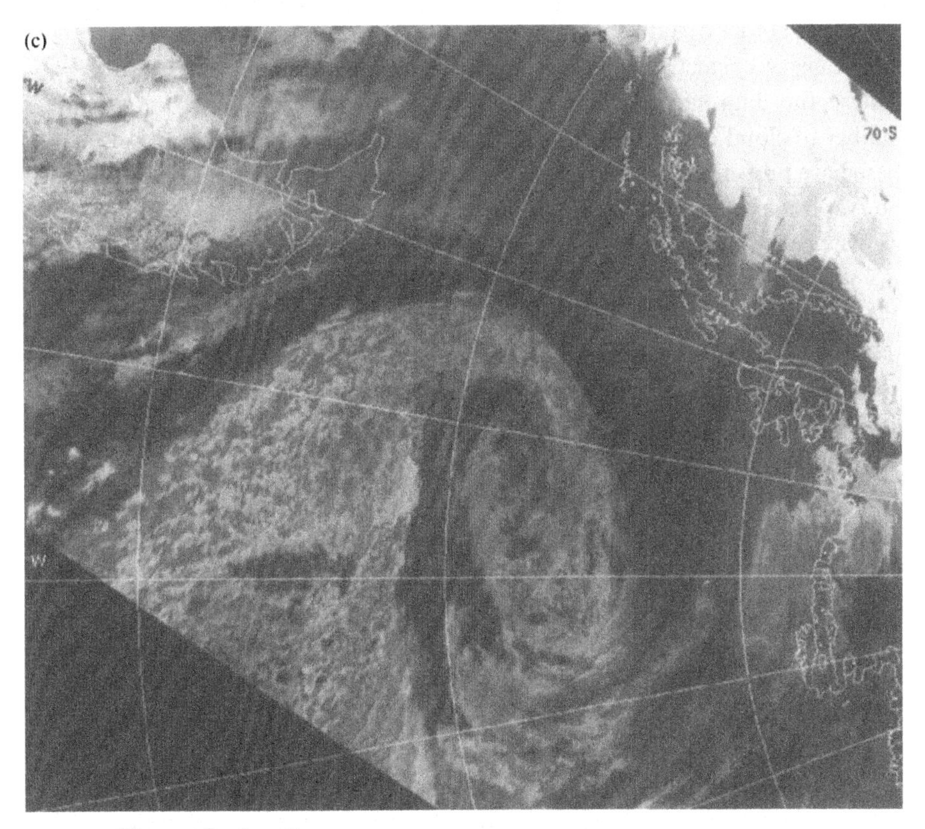

Figure 2.8 (*cont.*)

the 1960s the relatively coarse-resolution imagery available at that time was used to produce the first climatology of synoptic-scale weather systems over the Southern Ocean (Streten and Troup, 1973) and this work allowed the major regions of cyclogenesis and cyclolysis to be determined. More recently, a climatological investigation of the regions where mesocyclones occur was carried out using southern hemisphere composite TIR images for several winter seasons (Carleton and Carpenter, 1990), showing that even mesoscale phenomena can be routinely observed in this imagery. Although all the early investigations with satellite imagery were carried out through the manual analysis of hardcopy prints, during the last few years there has been a growing shift towards the automatic processing of satellite data to produce climatological fields, such as cloud cover, sea surface temperature and sea ice extent. One of the largest projects of this kind is the International Satellite Cloud Climatology Project (ISCCP), which is producing fields of mean cloud extent and type for use in climate studies and for the validation of general circulation model parameterisation schemes (Schiffer and Rossow, 1983). Producing the fields required for ISCCP over the Antarctic continent has proved to be a particularly challenging task (Raschke, 1987) and work is still required on algorithm development before we can be confident in the products generated by automatic processing in the Antarctic. Under certain conditions, it is difficult for a trained analyst to interpret imagery correctly, so the problems in developing computer software to analyse the data successfully can be imagined. Over the unfrozen ocean a number of cloud detection schemes have been developed (Saunders, 1986) and these have proved effective at detecting most types of cloud at mid-latitudes. In the Antarctic, most difficulties are experienced over the sea ice and especially over the coastal region of the continent where the topographic height changes very rapidly. In these regions the surface albedo and temperature are comparable to those of the cloud and the determination of cloud-covered and cloud-free areas is by no means simple. The cloud can usually be detected through visual inspection of the visible imagery because the cloud banks often have a brighter band on the sun-facing side and a shadow on the opposite side. Isolated areas of low cloud over the continent can also often be detected by their slightly lower albedo and their textural appearance, although this is difficult to quantify in computer software. A number of attempts have been made to develop automatic cloud detection schemes and software that can differentiate the different types of surface found in the Antarctic. Yamanouchi *et al.* (1987) made use of the differences between 3.7, 11 and 12 μm radiances in the absence of solar radiation to detect cloud over ice surfaces and to obtain information on the particle size and thickness of the cloud. Their scheme used differences between 11 and 12 μm radiances to detect thin cloud and differences between 3.7 and 11 μm radiances to isolate thick cloud. Ebert (1988; 1989; 1992) used a pattern recognition algorithm to segment scenes into 18 surface and cloud types using ten spectral and textural features and was able to obtain information on fractional cloud cover and albedo. However, the limited number of channels on the present generation of imagers and the large range of surface types and cloud mean that, even during periods when sunlight is available, problems can be experienced in correctly classifying all features in the satellite imagery (Zibordi and Meloni, 1991).

Although single satellite images can show the distribution of clouds in the Antarctic, on many occasions the availability of sequences of images greatly aids their detection and their differentiation from the ice surface. Since the polar orbiting satellites make passes over the continent about every 100 min and two satellites from each of the major satellite series are usually in operation at any time, it is possible to make short 'movie loops' covering periods of interest by re-mapping the data onto a common projection. These short loops can greatly aid interpretation of the imagery by allowing the evolution of the cloud to be observed against the background pattern of the surface and so help the discrimination of glaciological and meteorological information. This technique is also very useful for determining the cloud that is fixed in relation to the topography and not associated with the major weather systems. An extension of this technique is the extraction of cloud motion vectors, and hence upper winds, at cloud level from these movie loops (Turner and Warren, 1988a) by determining the displacement of the cloud elements between successive pairs of images.

Satellite imagery available on a routine basis

The most commonly used imagery for forecasting and research in the Antarctic is that from the Advanced Very High Resolution Radiometer (AVHRR) on the USA's NOAA series of weather satellites. These platforms are the operational civilian polar orbiting satellites operated by the National Oceanographic and Atmospheric Administration (NOAA), who usually maintain two spacecraft in orbit at all times. Although they are launched primarily to provide data for weather forecasting and monitoring the oceans, the data from these platforms have found applications in many areas of Antarctic science, including climatology, glaciology and geology. The satellites are in a low-altitude (approximately 800 km) orbit and make just over 14 orbits of the Earth each day, so they provide frequent coverage of the polar regions. The AVHRR is a useful instrument since it has such a good data coverage by virtue of its medium horizontal resolution of 1 km and a wide, 3000 km swathe of data. With the five visible and infra-red channels available (Table 2.2) the AVHRR is a very valuable source of data for monitoring cloud, the oceans, sea ice and land ice. Data from the instrument are broadcast continuously by the satellites in real time in two forms. First, the full resolution five-channel data are broadcast as part of the High Resolution Picture Transmission (HRPT) data stream at a rate of 665 kbps for collection by stations with a steerable antenna. Secondly, a reduced resolution (4 km) Automatic Picture Transmission (APT) broadcast is made of one visible and one infra-red channel at a data rate that can be taken by very simple receivers with only an omnidirectional, helical antenna.

The Defense Meteorological Satellite Program (DMSP) series of satellites are the military equivalent of the NOAA spacecraft and are operated by the USA Department of Defense. As with the NOAA series, there are usually two spacecraft operational at any time at an orbital height of about 800 km. The imager on these spacecraft is the Operational Linescan System (OLS), which has a horizontal resolution of about 0.5 km and a swathe width of over 3000 km. The OLS differs from the AVHRR in having only two channels – a broadband visible channel and a thermal infra-red channel (Table 2.3).

Satellites of the Russian Meteor series of polar orbiting weather satellites began to be launched in the 1970s by the USSR and provide a useful supplement to the data from the USA's spacecraft. They are at a relatively high altitude of approximately 1200 km and there are usually two spacecraft in orbit at any time. Although the spacecraft carry a high-resolution (1.5 km) radiometer and, occasionally, an Earth resources instrument, it is the APT transmissions, which are in a similar format to the American broadcasts, that are usually received in the Antarctic.

Table 2.2 *The channels of the Advanced Very High Resolution Radiometer (AVHRR) on the USA's NOAA series of satellites and typical applications in Antarctic meteorology.*

Channel number	Central wavelength	Applications
1	0.6 μm	Monitoring cloud and sea ice during the day
2	0.9 μm	Similar to channel 1. The difference between channels 1 and 2 is valuable for minimising the effect of cloud in sea ice observation.
3	3.7 μm	Detection of water droplet clouds over ice during the day. Used in night-time SST algorithms.
4	11 μm	Year-round observing of cloud. Computation of SST, ice surface temperature and cloud top temperature.
5	12 μm	Similar to channel 4. The difference between channels 4 and 5 is useful in detecting semi-transparent cirrus. Used with channel 4 in SST algorithms.

Table 2.3 *The channels of the Operational Linescan System (OLS) on the DMSP satellites.*

Channel number	Wavelength (μm)	Applications
1	0.4–1.1	A broadband visible channel for observing cloud, sea ice and land ice.
2	10.5–12.6	A thermal infra-red channel for monitoring of cloud and the surface.

2.6.2 Satellite sounder data

The satellite imagery described above is very valuable for observing the clouds associated with weather systems but cannot provide any direct information on the vertical temperature structure of the atmosphere. Yet it is this information which is so vital in determining future atmospheric developments and which is required as input to numerical weather prediction systems. With there being so few radiosonde ascents over the Antarctic continent and Southern Ocean, other means have had to be found to obtain information on the thermal field at upper levels. This requirement has resulted in the development of techniques for determining temperature profiles from satellite radiance measurements. It is possible to derive atmospheric profiles by making observations of the upwelling radiation in non-window regions of the spectrum within which most of the radiation has been emitted by gases in the atmosphere rather than by the surface. By using a number of different wavelengths with varying degrees of atmospheric absorption, radiation can be received from the whole atmospheric profile. With suitable mathematical inversion procedures it is possible to convert this radiance data into temperature profiles of the atmosphere extending from the surface to the middle stratosphere.

By far the most commonly used form of satellite sounder data is measurements from the Tiros Operational Vertical Sounder (TOVS) on the NOAA series of satellites. The TOVS has been in use since the 1970s and, as shown in Table 2.4, consists of three instruments – the High Resolution Infra-red Spectrometer (HIRS), the Microwave Sounder Unit (MSU) and the Stratospheric Sounder Unit (SSU) (Schwalb, 1978; 1982). These instruments are for obtaining soundings under cloud-free conditions, for obtaining soundings where there is thick cloud and for sounding in the stratosphere respectively. NOAA have been processing these data operationally on a world-wide basis for many years and distributing the temperature and humidity profiles via the GTS as SATEM messages. The transmitted data originally had a horizontal resolution of 500 km but recently this has been improved to 250 km, so giving better resolution of the

Table 2.4 *Characteristics of the three instruments constituting the Tiros Operational Vertical Sounder (TOVS).*

Instrument	Channels	Applications
High Resolution Infra-red Spectrometer (HIRS)	12 infra-red channels in the 4.3 and 15 μm CO_2 and N_2O bands. Three channels in the 6–8 μm H_2O band. Four window channels for cloud detection and surface observing.	Temperature and humidity sounding under partly cloudy and cloud-free conditions
Microwave Sounder Unit (MSU)	4 channels in the 5.5 μm oxygen band.	Temperature sounding under cloudy conditions
Stratospheric Sounder Unit (SSU)	3 channels in the 15 μm CO_2 absorption band.	Obtaining temperatures in the middle and lower stratosphere.

thermal structure of weather systems. The SATEM messages are primarily used in numerical weather prediction systems for defining the initial conditions over the ocean where there are few radiosonde ascents. Over the Southern Ocean they are particularly valuable because the only upper air data in this region come from radiosonde ascents made from isolated islands and from the occasional research ship. An example of the thermal structure that can be revealed by TOVS soundings is shown in Figure 2.9, which depicts the TOVS 1000–500 hPa thickness field for the time of the imagery in Figure 2.8.

In the coastal region of the Antarctic, TOVS retrievals are a valuable research tool because they can provide information on the thermal structure of the atmosphere that cannot be resolved by the widely spaced research stations that make upper air soundings (Turner *et al.*, 1993a). When working with the raw TOVS data, soundings can be produced every 40 km, which gives improved resolution over the SATEM messages. However, there are a number of problems in processing the TOVS data in the Antarctic, including the detection of cloud over ice, obtaining a good first-guess profile for the retrieval process and dealing with the high topography of the continental interior and the very stable near-surface layer. The statistical regression procedures used for operational retrievals have presented problems in the polar regions (Tanaka *et al.*, 1982), so a number of workers have adapted the techniques to Antarctic conditions. Lachlan-Cope (1992) investigated the performance of a statistical scheme (Turner *et al.*, 1985)

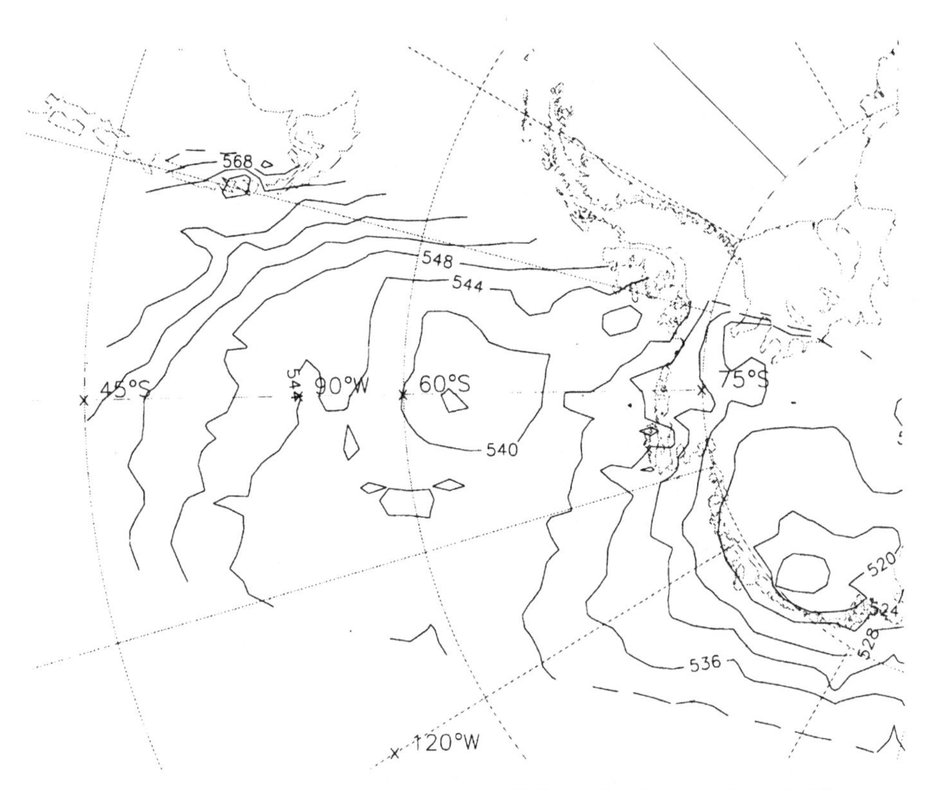

Figure 2.9 The field of 1000–500 hPa thickness for the case shown in Figure 2.8, determined from TOVS data.

and a simultaneous physical inversion scheme (Smith *et al.*, 1985) in the Antarctic coast region and compared the temperature profiles generated by both schemes with data from radiosonde ascents. He found that the statistical scheme had slightly smaller temperature differences when compared with radiosonde ascents and obtained an RMS error of about 2°C up to 20 hPa. Lutz and Smith (1988) and Lutz *et al.* (1990) examined the performance of the University of Wisconsin physical simultaneous TOVS retrieval scheme over the high interior parts of the continent where the ability of the scheme to take into account the high elevation should have given significant improvements over the statistical inversion techniques. However, they identified a number of problems in using the raw data over the Antarctic, including contamination of the 4.5 μm channels by reflected sunlight during the day and errors in the calibration of certain channels. They also found that the selection of the first-guess profile was very important in the Antarctic and they produced a set of Antarctic climatological first-guess profiles that was found to give much better results than did those available in the scheme that had been used previously. Their modified inversion procedure gave improved profiles when compared with collocated radiosonde ascents, although problems still remained in the detection of cloud over the ice surface. The differences between the temperature profiles produced by the modified scheme and radiosonde ascents were generally less than 2°C throughout the troposphere, with values increasing to 3°C in the stratosphere.

The raw TOVS data also allow the computation of humidity profiles through the troposphere, but the availability of only three water vapour channels on the HIRS means that the retrievals are poor, with little vertical structure. Nevertheless, NOAA routinely produce global humidity profiles that are distributed via the GTS and these provide the only upper air humidity data for the Southern Ocean, besides the radiosonde data from the isolated island meteorological stations. Over the continent itself, the low humidity levels preclude the calculation of realistic humidity profiles with the existing sounding instruments but the raw HIRS data have proved to be of value when used as coarse-resolution imagery. Turner and Ellrott (1992) used Antarctic radiosonde ascents and a radiative transfer model to determine the levels of the atmosphere that were being observed by the 6.7 and 7.3 μm channels of the HIRS. They showed that the imagery was capable of indicating areas of dry air descending from the upper troposphere in association with frontal bands. Such data should be of value in investigating synoptic and mesoscale weather systems over the high ground where there are no radiosonde ascents.

2.6.3 Passive microwave imagery

Although the early satellite imagery was at visible and TIR wavelength, the rapidly evolving space technology soon allowed reception of naturally upwelling radiation at longer wavelengths in the microwave region of the spectrum. Here clouds are far more transparent to upwelling radiation than they are at shorter wavelengths, allowing continuous monitoring of the surface, regardless of the meteorological conditions above. However, the Earth has its peak of emission close to 11 μm and the microwave region is in the tail of the emission spectrum, so that the energy available at these wavelengths at the top of the atmosphere is very small. This means that microwave instruments must observe much larger areas in order to collect enough radiation for the signal-to-noise ratio to be acceptable. Therefore, although instruments observing in the visible and TIR regions can have a horizontal resolution as small as a few tens of metres, in the microwave region it has been necessary to have a resolution of around 20–50 km, although the latest instruments now have channels that can observe at 12.5 km. This coarse resolution has not been a problem for most applications, which have been concerned with observing the broad-scale features of the continent and the surrounding ocean, although, when attempting to examine precipitation structure within weather systems, a resolution of around 10 km or better is desirable.

Passive microwave instruments have been flown continuously since the early 1970s with the most important application of the data since that time being the

Table 2.5 *Some forms of passive microwave data encountered in Antarctic research.*

Satellite	Instrument	Period of operation	Wavelengths	Horizontal resolution
Nimbus-5	Electrically Scanning Microwave Radiometer (ESMR)	1972	19.35 GHz	30 km
Nimbus-7	Scanning Multi-channel Microwave Radiometer (SMMR)	1978–87	6.63, 10.69, 18.00, 21.00 and 37 GHz and horizontal and vertical polarisations	27 by 32 km at 37 GHz to 148 by 151 km at 6.63 GHz
Seasat	SMMR	1978	As above	As above
DMSP	The Special Sensor Microwave/Imager (SSM/I)	1987–	19.35, 22.235, 37 and 85.5 GHz. All horizontal and vertical polarisation except 22 GHz which is vertical only.	25 km except 85 GHz (12.5 km)

routine monitoring of the extent of the sea ice. The large difference in emissivity between open water and ice in the microwave region (Gloersen *et al.*, 1973) has made this possible even with the early, single-channel Electrically Scanning Microwave Radiometer (ESMR) (see Table 2.5). The availability of the more advanced Scanning Multi-channel Microwave Radiometer (SMMR) from 1978 and the current Special Sensor Microwave/Imager (SSM/I) instrument, with multi-channel and multi-polarisation capability, has allowed the development of algorithms that give accurate estimates of ice extent and also allow the classification of ice into first-year and multi-year ice and give improved estimations of ice concentration (Cavalieri *et al.*, 1984).

Over the Antarctic continent itself the main application of passive microwave data has been for the investigation of accumulation rates of snow. This work began through the examination of the brightness temperatures from early passive microwave instruments, which suggested that they were lower than was to be expected from theory and that scattering of radiation was taking place within the upper layers of the snow (Gloersen *et al.*, 1974). Several workers have attempted to model the scattering properties of the snow (Chang *et al.*, 1976; Tsang and Kong, 1977), which are functions of the temperature, density and particle size of the snow. Zwally (1977) related the observed emissivities to snow particle size and computed snow accumulation rates for selected areas in Greenland and the Antarctic. This work was extended by Rotman *et al.* (1982), who calculated long-term snow accumulation rates for parts of the Antarctic and Greenland. Good agreement was found with *in situ* data, except where remelting was thought to be taking place. The determination of accumulation rates from microwave emissivities is not simple but these techniques have great potential for monitoring precipitation in remote parts of the continent where *in situ* data are not available.

The main meteorological application of passive microwave data is in the estimation of areas of precipitation over the ice-free ocean. Work in the tropics has shown that realistic estimates of precipitation extent and intensity can be made (Petty and Katsaros, 1990) and such techniques have been validated against radar data. In the polar regions the use of such techniques is much more difficult because the precipitation rates are usually quite low. However, the first attempts to examine precipitation associated with mesocyclones have been made (Carleton *et al.*, 1993) and this subject will undoubtedly receive much more attention in coming years.

2.6.4 Wind data over the ocean

Active microwave instruments (radars) on the polar orbiting satellites are an important means of obtaining data on the near surface wind field over the ice-free ocean areas around the Antarctic for climatological investigations and studies of weather systems. These instruments are of two types, wind scatterometers and radar altimeters. Both work by transmitting pulses of microwave radiation towards the surface and measuring the signal returned from the surface.

Radar altimeters

Radar altimeters are nadir-viewing instruments that make measurements of backscattered radiation along the sub-satellite track only, so that they obtain a relatively small amount of data and there are large gaps between tracks. With one measurement from the surface it is possible to obtain the wind speed near the surface (Glazman and Greysukh, 1993) and information on the sea state, but not the wind direction. This is a major limitation when using the data in case studies of individual weather systems, although these data have proved to be of value in the polar regions when combined with other satellite data (Claud *et al.*, 1993). The lack of wind direction data is also not a problem when the data are

Table 2.6 *Satellite instruments capable of providing surface wind data over the ice-free ocean.*

Satellite	Period of operation	Wind measuring instrument	Frequency	Polewards limit of data coverage
Seasat	June–October 1978	Scatterometer	14.595 GHz	74° S
ERS-1/2	1991–	Scatterometer	5.3 GHz C-band	79° S
GEOS-3	1975–78	Altimeter	13.0 GHz	65° S
Seasat	June–October 1978	Altimeter	13.5 GHz	72° S
Geosat	1985–90	Altimeter	13.5 GHz	72° S
Nimbus 5	1972–82	ESMR	See Table 2.5	90° S
Seasat	1978	SMMR	See Table 2.5	72° S
DMSP	1987–	SSM/I	See Table 2.5	90° S
Nimbus 7	1978–87	SMMR	See Table 2.5	90° S
ERS-1/2	1991–	Altimeter	13.8 GHz	81.48° S
TOPEX/ POSIEDON	August 1992–	Altimeter	5.3 and 13.65 GHz	63° S

being used in the construction of climatological fields of the wind speed over the ocean or significant wave height when the data from many orbits of the satellite can be accumulated to get very good coverage (Tournadre and Ezraty, 1990; Carter, 1993). The wind speed data can also be valuable for assimilation into models and have been found to improve analyses and forecasts of atmospheric and sea state models, especially in the Southern Ocean. As can be seen in Table 2.6, the availability of altimeter data was rather sporadic throughout the 1970s and 1980s but continuous operation of at least one altimeter should be guaranteed from now on via the ERS series, TOPEX/POSEIDON and the planned launches discussed below.

Wind scatterometers

Scatterometers differ from altimeters in that they view areas of about 2000 km^2 across a swathe to the side of the sub-satellite track. They also view each scan spot two or three times from different angles, which allows the determination of the wind direction for each spot as well as the wind speed.

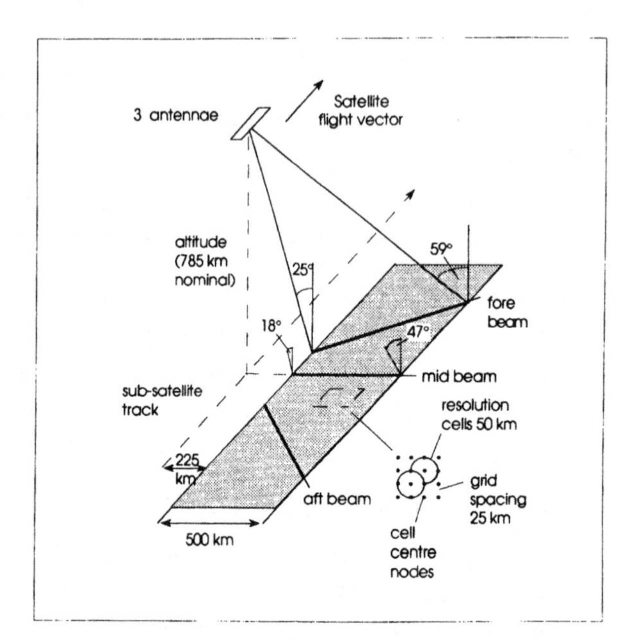

Figure 2.10 The viewing geometry of the wind scatterometer on the ERS-1 satellite.

This is accomplished by a mathematical model which relates the backscattered radiation, wind speed, wind direction and incidence angle of the observation (Offiler, 1990). This can be seen in the viewing geometry of the ERS-1 scatterometer, which is shown in Figure 2.10. This instrument views overlapping 50 km diameter cells across a 500 km wide swathe set some 225 km to the right of the sub-satellite track. The motion of the satellite is used to obtain three views of each cell to give sufficient information for the calculation of the wind direction. The result after processing of the raw data is a continuous 500 km swathe of wind vectors over the ice-free oceans that have an accuracy of 20° in direction and 2 m s^{-1} or 10% in wind speed over the range 4–24 m s^{-1}. An example of wind vectors from ERS-1 can be seen in Figure 2.11, which shows data for 06:14 GMT 4 March 1993 when the satellite was crossing a small synoptic-scale depression over the southern Bellingshausen Sea. The cyclonic circulation is clearly seen in the vectors, as is a sharp veering of the winds to the east of the low centre across a weak frontal band. This pass of data also shows two problems associated with scatterometer data. First, at the southern end of the pass the winds are chaotic since there is sea ice in this area close to the coast and the backscattered radiation from areas such as this does not allow the calculation of realistic vectors. Secondly, at the western end of the front, close to the centre of the low, some of the vectors have directions 180° different from that which would be expected. This is a result of there not being a unique solution in the inversion of the radar backscatter values into a wind vector and of the problems involved in selecting a solution when the wind field is changing rapidly, such as in the centre of lows and near fronts.

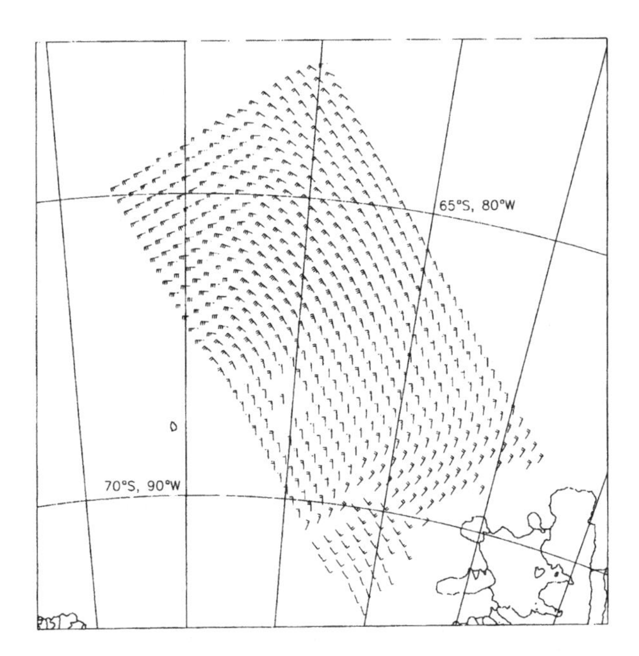

Figure 2.11 Scatterometer winds over the southern Bellingshausen Sea at 6:14 GMT 4 March 1993.

The first scatterometer was flown on Skylab in 1973 but it was the Seasat mission in 1978 that provided the first opportunity to collect large amounts of wind vector data over the world's ocean and to assess its value. Subsequent research has shown that the data can improve the numerical analyses (Ingleby and Bromley, 1991) and are also of value in research activities (Allan and Guymer, 1984). After Seasat there was a long gap before a further scatterometer was flown on the ERS-1 satellite but one can now be hopeful that continuity of scatterometer data will be provided by future missions.

Passive microwave instruments

It is also possible to estimate surface wind speed, but not wind direction, from the upwelling radiation measured by passive microwave instruments, such as the SMMR and the SSM/I. These data have a relatively coarse resolution, but can still be used to study mesoscale weather systems in the polar regions (Claud *et al.*, 1993). One significant limitation is that the winds can only be derived in the absence of precipitation. Extensive information on the determination of geophysical products from passive microwave data has been provided by Petty (1990).

2.6.5 Instruments providing climatological data

Although data from many of the instruments described above have been used in climatological investigations, some instruments have been launched specifically for this purpose. Here we will briefly consider two such instruments.

The Along Track Scanning Radiometer (ATSR)

Although data from the AVHRR have been used for many years to obtain global fields of sea surface temperature (McClain *et al.*, 1983), this instrument does not have an especially low level of noise and the data are occasionally severely affected by dust in the stratosphere after volcanic eruptions. ATSR, which was launched in 1991 on ERS-1, has essentially the same infra-red channels as AVHRR but uses a Stirling cycle cooler to keep the detector at a low temperature and obtain measurements with very low noise levels. It also uses a dual-view (nadir and 55° forward) technique to aid in the computation of the atmospheric correction and so produce SSTs with an accuracy of 0.3 K for use in climate studies. Recent studies have also suggested that the ATSR data can be used to compute very accurate surface skin temperatures over the Antarctic continent (Bamber and Harris, 1994). The narrow 500 km swathe limits use of the data for many meteorological applications but data from this instrument provide a valuable supplement to AVHRR imagery.

The Earth Radiation Budget Experiment (ERBE)

Satellites are above most of the atmosphere and are therefore able to make measurements of the solar radiation that has been reflected from, and the long-wave radiation emitted by, the clouds, atmospheric gases and the surface, so providing far greater coverage of radiation data than is possible from *in situ* instruments. The channels of most satellite instuments cover relatively small regions of the spectrum in order to allow the determination of cloud and surface properties. However, for radiation studies, broad regions need to be encompassed. The ERBE instrument (Barkstrom and Smith, 1984) is designed specifically for such investigations and has a coarse (40 km) resolution and channels covering the visible and infra-red parts of the spectrum. The instrument has been flown on the NOAA series spacecraft to obtain data concerning radiative fluxes and albedo at the top of the atmosphere for use in radiation budget studies. Data from ERBE have recently started to be used in the study of the effects of cloud and sea ice on the Antarctic radiation budget (Yamanouchi and Charlock, 1994) and will undoubtedly find further applications in the future.

2.6.6 Receiving stations

The APT imagery broadcast by the NOAA and Meteor satellites has been received at Antarctic research stations and on ships for many years, for the receivers required are simple and compact and only require an omnidirectional helical antenna. Such receivers can usually store several images and can provide hardcopy output and archive the data on computer disc for later analysis. Although APT data are very valuable for weather forecasting and ice analysis,

Table 2.7 *Antarctic research stations collecting high-resolution, digital data from the operation weather satellites.*

Station	Data received	Period of data collection
McMurdo	HRPT and DMSP	October 1985–
Palmer	HRPT and DMSP	January 1990–
Syowa	HRPT	February 1980 – May 1991
Casey	HRPT	December 1991–
Rothera	HRPT	February 1993–

they have a number of disadvantages in comparison with the full digital HRPT data in research investigations. The imagery is only available at a coarse (4 km) resolution, only two of the five AVHRR channels are included and no ARGOS or TOVS data are available. For these reasons a number of nations have installed receivers for the HRPT broadcasts so that data can be collected for research studies. To date, five HRPT receivers have been deployed (Table 2.7) and, as can be seen in Figure 2.12, complete coverage of the continent can be obtained from these systems. Receivers for the high-resolution digital data streams require a steerable antenna because the data rates broadcast by the satellites are high; for example, the HRPT data are transmitted at a rate of 665 kbps, producing around 100 Mb of data for a typical 20 min overpass. Nevertheless, modern digital storage devices can handle these quantities of data and substantial archives of HPRT and DMSP (the latter from Palmer and McMurdo only) data have been established (see Section 2.8).

Although HRPT receivers are substantial pieces of equipment and are usually installed permanently at the research stations, one system has been mounted on the German research vessel Polarstern. This is capable of being operated when the ship is in sea ice or when the waves and swell are limited and the slight motion of the vessel can be compensated for by gimbals.

2.6.7 Future developments

During the first 30 years of satellite remote sensing of the polar regions most of the spacecraft which provided data for operational and research activities were launched by individual countries as part of their national space programmes. However, the high cost of launching satellites has meant that there is now a trend towards multi-national initiatives and a greater degree of co-operation in the planning of future missions. The first results of this were seen with the launch of ERS-1 in 1991 by the European Space Agency (ESA) on behalf of its member states. ERS-1 is the first of a new generation of Earth observation satellites and

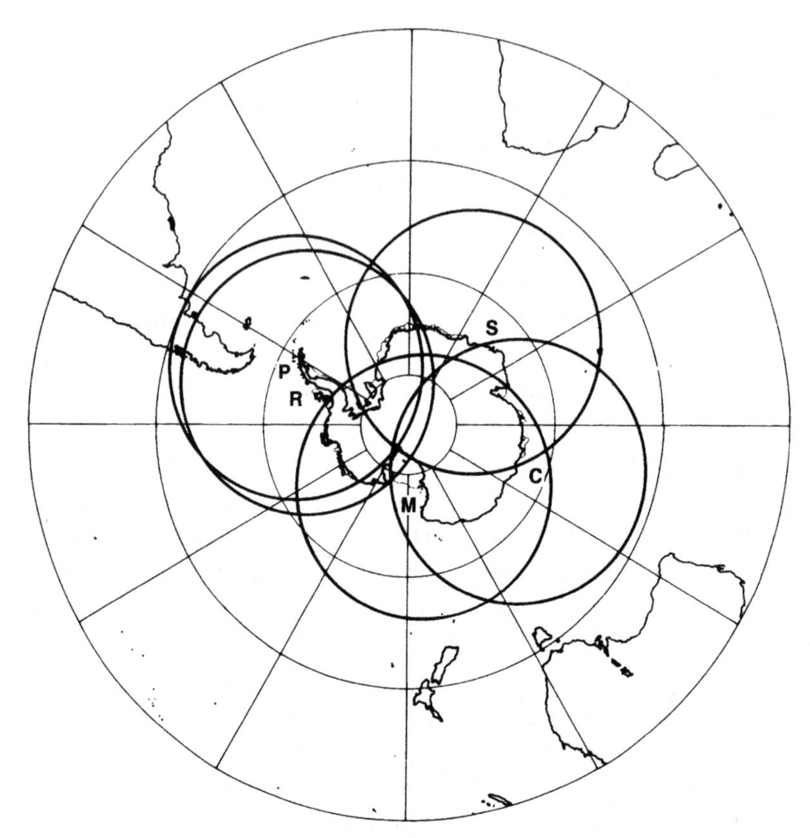

Figure 2.12 The approximate areal coverage obtained from the five HRPT satellite receivers that have been operated at the research stations. The letters refer to Casey (C), McMurdo (M), Palmer (P), Rothera (R) and Syowa (S).

its mission is to provide data on the oceans and polar regions using a suite of active and passive instruments. The continuity of data from these instruments is now assured with the launch of ERS-2 in 1995 and the planned ENVISAT satellite in 1998.

The TIROS-N/NOAA series of civilian weather satellites launched by the USA has been one of the mainstays of polar meteorology and climatology for almost 20 years and the data have found applications in a wide range of disciplines. However, the limitations of some of the instruments are now apparent and, over the coming years, they will be replaced by state-of-the-art sensors. However, there is also a need to ensure that there is continuity in the data record for climatological investigations so that intercomparisons with the data produced by the present generation of instruments will also be necessary. The first change will be the replacement of the MSU with the Advanced Microwave Sounder Unit (AMSU), a 20-channel instrument with 16 channels between 31 and 89 GHz for temperature sounding (AMSU-A) and four channels between 90 and 190 GHz for humidity sounding (AMSU-B). The introduction of AMSU on the NOAA spacecraft will probably take place in 1997 and will result in vastly improved sounding capability under cloudy conditions, such as those found around the coast of the Antarctic. However, the major change in the series will not take place until the year 2000, when the NOAA spacecraft will be replaced by new satellites, which are often referred to as the polar platforms. The goal will be to maintain two spacecraft in sun-synchronous orbits at all times, with the morning satellite (METOP) being launched and operated by EUMETSAT on behalf of its member states and the corresponding afternoon satellite by NOAA. The payloads of these satellites will include AMSU, a scatterometer, the Multifrequency Imaging Microwave Radiometer (MIMR) and an improved imager to replace AVHRR.

Over the next two decades there will also be new satellite instruments introduced that will be of great value in the data-sparse Antarctic. At the present time, data on the upper air wind field can only be obtained from radiosonde ascents or indirectly from cloud track winds. However, lasers mounted on satellites will be able to provide data on the backscattered radiation from aerosols in the atmosphere, which could be turned into wind direction and speed information. These data will be of value for assimilation into atmospheric models providing much needed upper air data for areas with few radiosonde stations, such as the Antarctic. Even though such systems require considerable power and will be expensive to render operational, by early in the next century the technology should have advanced to the point at which they can be considered for use on the operational weather satellites.

2.7 The station network and communications

The Antarctic meteorological observing network as we know it today was really established during the International Geophysical Year (IGY) of 1957/58. Before then, there were few observations available from Antarctica, mostly short-term records from expeditions that only spent one or two winters on the continent (Section 1.2). Many of the stations established during the IGY continued to function as permanent bases and, during the 1960s and 1970s, further stations were established. Since the early 1980s, the number of manned stations making year-round meteorological observations has declined slightly as many countries have rationalised their Antarctic programmes. Most notably, a number of stations formerly operated by the USSR have closed, including Vostok, which was one of the few meteorological stations on the high plateau of East Antarctica. The closure of manned stations has been offset to some extent by the rapid growth in the number of automatic weather stations deployed. However, AWS observations are not as reliable or as complete as those made at manned stations and there are still large gaps in the station coverage, particularly in West

Antarctica other than the Antarctic Peninsula and over the interior of East Antarctica. A list of all meteorological stations currently operating in the Antarctic and all sites where observations have previously been made for at least two winters is given in Appendix A. Maps showing the station locations are included as Figure 2.3.

A census of Antarctic meteorological stations operating during the winter of 1993 was undertaken as part of the FROST (First Regional Observing Study of the Troposphere) project. At that time, 35 manned stations were making surface observations. Of these, 14 were on the coast of East Antarctica, three were in the interior of East Antarctica, seven were in the Antarctic Peninsula and nine were on islands south of 60° S (of these nine, six were on King George Island in the South Shetland Islands). These observations were supplemented by about 55 automatic weather stations, mostly in East Antarctica but with six deployed in West Antarctica and the Antarctic Peninsula and a further three on islands south of 60° S. Upper-air observations were made at even fewer stations – eight on the coast of East Antarctica, one (Amundsen–Scott) on the East Antarctic plateau, one (Marambio) from the Antarctic Peninsula and one from an island (Bellingshausen on King George Island).

Observations from the Southern Ocean south of 60° S are equally sparse. During the summer months surface observations are available from ships supplying the Antarctic bases or engaged in oceanographic research and a number of these vessels also launch radiosondes. In winter the pack ice prevents ships from operating in Antarctic waters and the only observations available are those from drifting buoys. Since 1985 there have been 5–10 drifting buoys deployed each year in the Weddell Sea, with similar numbers operating in the Pacific and Indian Ocean sectors of the Southern Ocean.

As well as finding application in climatological research, Antarctic meteorological data are required for the preparation of local weather forecasts and as initial data for global numerical weather prediction models. For the former task it is necessary for observations to be exchanged between stations in the Antarctic in a timely fashion whereas the latter task requires Antarctic observations to be passed to numerical weather prediction centres remote from the Antarctic without delay. The World Meteorological Organisation (WMO) has responsibility for overseeing the collection and distribution of observations from Antarctic stations and has designated certain stations as 'collecting centres'. These centres collect observations from surrounding stations, field parties, aircraft and ships, generally using short-wave radio, and pass them on to national meteorological centres outside the Antarctic, often using satellite communication links for this purpose. Some collecting centres also re-broadcast their bulletins by short-wave radio for use by forecasters at other Antarctic stations. The national meteorological centres inject the observations received into the GTS of the WMO, making them available to forecasting centres worldwide. An increasing number of

Antarctic stations are equipped with satellite data collection platforms and can thus transmit their observations directly to a satellite ground-station connected to the GTS, removing the need to go through an Antarctic collecting centre. Observations from automatic weather stations and drifting buoys that use the ARGOS system carried on polar-orbiting satellites are received at one of the three ARGOS Global Processing Centres and are then forwarded to a national meteorological centre for insertion into the GTS.

2.8 Data sets and data centres

The record of Antarctic meteorological observations extends back to the earliest expeditions of the seventeenth and eighteenth centuries. However, data collected during these early voyages south are scattered across many archives and libraries and there is clearly no continuous series of observations for any one location. It was only with the establishment of permanently manned stations during the early part of the twentieth century that extended series of data could be assembled and used for climatological investigations. With the introduction of Earth-observing polar orbiting satellites in the early 1960s very large quantities of observations of the Antarctic became available, requiring a much higher degree of organisation. This has been aided in recent years by the advances in computer technology which allowed such data to be held on computer data bases and often accessed remotely. The result has been the establishment of a number of data centres that now hold long, comprehensive series of observations of interest to research scientists.

It is stated under the terms of Article III of the Antarctic Treaty that, 'In order to promote international co-operation in scientific investigation in Antarctica ... the Contracting Parties agree that, to the greatest extent feasible and practicable: scientific observations and results from Antarctica shall be exchanged and made freely available'. Exchanges of reasonable quantities of meteorological data therefore take place quite frequently between nations. However, the practicalities of copying large amounts of data, when some of the synoptic observations may be in hardcopy format, can be considerable, so it is preferable for the data to be held in machinable form within internationally agreed data centres that have the resources to disseminate the data.

In this section we will consider the data sets that are available to those conducting research into Antarctic meteorology and the data centres that hold reasonably long data sets of Antarctic observations that may be of value in process studies and investigations into change. When possible the full address of the data centre is provided so that contact can be made and questions posed regarding the availability of data. With the limited space available the list of data centres clearly cannot hope to be comprehensive, so the emphasis is on the centres with major data holdings. However, we also provide suggestions about where further enquiries can be made to help locate particular data. Our approach has been to highlight international data centres rather than national archives, but if it is necessary to contact the Antarctic operators of a particular country a good starting point is the national committee of the Scientific Committee on Antarctic Research (SCAR). Details of national representatives can be obtained from the Executive Secretary, Scientific Committee on Antarctic Research, Scott Polar Research Institute, Lensfield Road, Cambridge, CB2 1ER, UK, tel +44 (1223) 362061, fax +44 (1223) 336549.

SCAR, in association with the Council of Managers of National Antarctic Programmes, is developing and promoting an Antarctic Data Directory System (ADDS) in order to improve access to and comparability of Antarctic scientific data. The ADDS is composed of a network of National Antarctic Data Centres (NADC) with knowledge of Antarctic scientific data within any one nation, linked to a central Antarctic Master Directory (AMD). The AMD acts as a point of contact for information on Antarctic scientific data sets. For further information about the ADDS, contact the SCAR Secretariat. For information about the AMD, contact its host, the International Centre for Antarctic Information and Research, Orchard Road, P. O. Box 14199, Christchurch, New Zealand, or http://icair.iac.org.nz/.

2.8.1 Conventional surface and upper-air data

Observations from early expeditions are usually still held in hardcopy format and are scattered across many archives, research centres and libraries. In order to locate data collected by a particular country it is suggested that the current Antarctic research organisation in that country be contacted, together with the national committee of SCAR. They should be able to advise on data holdings and how access can be obtained.

Data availability has been much improved since the IGY of 1957/58 and many of the observations collected by the research stations established during that experiment have been published. The data collected by the stations that were formerly operated by the USSR have been published in detail by Dolgina (1962).

Although large amounts of synoptic data are transmitted routinely across the GTS for use in operational weather analysis and forecasting, few centres have archives of these data extending back for long periods. Today, most national meteorological services keep the observations that they receive on a permanent basis to allow model re-runs to be carried out for cases of interest. Unlike many organisations, the European Centre for Medium Range Weather Forecasts (ECMWF) has retained all observations since they began producing routine forecasts in 1979, so assembling one of the largest archives of Antarctic observations distributed on the GTS in recent years.

Climatological means of the main meteorological parameters are held by a number of centres including the following.

(i) The Climate Research Unit, University of East Anglia, Norwich, UK. They have a data base of climatic data (pressure and temperature) from Antarctic stations that have been operating for long periods. A selection of these data have been published by Jones and Limbert (1987), including monthly mean surface pressure and temperature data for 29 manned stations for the period 1957–86.
(ii) National Climate Data Centre, Federal Building, Asheville, NC 28801–2733, USA. They hold data from the USA's stations and some other sites.

A number of routinely published documents give Antarctic observations and show mean fields and anomalies. These include the following.

(i) The monthly *Climate Diagnostics Bulletin* produced by the Climate Analysis Center, NOAA/NWS/NMC, World Weather Building, 5200 Auth Road, Washington, DC 20233, USA. This publication provides a summary of climatic conditions across the globe, but gives data of interest to Antarctic investigations, including Southern Hemisphere maps centred on the South Pole. Data given include mean fields and anomalies.
(ii) The monthly *Climate Monitoring Bulletin Australia* (formerly the *Climate Monitoring Bulletin, Southern Hemisphere*), published by the National Climate Centre, Bureau of Meteorology, P. O. Box 1289K, Melbourne, 3001, Australia. These contain much valuable information on climatic conditions for each month, together with oceanographic and ozone data.

2.8.2 Automatic weather station observations

The University of Wisconsin-Madison has maintained a number of AWSs around the coast of the continent and in the interior since 1980, with the number of systems in the network steadily increasing during this period (see Section 2.3.2). The data held consist of wind speed, wind direction, temperature, pressure, relative humidity and 3.0–0.5 m temperature difference, with observations being made every 10 or 20 min. The individual measurements at this frequency are held together with extremes and averages for individual months and years. Data from the stations are stored in a computer archive in Madison, with the most recent data being available for access from remote computer systems. Until 1995, a hard-copy monthly summary of three-hourly observations and monthly means and extremes for the stations funded by the USA's antarctic programme was published. The monthly summaries have recently been made available on floppy disc to allow easier manipulation of the data. Finally, the annual summaries are collected into annual reports, which can also be obtained from Madison, e.g. Keller *et al.* (1990). For further details contact Professor C. R. Stearns, Department of Atmospheric and Oceanic Sciences, University of Wisconsin, Madison, Wisconsin 53706, USA.

In recent years other nations have installed AWSs in various parts of the Antarctic so that the total number of observations on the GTS from these platforms is now greater than that from manned stations. However, the AWSs used are of different types so that care must be taken when using data from different systems. Australian data have recently become available on the GTS and details of these systems and the availability of the full data set can be obtained from Dr I. Allison, Co-operative Research Centre, University of Tasmania, G. P. O. Box 252C, Hobart, Tasmania 7001, Australia.

2.8.3 Drifting buoy observations

Drifting buoys were first deployed extensively during the First GARP Global Experiment in 1979, when their observations were incorporated into the operational analysis schemes. Since that time buoys have been deployed in association with particular experiments and no network of buoys has been maintained.

Some observations from drifting buoys are transmitted over the GTS and can be taken at an appropriate node of the network. Once the International Programme for Antarctic Buoys (IPAB) is fully operational, it is planned that the data will also be archived at the Marine Environmental Data System in Canada and at the National Snow and Ice Data Center (NSIDC) in Boulder, Colorado, USA.

At present, the best means of obtaining buoy data is via the national centres responsible for the buoy programmes. These include the following.

(i) Australia. Dr I. Allison, Co-operative Research Centre, University of Tasmania, G. P. O. Box 252C, Hobart, Tasmania 7001, Australia.

(ii) Germany. Dr C. Kottmeier, Alfred-Wegener-Institut für Polar und Meeresforschung, Postfach 12 01 61, Columbusstraße, D-2850 Bremerhaven, Germany.

2.8.4 Raw satellite data

Satellite observations of the Antarctic have been available since the early 1960s, when the first low-resolution imagery data became available. However, many of these early data are difficult to use because of the low horizontal resolution and the limited number of grey scale levels. Digital imagery has been available since the 1970s but many of these data have been lost because of the costs that would have been involved in transferring the data on to new high-density storage media.

NOAA series data

NOAA maintain an archive of data from the TIROS-N/NOAA series of satellites (see Section 2.6.1) extending back to 1978. Imagery is available for the whole globe as five-channel, 4 km horizontal resolution global area coverage (GAC) data. Full 1 km resolution AVHRR imagery can be obtained in local area coverage (LAC) form if ordered in advance from NOAA. An archive is also maintained of global TOVS sounder data. Information on obtaining these data can be obtained from NOAA/NESDIS, World Weather Building, Washington DC 20233, USA.

The USA's NSIDC has recently begun collecting full 1 km resolution AVHRR data for use in studies of polar cloud and ice/snow. They should be contacted at the address in Section 2.8.3 for the status of this work.

As described in Section 2.6.6, a number of nations operate receivers that take the digital, high-resolution broadcasts from the NOAA satellites. These data are archived at a number of centres outside the Antarctic including the following.

(i) USA. HPRT and DMSP data from McMurdo and Palmer Stations are archived at the Scripps Institute of Oceanography in California. Further details can be obtained from the Antarctic Research Center (ARC), Ocean Research Division, A-014, Scripps Institute of Oceanography, La Jolla, California 92093, USA.

(ii) Japan. HRPT data have been collected at Syowa Station. Details can be obtained from the National Institute of Polar Research, 1-9-10 Kaga, Itabashi-ku, Tokyo 173, Japan.

(iii) Australia. HRPT data are taken at Casey Station. Information can be obtained from the Bureau of Meteorology, 150 Lonsdale Street, Melbourne, Victoria, Australia 3000.

(iv) UK. Rothera Station has collected HRPT data since February 1993 and the data are archived at the British Antarctic Survey in Cambridge. Information on the data available can be obtained from the Ice and Climate Division, British Antarctic Survey, High Cross, Madingley Road, Cambridge, CB3 0ET, UK.

(v) Germany. A HRPT receiver has been operated on the vessel *Polarstern* during periods when the ship was in sea ice or when the sea was relatively calm. Data from these cruises are held at the Alfred-Wegener-Institut für Polar und Meeresforschung, Postfach 12 01 61, Columbusstraße, D-2850 Bremerhaven, Germany.

DMSP data

DMSP broadcasts received in the Antarctic at Palmer and McMurdo Stations are archived at the Scripps Institute. The address given above should be contacted for information on the data holdings.

Twice daily infra-red and once daily visible mosaics of DMSP imagery are held by the National Snow and Ice Data Center, CIRES, Campus Box 449, University of Colorado, Boulder, Colorado 80309, USA. This data set extends back until at least 1977.

Scatterometer winds

Winds vectors generated from measurements of the scatterometer on board the ERS-1/2 satellites since February 1992 are available. Information on these data can be obtained from the ERS Help Desk, ESRIN, Via Galileo Galilei, 00044 Frascati, Italy.

SSM/I data and products

Passes of raw SSM/I data can be obtained from a commercial supplier. Contact Seaspace, 5360 Both Avenue, San Diego, California 92122, USA. These data are suitable for use in studies of precipitation, surface wind speed etc., which require single-pass data. Daily mean SSM/I brightness temperatures, which are more appropriate for investigations of sea ice, can be obtained on CD-ROM from the NSIDC at the address above.

2.8.5 Surface and upper-air analyses

In the pre-satellite era the very limited number of synoptic observations made it very difficult to draw surface or upper air charts, but areas with a greater concentration of observations, such as the Antarctic Peninsula, allowed charts to be prepared for limited regions of the continent. Surface charts for the peninsula were drawn by British meteorologists during the late 1940s, although observations were very scarce in the southern part of the area. In the mid-1950s the South African weather service began to prepare surface and upper air charts regularly for the whole of the Southern Hemisphere and monthly mean and seasonal mean charts produced from these were published in the journal *Notos*. During the 1960s *Notos* extended the charts published to include daily surface analyses with frontal data. The South African charts were also used to prepare upper level climatological mean fields, which are available in the volume by Taljaard *et al.* (1969). These are obviously now somewhat out of date but they do provide a convenient summary of upper air conditions for the IGY period.

By the 1970s satellite imagery and sounder data were of a sufficiently high quality for analyses to be prepared on a global basis. The longest series available is that prepared by the Australian Bureau of Meteorology beginning in 1971. These fields are available in digitised form and consist of surface pressure and 500 hPa heights both for 00 and for 12 GMT. The data are on a 47 by 47 point polar stereographic grid covering the Southern Hemisphere. The reliability of many of the early surface pressure analyses has been assessed by Jones (1991).

The other long series of analyses is from the European Centre for Medium Range Weather Forecasts (ECMWF), who have retained all the products generated since 1979. A re-analysis has recently been completed of all the data received since 1979 using the current operational assimilation scheme. This has produced a unique series of consistent analyses that will be of great value in studies of climate variability. The Data Support Section (DSS) of the Scientific Computing Division of the National Center for Atmospheric Research (NCAR) was involved in the collation of data sets to support the re-analysis. The DSS has both an extensive data archive and knowledge of sources and availability of data to support climate research. A major strength of NCAR's data archive is the variety and versatility of its data collection. Further details are available from NCAR, P. O. Box 3000, Boulder, Colorado 80307–3000, USA, or http://http.ucar.edu/dss/index.html.

2.8.6 Sea ice data

Information on sea ice extent in the pre-satellite era is very limited and based mostly on observations from stations on islands or near the coast. These data are usually held by the organisations responsible for the maintenance of the research stations. Since the introduction of passive microwave imagers on polar orbiting satellites in 1973, it has been relatively easy to prepare routine maps of ice extent for both polar regions. Such maps have been prepared since 1973 by the NOAA/US Navy Fleet Weather Facility (FLEWEAFAC) National Ice Center (formerly the Joint Ice Center (JIC)) from the available satellite data, aircraft reports and ship observations. Information on obtaining their weekly sea ice maps can be obtained from National Ice Center, 4251 Suitland Road, Washington DC 20395, USA.

The maps have been processed to give additional data including

(i) the latitude of the ice edge every 10° of longitude (Jacka, 1983) and
(ii) the total area of ice (Naval Oceanography Command Detachment, 1985).

The data are also available digitally, coded in SIGRID form from NSIDC.

Monthly maps of Antarctic sea ice extent and anomalies based on passive microwave satellite data are available for the period 1973–87 in two atlases produced by NASA. These cover the ESMR period of 1973–76 (Zwally *et al.*, 1983a) and the SMMR years of 1978–87 (Gloersen *et al.*, 1992). In these volumes colour-coded maps showing ice concentrations are provided, together with data for various sectors around the continent.

The NSIDC in Boulder, Colorado makes available digital sea ice data produced from several satellite missions together with raw data for research into retrieval techniques. Data available include the following.

(i) Monthly averaged Southern Ocean ice concentrations derived from the ESMR data for the period 1973–76.
(ii) Brightness temperatures and ice concentrations for every other day during the period of operation of the SMMR instrument on the Nimbus 7 satellite between 1978 and 1987. Data are made available on a set of 12 CD-ROMs. A further CD-ROM containing monthly mean ice extents will also soon be available.
(iii) Brightness temperatures and sea ice concentration maps from the SSM/I instrument since its launch on 19 June 1987. The data are available to the research community on CD-ROM.

2.8.7 Sea surface temperatures

There is a relatively long record of SST measurements over the Southern Ocean because of the large number of observations that were made by commercial vessels, many of which were concerned with whaling activities during the nineteenth century and the early part of the twentieth century. Many of these observations have been assembled into global SST data sets that have been analysed to produce monthly mean SST fields and anomalies from the long-term means. Some of the SST data sets include the following.

(i) The COADS (Comprehensive Ocean–Atmosphere Data Set) was jointly developed by several institutions in the USA, including NOAA, the National Climatic Data Center, the Cooperative Institute for Research in Environmental Sciences and the National Center for Atmospheric Research (Woodruff *et al.*, 1987). It contains a number of marine fields including sea surface temperatures computed from ship observations extending back to 1854. The data set contains 63 million non-duplicate SSTs for the period up to 1979 (Slutz *et al.*, 1985) and monthly mean SSTs for 2° by 2° latitude by longitude boxes. The data set can be obtained from the USA's National Climatic Data Center.
(ii) The USA Navy Fleet Numerical Oceanography Center (FNOC) at Monterey, California has produced the SST Consolidated Data Set (CDS). This is based on about 35 million sets of SST observations (Hsiung and Newell, 1983).
(iii) The UK Meteorological Office have assembled the Meteorological Office Main Marine Data Bank (MOMMDB) (Shearman, 1983) from a number of data sources, so that it contains about 56 million SST observations for the period up to 1989. In a combined project, the UK Meteorological Office and the Massachusetts Institute of Technology published an atlas of global SSTs based on their combined data holdings (Bottomley *et al.*, 1990). This contains monthly mean SSTs for the period

1951–80 and monthly anomalies from the 30-year means for the periods 1968–77 and 1982–83. Decadal seasonal average SST anomaly fields are also provided for 1866–1985 based on SST values corrected for the use of uninsulated buckets.

In the USA, NOAA produce weekly mean SST fields that have global coverage. These are also available in digital form and NOAA should be contacted at the above address for more details.

Weekly and monthly analyses of sea surface temperatures on a 1° latitude by longitude grid are also produced by the Australian National Meteorological Centre, Bureau of Meteorology (Warren, 1994).

2.8.8 Ocean data

The COADS data set described above contains some oceanographic data, but information on ocean currents is much more difficult to obtain, for such data are not usually collected on a routine basis by commercial vessels. We therefore have to rely heavily on data collected from oceanographic cruises.

A major advance was recently made in the understanding of ocean currents around the Southern Ocean through the development of the Fine Resolution Antarctic Model (FRAM). This model had six years' worth of data from oceanographic cruises assimilated into it, producing high-resolution fields of oceanographic data. The results from the FRAM integrations have been published as an atlas (Webb *et al.*, 1991), which contains maps of velocity, pressure and salinity for a number of levels.

Chapter 3

Physical climatology

3.1 Radiation

Short-wave or solar radiation drives the general circulation of the atmosphere and the weather systems that are observed on a day-to-day basis. Most of this energy is absorbed in the tropics and mid-latitude areas, with the polar regions receiving much less as a result of the low angle of the sun at these latitudes. The high albedo (reflectivity) of the snow and ice surface also results in much of the incoming solar radiation being returned to space so that the fraction absorbed is much less than that which is absorbed in the extra-polar regions. Nevertheless, radiation is still extremely important in determining the surface energy budget and also affects many aspects of the climate of the Antarctic, including the nature of the low-level temperature inversion, the katabatic wind regime and the stability of the atmosphere. In this section we will examine the various components of the surface radiation budget and consider their climatological values at three contrasting sites in the Antarctic. These are Faraday Station (65.3° S, 64.3° W) on the Argentine Islands, Antarctic Peninsula, which represents a relatively northerly location. Halley (75.5° S, 26.6° W), a coastal station on the eastern side of the Weddell Sea and Vostok (78.5° S, 106.9° E) on the Antarctic plateau, which is one of only two interior stations with a reasonably long record of radiation measurements. The data for Faraday and Halley cover the period 1963–82 and are taken from Gardiner and Shanklin (1989) whereas the Vostok observations cover the period 1963–73 and are from Dolgina *et al.* (1976). Although the periods covered by these two data sets are different, they are both sufficiently long to provide a broad indication of the conditions at all three sites.

3.1.1 The components of the radiation budget

The net radiative flux, Q, on a horizontal reference area is given by

$$Q = G - R - L_T + L_S \tag{3.1}$$

where G is the downwelling, global solar radiation received from a solid angle of 2π steradian. This consists of two components – a direct element, which arrives from the solid angle subtended by the sun's disc and a diffuse component, which arrives from the remainder of the sky. R, the reflected solar radiation, is the upwards shortwave radiation that has been reflected by the Earth's surface and diffused by atmospheric particles in the layer between the surface and the observation point. When measuring R close to the ground the second term is very small.

The albedo (a) of the surface is the ratio of R to G and is clearly important in determining what proportion of downwelling solar radiation is absorbed at the surface. This quantity varies considerably across the Antarctic and according to season. The amount of solar radiation absorbed by the surface is called the effective short-wave radiation and is defined as the global radiation less the reflected short-wave element:

$$E_S = G(1 - a) \tag{3.2}$$

L_T is the emitted terrestrial long-wave radiation. This can be computed from the Stefan–Boltzmann law:

$$L_T = \epsilon\sigma T^4 \tag{3.3}$$

where σ is the Stefan–Boltzmann constant (5.67×10^{-8} W m^{-2} K^{-4}), T the radiative temperature of the surface in kelvins and ϵ the surface emissivity. It should be noted that the presence of a strong surface temperature inversion can lead to large differences in temperature between that of the radiating top few micrometres of the surface and that at the standard meteorological screen height of 2 m in the Antarctic (see Section 3.2.3) and care must be taken when using screen temperature to calculate the long-wave flux. Typical long-wave emissivities for surfaces found in the Antarctic are given in Table 3.1. The experimentally determined emissivities for snow are in general agreement with modelled values estimated by Dozier and Warren (1982). However, some reports of lower emissivities for snow do exist, e.g. that by Rees (1993), but the reasons for such variations are not known. Warren (1982) considered the factors affecting the emissivity of snow and found that it was very insensitive to its physical properties, such as grain size, impurities, snow depth, liquid water content or density. However, it does vary as a function of viewing angle (Dozier and Warren, 1982), which affects the derivation of surface temperatures from satellite measurements (Bamber and Harris, 1994).

Table 3.1 *Typical long-wave emissivities for surfaces found in the Antarctic.*

Surface type	Emissivity	Reference
Snow	0.97–0.98	Carroll (1982)
		Kondo and Yamazawa (1986)
Sea water	0.94–0.98	Masuda *et al.* (1988)
Sea ice	0.92–0.97	Arya (1988)

L_S is the downwelling long-wave radiation component emitted by clouds and atmospheric gases. This is dependent on the vertical distribution of temperature and water in the atmosphere and the optical properties of any clouds present. In the absence of cloud, L_S is dominated by emission from atmospheric water vapour and carbon dioxide. Even in the relatively dry Antarctic atmosphere there is sufficient water vapour for the dominant contribution to L_S to come from atmospheric emission at relatively low levels. Under conditions of total low cloud cover, L_S will be dominated by emission from cloud water droplets and ice particles. Examination of monthly mean values of downwards long-wave radiation at four Antarctic stations (two in the coastal region and two on the high plateau) shows that L_S varies with the fourth power of the surface air temperature, T_A, and the sixth power of the temperature T_M at the top of the surface inversion layer. To a reasonable approximation

$$L_S = 0.49 + 4.70 \times 10^{-8} T_A^4$$
$$L_S = -5.08 + 6.68 \times 10^{-13} T_M^6 \qquad (3.4)$$

where L_S is in W m^{-2} and the temperatures are given in kelvins. At Neumayer, a coastal station, König-Langlo and Augstein (1994) showed that hourly values of L_S were well described by the relationship

$$L_S = (0.765 + 0.22c^3)\sigma T_A^4 \qquad (3.5)$$

where c is the fractional cloud cover. This parameterisation is unlikely to work well over the interior of the continent, where the clouds are generally composed of ice crystals, in contrast to the dense water clouds found in the coastal regions.

The two long-wave radiation terms are often combined into an effective long-wave radiation (E_L) that is the sum of the upwards terrestrial component (negative) and the downwards (positive) radiation emitted by the clouds and gases. The upwards component is usually greater than the downwelling flux, so this quantity is generally negative.

Each of these terms represents a radiative energy flux that is normally measured in W m^{-2}. However, when considering the total radiation received or lost over periods of a month or more it is convenient to express these quantities in units of millions of joules per square metre (MJ m^{-2}).

Here we will consider the climatology and variability of these components and the properties of the surface that affect their magnitude. The role of radiative fluxes in the heat budget of the Antarctic atmosphere is not discussed since this is covered in Section 4.2.

3.1.2 Solar radiation

The spectrum of solar radiation, as measured above the top of the Earth's atmosphere by satellites and rockets, extends from the X-ray region through the ultraviolet, visible and infra-red to the radio wave region. However, there is absorption of selected wavelengths in the upper atmosphere and radiation measured at the ground almost all lies between 0.29 μm in the ultraviolet and 5 μm in the infra-red. The amount of radiation that arrives at the ground is a function of the following.

(i) The latitude and time of year, which determine the radiation arriving at the top of atmosphere at a given location.

(ii) The elevation of the surface, since radiation will have to pass through a shorter atmospheric path to reach the ground for higher locations.

(iii) The water vapour content of the atmosphere because radiation is scattered and absorbed by water vapour.

(iv) The amount of ozone in the atmosphere. The decrease in springtime stratospheric ozone over Antarctica that has been observed since the early 1980s (Farman *et al.*, 1985) is now large enough to have a significant effect on solar radiation at ground level.

(v) The amount of cloud.

(vi) The turbidity of the atmosphere, i.e. the concentration of aerosol particles, such as dust, which may scatter and absorb solar radiation.

In the Antarctic, aerosol levels are usually very low and the solar elevation is a function of the time of year so that the inter-annual variability of solar radiation is dominated by variations in the amount of cloud.

Figure 3.1 shows the mean monthly global solar radiation received at Faraday, Halley and Vostok, with annual totals of radiative flux being given in Table 3.2. It can be seen that the annual cycle of global radiation received at these stations is strongly dependent on latitude with Vostok receiving almost twice the radiation of Faraday in January whereas during the winter it experiences a long period of total darkness. In fact, on the plateau during the summer, when it is the coldest place in the Southern Hemisphere, more solar radiation is received

Table 3.2 *Mean annual sums of radiative energy fluxes (MJ m⁻²) and mean annual albedo (%) for Faraday (1963–82), Halley (1963–82) and Vostok (1963–72).*

Quantity	Faraday	Halley	Vostok
Global radiation (G)	3156	3513	4664
Effective short-wave radiation (E_S)	757	632	700
Effective long-wave radiation (E_L)	−903	−937	−831
Net radiation (Q)	−152	−302	−87
Albedo (a)	76	82	85

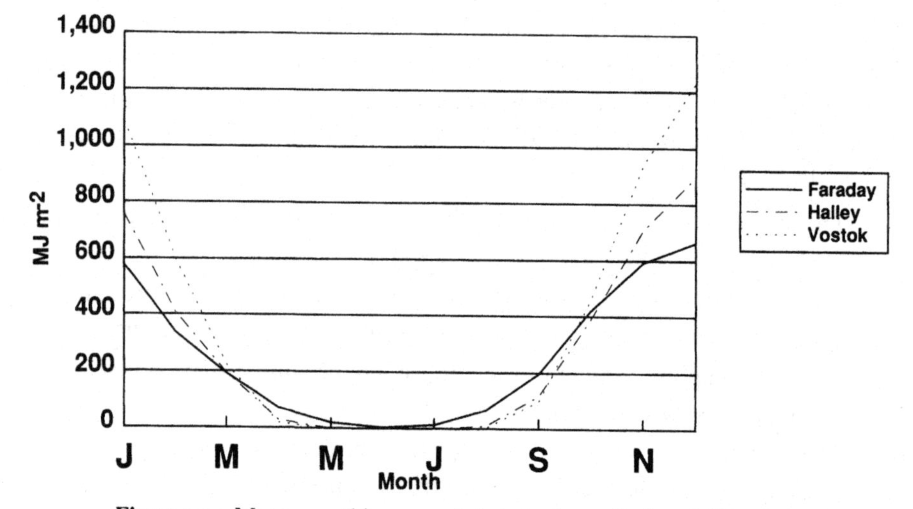

Figure 3.1 Mean monthly sums of global solar radiation (MJ m^{-2}) for Faraday (1963–82), Halley (1963–82) and Vostok (1963–73).

than anywhere else on Earth (Zillman, 1967). Faraday, on the other hand, is north of the Antarctic Circle and has no days of complete darkness in the middle of the winter. The mean annual totals of global radiation show that, despite the long period with no incoming solar radiation experienced on the Antarctic plateau, Vostok receives about 50% more radiation over the year than does Faraday. This is a result of the limited amount of cloud found on the plateau (see Section 3.4.1), the short atmospheric path that the radiation has to pass through and the very low levels of aerosol and atmospheric water vapour found in the atmosphere. Under ideal conditions the amount of solar radiation received at the surface on the plateau can be about 90% of that arriving at the top of the atmosphere (Kuhn *et al.*, 1977a) although, as discussed in Section 3.4.1, the presence of a thin veil of ice crystal cloud often reduces the fraction of the incident solar radiation which reaches the surface.

Of the global radiation arriving at the surface, about one third comes as direct radiation with the remainder arriving as a diffuse component that has been scattered by molecules and aerosol in the atmosphere. The proportion received directly varies throughout the year, with more arriving this way during the summer than at other times of the year. This is a result of the higher elevation of the sun at this time of year and the correspondingly shorter path through the atmosphere resulting in there being a smaller chance of the radiation being scattered. At more southerly latitudes, and especially on the Antarctic plateau, the atmosphere is very clear and the ratio of direct to diffuse radiation is higher.

The amount of solar radiation absorbed by the surface and therefore able to play a role in surface processes is dependent on the albedo. In the Antarctic this will generally be high as a result of the extensive snow and ice cover, although the summer melting of the snow in some coastal areas gives a large annual cycle in the albedo. The albedo of snow was investigated by Warren (1982), who used observational results and model calculations to assess the albedo across the visible and near infra-red parts of the spectrum. He found that the albedo in the visible is a function of the thickness of the snowpack, the amount of impurities it contains and the grain size, which increases as the snow ages, giving a reduction in the albedo. Snow albedo is much lower in the infra-red than in the visible region of the spectrum, it being sensitive both to snow grain size and to solar zenith angle.

Figure 3.2 shows the monthly mean albedo for the three stations considered earlier. Both Halley and Vostok have snow cover throughout the year and their albedo is always in the range 80–90%. This is consistent with the albedo of 85% found by Dalrymple *et al.* (1966) for the South Pole, showing the uniform nature of conditions across the plateau. At both Halley and Vostok the albedo is highest

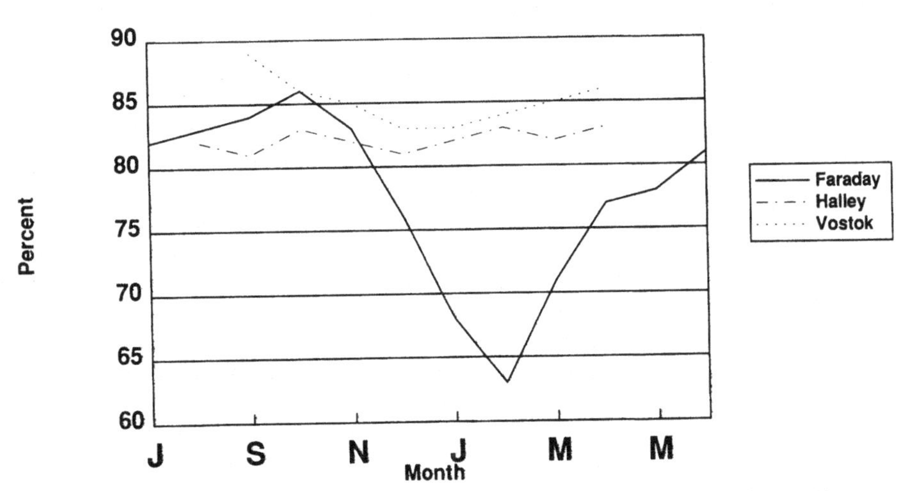

Figure 3.2 Monthly mean albedo (per cent) for Faraday (1963–82), Halley (1963–82) and Vostok (1963–73).

during the intermediate months and drops during the summer as the elevation of the sun above the horizon increases. This increase in albedo with greater solar zenith angle has been noted both for dry and for melting snow. Faraday, on the other hand, has a high albedo throughout the winter and spring, but experiences a marked drop over the summer as the snow disappears and exposes bare rock during the relatively warm conditions experienced at this northern site. However, as discussed in Section 7.1, the climate on the western side of the Antarctic Peninsula is very variable and the snow cover found at Faraday in February and March has a high interannual variability, the standard deviation of the albedo being 12% for these two months; the largest values over the year. At sites such as Faraday, where the rock is snow-free during the summer months, a large amount of heat is absorbed by the surface, resulting in warming of the lowest layers of the atmosphere and shallow convection taking place, resulting in cumulus cloud formation on many occasions.

The coastal ablation zone around the Antarctic is quite narrow but here persistent katabatic winds can scour the surface, creating snow-free areas of blue ice. This has a lower albedo than the surface of the plateau and a value of 69% has been estimated by Weller (1980) with the value thought to change but little throughout the year. The albedos of surfaces found in the Antarctic are summarised in Table 3.3.

The effective shortwave radiation E_S combines the information on global radiation and albedo to show the amount of radiation that is available at the surface to take part in the heat budget of the surface layer. This is a very useful parameter and the monthly mean values of E_S for the three stations are shown in Figure 3.3. Although the global radiation at Vostok was shown to be almost double that of Faraday in January, the differences in albedo mean that the

Table 3.3 *Typical albedos of surfaces found in the Antarctic.*

Surface type	Albedo (%)	Reference
Snow on the plateau	80–90	Gardiner and Shanklin (1989)
		Dolgina *et al.* (1976)
		Dalrymple *et al.* (1966)
Blue ice	69	Weller (1980)
Sea water	10–15	Lamb (1982)
Sea ice (concentration>85%)	75–80	Weller (1980)
		Grenfell (1983)
Sea ice (concentration 15–85%)	54	Weller (1980)
Bare ground (Bunger Oasis)	15–20	Solopov (1969)

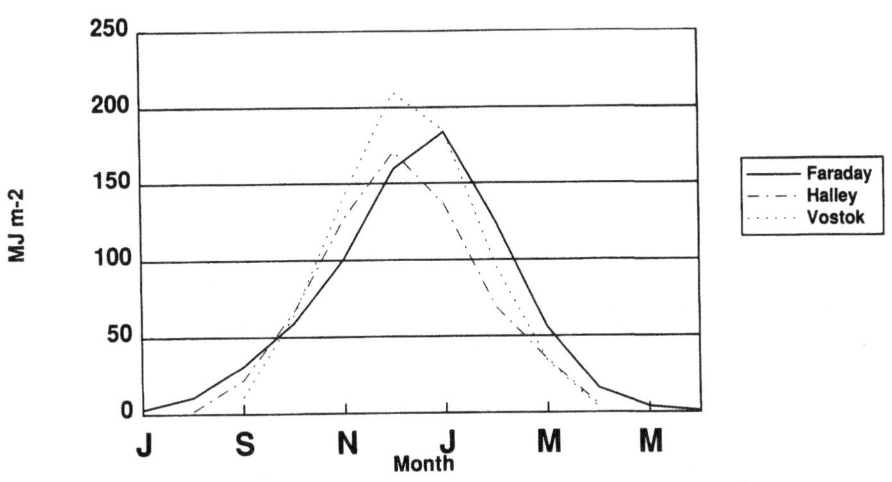

Figure 3.3 Monthly mean effective shortwave radiation (MJ m^{-2}) for Faraday (1963–82), Halley (1963–82) and Vostok (1963–73).

summer time values of E_S are very similar at these two sites. This comes about because much of the radiation received at Vostok is returned to space from the very reflective snow surface whereas the lower albedo at Faraday allows much of the solar radiation to be absorbed at the surface. Figure 3.3 shows that the form of the E_S curve is very similar at all three sites, although the data for Faraday exhibit a peak later in the summer and higher values in the autumn in comparison with the other sites, as a result of the minimum in the albedo being in February. Of course, the effective short-wave radiation over the unfrozen ocean will be greater than for any part of the continent since, despite the large amounts of cloud, the albedo is very low and the northerly location will ensure a large amount of incoming solar radiation.

3.1.3 Long-wave radiation

The monthly mean effective long-wave radiation flux, $E_L = L_T + L_S$, for Faraday, Halley and Vostok is shown in Figure 3.4. Here the negative values indicate a net flux away from the surface, which is the condition found at all three sites throughout the year. Figure 3.4 shows that E_L is a minimum during the winter and a maximum during the summer at all three sites, but with a larger annual cycle at the more southerly locations. The upwelling component L_T is a function of surface temperature and is therefore small on the plateau during the winter months. With the strong temperature inversion found over the plateau during the winter the cloud and water vapour, which emit long-wave radiation downwards, will be relatively warm, thus offsetting much of the radiation loss from the surface. On the other hand, during the summer the snow surface is much warmer so that the upwelling radiation is much greater whereas there is little increase in

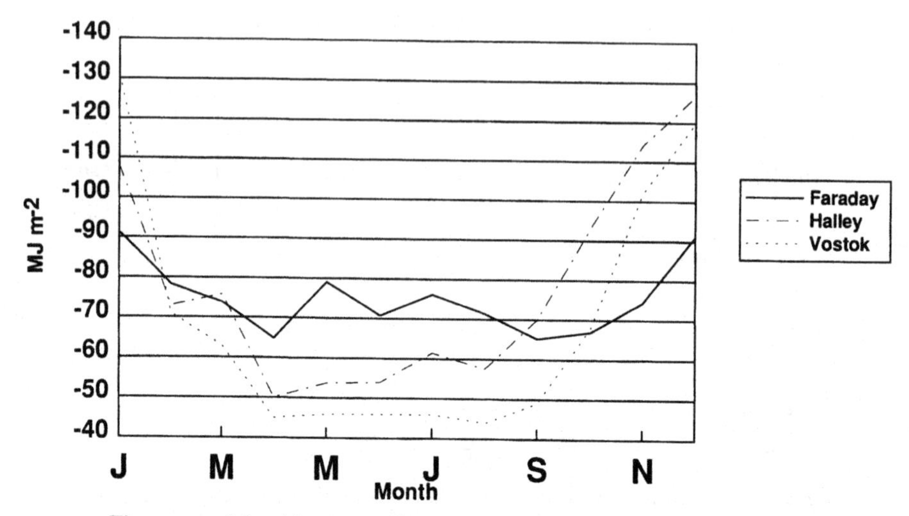

Figure 3.4 Monthly mean effective long-wave radiation (MJ m^{-2}) for Faraday (1963–82), Halley (1963–82) and Vostok (1963–73).

the downwards component, resulting in large, negative values of effective long-wave radiation. In contrast to this situation, Faraday has extensive low cloud throughout the year and relatively high surface temperatures. The upwelling radiation will therefore always be quite large but the cloud will provide a strong downwelling component throughout the year. As the annual cycle of temperature is greater close to the ground than it is at the level of much of the cloud, so the effective long-wave radiation has a smaller magnitude during the winter season. We can summarise most of the above discussion by saying that the net long-wave radiative loss is greater when the lapse rate is steeper.

The mean annual sums of effective long-wave radiation shown in Table 3.2 indicate that similar values are recorded at all three sites despite the annual cycles being of different amplitude. Halley has the largest annual value since the surface temperatures are significantly warmer than on the plateau yet there is less cloud throughout the year than at more northerly sites such as Faraday.

3.1.4 Net radiation

The net radiation, or radiation balance, is the sum of the effective short-wave and long-wave components. It indicates the extent to which a surface is gaining energy through a surplus of incoming solar radiation compared with emitted terrestrial radiation or, when there is little short-wave flux, the degree to which energy is being lost from the surface. Figure 3.5 shows the mean monthly net radiation for the three sites discussed earlier. In Figure 3.5 positive values of Q indicate net energy gain from short-wave radiation and negative values show net loss through long-wave emission. All three sites have a similar annual cycle with

Table 3.4 *Monthly sums of mid-winter net radiation at various Antarctic sites.*

Location	Net radiation (MJ m^{-2})	Reference
South Pole	−46	Carroll (1982)
Maudheim	−57	Liljequist (1956)
Mawson	−111	Weller (1968a)
Coastal sea ice near Mawson	−145	Weller (1968b)
Open ocean close to the sea ice edge	−31	Zillman (1972)

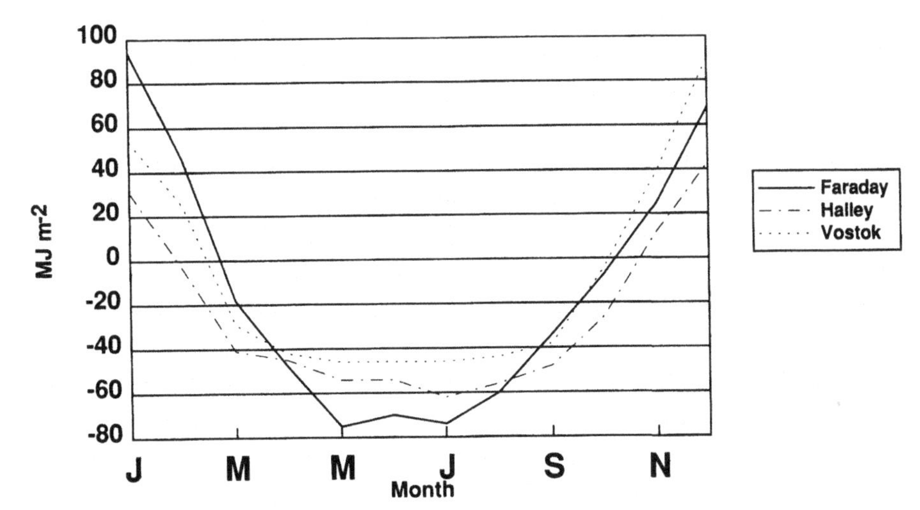

Figure 3.5 Monthly mean sums of net radiation (MJ m^{-2}) for Faraday (1963–82), Halley (1963–82) and Vostok (1963–73).

a net loss during the winter months when there is little or no incoming solar radiation and net gain during the summer. The period of net radiation loss varies from eight months at Faraday to over nine months at Halley. The differences between the annual cycles of net radiation at the three sites are dominated by the effective long-wave radiation since, as shown in Figure 3.3, the effective short-wave radiation is similar for all the locations. The greatest range of mean monthly values is at Faraday, where the limited net long-wave loss from the surface during the summer (because of extensive cloud) and relatively large long-wave cooling during the winter (because of fairly high surface temperatures compared with those at the other sites) result in a large annual range of net radiation values. The picture of the smallest losses of radiation over the centre of the continent during mid-winter with increasingly large losses at more northern latitudes is supported by studies carried out since the 1950s, which are summarised in Table 3.4 and adapted from Weller (1980). Here the net radiation losses are greater over the coastal ablation zone and over blue ice than they are in the interior of the continent, with even greater values being found over the ocean, where the surface temperature is relatively high because of heat conduction from below the surface. There have been few mid-winter measurements of net radiation over the unfrozen ocean but Zillman (1972) argues that net losses are small because of the large amounts of cloud over the ocean and the consequent large downwelling component of long-wave radiation.

The mean annual sums of net radiation shown in Table 3.2 are all quite small compared with the effective long- and short-wave components indicating the extent to which the Antarctic is close to a radiative balance, especially on the plateau. However, at all three sites there is a net loss of energy over the year, with Halley having the largest value. This comes about because of the limited amount of solar radiation received during the short summer and the large emission of long-wave radiation from a site where winter temperatures are not exceptionally cold and there is little cloud and atmospheric water vapour to radiate downwards. Further south, on the Antarctic plateau, the long summer period of perpetual sunlight provides greater input of solar radiation and the long-wave radiation lost to space is less because the surface temperatures are very cold during the winter. Here the net radiation loss is small compared with the more northerly parts of the Antarctic. At Faraday the net radiative loss is smaller than at Halley because of the greater input of solar radiation as a result of the low surface albedo in the summer months and the limited loss of long-wave radiation because of large amounts of cloud. The implications of this net radiative heat loss on the heat budget of the Antarctic atmosphere are discussed in Section 4.2.

Clouds and net radiation

From the above discussion it is clear that clouds affect the surface radiation balance in two main ways. First, they reflect some of the incident solar radiation, thus reducing the net energy gain by the surface. However, they also increase the downwards component of long-wave radiation since clouds are more effective radiators than clear air. Over high-albedo surfaces, such as the Antarctic ice sheets and sea ice, the second effect outweighs the first since the fraction of solar radiation absorbed at the surface is already small. Increased cloud cover thus, somewhat paradoxically, *increases* the net radiative warming of the surface. This effect was first noted in observations of net radiation over the Greenland ice sheet by Ambach (1974). Over the open ocean, where much of the incident solar radiation is absorbed at the surface, the short-wave effects of cloud dominate and increased cloudiness has a cooling effect on the surface.

3.1.5 Radiation in the sea ice zone

Because of the difficulties involved in making *in situ* measurements over the sea we have far less knowledge of the radiation regime here than over the continent itself and clearly there are no long series of data available. However, some data have been collected from ships and satellite data can provide some information on surface conditions. We know from visible satellite imagery that the albedo of sea ice is much more variable than that for the interior parts of the continent and values can range from the very high, when the sea ice is covered in deep, new snow, to very low values when there is extensive surface melting and pools of standing water form. The albedo of sea ice is considered further in Section 3.5.1 but here we can assume a range of values of 40–90%. No long series of instrumental values of the components of net radiation budget are available but we can assume that, because of the ice-free conditions during the summer and the leads that are present during the winter, the surface temperature is greater than that over land areas at the same latitude with a correspondingly large loss of long-wave radiation. Weller (1980) estimates that the midwinter loss of radiation is large and exceeds that of nearby land areas by about 30%. We can also expect the net radiation loss to show less of an annual cycle than it does over the interior parts of the continent, although the range in the sea ice zone will be greater than over the ice-free ocean. During the summer the relatively low albedo of the surface will result in the absorption of more solar radiation than over the land but no instrumental data are available to indicate the exact radiation balance.

3.1.6 Interannual variability

Some of the Antarctic stations have been making radiation measurements for many years, allowing the interannual variability of the various components of the radiation budget to be examined. Here we will consider the 20-year records from Faraday and Halley because these provide data from a coastal station and a more northerly site. Figure 3.6 shows the annual sum of global solar radiation

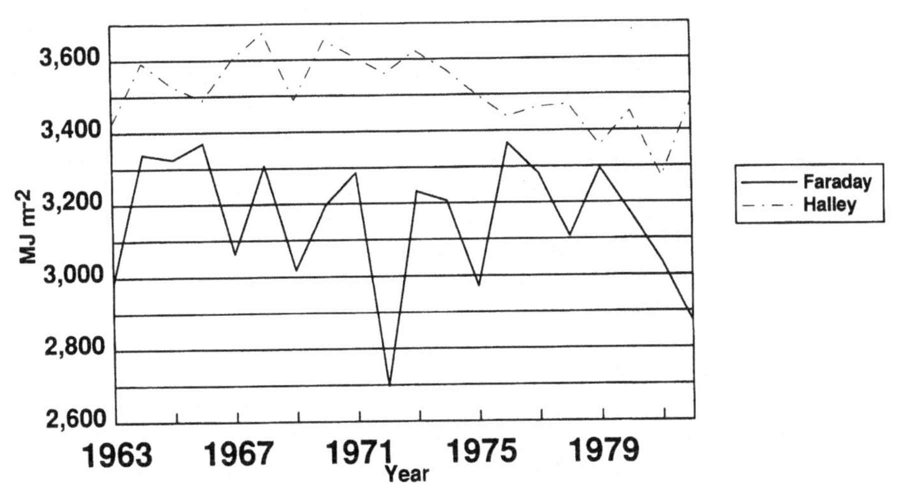

Figure 3.6 Annual sums of global solar radiation (MJ m^{-2}) for Faraday and Halley for the period 1963–82.

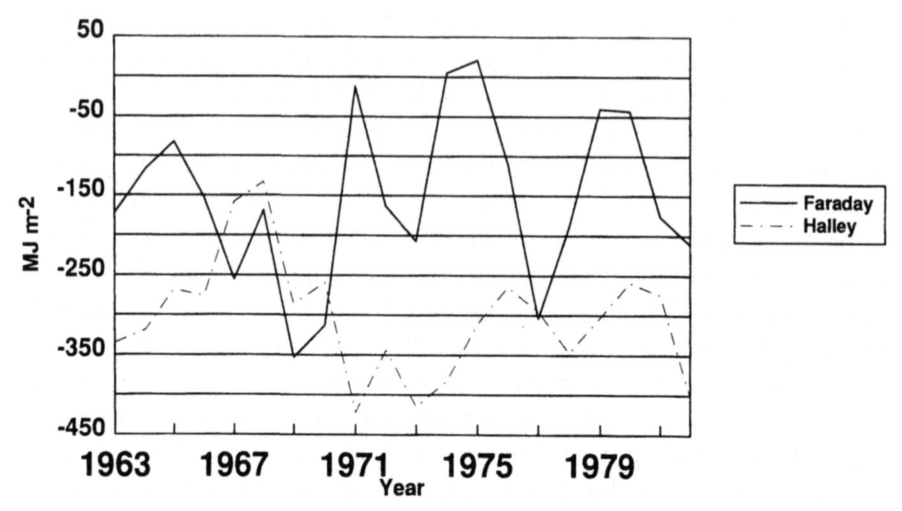

Figure 3.7 Annual sums of net radiation (MJ m^{-2}) for Faraday and Halley for the period 1963–82.

for 1963–82. The global solar radiation measured at a site is mainly determined by the amount and form of cloud that is present so that the high degree of variability in G at Faraday (SD 183 MJ m^{-2}) is a result of its location at the latitude of the circumpolar trough. Here the interannual variability of depression activity is very large and this affects the measurements of cloud amount and solar radiation. Halley, on the other hand, is south of the circumpolar trough and has much less cloud, so the interannual variability of this and the global radiation are correspondingly reduced (SD 101 MJ m^{-2}). On the Antarctic plateau, where there is even less cloud, the variability of G is smaller still. The total number of hours of sunshine and G are closely related at both sites.

The 20-year series of annual totals of net radiation for Faraday and Halley are shown in Figure 3.7. Compared with the total global solar radiation, the net radiation is relatively small and is only about 10% of G at Faraday and Halley. Because Q is the difference between two large quantities (the effective long-wave and effective short-wave totals for the year) the totals of net radiation show little correlation with G. Again, it can be seen that the interannual variability of Q is greater at Faraday than it is at Halley because of the greater variability of cloud cover at the former site.

3.1.7 Future research needs

Compared with many areas of Antarctic meteorological and climatological research, radiation studies are fortunate in having comparatively good data with which to study processes and variability with, in many cases, records extending back more than 20 years. This allows us to have a fairly good picture of the radiation regimes across the continent, although the data for the interior are much poorer than those for the coastal region. In the future, satellite data will provide a powerful tool for studying the details of some aspects of the radiation balance and of how the components are changing, although more work needs to be carried out on comparing *in situ* and satellite data sets. Satellite data will be of particular value in the sea ice zone where there have been few *in situ* measurements, especially in winter, and where it will never be possible to operate surface-based instrumentation for long periods.

3.2 Temperature and humidity

3.2.1 Geographical variation of annual mean temperature

The sparse network of climatological observing stations in Antarctica is sufficient to reveal only the broadest details of how temperature varies over the continent. In particular, there are very few records from the high interior plateau and much of West Antarctica is entirely devoid of observations. In recent years the installation of automatic weather stations has improved the situation somewhat but the records from these stations are still short and are not always sufficiently reliable or complete to use for climatological purposes.

Direct measurements of air temperature can be supplemented by two techniques that can provide improved geographical coverage. First, snow surface temperatures can be estimated from infra-red radiometric observations made from polar orbiting satellites. Provided that the emissivity of the snow surface and the correction required to compensate for atmospheric transmission are well known, measured radiances can be converted into snow surface temperatures. Bamber and Harris (1994) have demonstrated that, in the dry Antarctic atmosphere, uncertainties in snow surface emissivity are a greater source of potential error than are atmospheric effects. Satellite infra-red measurements can provide good spatial and temporal coverage, although measurements can only be made under cloud-free conditions and the detection of cloud over snow and ice is a major problem in the Antarctic (see Chapter 2, Section 6.1). A series of maps of monthly and annual average Antarctic surface temperatures has been produced by Comiso (1994) using data from the Temperature Humidity Infrared Radiometer on the Nimbus 7 satellite.

A second technique involves making measurements of temperature profiles in the snow pack. In principle, a snow temperature profile can be inverted to give a snow surface temperature history, thus the form of the annual temperature cycle can be deduced from measurements made on a single day. In practice, complete inversion can be difficult. However, analysis of the heat conduction equation shows that high-frequency surface temperature variations will be attenuated with increasing depth more rapidly than will low-frequency variations. At a depth of 10 m, the amplitude of the annual temperature wave is attenuated to about 5% of its surface value. Thus, the snow temperature measured at 10 m depth remains close to the annual average surface temperature throughout the year. A large number of 10 m snow temperature measurements have been performed on traverses, giving sufficient spatial coverage to produce reasonable maps of annual mean surface temperature, such as that shown in Figure 3.8. It should be remembered that this technique (like the satellite measurements discussed above) produces a value for annual mean *surface* temperature. On the Antarctic plateau, surface temperatures can be 2°C or more colder than air temperature measured at the standard 1.5 m height as a result of the strong stable stratification of the near-surface air.

Figure 3.8 reveals the main features of surface temperature variation across the continent. The correlation of temperature with surface elevation appears to be strong but not exact – note the double 'pole of cold' on the East Antarctic plateau, somewhat displaced from the single maximum of elevation. Also, the major ice shelves appear to be anomalously cold for their low elevation. A number of attempts have been made to develop regression models that relate surface temperature to location and elevation. Fortuin and Oerlemans (1990) found that simple linear regression with just elevation and latitude as predictor variables could provide a good fit to the observations provided that separate models were used for the continental interior, the coastal escarpment region and the ice shelves. Their regression coefficients are shown in Table 3.5. In the interior region, the elevation coefficient indicates that the surface lapse rate is

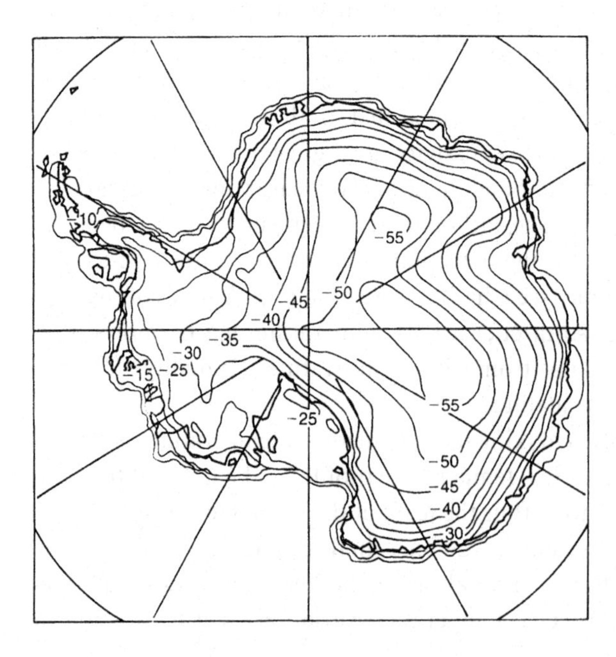

Figure 3.8 Annual mean surface temperatures over Antarctica, deduced from 10 m snow temperature measurements. Reproduced with permission from Connolley and Cattle (1994), *Antarctic Science*, **6**, 115–22 (After Radok *et al.* (1987).)

Table 3.5 *Regression parameters for surface temperature in three different regions of Antactica. Surface temperature has been modelled as $T_S = Ah + Bl + C$, where T_S is the surface temperature in degrees Centigrade, h is the surface elevation in kilometres and l is the latitude in degrees south. The final column gives the percentage of the observed variance of surface temprature in the region explained by the regression model (Fortuin and Oerlemans, 1990).*

Region	A	B	C	Variance explained (%)
Interior	−14.285	−0.180	7.405	85
Escarpment	−5.102	−0.725	36.689	59
Ice shelves		−0.943	49.642	82
Whole ice sheet	−9.140	−0.688	34.461	81

significantly superadiabatic. This is somewhat surprising since we shall see that the continental interior is characterised by katabatic winds draining outwards from the highest parts of the ice sheets and one might expect the near-surface air in such flows to warm adiabatically as it flows down the ice slopes. The superadiabatic value points to the importance of the entrainment of warmer overlying air into the katabatic flow (see Chapter 6). In the escarpment region, the regression model is less successful at describing the temperature variation. This may partly be a problem of analysis – collocation of topographic and temperature data is more difficult in this region of high slope – but probably also reflects the increased complexity of the katabatic flow in this region. The elevation coefficient is significantly subadiabatic, reflecting the stable stratification of the free atmosphere and suggesting that surface cooling more than outweighs warming by entrainment here. The surface elevation of the ice shelves varies so little that it is not possible to compute meaningful lapse rates for these regions. However, the decrease in temperature with increasing latitude appears to be more rapid here than in the interior or over the coastal escarpment. This may to some extent reflect the high correlation between latitude and surface elevation over the continent.

Of course the broad-scale temperature variations described by the simple regression models do not necessarily hold on smaller scales and there are parts of the continent where the spatial variation of temperature is more complex. The Antarctic Peninsula is a case in point. Station observations and 10 m snow temperature measurements show that the mountainous spine of the Peninsula, which reaches an elevation of 2000 m over most of its length, forms a sharp divide between the relatively mild maritime climate of the west coast and the east coast, where the climate is influenced by much colder continental air flowing west and north (Figure 3.9). Annual average temperatures on the east coast are

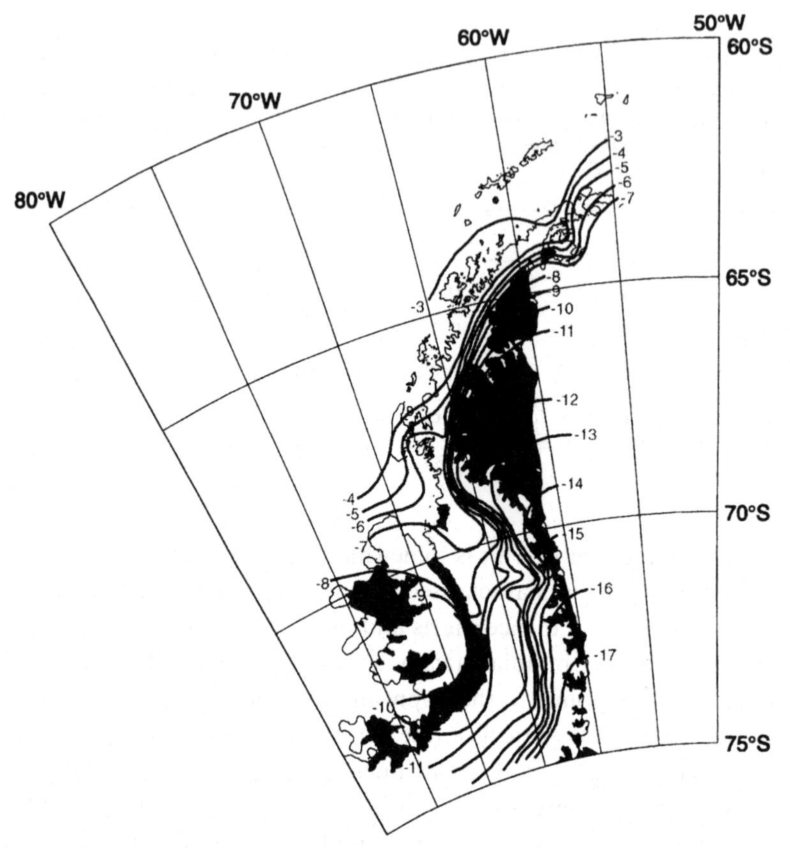

Figure 3.9 Contours of annual mean surface temperature (adjusted to sea level using observed regional lapse rates) deduced from 10 m snow temperature measurements and station observations in the Antarctic Peninsula. After Reynolds (1981).

some 7°C colder than at the same latitude on the west coast and the climatic divide appears to lie to the east of the Peninsula spine and close to the crest of the east coast escarpment (Martin and Peel, 1978; Reynolds, 1981). Surface lapse rates are also different on either side of the topographic divide. On the west side, the lapse rate is close to adiabatic at $-8.2°C\ km^{-1}$ whereas to the east a lapse rate of $+0.08°C\ km^{-1}$ reflects the strong stable stratification of the free atmosphere in this region (Morris and Vaughan, 1994).

3.2.2 Seasonal variation of surface temperature

The annual range of surface air temperatures varies considerably across the Antarctic continent. Monthly mean temperatures for selected stations are shown in Figure 3.10. Only in the northern part of the Antarctic Peninsula in summer do monthly mean temperatures exceed freezing; hence, over the greater part of the Antarctic ice sheets, there is little or no direct ablation of the snow surface. In

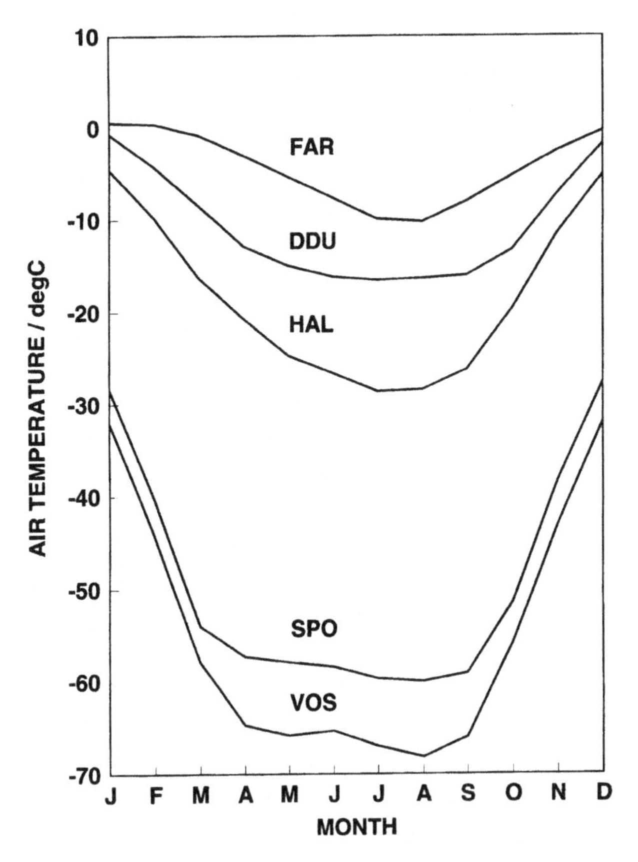

Figure 3.10 Monthly mean temperatures at Faraday (FAR, Antarctic Peninsula), Halley, Dumont d'Urville (HAL, DDU, both on the coast of East Antarctica), South Pole and Vostok (SPO, VOS, both on the plateau of East Antarctica)

the Antarctic Peninsula and around the coast of East Antarctica, the annual cycle of temperature takes a familiar form with a broad summer maximum and a minimum in July or August. However, moving southwards onto the polar plateau, the form of the cycle changes to a short, 'peaked' summer season and a 'coreless' winter during which the temperature varies relatively little. A number of factors contribute to this behaviour. First, stations south of the Antarctic Circle experience an abrupt change in solar radiation at the beginning and end of the period of winter darkness. During the dark period, the surface radiation balance is determined only by the difference between surface long-wave cooling and incoming long-wave radiation. Secondly, the annual cycle of advection of warm air to Antarctica has a strong semi-annual component, as revealed by the dominant semi-annual signal in surface pressure (see Section 3.3). Finally, the snow pack acts as a heat reservoir, smoothing out the extremes of temperature variation.

The unusual nature of the annual cycle of temperature in Antarctica means that the standard three-month seasons used elsewhere for climatological analysis may not be entirely appropriate. Périard and Pettré (1993) applied cluster analysis to ten-day temperature averages in order to produce an objective definition of the seasons at Dumont d'Urville. This suggested the following division: Summer – late November to the end of February, Autumn – March and April, Winter – May to early October and Spring – mid-October to mid-November. The transitional seasons are very short, reflecting the rapid changes occurring at the beginning and end of winter. This classification may not be appropriate for all Antarctic stations and, in practice, the standard three-month seasons are frequently used by climatologists.

3.2.3 Upper-air temperatures

The surface inversion

Probably the most notable feature revealed by upper-air soundings in Antarctica is the strong surface temperature inversion which is maintained by the radiative cooling of the surface and lower atmosphere. Although the inversion is at its strongest in the interior of the continent during the winter (Figure 3.11), when the cooling of the surface is at a maximum, significant inversions occur for much of the year even in the coastal regions. The spatial variation of inversion strength is difficult to determine since only three stations have ever made year-round upper-air soundings in the interior of the continent. However, a map showing the probable variation of inversion strength (Figure 3.12) has been produced by Phillpot and Zillman (1970), using the difference between summer and winter surface temperatures as a predictor of inversion strength. The inversion strength is usually defined as the difference between the

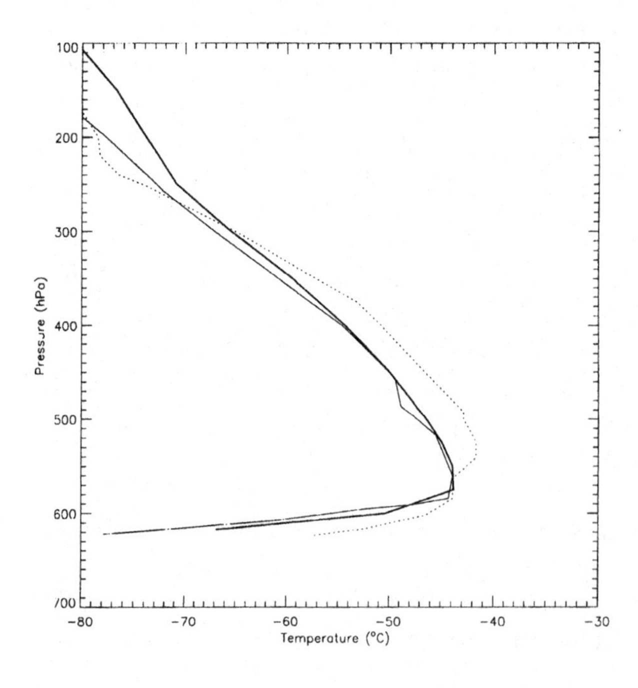

Figure 3.11 Temperature profiles for Vostok, July 1989, showing the strong surface temperature inversion. The bold line indicates the mean temperature profile, with the broken and light lines showing the soundings with the weakest and strongest inversions respectively.

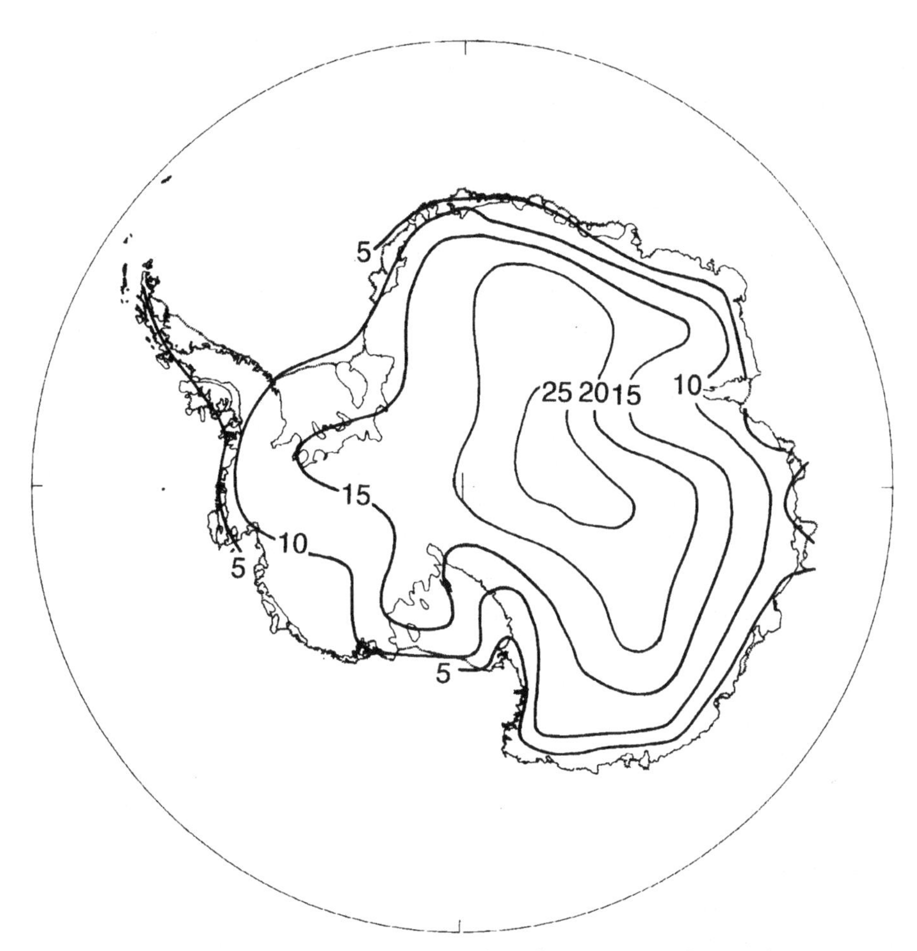

Figure 3.12 The geographical variation of the strength of the surface inversion (in degrees Centigrade) during winter. After Phillpot and Zillman (1970).

surface temperature and the highest temperature observed in the lower troposphere. An alternative way to model inversion strength is simply to relate it to surface temperature through a regression model (Jouzel and Merlivat, 1984), yielding

$$I = -0.33T_S - 1.2 \tag{3.6}$$

where I is the annual mean inversion strength and T_S the annual mean surface temperature, both in degrees Centigrade.

The seasonal variation of inversion strength and depth at an interior station (Vostok) and a coastal station (Halley) is shown in Figure 3.13. At both stations the inversion strength closely follows the seasonal variation in surface temperature and the inversion nearly disappears at Halley during the summer months. The inversion layer is deepest at both stations during winter but the seasonal variation in depth is much greater at Halley than it is at Vostok. The maintenance of the surface inversion is dicussed in Section 4.2.6.

Seasonal variation of tropospheric temperatures

The extreme static stability of the surface inversion is often remarked on and has rather overshadowed the observation that the lower part of the Antarctic troposphere is much more strongly stratified than its mid-latitude counterpart is. Tropospheric and lower stratospheric temperature profiles for Halley are shown in Figure 3.14. The troposphere is stably stratified during all seasons but the stability is strongest below 4 km during the winter when it is maintained by a combination of strong radiative cooling and subsidence (see Section 4.2.6). A tropopause is clearly evident in the summertime temperature profiles but becomes very indistinct in winter as the stratosphere cools rapidly.

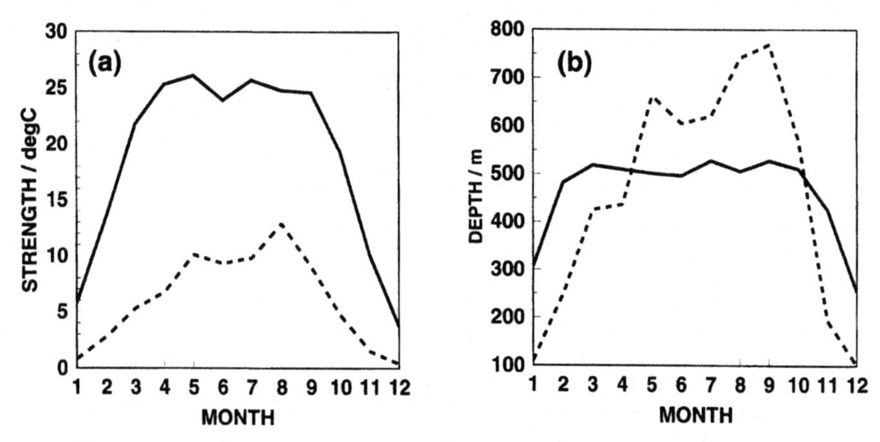

Figure 3.13 The seasonal variation of (a) inversion strength and (b) inversion depth at Vostok (solid line) and Halley (broken line).

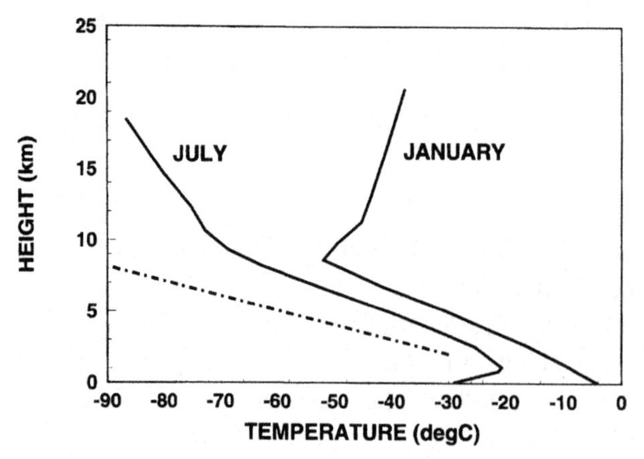

Figure 3.14 Mean tropospheric and lower stratospheric temperatures at Halley during January and July for the period 1957–93. The dashed line indicates the dry adiabatic lapse rate.

3.2.4 Humidity

Accurate determination of humidity at the low temperatures experienced in the Antarctic is extremely difficult and the problems associated with such measurements were discussed in Chapter 2. Over the continental ice sheets and ice shelves the near-surface air is close to saturation with respect to ice for most of the year, although, during the warmer summer months, significant subsaturation may occur, leading to enhanced surface evaporation rates. The frequent occurrence of ice crystal precipitation from clear skies over the high plateau (Section 3.4.2) indicates that air in the near-surface layers must be significantly *supersaturated* with respect to ice at times, otherwise the precipitating ice crystals could not form and grow. Calculations (Schwerdtfeger, 1984, p. 192) indicate that supersaturations of over 20% with respect to ice are required to grow crystals of the size observed in a reasonable length of time. The humidity of the free atmosphere is discussed in Section 4.4.

3.3 Pressure, geopotential and wind

3.3.1 Surface pressure and its seasonal variation

Once again, the sparsity of observations over the Antarctic continent and surrounding oceans makes the compilation of climatological pressure charts a difficult task. A further complication arises because the extreme elevation of Antarctica makes it almost impossible to reduce interior station level pressures to mean sea level (MSL) values in a systematic way (see Chapter 2). MSL pressures are often plotted over the interior of Antarctica but it must be remembered that this is an essentially fictitious quantity where the surface pressure is 600 hPa or less.

Seasonal mean MSL pressure charts for Antarctica and the Southern Ocean are shown in Figure 3.15. These fields exhibit three main features.

(i) The circumpolar trough of low pressure, extending around Antarctica at a mean latitude of about 66° S. This feature has no counterpart in the Northern Hemisphere and reflects the lack of barriers to zonal flow at high southern latitudes. It results from the high level of cyclonic activity around the coast of Antarctica (see Chapter 5).

(ii) A weak surface anticyclone over the Antarctic continent. As stated above, difficulties with reducing pressures to sea level values mean that the MSL pressure field depicted over the continent should not be regarded as more than a qualitative indication of the circulation in this region.

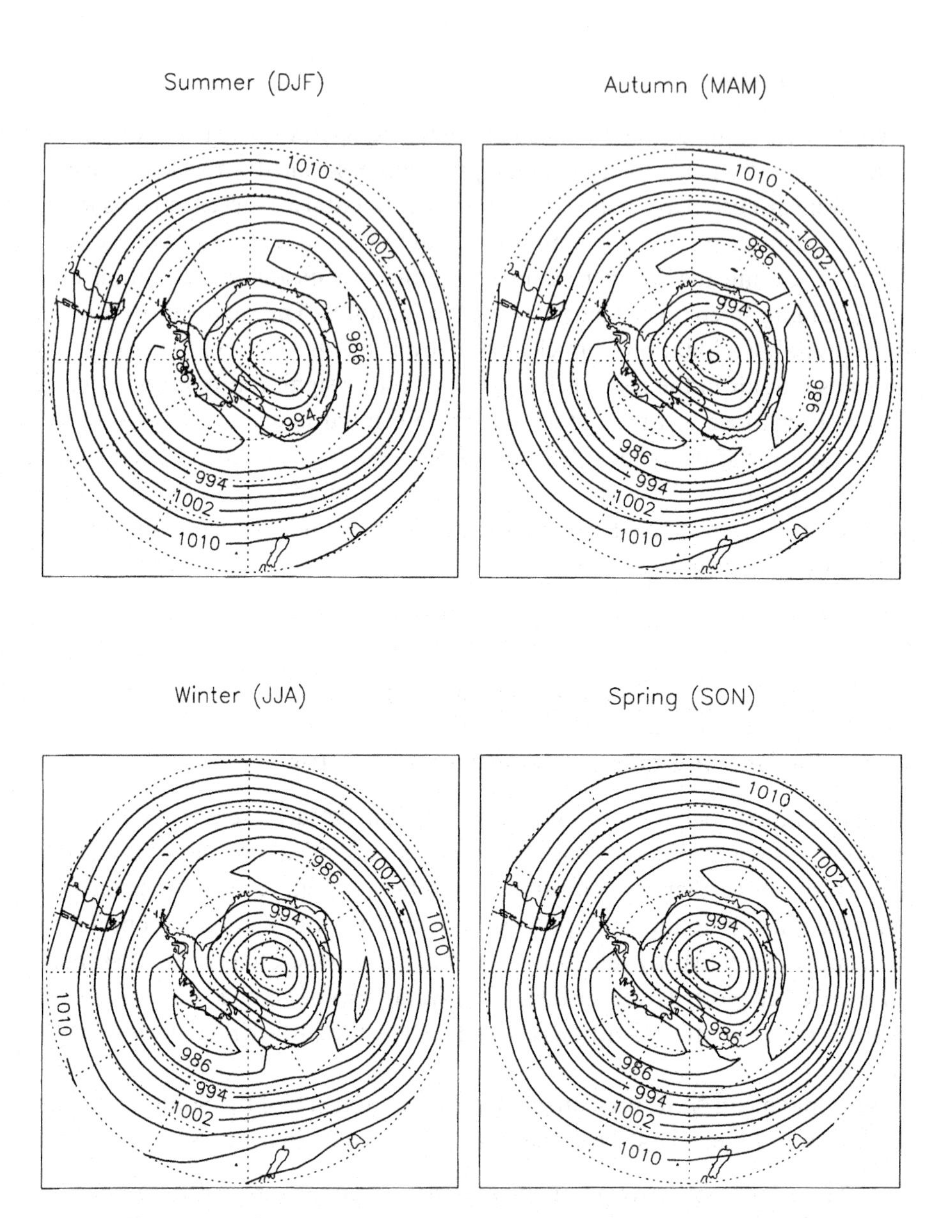

Figure 3.15 Seasonal average MSL pressure fields, from analyses produced by the Australian Bureau of Meteorology for the period 1972–91. Contours are at 4 hPa intervals.

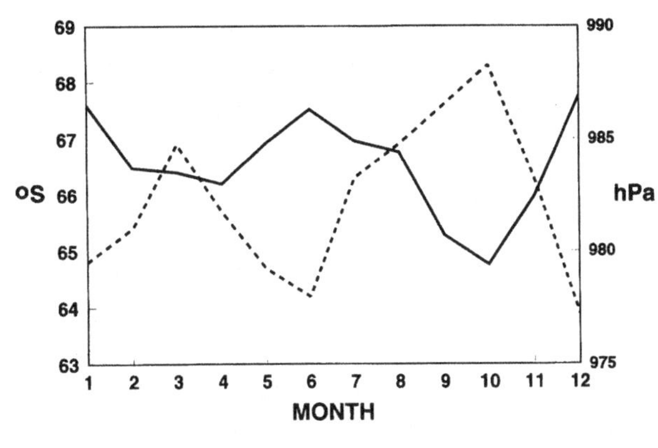

Figure 3.16 The annual cycle of zonally averaged pressure (solid line) and zonally averaged latitude (broken line) of the circumpolar trough. After van Loon (1972).

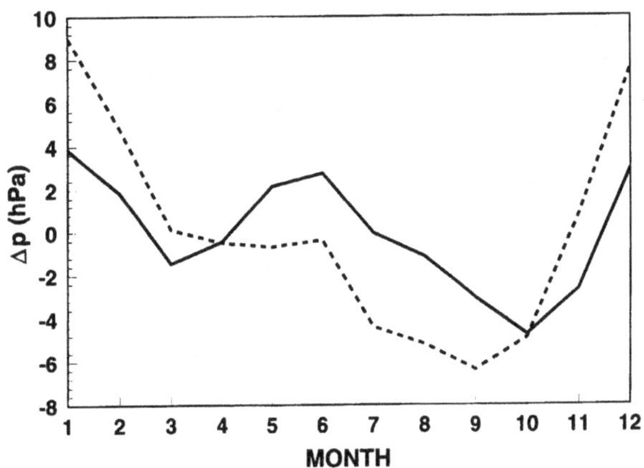

Figure 3.17 Monthly mean surface pressures, shown as a departure from the annual mean, at Halley (solid line) and South Pole (broken line).

(iii) Some departure from zonal symmetry within the circumpolar trough. Three climatological low pressure centres are seen in all seasons, located at approximately 20° E, 90° E and 150° W.

The circumpolar trough undergoes a marked semi-annual oscillation both in position and in strength (Figure 3.16). It is deepest in the equinoctial seasons and is also closest to the Antarctic continent at these times. The strength of the circumpolar westerlies to the north of the trough is largely determined by the depth of the trough, giving rise to the well-known 'equinoctial gales' of the Southern Hemisphere mid-latitudes.

The semi-annual oscillation of the circumpolar trough has been thoroughly documented in the work of van Loon (1967, 1972). It dominates the seasonal variation of pressure at Antarctic coastal stations and is apparent to a lesser extent in surface pressure records from the high plateau (Figure 3.17). Surface pressure records show well-defined maxima around the summer and winter solstices, when the circumpolar trough is at its furthest north. In the mid-latitudes of the Southern Hemisphere, to the north of the circumpolar trough, the phase of the semi-annual oscillation is reversed, with pressure maxima seen around the equinoxes. The change in surface pressure over Antarctica implies that there is a net transport of air into high latitudes in early winter and in spring, with corresponding export of atmospheric mass to lower latitudes in late winter and summer. This mass transport must reflect seasonal differences in the energy balance of the mid- and high-latitude Southern Hemisphere atmosphere.

3.3.2 Geopotential

Summer and winter mean 500 hPa geopotential height fields are shown in Figure 3.18. The 500 hPa surface is the lowest standard pressure level which is everywhere above the surface of the Antarctic continent and the flow at this level takes the form of a weak cyclonic vortex. The vortex is not, however, centred on the geographic pole but rather is displaced towards the Ross Ice shelf. At 300 hPA (Figure 3.19), the vortex is both stronger and more symmetric about the South Pole. The organized structure of the upper vortex only becomes apparent when the fields are averaged over many days (as in Figs. 3.18 and 3.19). On any given day, the flow will be highly distorted by individual weather systems. The structure and maintenance of the upper-level circulation is discussed further in Section 4.3.

3.3.3 Wind

Surface Winds

Surface winds are possibly the most intensively studied climatological element in Antarctica. Early explorers remarked on the strength and persistence of the

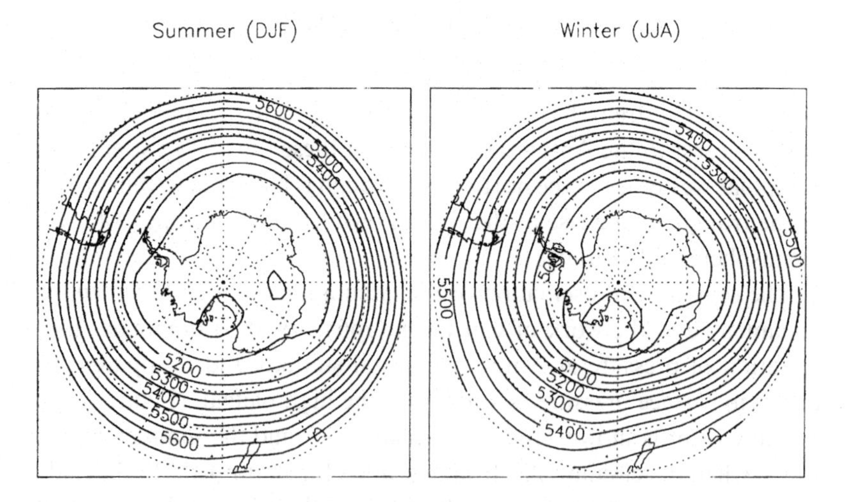

Figure 3.18 The average 500 hPa geopotential height in summer (DJF) and winter (JJA). From analyses produced by the Australian Bureau of Meteorology, 1972–91. Contours are in metres at 50 m intervals.

winds which characterise much of the continent and subsequent scientific studies attempted to determine how such winds were maintained. It is now known that the surface winds over the ice sheets are largely katabatic in origin, resulting from the drainage of cold, dense air in the boundary layer from the high interior of the continent outwards and downwards towards the coast. The forces controlling this drainage flow are discussed in Section 4.3.

Wind speed and direction measurements are available for most manned stations in Antarctica and, somewhat less reliably, from automatic weather stations. However, since surface winds are strongly affected by local topography, these measurements alone may not give a good picture of the broad-scale surface wind field. Over the surface of the ice sheets, the orientation of the sastrugi (erosional features which are aligned with the predominant wind direction) can give some indication of the prevailing wind direction. Mather and Miller (1967) produced a map of surface streamlines over Antarctica by combining station wind records with observations of sastrugi made on traverse expeditions (Figure 3.20). This clearly shows the katabatic flow draining from the high plateau, turning to the left under the action of the Coriolis force and merging with the coastal polar easterlies, which are forced by the presence of the circumpolar trough to the north of the continent. The near-surface flow thus takes the form of an anticyclonic vortex with an outflow of cold surface air from the continent.

Annual mean wind speeds for selected stations are shown in Table 3.6. Also shown are values of the annual mean directional constancy, q, defined as the ratio of the vector mean wind speed to the scalar mean. Over the interior of Antarctica, surface winds exhibit high directional constancy, indicating that the

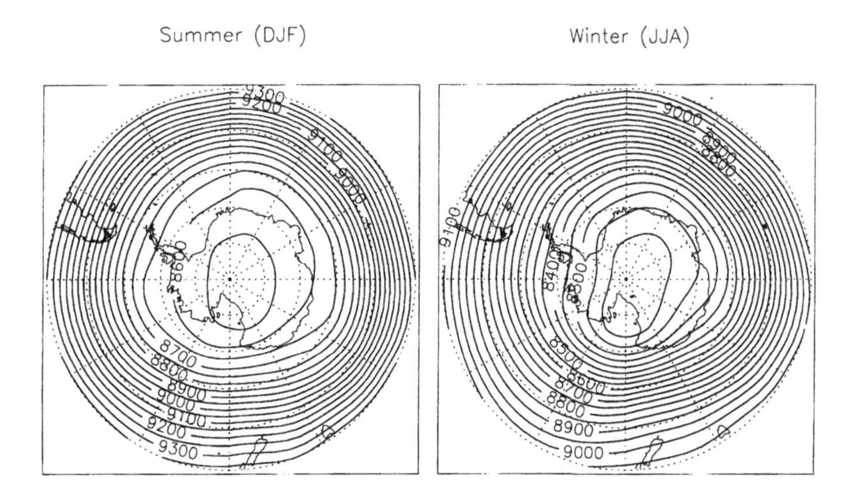

Figure 3.19 The average 300 hPa geopotential height in summer (DJF) and winter (JJA). From analyses produced by the Australian Bureau of Meteorology, 1972–91. Contours are in metres at 50 m intervals.

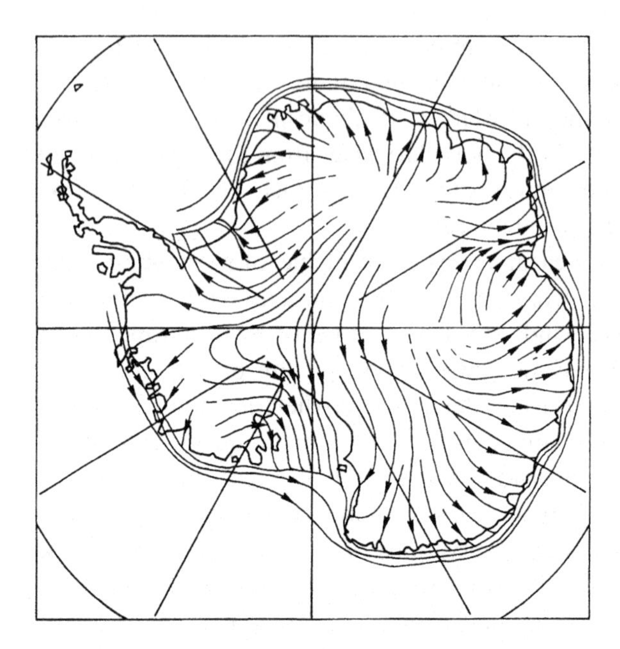

Figure 3.20 Surface wind streamlines over Antarctica. Reproduced with permission from Connolley and Cattle (1994), *Antarctic Science*, **6**, 115–22. After Mather and Miller (1967).

flow is largely controlled by the local topography through katabatic forcing and that forcing by synoptic-scale weather systems is relatively weak. Wind speeds are closely related to the local topographic slope, with the strongest winds seen at stations on the coastal escarpment (e.g. Mizuho) and the weakest on the flattest areas of the high plateau (e.g. Dome C). Coastal stations exhibit a wider range of directional constancy, reflecting the importance both of katabatic winds and of synoptically forced winds in determining the surface wind regime in this region. Along the coast of Adélie Land (represented by Dumont d'Urville and Cape Denison in Table 3.6), local topography channels the katabatic outflow onto a small stretch of coast (see Section 6.1), causing intense and persistent winds with a very high directional constancy. At Cape Denison, Mawson's 1912–13 expedition not only recorded a world record annual mean wind speed of 19.4 m s^{-1} but also experienced gale-force winds on all but one of 203 consecutive winter days. Such extremes occur only in a few regions where the local topography favours convergence of the katabatic flow; around much of the Antarctic coast annual mean wind speeds are typically in the range 5–10 ms^{-1}

Figure 3.21 shows the seasonal variation of mean wind speed at three representative stations. At Mizuho, situated on the steep coastal escarpment, there is a pronounced annual cycle in wind speed. The surface wind at this station is largely forced by katabatic drainage down the steep slopes and the strength of the katabatic forcing (see Chapter 4) will follow the annual cycle in inversion strength and surface temperature. In contrast, the wind speed at Halley, situated on a coastal ice shelf, has a significant semi-annual component. Winds at coastal stations are strongly influenced by synoptic forcing, the strength of which varies in a semi-annual fashion

Table 3.6 *Annual scalar mean wind speed (V) and directional constancy, (q), at selected Antarctic stations. Station locations may be found in appendices A and B.*

Region	Station	V (m s^{-1})	q
Continental interior	Amundsen–Scott	5.8	0.79
	Vostok	5.1	0.81
	Mizuho	11.6	0.96
	Dome C (AWS)	2.9	0.60
	D-80 (AWS)	6.1	0.91
	Siple (AWS)	5.3	0.57
Coast	Dumont d'Urville	9.4	0.91
	Cape Denison	19.4	0.97
	Mirny	10.8	0.90
	Mawson	10.5	0.93
	Syowa	5.8	0.78
	Halley	6.7	0.59
Antarctic Peninsula	Rothera	6.2	0.56
	Faraday	4.3	0.24

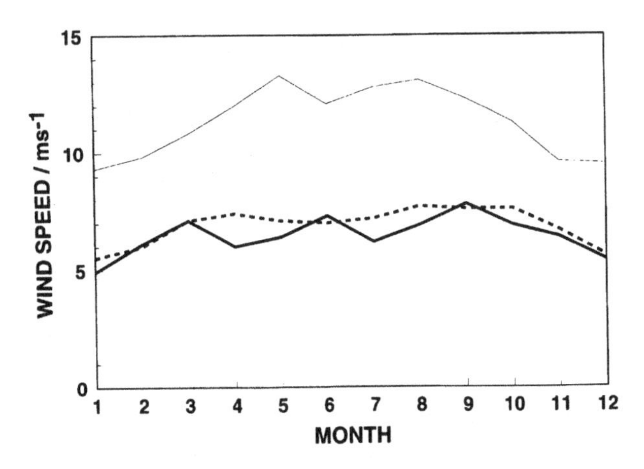

Figure 3.21 The annual variation of mean surface wind speed at South Pole (heavy solid line), Mizuho (light solid line) and Halley (broken line).

with the movement of the circumpolar trough. At the South Pole, topographic slopes are smaller than they are at Mizuho so katabatic forcing of the wind is smaller. The rather complex seasonal variation of surface wind speed suggests that both katabatic and synoptic forcing are important at this station.

Upper-level winds

The wind field in the free troposphere broadly follows the pattern expected from the pressure and geopotential fields presented in the previous section. Annual average upper wind vectors measured at radiosonde stations are shown in Figure 3.22. The flow at 850 hPa broadly follows the anticyclonic pattern seen at the surface (Figure 3.20), although the outflow of cold air from the continent is already substantially weaker at this level. At 500 hPa, the flow is rather weak but is generally cyclonic, with inflow replacing outflow at many locations. At 300 hPa there is a well-developed cyclonic vortex.

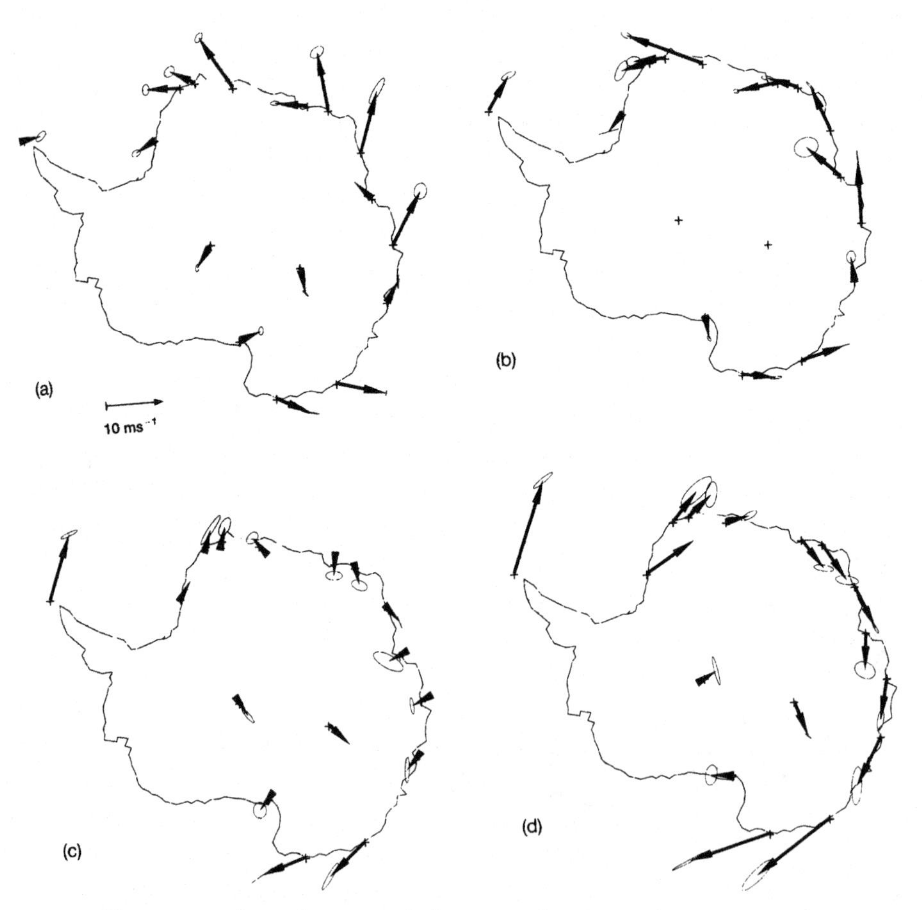

Figure 3.22 Annual average winds measured at Antarctic radiosonde stations: (a) surface, (b) 850 hPa, (c) 500 hPa and (d) 300 hPa. The ellipses at the ends of the arrows indicate interannual variability. From Connolley and King (1993).

Seasonal wind profiles at Molodeznaja are shown in Figure 3.23. During the summer, both the low-level winds and the upper cyclonic flow are weak. As the surface and atmosphere cool during autumn, the katabatic outflow at low levels intensifies and the upper vortex accelerates. This stronger flow persists through the winter, weakening again once the spring warming starts.

3.4 Clouds and precipitation

3.4.1 Clouds

Clouds are a very important component of the Earth's climate system by virtue of their role in radiative processes and the hydrological cycle. Because they have a relatively high reflectivity or albedo they return a large proportion of the incident solar radiation back to space and reduce the short-wave radiation received

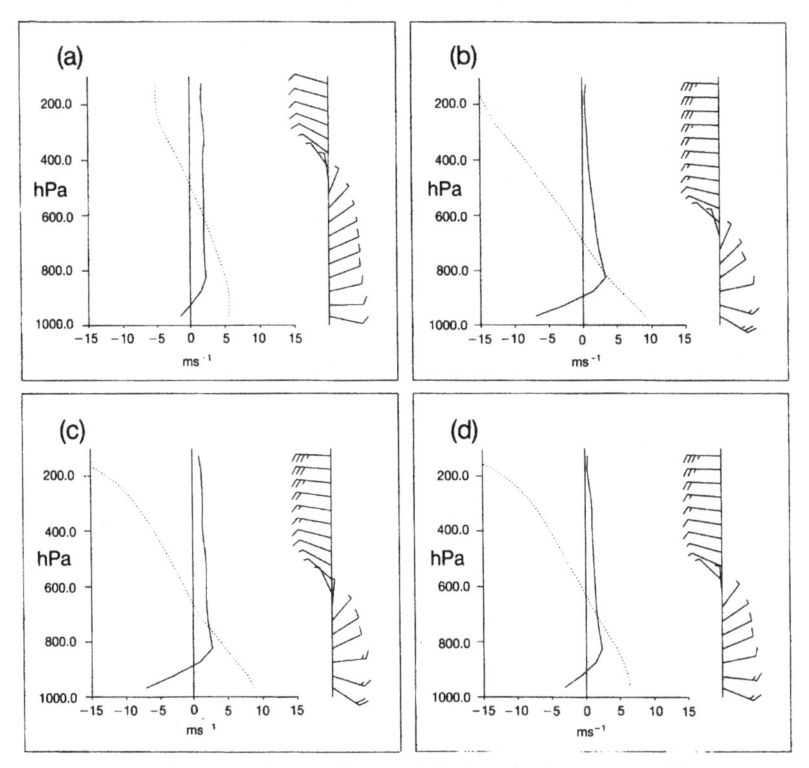

Figure 3.23 Seasonal wind profiles at Molodeznaja: (a) summer, (b) autumn, (c) winter and (d) spring. The solid line is the N–S wind component, northerly positive; the dotted line is the E–W wind component, easterly positive. Wind arrows: full barb 5 m s^{-1}, short barb 2.5 m s^{-1}. From Connolley and King (1993).

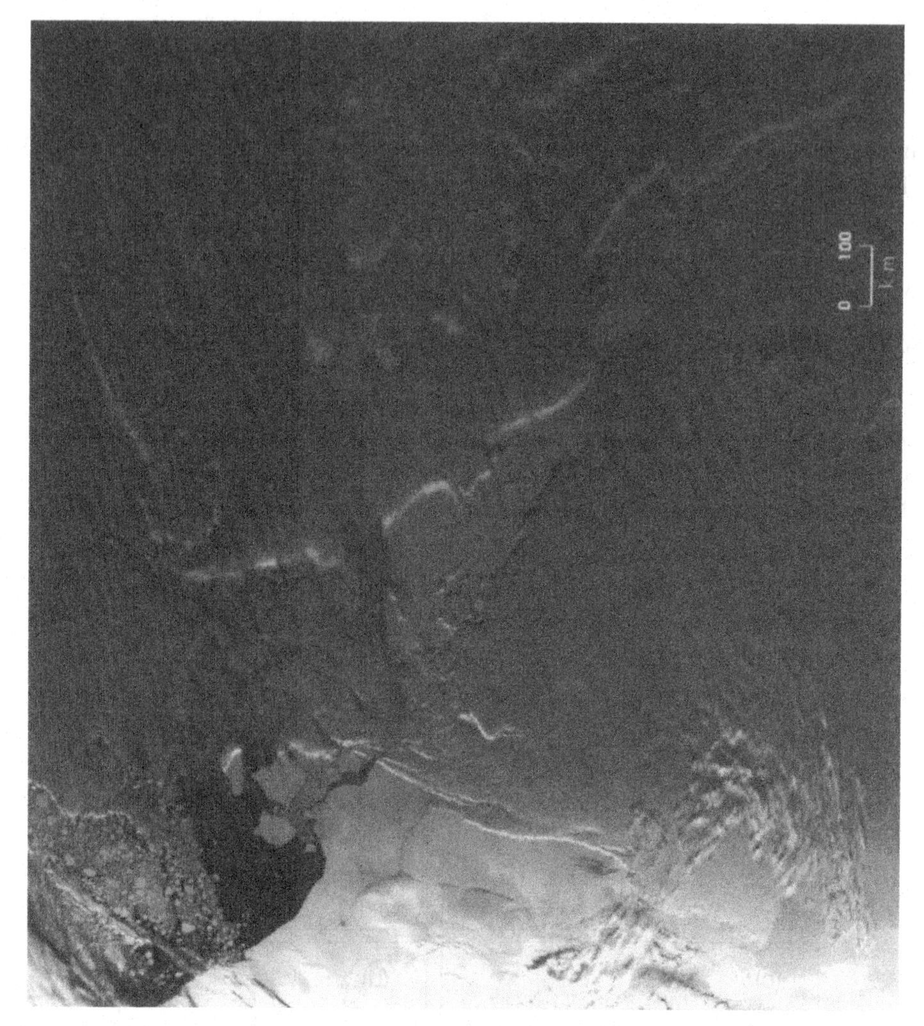

Figure 3.24 A visible wavelength satellite image of part of the interior of the Antarctic, showing clouds over the snow-covered surface.

at the ground. When clouds form over the ice-free ocean, which has a low albedo, they have a significant effect on the radiation balance at the top of the atmosphere by reflecting most of the radiation back into space. On the other hand, clouds over the Antarctic continent itself have less of an effect on the short-wave radiation balance because they have an albedo similar to that of the snow-covered surface. Satellite imagery, such as that in Figure 3.24, shows that the albedo of the cloud over the continent can be slightly higher or lower than that of the surface depending on the microphysical properties of the cloud, but is never significantly different from that of the ice and snow. The effects of clouds on the surface radiation budget are discussed in detail in Section 3.1.

The second major role of cloud is in the production of precipitation. In the extra-polar regions virtually all precipitation comes from moderately deep convective or layer cloud. Within these clouds, droplets grow until they are heavy enough to fall out of the cloud and reach the surface as rain or snow. In the Antarctic coastal region the few studies carried out to date suggest that precipitation comes mostly from relatively thin cloud. Many of these precipitating clouds are associated with major low-pressure systems, although shower clouds and isolated areas of layer cloud can bring significant amounts of precipitation. However, in the interior of the Antarctic, observers have noted that many suspended ice crystals are found in the lower layers of the atmosphere and that these gradually fall to the surface as 'clear sky' precipitation. This very thin veil of cloud obviously causes difficulties in determining what to report in the surface meteorological observations and this problem will be discussed later in this section.

From the preceding discussion it will be clear that knowledge of the amount and types of clouds found in the Antarctic is required for climatological investigations of the radiation budget of the continent, climate monitoring and the parameterization of cloud in numerical models. However, after a number of decades of investigation we still have a very poor understanding of the cloud distribution in the Antarctic and its variability on a range of time scales. The use of surface observations and satellite imagery to produce statistics on cloud cover presents many problems (Hughes, 1984) but the development of an accurate cloud climatology for the Antarctic is a high priority for many climatological and modelling studies and work towards this goal is currently taking place. In this section we will examine some of the problems that stand in the way of this goal and consider the progress made and the currently available data sets.

Observations of cloud amount from conventional surface observations

The difficulties associated with making accurate and reliable cloud observations in Antarctica were discussed in Chapter 2. Despite these problems, synoptic observations of cloud type and amount have been made since the earliest expeditions, enabling climatological summaries to be compiled. The most extensive study of this kind was carried out by Warren *et al.* (1986, 1988), who examined cloud cover world-wide via the cloud reports included in the surface synoptic reports from land stations and ships. The data used consisted of observations from land stations for the period 1971–81 and reports from ships for 1930–81. The ship reports are obviously concentrated along the major shipping lanes of the world so they do not provide a great deal of information on the spatial distribution of cloud across the Southern Ocean. However, the availability of some ship reports from the vessels involved in whaling during the middle part of the century does at least provide some coverage of Antarctic waters. Over the Antarctic continent most of the manned stations are around the coast so that data for the interior are based on only a few stations. At these sites problems can arise over the classification of clouds because of the very high altitude of the stations and the consequent interpretation of the terms low, medium and high cloud. For example, at South Pole Station, cloud that had a base at less than 1000 m was, until 1986, always reported as being at medium level, reflecting its absolute elevation above sea level rather than the height above the ground. This resulted in liquid water clouds in the lowest kilometre of the atmosphere being reported as altocumulus or altostratus, rather than stratocumulus or stratus (Warren *et al.*, 1986). However, this was not the case for the USSR's stations on the plateau (Rusin, 1961).

Satellite observations of cloud

More recently, attention has focused on the use of satellite data to provide information on global cloud cover by using imagery from the geostationary and polar orbiting satellites. This technique has the advantage that the satellites provide true global coverage so that the problem of the very uneven distribution of the observing network across the world that plagues surface-based studies is overcome. It is also possible using satellite data to derive information on cloud micro- and macro-physical properties such as albedo, optical depth and droplet size, although this information is much easier to determine in the extra-polar regions than it is in the Antarctic. Unfortunately, there are a number of problems in estimating cloud amount and type from space and the difficulties that multi-layered and very thin cloud present make the correct interpretation of the imagery very difficult. This is especially so in the Antarctic, where visible imagery is not available for long periods of the winter, much of the cloud has a cloud top temperature that is similar to that of the ice surface and thin cirrus is often present. The early satellite-based cloud climatologies were assembled by the manual analysis of many individual images. This was an extremely time-consuming task and later work used automatic techniques based on the assessment of the data by computer systems. As discussed in Section 2.6.1, some progress has been made in recent years on the question of automatically detecting cloud in imagery so that it is now possible to consider the production of climatological fields of cloud cover from long series of satellite images. The main focus for this work has been the International Satellite Cloud Climatology Project (ISCCP) of the World Climate Research Programme (Schiffer and Rossow, 1983) that came into being during 1982. This project undertook the collection of global radiance data from the operational satellites since 1 July 1983 for the production of climatological data on cloud occurrence and properties. The data collected consist of visible and 11 μm thermal infra-red imagery from the polar orbiting and geostationary satellites sub-sampled to a resolution of 25–30 km. The processing consists of using the radiances, in conjunction with information on the surface conditions, to detect clouds and produce statistics on a coarse, 280 km grid. In the Antarctic, the similarity between the cloud-top and surface temperatures and albedos means that the ISCCP processing algorithms may miss significant amounts of cloud (Rossow, 1993), although testing of an early version of the operational algorithm on a limited polar data set found that too much cloud was identified (Rossow, 1987). Because of concerns about the polar element of ISCCP, a workshop was held to examine possible algorithms and it recommended that, for a reliable cloud climatology of the Antarctic to be prepared, the full five-channel AVHRR data need to be used (Raschke, 1987).

In the following sections we will examine the climatological occurrence of cloud in the Antarctic and the distribution of various cloud types through the results of the studies that used surface observations and satellite imagery. Where possible we will attempt to resolve discrepancies between the different data sets in terms of the characteristics of the data and the processing and interpretation that has taken place.

Distribution

Composite satellite images, such as that shown in Fig 5.15 later, indicate that the zone of greatest cloud cover in the high southern latitudes occurs over the ocean area north of the edge of the continent, where the synoptic-scale weather systems are most active. North and south of this band the amount of cloud decreases, relatively little cloud occurring over the interior of the continent. This observation is confirmed by the results of Warren *et al.* (1986) and their data on zonally averaged total cloud amount for land stations for the four seasons are shown in Figure 3.25. Although there are only a few land stations along the 60° S line of latitude the results from ship observations (Warren *et al.*, 1988) are very similar, with a peak of 85–90% cloud cover close to this latitude. The main difference between the land and ship cloud statistics is a more gradual decrease of total cloud amount to the south in the ship data.

Figure 3.25 shows that the zone covering the Southern Ocean near 60° S is the cloudiest place in the Southern Hemisphere, with around 85–90% cloud cover throughout the year. Here the variation between seasons is very small with only a slight decrease in cloud amount during the winter months. This is probably a result of the extensive sea ice around the Antarctic in this season reducing the flux of water vapour into the lowest layers of the atmosphere. The latitude of maximum cloud amount is north of the belt of lowest surface pressure associated with the circumpolar trough, probably as a result of most of the cloud associated with synoptic-scale depressions lying equatorwards of the actual low-pressure centres of individual systems. However, the zone of maximum cloudiness is south of the belt of maximum westerly wind. The position of the cloud maximum found by Warren *et al.* is in agreement with that given by Schwerdtfeger (1970), who examined the fractional cloud cover at stations over a range of latitudes and found that Orcadas, near 60.7° S in the South Orkney Islands, had the highest annual mean coverage, 87%. The data quoted above are

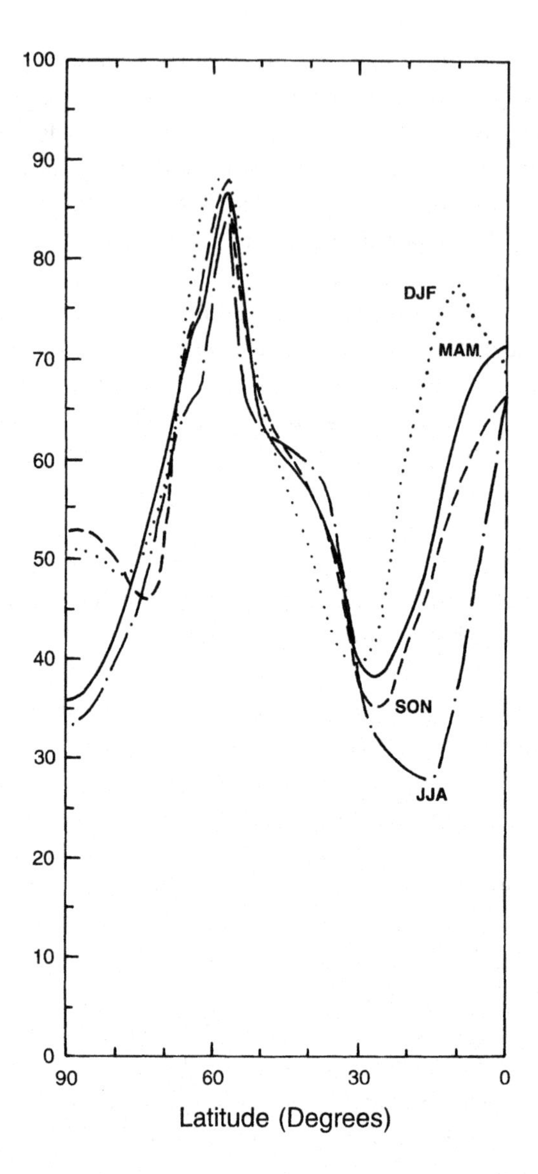

Figure 3.25 Zonal average total cloud cover (per cent) over land as a function of latitude for the four seasons. After Warren *et al.* (1986).

also in agreement with the ISCCP results, which are presented as maps of Southern Hemisphere mean cloud amount in Figure 3.26. These show a similar belt of high cloudiness around 60° S with an annual mean cloud amount of 80–90%. It shows some minor variations around the Southern Ocean with cloud amounts in excess of 90% south of Africa and to the east of South America; both are areas of extensive cyclonic activity (see Section 5.3.2).

In the coastal region of the Antarctic, near 70° S, the surface observations show a total cloud cover of around 45–50% with little variability throughout the year and only a small decrease during the winter months. The ISCCP data show a similar decrease in cloud between 60° S and the coast of the Antarctic with the

(a)

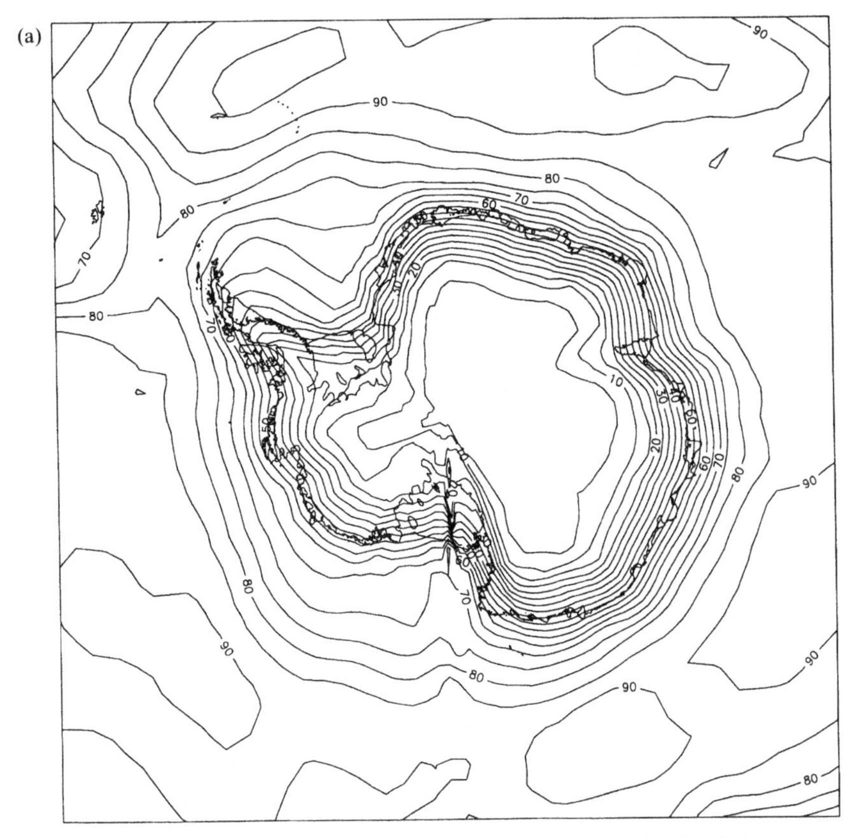

Figure 3.26 Mean annual cloud amount (per cent) from ISCCP: (a) summer and (b) winter (overleaf).

coastal area having 40–60% cloud cover. The ISCCP data indicate little variability of cloud cover around the coast of the continent, although the Antarctic Peninsula and the western Weddell Sea, in the lee of this north–south topographical barrier, have less cloud than do the ocean areas at these latitudes. South of the Peninsula, the mean cloud amount is higher than that in other areas at a similar distance into the interior of the continent, reflecting the number of depressions that pass between the southern Bellingshausen and southern Weddell Seas by this route. The belt of higher cloud amount that is located in mid-latitudes extends closer to the coast of the continent where there is greater cyclonic activity, such as around the coast of East Antarctica and over the Bellingshausen Sea. Synoptically quiet regions, such as the Weddell Sea, correspondingly have less cloud on average.

The amount of cloud over the continent decreases rapidly inland from the coast in line with the decrease in precipitation and the occurrence of active weather systems. However, the seasonal variation in the interior is much larger than that over the ocean or the coastal region and changes from about 35% in

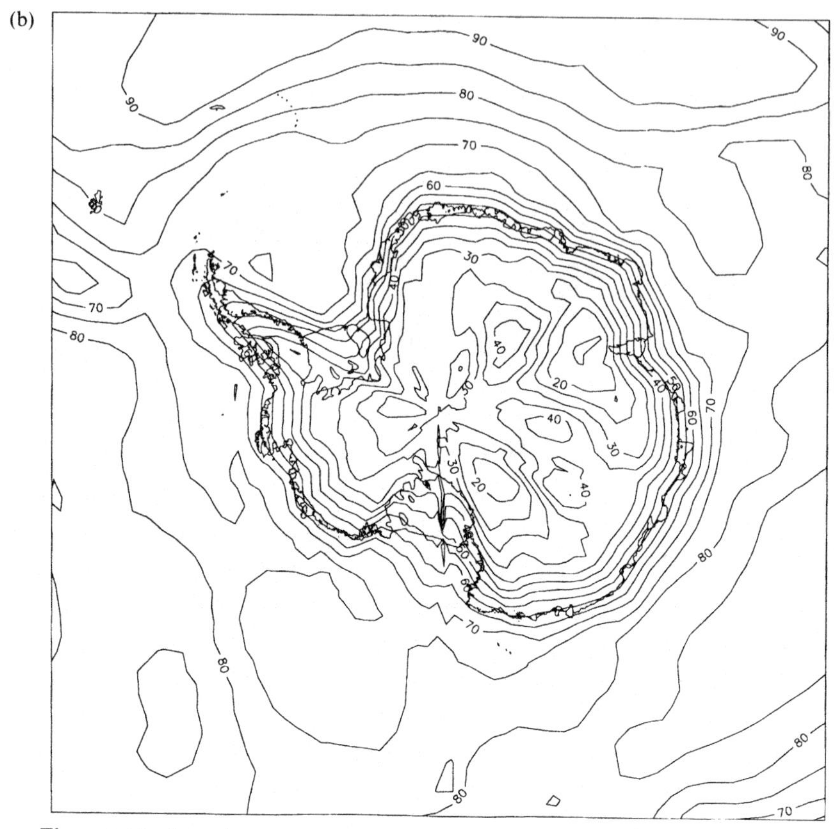

Figure 3.26 (*cont.*)

autumn and winter to close to 50–55% in spring and summer, according to the surface observations. The greatest seasonal variation is in the 80–90° S band, with the cloud amount becoming much more seasonally independent towards the coast. Because of the problems in detecting cloud during the period of winter darkness it is not known whether this seasonal cycle in the interior is real or rather a result of the observational problems when there is no sunlight. The spatial variation of cloud amount over the continent is difficult to determine with a high degree of confidence because of the limited number of manned stations and the problems in automatically interpreting the satellite imagery. However, manual examination of satellite imagery and the first results of ISCCP are in agreement in suggesting that West Antarctica, with its lower topographic height, has more cloud than does high East Antarctica. At the South Pole the surface observations indicate that the annual average cloud amount is about 45% but the ISCCP results have a lower estimate of about 20–30%. This discrepancy is probably a result of the difficulties in detecting the often thin ice clouds over the continental interior and suggests that further development of the cloud detection algorithms is needed for use at high southern latitudes. The amount of cloud over

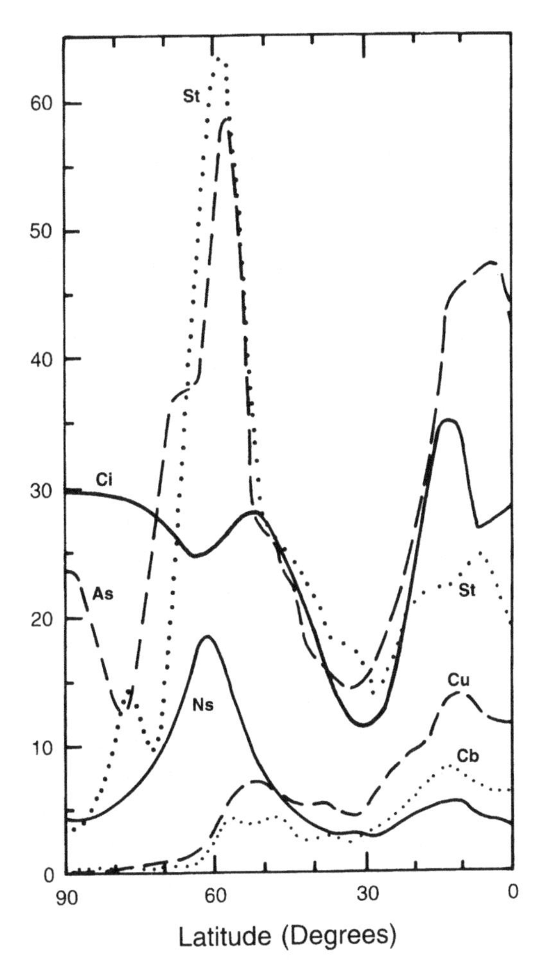

Figure 3.27 Zonal average amounts (per cent) of each cloud type over land, for December–February, as a function of latitude. After Warren *et al.* (1986).

the continent is strongly correlated with topography and this was found by Schwerdtfeger (1970), who examined the synoptic reports from two plateau stations with elevations between 2400 and 2700 m and from two in the elevation range 3500–3600 m. Although the records for these stations only covered periods of a few years, they were in agreement with the general picture of limited cloud in the interior, with annual mean cloud cover totals of 53 and 36% at the lower and higher stations respectively.

Cloud type

The distribution of different cloud types is illustrated by the graph of zonally averaged cloud amount for summer produced from surface observations over the land by Warren *et al.* and shown in Figure 3.27. Reference will also be made to the corresponding oceanic data derived from ships' observations that are not shown here. The most striking feature of the cloud distribution is the large amount of stratus close to 60° S, which is a feature both of the land and of the ship data. This is associated with the large number of depressions at this latitude and is a feature of all seasons, with only a slight drop during the winter months. This extensive low cloud obscures cloud at high levels from the surface observer but there are also large amounts of altostratus reported, especially over the land. The amount of stratus reported drops rapidly towards the south and, at the more southerly sections of the Antarctic coastline, the amount of stratus is exceeded by that of cirrus and altostratus. At 70° S, cirrus and altostratus are found in about equal amounts and are included in 25–30% of routine reports, stratus being found in about 15%. There is little variation in these values throughout the year.

From the oceanic cloud type statistics of Warren *et al.* cumulus cloud would appear to be quite rare and is only reported in about 5% of ship reports at 60° S. However, satellite imagery shows extensive low-level convective cloud in southerly outbreaks of cold, continental air over the ice-free ocean. Sometimes this cloud can extend up to medium levels, but the deep cumulonimbus clouds found in cold air outbreaks in the Arctic are not found in the Antarctic, although such clouds may occur north of 60° S.

Over the interior of the Antarctic, cirrus is the most commonly reported type of cloud, with most being found in the spring and the least during the winter months. Here the problem of detecting the cloud when there is no sunlight probably means that the figures for the winter are an underestimation, especially of thin cirrus. Altostratus is the next most commonly reported cloud type, although, as discussed earlier, this will include some cloud that has a cloud base of less than 1000 m above the surface. Although some reports of cumulus over the interior of the Antarctic have been recorded (Kuhn, 1969), these events are rare because the great stability of the atmosphere favours the formation of stratiform cloud.

Cloud micro- and macro-physical properties

Compared with the extensive research that has been carried out into clouds in the tropics and mid-latitude areas, there has been very little work on the microphysics of clouds in the Antarctic. This is mainly because of the difficulties of operating the well-equipped research aircraft that are the main platform for cloud physics research from small airstrips on the continent. Without aircraft data, cloud investigations are limited to that which can be achieved using remote sensing from satellites and the surface, and the collection of *in situ* data from tethered balloons and instruments operated from high ground that penetrates some clouds. As discussed earlier, satellite remote sensing still has problems in detecting some clouds over ice-covered surfaces and it is not possible to estimate properties such as optical depth and droplet size with a high degree of confidence at the present time. Although some satellite imagery, such as the AVHRR 3.7 μm imagery from periods with solar illumination, can show considerable structure in cloud tops that would appear to be related to droplet size, as well as help differentiate between glaciated and non-glaciated clouds (see Section 2.6.1), the value of these data is limited until we have better *in situ* data to help in their interpretation. Work on Antarctic cloud properties has therefore mostly been carried out using surface-based remote sensing instruments, a number of lidars having been deployed in recent years. One such instrument was operated at the French Dumont d'Urville station for a year to determine the optical and radiative properties of clouds above that location (del Guasta *et al.*, 1993). This work was aided by the coincident radiosonde data that were obtained each day and analysed in conjunction with the lidar observations. Dumont d'Urville was often found to have a temperature inversion between 1000 and 3000 m that resulted in stratiform low cloud forming on many occasions. The tropopause over the year varied between 8 and 9 km in height and cirrus was often found to extend up to this level. The lidar data allowed the calculation of statistics of cloud base and cloud top height together with mid-cloud temperature and height. At this location the high clouds were thicker than those found at mid-latitudes with a thickness of around 3000 m. The uniformity of depolarization of the lidar returns suggested that the ice crystal form and size distribution were also very similar in the deep clouds that had temperatures below −40°C.

In the coastal region clouds are relatively warm and contain significant concentrations of liquid water droplets. However, on the plateau (and especially during the winter season) clouds are very cold and composed primarily of ice crystals (Morley *et al.*, 1989). The investigations by del Guasta *et al.* (1993) therefore provided information on the properties of clouds in the coastal region but the results are not applicable to cloud found on the plateau.

Studies of cloud properties have been carried out at South Pole Station by Stone (1993) using radiometersonde data in conjunction with ancillary meteorological measurements. This study was concerned with winter season clouds associated with large-scale synoptic disturbances that reached the Pole, often from the Weddell Sea. These clouds were found to be moderately thick and to extend typically to heights of more than 3 km with the cloud base being at the top of the surface temperature inversion. The clouds were found to be optically thin and non-black with a mean effective emissivity of about 0.6. Because the temperature was less than −40°C, the clouds were composed of small, non-spherical ice crystals that usually had the shapes of bullets, plates and prisms. The total ice content was low and the clouds contained little or no liquid water. The study concluded that, overall, the clouds at the Pole were similar to the high-level cirrus found at mid-latitudes.

Future research needs

Although the ISCCP has already had great success in producing the most reliable climatology of cloud cover from satellite data, the areas where there are most problems are in the polar regions. Because of this a special working group was established to consider how the polar element of the work could be improved (Raschke, 1987). They concluded that to produce reliable cloud cover data at high latitudes additional satellite data beyond the visible and thermal infra-red imagery would be required. The polar element of ISCCP needs to be further developed by using improved algorithms for the detection of cloud and the determination of cloud type.

There have been very few *in situ* observations made of the microphysical properties of tropospheric clouds at high southern latitudes and a pressing need is for campaigns using instrumented aircraft, tethersondes and ground-based remote sensing. These data will be of value in a number of areas including the interpretation of satellite imagery, the representation of clouds in numerical models and in precipitation studies.

3.4.2 Precipitation

Knowledge of precipitation formation mechanisms, snowfall distribution over the Antarctic continent and the synoptic origins of the precipitating air masses is important for the investigation of whether the ice sheets are growing or shrinking (mass balance), how accumulation may change in an environment of higher mean atmospheric temperature and for the study of past climate through the analysis of ice cores. It has been estimated that about 2300 km^3 of ice falls on the continent each year (Fortuin and Oerlemans, 1990), but it is extremely difficult to make measurements of precipitation in the Antarctic, not least because of the problems in differentiating between blowing snow and falling precipitation. With the strong winds experienced in many parts of the continent, conventional snow gauges have been found to give poor results so that other means of determining the accumulation have had to be employed. These have included the use of snow stakes, stratigraphy, a knowledge of snow drift density and radio-isotope analysis of snow and ice retrieved from pits and ice cores. As discussed in Section 4.4, it is also possible to estimate precipitation on a continental scale from water vapour fluxes determined from radiosonde ascents made at coastal stations around the continent. Although stake and pit measurements can give an accurate estimate of the snow that has built up on the surface, care must be taken when interpreting these data if the goal is to determine precipitation, because they will include contributions from precipitation that fell at the site and from snow that blew onto the area from upwind. A similar problem is experienced when the accumulation is determined from passive microwave imagery (Zwally, 1977), which also gives an estimate of the combined deposition as a result of precipitation and blowing snow.

Following Bromwich (1988), the snow accumulation rate (B) at a site can be written as:

$$B = P - E - \nabla \cdot Q - R \qquad (3.7)$$

where P is the precipitation rate, E the net evaporation (evaporation minus deposition of hoar frost) rate, Q is the horizontal flux of drifting snow (see Section 6.4 for further discussion of this term) and R is the rate of run-off of melt-water.

The precipitation term indicates the input through snowfall and this dominates in the more sheltered locations in the coastal region where there is a relatively high annual accumulation and limited loss of snow blowing across the coast. In the interior of the continent the precipitation is approximately equal to net snow accumulation because there is little loss through blowing snow or evaporation.

Net evaporation is made up of two elements; the evaporation of snow from the surface and the build-up of ice on the surface as hoar frost. The amount of evaporation depends on the surface wind speed and is greatest where persistent katabatic winds are found at the base of valley glaciers in the coastal region. As discussed in Section 4.4, evaporation during the summer months can remove up to 45% of the accumulation at some coastal sites while during the winter it can be almost zero (Loewe, 1962). Growth of ice on the surface by hoar frost occurs during the winter months, but is not thought to contribute significantly to the mass balance over most of the interior. Annual net evaporation decreases away from the coast and falls almost to zero about 150 km from the coast (UNESCO, 1978) (see Section 4.4 for a discussion of observations of evaporation).

The blowing snow term represents the mass of snow that is advected away from or on to a location during periods of strong winds. In coastal valleys, subject to strong, persistent katabatic winds, the bulk of the precipitation may be blown away, with most being advected out to sea. In areas such as Adélie Land, where the wind speeds increase towards the coast, net loss may occur. Well away from the coast, where there are few major weather systems and no steep topographic gradients, advection of snow is minimal.

During the summer months, when air temperatures rise above freezing point, there will be melt-water run-off from ice- and snow-covered coastal areas into the ocean. This becomes more important in the more northerly parts of the continent, such as the Antarctic Peninsula, where summer temperatures are higher and where melting may result in areas of rock being exposed.

In the following sections we will examine the processes behind Antarctic precipitation and see how these influence accumulation rate over the continent. The link between synoptic activity and precipitation will also be examined, together with seasonal variations.

The nature and formation mechanisms of precipitation

The Antarctic is always envisaged as the land of ice and snow, and solid precipitation is indeed the most common form of deposition over the continent.

However, rain does fall, although it tends to be confined mainly to the western side of the Antarctic Peninsula, where mean temperatures during the summer months rise to several degrees above freezing. At these relatively northerly latitudes there is a large degree of variability in the circulation and strong north or northwesterly flow can become established even in the middle of winter, when rain can fall as far south as the middle of the Peninsula. Rothera Station, on the western side of the Peninsula, had solid, mixed and liquid precipitation on 88, 5 and 7% of occasions according to the 1977–93 statistics, but the frequency of rainfall decreases rapidly towards the south and at Wilkes Station (66.3° S, 110.5° E) it makes up only 2% of the precipitation reports. In the interior parts of the continent precipitation is almost always in the form of snow because intrusions of very mild air beyond the coastal zone are rare. On the high Antarctic plateau liquid precipitation is never recorded.

The sources of precipitation are quite different in the coastal and interior parts of the continent. Near the coast, frontal cyclonic systems are responsible for most of the snow whereas further inland most precipitation is in the form of ice crystals falling from thin, isolated cloud or even from an apparently clear sky ('clear-sky' precipitation or 'diamond dust'). In the coastal region, the depressions providing the precipitation are usually weak or declining so that most of the precipitation is reported to be of slight intensity, which means a water equivalent of no more than 0.5 mm $hour^{-1}$ or snow accumulation not exceeding 5 mm $hour^{-1}$ (Meteorological Office, 1982). The tendency for Antarctic precipitation to be of lower intensity is also a result of the limited capacity of air to hold moisture at temperatures close to and below freezing. Reports of moderate or heavy precipitation are usually indicative of more active fronts reaching the coastal region at the leading edge of warm air masses being advected southwards. During such periods, when the atmospheric long waves are amplified, a strong meridional temperature gradient can become established near the coast, often giving precipitation in the form of rain or a mixture of rain and snow.

Near the coast the most important mechanism for the production of precipitation is the adiabatic cooling of the air as it rises up the steep topography that characterises many coastal areas. An onshore airstream, such as occurs on the eastern side of a major low-pressure system, will be most effective in giving precipitation at the edge of the continent since the air arriving from over the ocean will contain a large amount of moisture and the east–west pressure gradient will provide sufficient energy to lift the air up the steep topographic barrier. The north–south topographic barrier of the Antarctic Peninsula has a profound effect on the precipitation experienced in that sector of the continent, with the western side coming under the influence of many moist, maritime airstreams and the eastern side having a more continental climate (see also Section 3.2). Satellite imagery frequently shows a banner of cloud extending eastwards from the high central plateau and this is evidence of the lifting of air as it rises and is cooled on the western flank. This will result in some topographically induced precipitation on the west-facing slopes, although, as discussed later in this section, it is known from ice cores that the most precipitation occurs near the west coast, where the maritime influence is greatest. Although the frequent banner clouds extending towards the east will give some precipitation on the eastern side of the Peninsula, the lack of maritime influence predominates and there is less precipitation here than there is over the western and central areas.

Although most precipitation in the coastal region comes from organised weather systems significant amounts fall from cloud well removed from the major depressions and in synoptically quiet or high-pressure situations. In their one-year study of Rothera precipitation Turner *et al.* (1995) found that almost 20% of the precipitation reported over the period fell into this category, although all but one of the falls were classed to be of slight intensity. Sometimes these precipitation events could be ascribed to weak fronts that had penetrated into the centre of anticyclones but most were from extensive areas of stratus or stratocumulus cloud. In these situations the snow was a result of the gradual coalescence of cloud droplets over an extended period and the slow falling out of ice crystals in the form of slight snow. Because ice crystals are very slow to evaporate, most of this precipitation reaches the ground as a light fall of snow.

The amount of cloud over the Antarctic decreases rapidly away from the coast because most major weather systems do not penetrate into the interior. Consequently the nature of the precipitation on the plateau also changes and, although some precipitation does come from organised cloud, most deposition occurs from isolated cloud or when no optically thick cloud is present. Precipitation falling from a clear sky is almost a daily occurrence on the plateau (Rusin, 1961) and seems to be a feature of all seasons. This was confirmed by a one-year study of precipitation at Plateau Station carried out by Radok and Lile (1977), who found that 87% consisted of ice crystals and that most of these had fallen under cloud-free conditions. The accumulation rates for such clear-sky precipitation are very low but have been estimated by Sato *et al.* (1981) for the summer season at South Pole Station to be 0.1 mm hour^{-1} by collecting ice crystals on Formvar-coated glass slides. Although the precipitation at inland stations is usually recorded as 'trace', it can occasionally achieve a depth of 3 mm in a few hours (Hogan, 1975). Bromwich (1988) has suggested that, at levels above 3000 m, radiative cooling is the primary means by which saturation of the air is maintained and which results in the formation of ice crystals. However, orographic lifting of maritime air is also thought to be important in the production of clear-sky precipitation (Bromwich, 1988). When precipitation is observed to fall from a clear sky there is still a 'cloud' of ice crystals suspended in the air, but this is optically too thin to be seen either from the ground or in satellite imagery. The nature of the crystals falling at the South Pole during one summer season was investigated by Hogan (1975), who found that most were of columnar form, although smaller 'diamond dust' crystals were also found. The lidar studies of Smiley *et al.* (1980) observed additional ice crystal types including prisms, plates, bullets and clusters. Hogan found that the crystals formed under supersaturated conditions when there were cirrus bands at higher level with, it is assumed, the crystals growing as they fell through the moist near-surface layer. Smiley *et al.* observed crystals growing at the top of the surface inversion at temperatures between −30 and −50°C. Occasions of persistent supersaturated conditions with temperatures below −30°C and no precipitation suggest that the air at the South Pole is free of heterogeneous freezing nuclei.

As discussed in Section 5.3.4, Phillpot (1968) examined three years of surface observations from Vostok to assess the penetration of depressions onto the plateau. Over this time he found that 42 depressions had affected the station, with many giving precipitation. There is, however, no evidence to support the suggestion of Rubin and Giovinetto (1962) that precipitation over the plateau comes from a few very active depressions. Nevertheless, studies of surface observations of precipitation in conjunction with long sequences of satellite images are needed to assess fully the role of depressions in bringing precipitation to the high plateau area.

Synoptic origins

The circumpolar trough, which rings the continent between 60 and 70° S, is the result of the many depressions that are found just north of the edge of the continent and these lows are responsible for much of the precipitation that falls in the coastal region. Turner *et al.* (1995) investigated the precipitation that fell at Rothera Station in the middle of the Antarctic Peninsula during a one-year period using the synoptic observations from the station, satellite imagery and routine synoptic charts. Over the year, 80% of the precipitation was found to be associated with cyclonic disturbances, with most of these located in the Bellingshausen Sea to the west of the station. This situation results in the Peninsula being under a mild, moist northwesterly airstream, which Schwerdtfeger *et al.* (1959) also linked to precipitation at Melchoir Station at the northern end of the Peninsula. Such a feed of warm, moist air onto the western side of the Peninsula strengthens the thermal gradient in the north–south direction and most of the moderate and heavy precipitation is experienced in this situation. Although the Antarctic coastal region has often been considered to be affected by declining depressions that have moved south from mid-latitudes, it was found that, during the one-year study, 49% of the depressions that gave precipitation at Rothera had developed south of 60° S, indicating that the coastal area is a far more active region than was originally thought. Many of these lows had developed well to the west of the Antarctic Peninsula and tracked a long distance around the Southern Ocean before stagnating in the Bellingshausen Sea.

Although many of the lows in the coastal region are past their most active phase, many still have fronts associated with them, as can be seen on satellite imagery. Most of these are rather weak, indistinct feature but many nevertheless can still give snow or rain at the coastal stations, although most is only of slight intensity. In the Turner *et al.* study described above, 48% of the precipitation recorded at Rothera during one year was attributed to fronts, most of which were warm fronts arriving from the northwest in mild airstreams from the South Pacific.

Precipitation at coastal stations can also originate under quiet synoptic conditions or even from high-pressure systems when there is extensive low cloud. At Rothera Station Turner *et al.* (1995) found, during the one year they examined that there were 26 reports of precipitation from anticyclones or ridges. This came mostly from extensive stratus or stratocumulus cloud or when a active front had penetrated into the circulation of an anticyclone.

Reports of showers are rare at Antarctic stations because the conditions which give intermittent precipitation, consisting of fields of cumulus cloud, are usually found in southerly air streams where the shower clouds are being carried to the north. For the showers to reach the continent the unstable air mass must be carried around the circulation of a deep low so that it is moving towards the south or southeast. Stations that receive occasional showery air streams are usually on the west-facing side of a promontory, such as the Antarctic Peninsula.

Mesocyclones, or sub-synoptic-scale low-pressure systems, are a common occurrence over the Southern Ocean and in the Antarctic coastal region, where they can bring strong winds and low surface pressure. Little is known, however, about the precipitation that is associated with these systems because they are difficult to investigate using surface observations owing to their small horizontal scale. A further problem is that satellite techniques for precipitation determination are still not well developed for the polar regions. It has been shown that occasional larger, more long-lived mesocyclones can bring significant amounts of precipitation to coastal stations (Turner *et al.*, 1993b) but these are exceptional systems and we know from climatological investigations that the majority of mesocyclones are found over the ice-free ocean in southerly airstreams to the west of synoptic lows. However, recent work by Rockey and Braaten (1995) suggests that mesocyclones are responsible for over a third of the precipitation falling at McMurdo Station and it will be important to determine how much precipitation these systems are giving and why they are so active in this region. Here mesocyclones could have glaciological consequences because many vortices have been observed on the ice shelf and found to exist for most of their lifecycle on the continent itself. Passive microwave imagery from instruments on polar orbiting satellites should soon tell us how many mesocyclones have precipitation associated with them, although we already know that this will not affect considerations of mass balance because most track northwards and do not make landfall.

Table 3.7 *Estimates of precipitation for various Antarctic research stations.*

Station	Elevation (m)	Annual precipitation (mm)	Period	Reference
Mizuho	2230	140	1980	Kobayashi (1985)
Mizuho	2230	230–260	1982	Takahashi (1985a)
Novolazarevskaya	87	303	1961–64	Schwerdtfeger (1970)
Molodezhnaya	42	712	1963–65	Schwerdtfeger (1970)
Mirny	30	625	1956–65	Schwerdtfeger (1970)
Melchior	8	1,189	1947–56	Schwerdtfeger (1970)
Orcadas	4	405	1904–06 1908–56	Schwerdtfeger (1970)
South Pole	2800	<70		Hogan (1975)
Deception Island	8	398	1948–56	Schwerdtfeger (1970)

Distribution and accumulation

In the Southern Hemisphere extra-tropics the zonally averaged precipitation maximum is between 40 and 60° S and is coincident with the zone of the most active extra-tropical frontal cyclones (Wallace and Hobbs, 1977). South of 60° S the depressions are generally declining and precipitation totals decrease sharply as the Antarctic continent is approached. This can be seen in the precipitation estimates for various Antarctic stations given in Table 3.7. These have been computed by a variety of means and the figures, particularly for the shorter periods, should be used with care considering the high variability of precipitation in the coastal region.

The precipitation over the continent has been estimated by Giovinetto and Bentley (1985) and their map of annual accumulation, as adapted by Bromwich (1988), is shown in Figure 3.28. The map shows net accumulation of snow at the surface and this has been equated to precipitation, which is a reasonable assumption for most of the interior. It can be seen that the largest accumulations are found slightly inland of the coast, where the steepest slopes are found and where air masses are lifted and cooled as they move polewards. The largest accumulations, which are over 1000 mm per year, are found along the coastal area of the southeastern Bellingshausen Sea, which is at the end of one of the major depression tracks from mid-latitudes and where many mature and declining lows are found. This is also an area of frequent cyclogenesis, with lows developing in the southern Bellingshausen Sea and moving northeastwards towards the tip of the

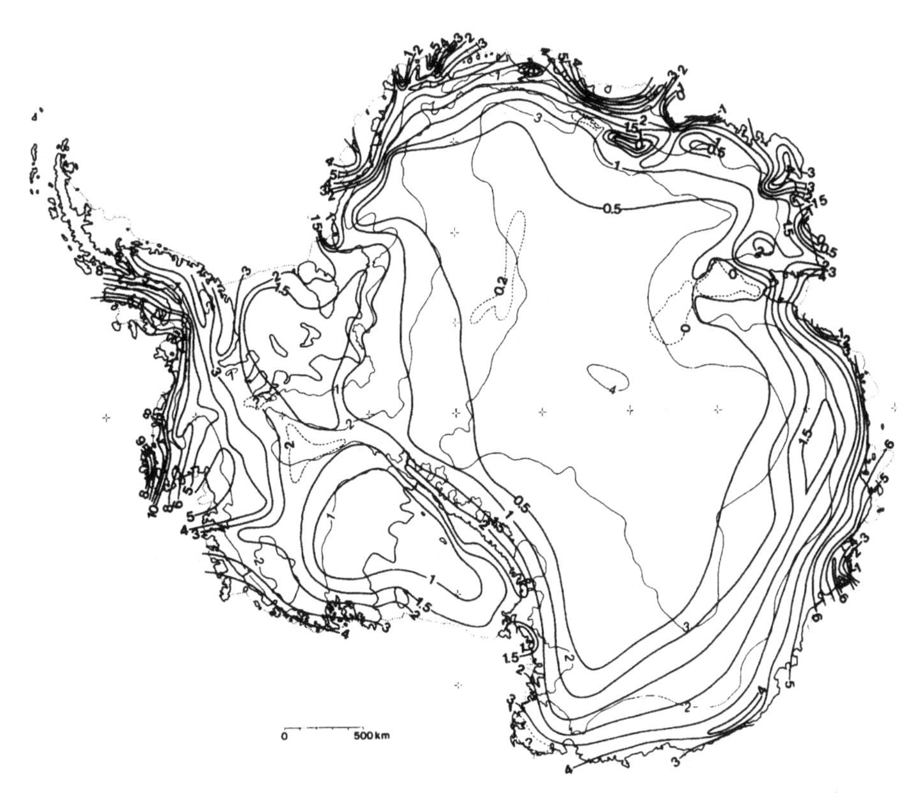

Figure 3.28. Precipitation over the Antarctic estimated from ice core data. From D. H. Bromwich, *Rev. Geophys.* **26**, 152, 1988, copyright by the American Geophysical Union.

Peninsula. Here many of the lows still have active fronts which bring precipitation to the coastal area as they move around the low centres. Precipitation is also supported in this area by the lack of sea ice in the Bellingshausen Sea during the summer months, which increases the fluxes of heat and water vapour into the lower layers of the atmosphere, giving moist unstable conditions.

Other areas of high annual precipitation around the coast coincide with regions of frequent cyclonic activity offshore, such as north of Enderby Land and along the coast of East Antarctica. The coastal areas with the lowest accumulation are the Ronne and Ross ice shelves, where there is no topographic lifting of the air masses to enhance precipitation. Inland of the immediate coastal area the precipitation totals drop very rapidly so that the bulk of East Antarctica, where the elevation is more than 3000 m, has less than 50 mm of accumulation a year.

The statistical relationship between accumulation and elevation has been investigated by a number of workers, including Muszynski and Birchfield (1985) and Fortuin and Oerlemans (1990). In this latter study, it was found using a linear multiple regression analysis that the surface slope, the surface shape and the saturation vapour pressure of the free atmosphere were the most significant predictor variables for accumulation. The link between the free air temperature and snow accumulation was first noted by Robin (1977), who found it to be similar to the Clausius–Clapeyron relationship between temperature and saturation vapour pressure. This can be explained as the saturation vapour pressure setting an upper limit on the amount of water vapour that is available for precipitation. The Fortuin and Oerlemans study found that the saturation vapour pressure was by far the most important parameter in determining the distribution of precipitation, particularly in the interior of the continent. Here, above 1500 m, 72% of the spatial variance in precipitation could be explained in terms of the saturation vapour pressure and the convexity of the surface.

Figure 3.28 does not show data for the northern half of the Antarctic Peninsula because the large variability in the topographic height over this area results in rapid spatial changes in the precipitation that cannot easily be mapped. However, the annual accumulation rates in this area have been estimated by Peel (1992a) from shallow (10 m) ice cores taken from various parts of the Peninsula where there is no summer melt-water run-off. The accumulation rates were estimated from seasonal stable isotope stratigraphy and the number of cores taken allowed relationships to be established with topographic height, the longitudinal location and annual mean temperature (using the 10 m snow temperature, which is a good indicator of the annual mean air temperature). These data are shown in Figure 3.29. It can be seen in Figure 3.29(a) that annual accumulation rates on the western side of the Peninsula are double or triple the value found to the east of the spine of the topographic barrier, where they are of the order of 500 mm per year. This is a result of the cyclonic activity on the western side, which does not usually extend across to the eastern half. Figure 3.29(b) indicates the strong relationship between accumulation and annual mean temperature, greater annual deposition being found at sites with higher mean annual temperature. However, the contrasting climatic regimes on the two sides of the Peninsula are shown by the different accumulation–temperature relationships indicated in Fig 3.29(b) which show 50% less accumulation on the east coast than occurs at sites on the western side with the same temperature. For both the eastern and western groups of sites the gradient of accumulation against temperature is close to that of the saturation mixing ratio. Figure 3.29(c) shows the relationship between topographic height and accumulation rate for all sites on the Peninsula. This provides further evidence of the different climatic regimes on either side of the barrier. On the western side the greatest accumulation is found at the low-level sites near the coast, with a progressive drop in accumulation at the higher locations which are more removed from the maritime influence. The east coast provides a contrasting relationship, with the greatest accumulation occurring at higher levels and the lowest annual deposition near the Weddell Sea, possibly reflecting the stable stratification of the atmosphere near the coast.

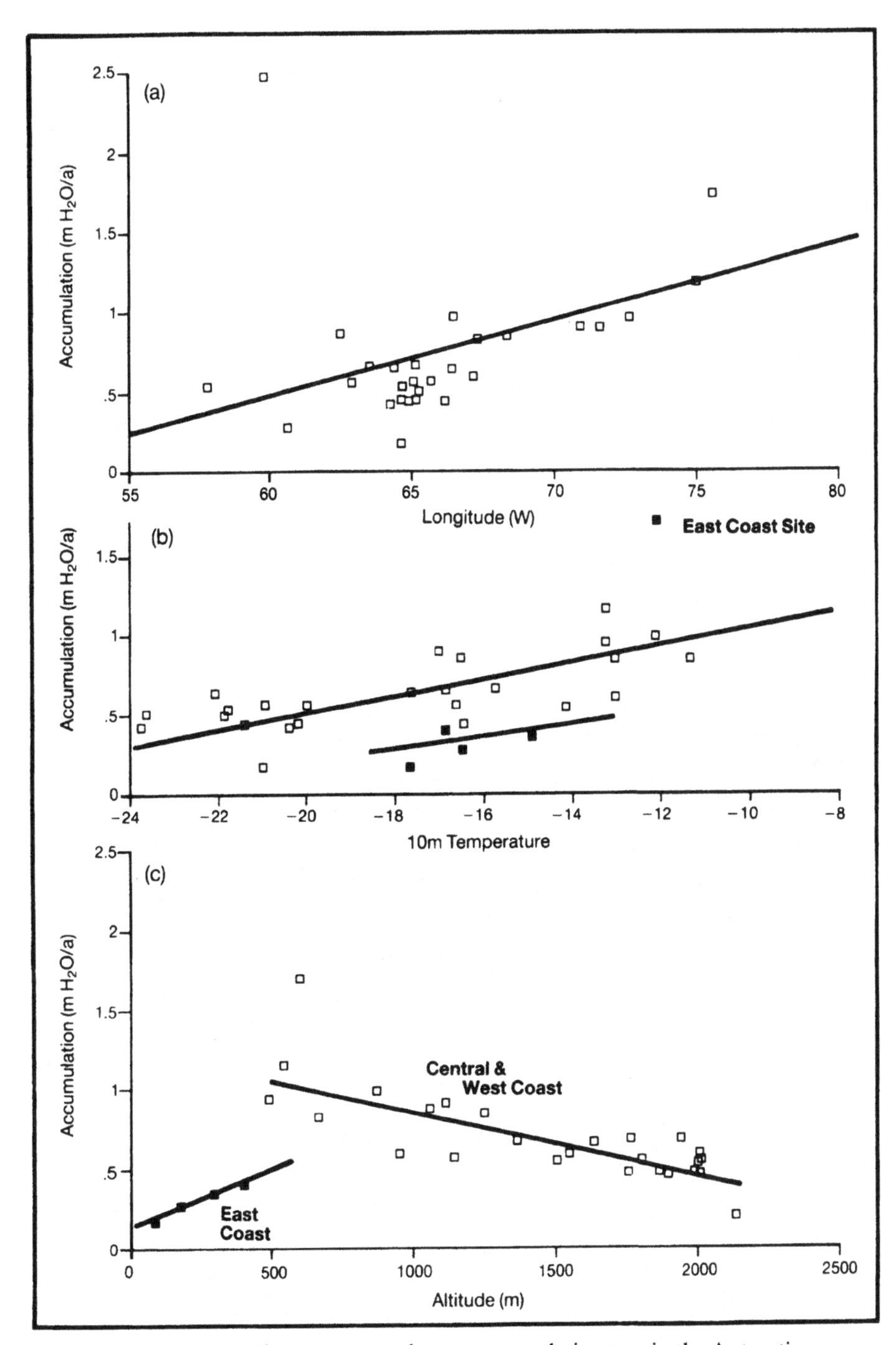

Figure 3.29 The mean annual snow accumulation rate in the Antarctic Peninsula as a function of (a) longitude, (b) 10 m temperature and (c) altitude. From Peel (1992a).

Table 3.8 *Estimates of precipitation over the Antarctic Peninsula (from Peel (1992a)).*

Zone	Present accumulation rate (mean) (mm water/year)
West coast	1260±390
Central	610±160
East coast	310±80

Typical accumulation rates for the east, central and western parts of the Peninsula are given in Table 3.8.

Seasonal variation of precipitation

The cycle of Antarctic precipitation throughout the year has been estimated at coastal sites using a range of direct observations and indirect estimates. In the northern part of the Antarctic Peninsula the surface synoptic observations from Faraday Station for the period 1977–92 show that the greatest number of precipitation reports come from the spring and autumn periods with a minimum in the summer. This is shown in Figure 3.30(a) together with the mean monthly surface pressure. The annual cycle of pressure at this station exhibits a semi-annual oscillation (see Section 3.3), which indicates that there is a maximum of cyclonic activity during the equinoctial months. The very strong correlation between the pressure cycle and the number of precipitation events confirms the dominant role of the extra-tropical cyclones in producing the precipitation over the western side of the Antarctic Peninsula. Figure 3.30(b) shows similar data for Rothera some 300 km to the south of Faraday. A similar relationship between pressure and precipitation events is apparent, although the amplitudes of the cycles are reduced because the station is further south of the centre of the circumpolar trough. Further south still, Kobayashi (1985) estimated precipitation rates at Mizuho Station in East Antarctica from snow drift density profiles that were processed to separate the snowfall and blowing snow elements. He estimated that, during 1980, there was a water equivalent of about 140 mm of precipitation; the monthly distribution of this total is shown in Figure 3.31. The maximum precipitation over this year occurred during the late winter months of July to September, with a minimum between November and January.

On intra-seasonal and inter-annual timescales the variations in precipitation experienced in the coastal region are mainly the result of changes in the location and intensity of the extra-tropical cyclones in the circumpolar trough, which is located just north of the edge of the continent. Frequent episodes of deep depressions near the coast result in greater precipitation, especially when the long waves

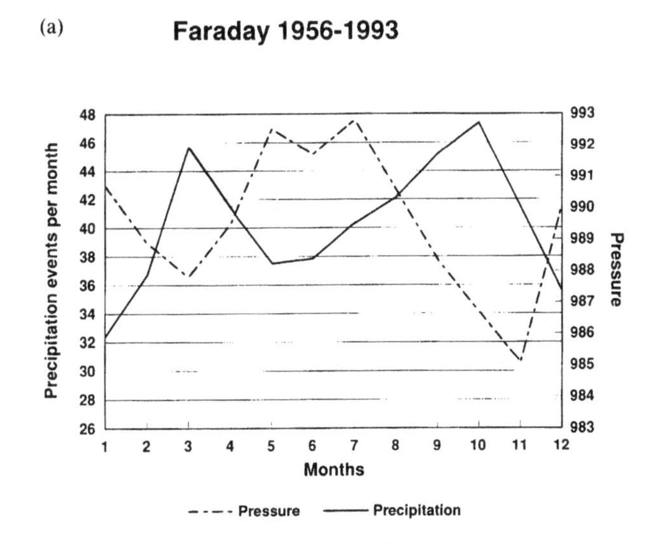

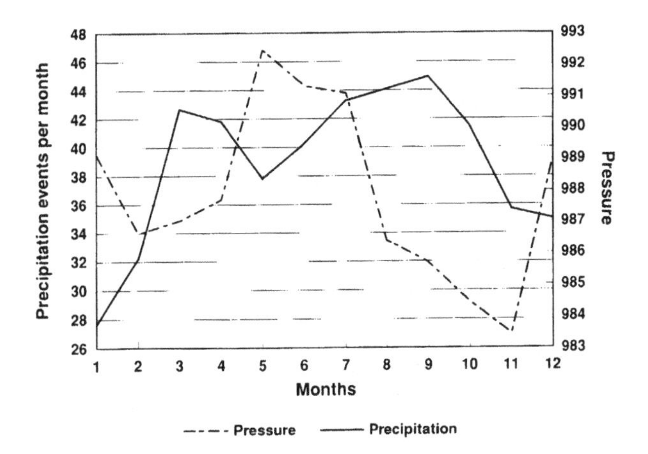

Figure 3.30 Monthly mean sea level pressures (hPa) and total number of precipitation events at (a) Faraday and (b) Rothera stations.

are amplified and there is advection of warm air to high latitudes. Conversely, blocking episodes give reduced precipitation, although there can still be some slight snow from anticyclones when they have extensive cloud. Because the inter-annual variability of synoptic-scale activity is large, it is not surprising that the precipitation also shows a high degree of variability.

Future research needs

We still have a very poor knowledge of precipitation processes in the Antarctic because of the lack of aircraft observations of cloud droplets and cloud micro-physical properties. However, surface-based remote sensing does offer us the opportunity of investigating some aspects of cloud and precipitation processes and the first investigations carried out in the Antarctic, such as those of del

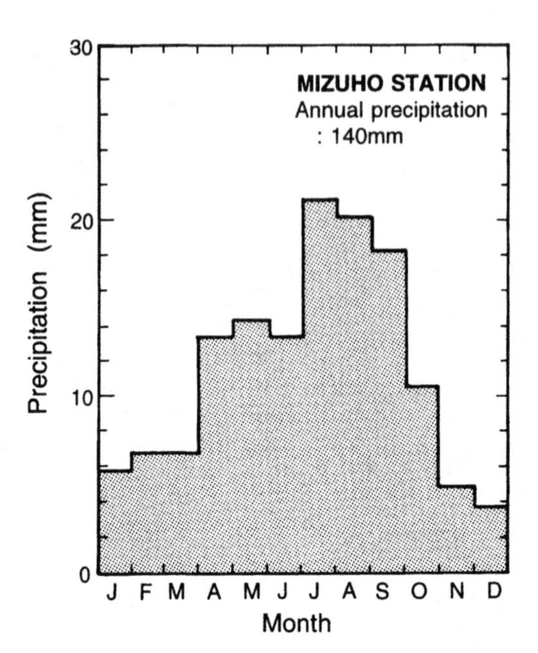

Figure 3.31 The distribution of monthly precipitation at Mizuho Station in 1980. After Kobayashi (1985).

Guasta *et al.* (1993), have shown that precipitation from clouds at a range of levels can be detected by a lidar. The collection of such data at a range of coastal and interior stations would be of great value in precipitation studies.

3.5 Sea ice and the Southern Ocean environment

3.5.1 Sea ice

One of the main features of the Antarctic maritime environment is the belt of sea ice which rings the continent and extends from the coast to, at its maximum in late winter, around 55° S at certain longitudes. At its greatest extent the ice covers about 19 million square kilometres – a larger area than the continent itself – and has a profound effect on the physical and biological conditions within the ocean. The majority of the sea ice in the Antarctic melts during the summer months so that, unlike in the Arctic, most of the ice encountered is new or first-year (FY) ice and relatively thin (Weller, 1980) with a thickness of less than 1 m and a high salinity. The main exception is the western part of the Weddell Sea, where significant amounts of sea ice persist over the summer, resulting in the largest proportion of multi-year ice. Sea ice plays an important role in many meteorological processes and has a major impact on the climatology of the maritime and coastal areas. In this section we will examine the part that sea ice plays in the climate system of the Antarctic and consider its temporal and spatial variability on a range of scales.

The role of sea ice in the climate system

Sea ice plays an important part in the climate system of the Antarctic through its modifying role in radiative, energy and mass exchange processes. One of its major effects is to alter the surface albedo of the upper layer of the ocean, so affecting the amount of solar radiation that can be absorbed. The albedo of the unfrozen ocean is typically around 10–15% (Lamb, 1982), causing the majority of incident shortwave radiation to be absorbed in the near-surface layer. However, the albedo of sea ice can be as high as 90%, when it is covered with fresh snow, so that most of the incoming solar radiation is reflected back from the surface. Much lower albedos are found when the ice is thin and when it contains a high proportion of leads. These are ice-free regions between floes that consist of either open water or very thin ice. On occasions when a high proportion of leads are present the areally averaged albedo can be almost as low as that of the ice-free ocean.

Sea ice has a further effect on the surface energy balance by capping the upper layer of the ocean and preventing the exchange of heat and moisture between the ocean and the atmosphere. In fact, during the winter months the sea ice is so efficient an insulator that the flux of heat into the atmosphere from the open ocean can be two orders of magnitude greater than that over the sea ice (Maykut, 1978). This affects the formation of cloud, the stability of the atmosphere and therefore precipitation. Studies using climate models (see Section 7.3) have suggested that a reduction in sea ice extent could produce marked changes in the climate of Antarctica. However, in numerical models the interactions among the atmosphere, ocean and sea ice are poorly represented at present and further developments are required before we can predict the outcome of a global warming on the Antarctic with any confidence.

During the winter, leads within the sea ice can account for 20% of the main sea-ice-covered region and can have an important effect on the surface exchanges of heat and moisture between the ocean and atmosphere. With the large air–sea temperature differences found in the Antarctic during winter there can be significant fluxes of heat into the atmosphere over the leads, although the relationship between the fraction of open water and heat flux is not linear. Nevertheless, it has been shown that, at ice concentrations of less than four tenths, the heat flux into the lowest layer of the atmosphere is almost equal to that from the open ocean (Worby and Allison, 1991).

Sea ice also affects the climate at coastal stations on a range of timescales. The opening up of coastal leads can significantly increase the temperature and humidity, especially during the winter months when fog may form and cause heavy riming of meteorological instruments, such as anemometers (Anderson, 1993). In some areas, such as the eastern Weddell Sea, coastal polynyas are found throughout the year (Kottmeier and Engelbart, 1992). On the interannual scale the extensiveness of sea ice in the Bellingshausen Sea has been found to affect temperatures along the western side of the Antarctic Peninsula (King, 1994; Weatherly *et al.*, 1991). The association between these two quantities is, however, less strong in other parts of the Antarctic coastal region (see Chapter 7).

Characteristics of Antarctic sea ice

Sea ice forms from the freezing of sea water under sub-zero conditions over the Southern Ocean. The exact freezing point depends on salinity, but is −1.9°C for water with a typical surface salinity of 35 parts per thousand by weight. When ice crystals form they consist of pure water, with the salt being rejected and so increasing the salinity and density of the surrounding water. This process of brine rejection initiates downward convection and promotes mixing between the deep ocean and the surface layer. It is also important in the formation of dense Antarctic bottom water (see Section 3.5.2) and for bringing nutrients to the upper layers.

Sea ice can take many forms and a full discussion of ice types is beyond the scope of this book. Those interested in the types of sea ice that can form should consult the *Marine Observer's Handbook* (Meteorological Office, 1995), in which the various types are illustrated. However, observations show that Antarctic sea ice is composed of a complex mixture of different ice types and that the nature of the pack ice is continuously evolving, even during the winter as a result of mechanical and thermal forcing. The ice concentration also varies considerably and it has been shown that, during the month of September, at the end of the winter, only half the ice pack has a concentration of over 85% (Zwally *et al.*, 1983a). In this study it was found that the remaining 50% of the sea ice had a concentration of between 15 and 85%, with the low concentrations being the result of extensive leads and polynyas. In the Weddell Sea it has been estimated that the areal coverage of the polynyas and leads is about 5% of the total area (Schnack-Schiel, 1987). Within the leads there is a continuous process of new ice formation, mechanical deformation resulting in rafting of the ice sheets and the maintenance of the leads.

There have been few measurements of sea ice thickness in the Antarctic, although some have been obtained by drilling holes through the ice during research cruises. The lack of data has meant that no conclusions can be drawn at present regarding climate-related changes. However, those measurements that have been made, e.g. Wadhams *et al.* (1987), suggest a modal thickness of about 0.5–0.7 m and relatively few areas with a thickness of more than 1 m. First-year ice is thought to be very thin and to have a thickness of about 0.6 m (Wadhams, 1994). Multi-year ice in the Weddell Gyre has been found to have a mean thickness of about 1.2 m when it is undeformed, roughly double that of first-year ice (Wadhams, 1994). Similar results were obtained by Lange and Eicken (1991) in their study of sea ice thickness in the Weddell Sea, with a modal thickness of about 1.5 m being found.

Another factor that affects the nature of sea ice and how we interpret remotely sensed data is the layer of snow on top of the ice. As noted above, this can affect the albedo of the surface layer and also change the thermal insulation properties of the sea ice/snow layer. A layer of snow on top of the sea ice, which can be very thick in the case of multi-year sea ice, can insulate it from cold air in the lowest atmospheric layer and suppress the growth of new ice. Alternatively, when the sea ice is thin, the top of the ice layer may be pushed below the surface of the water, causing the snow to become saturated with saltwater and to freeze into a new ice type, snow-ice, at the ice/snow interface (Lange *et al.*, 1989). Snow can also be converted into snow-ice by seawater infiltration and this has been found to be important in the Ross and Amundsen Seas (Jeffries *et al.*, 1994). The thickness of snow on sea ice is impossible to obtain over large areas but measurements made from ships in the Weddell Sea (Eicken *et al.*, 1994) found an average value of about 0.16 m for first-year ice in the central and eastern part of the area and about 0.53 m on second-year ice in the northwestern sector. In the Indian Ocean sector Allison *et al.* (1993b) found lower snow thicknesses of 0.05–0.1 m during the spring, suggesting less snow-ice formation. Areas of greatest cyclonic activity, such as the circumpolar trough, will have high annual snowfall and give the greatest rates of deposition on the sea ice.

The albedo of sea ice is very variable (Weller, 1980) and depends on the condition of the ice, the proportion of leads, the state of any covering snow and the amount of melting that is taking place on the surface. A number of estimates of the albedo of sea ice have been made. Grenfell (1983) estimated that the mean albedo is about 80%, which is in broad agreement with the figure of 75% given by Weller (1980) for a sea ice concentration of more than 85%. However, higher values can be found if the sea ice is covered with fresh snow, but much lower ones if the ice is forming or disappearing (Kukla and Robinson, 1980), is thin, contains a high proportion of leads or has extensive melting on the surface. For the outer pack ice zone, Weller gives an albedo of 54% for sea ice concentrations of 15–85%, but clearly large variations from this can occur. The variation of sea ice albedo as a function of snow depth on the surface was investigated by Weller (1968b), who found that values as low as 40% could occur at the beginning of summer.

Spatial and temporal variability

Accurate, broad-scale records of Antarctic sea ice extent have been available only since passive microwave imagers were first flown on the polar orbiting satellites in 1973. Since that time it has been possible to observe the change in sea ice extent almost on a daily basis and to produce accurate statistics on the total ocean area covered by ice. The edge of the pack ice can be located within an accuracy of about 30 km, although there is still debate over the ice concentration that should be taken to define the edge. Parkinson (1992) used a 30% threshold in her

(a)

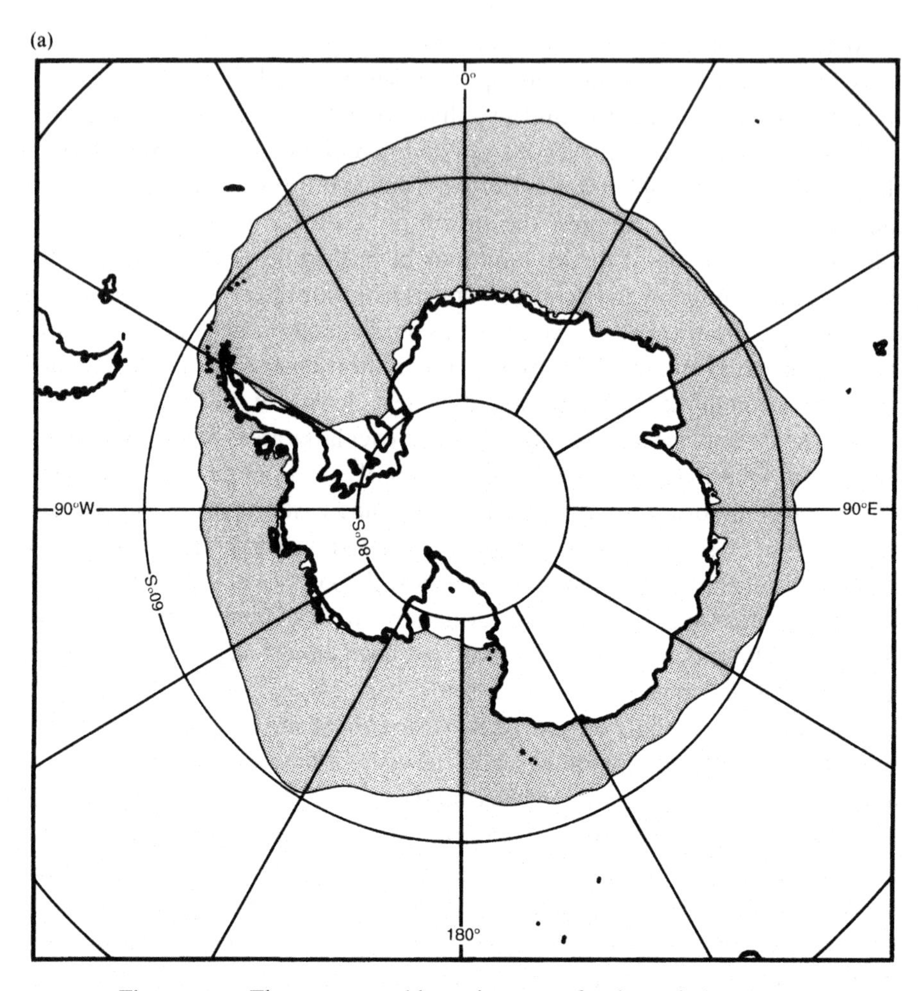

Figure 3.32 The mean monthly sea ice extent for the period 1978–87: (a) September and (b) February.

study of interannual variability and values in the range 10–15% have also been used. However, Zwally *et al.* (1983b) felt that the change in results from using different thresholds was small in that the concentration changes from 10% to more than 15% over less than 10 km, which is below the resolution of the imagery. Ice concentration and the area of first-year and multi-year ice can be estimated from passive microwave imagery with an approximate error margin of ±7.5%. The 20-year time series of sea ice data is not long compared with the meteorological records that are available from some Antarctic research stations. Nevertheless, it has allowed studies of the interannual variability of ice cover, the trend over this period and regional variability to be made.

The extent of the ice varies considerably throughout the year from a minimum in February of about 3.5×10^6 km^2 to a maximum in September of close to 19×10^6 km^2 (Gloersen *et al.*, 1992). The mean monthly ice concentration maps

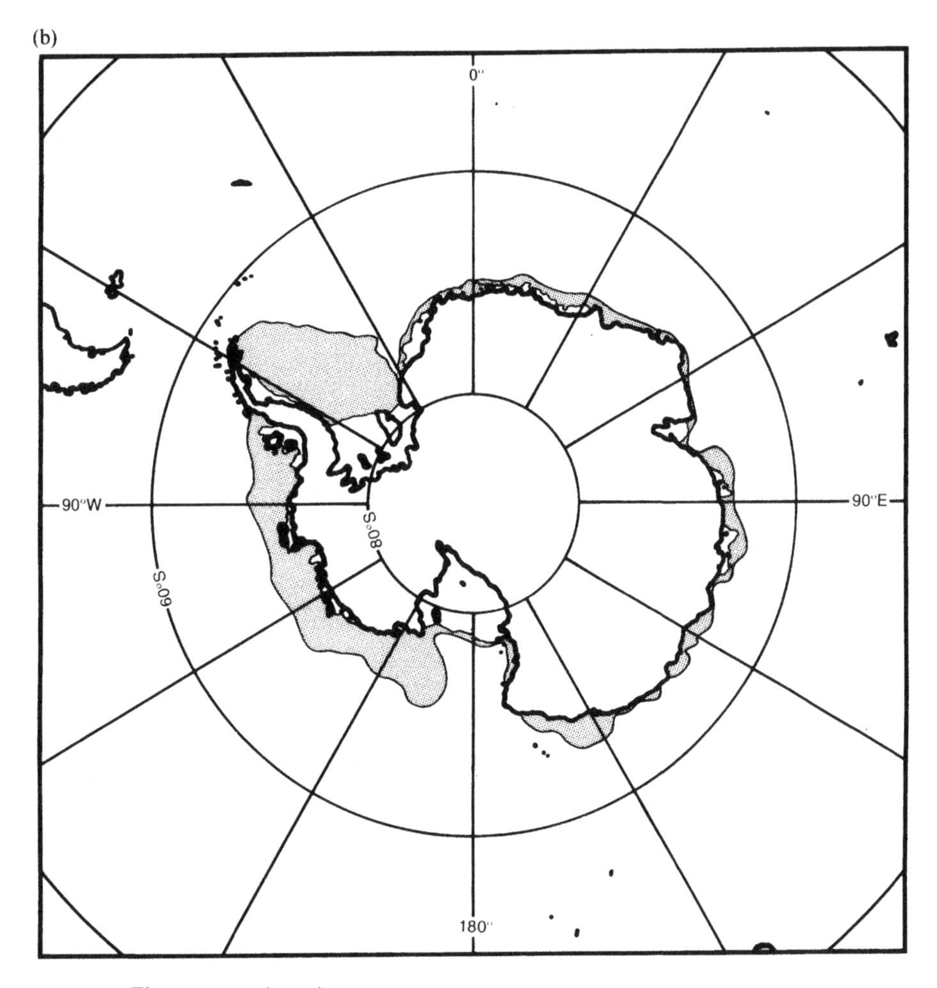

Figure 3.32 *(cont.)*

for these months over the period 1978–87 are shown in Figs. 3.32(a) and (b). This seasonal change of about 16×10^6 km^2 is much larger than that which occurs in the Arctic, where the distribution of land inhibits the growth of sea ice. On average the ice advance is most rapid in the autumn from April to June, but it continues to extend equatorwards until the late winter/early spring months of August and September. The period of ice growth is longer than that of its retreat so that the maximum ice extent is found about a month after the end of the winter season. Ice retreat is most rapid in the autumn/early summer period of October to December. As can be seen from Figure 3.32(b), the most extensive areas of sea ice remaining in February are in the western Weddell Sea and along the coast of the Bellingshausen and Amundsen Seas. The coast of East Antarctica extends further north than the western part of the continent so that the ice retreats back to the coast in most of this area by February. Since the northern ice edge reaches a similar latitude in this region to that in other sectors,

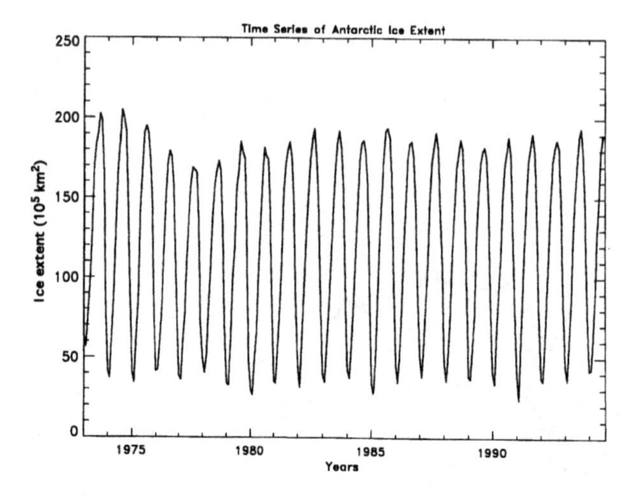

Figure 3.33 Time series of Antarctic sea ice extent for the period 1973–95.

both the maximum ice extent and the annual range in ice extent are smaller here than they are in other sectors of the Southern Ocean.

The passive microwave data first collected during the 1970s show that the Southern Ocean sea ice cover exhibits a large degree of interannual and regional variability during all seasons of the year, reflecting short-period variations and, possibly, long-term trends. However, some care needs to be taken when interpreting these data because different microwave instruments have been used over the period and sometimes there has been no overlap in their operation so that intercomparisons of ice data could not be performed. Nevertheless, here we will consider the total extent of the sea ice between 1973 and 1988 as determined by Zwally *et al.* (1983a) and Gloersen *et al.* (1992) and shown in Figure 3.33. The record is dominated by the large annual cycle of ice advance and retreat but also shows trends that have taken place during the 16 years of satellite data. In the early part of the record there was a decrease in the late winter maximum between 1973 and 1977 when the lowest value of 17×10^6 km^2 was recorded (Zwally *et al.*, 1983b). During the late 1970s there was also less ice at the summer minimum, with only about 2.5×10^6 km^2 in 1980 and 1981. However, the trend reversed after this time with the winter maximum and summer minimum both increasing by the early 1980s. Since that time the winter maximum and summer minimum values have shown less variability and no significant trend. It is difficult to draw definite conclusions regarding change in the extent of sea ice since 1973 because of the short period covered and the large interannual variability. However, at present there is no firm evidence to indicate that there is any long-term trend in Antarctic ice extent. Climatic changes may also be reflected in the total mass of Antarctic sea ice and not just in its extent. Detection of such changes will require a knowledge of ice thickness, which cannot be measured at present by satellite sensors and for which we only have spot measurements made from research vessels that have entered the sea ice.

In addition to this broad-scale picture the satellite imagery has also shown that there is a large regional interannual variability in the sea ice extent over some areas, such as the Weddell Sea. There are also decreases in some areas matched by increases in others over the period of the record. The Weddell Sea showed a large decrease in ice extent in the mid-1970s during the period that the Weddell Polynya (see below) was in existence, but since that time the ice extent in this area has recovered.

Interannual variations

Parkinson (1992) has examined the interannual variability of the spatial distribution of sea ice using monthly averaged data for the 13-year record covering 1973–76 and 1978–87 and her figures of February and September variability are shown in Figure 3.34. These show areas where ice was recorded every year in fine stipple and where ice was recorded in some years as coarse stipple. She found for the Antarctic as a whole that the greatest variability of ice cover occurred during the winter or spring months and the minimum in late summer. However, the area of interannual variability was much greater than the area of consistent ice coverage during that season. In winter the opposite is found to be true. Conditions in the Weddell and Ross Seas were different from those in other parts of the Antarctic insofar as the greatest variability was in December and occurred in connection with the late-spring opening over the eastern Weddell Sea and off the Ross Ice Shelf. The different sectors around the continent were found to have larger interannual variations than had the Antarctic as a whole, suggesting compensating mechanisms in the different regions.

Figure 3.34 shows that the only areas where sea ice was found in February of every year during the 13 years of data were over the western Weddell Sea and some parts of the Amundsen, Ross and Bellingshausen Seas. Along most of the coast of East Antarctica there were only very small areas where ice was always found at this time every year. However, by September sea ice is consistently found in all coastal regions of the Antarctic except the northwestern tip of the Antarctic Peninsula, which remains ice-free in some years. The variability occupies a 2–8° latitude band at the limit of the ice with the band at its narrowest in the western Ross Sea and along parts of the Indian ocean sector. The band of variability is largest in the eastern Ross Sea and north of the Antarctic Peninsula. One area of variability within the main mass of sea ice exists in the eastern Weddell Sea because of the Weddell Polynya of the mid-1970s, which is discussed below.

Cyclic fluctuations in Antarctic sea ice extent have also been observed. One of the most pronounced is a seven-year cycle in the sea ice maximum in the Weddell, Amundsen and Bellingshausen Seas. This is compensated for by a similar, but out-of-phase, cycle in the Western Pacific.

It has been suggested that sea ice could be very sensitive to changes in the global

(a)

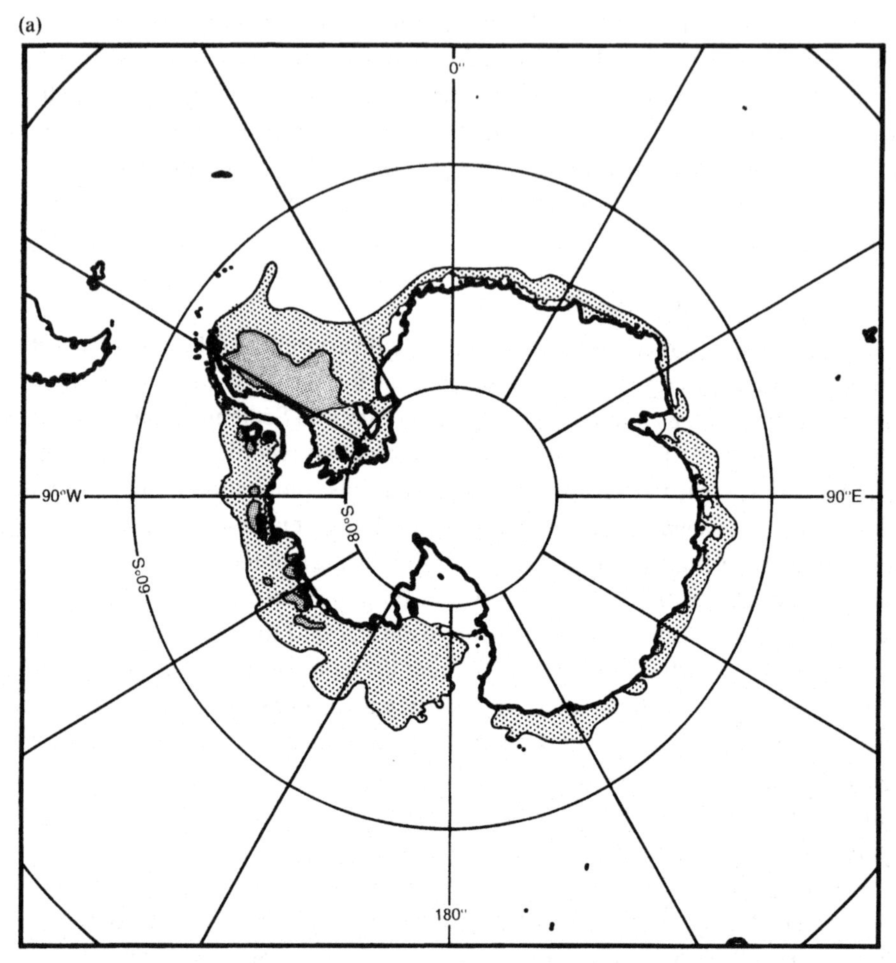

Figure 3.34 Spatial variability of monthly average sea ice for the period 1973–87. Fine stippling indicates sea ice coverage in every year and coarse stippling signifies coverage in some years, but not others: (a) February and (b) September. After Parkinson (1992).

climate system and may give an early indication of warming as a result of increased emission of greenhouse gases. However, the links between the sea ice and atmospheric and oceanic conditions are complex and at present we do not have a sufficiently good understanding of the reasons behind variations in sea ice extent to draw conclusions from the short series of sea ice data. The representation of sea ice in many climate models is also very crude and there are large differences in the response of such models to simulated changes in the climate system (Ingram *et al.*, 1989).

Polynyas

Polynyas are areas of open water that can develop at any time of year within the main area of sea ice or at the edge of the continent, where they are known as

(b)

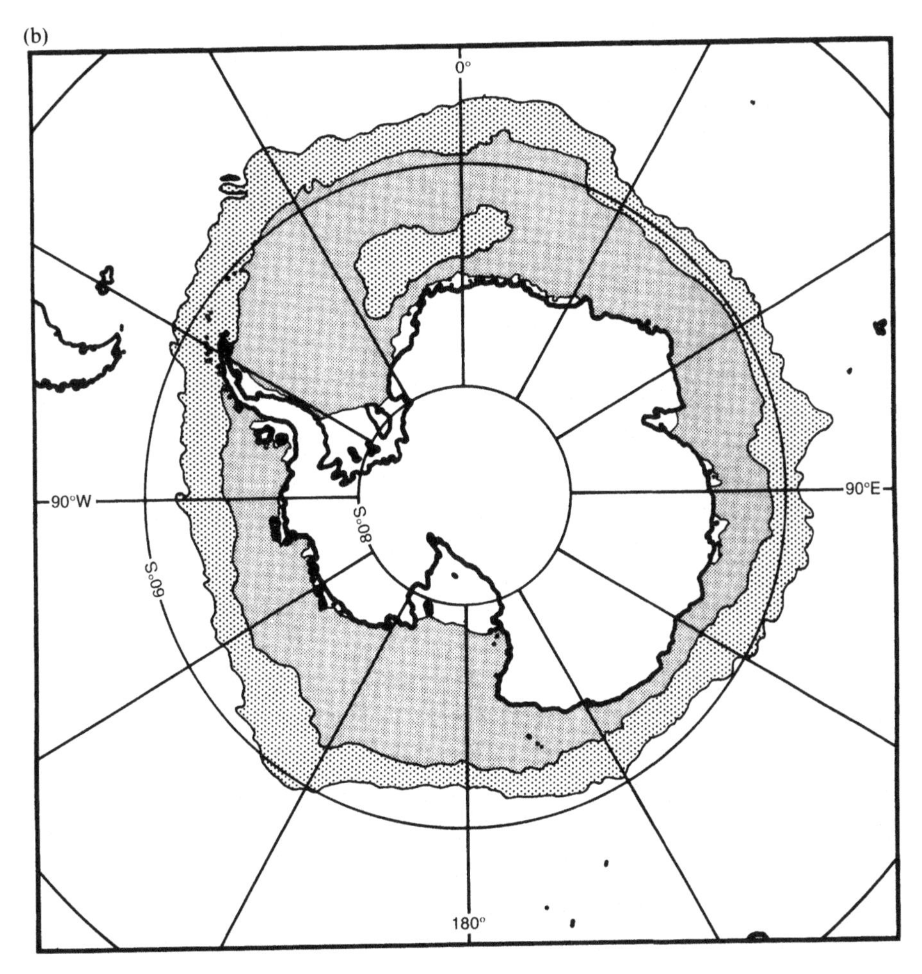

Figure 3.34 *(cont.)*

coastal polynyas. Their existence has been known since the early days of Antarctic exploration because they often provided access to the more southerly coastal areas at an early stage in the brief Antarctic summer. The coastal polynyas usually form as a result of the persistent katabatic drainage flow that is a feature of certain areas of the continent, forcing the sea ice away from the coast and thereby opening up areas of open water (Kurtz and Bromwich, 1983). This happens along the eastern coast of the Weddell Sea, which is subject to year-round easterly winds, so creating a narrow coastal passage that can be observed throughout the year. This coastal polynya was a factor in determining the location of Halley Station in that it allowed access to a relatively southerly location during December in most years.

The most famous polynya observed to develop in the main mass of sea ice was the Weddell Polynya, which was recorded during the winters of 1974, 1975 and 1976 over the eastern Weddell Sea (see Figure 3.35) (Gloersen *et al.*, 1974; Zwally

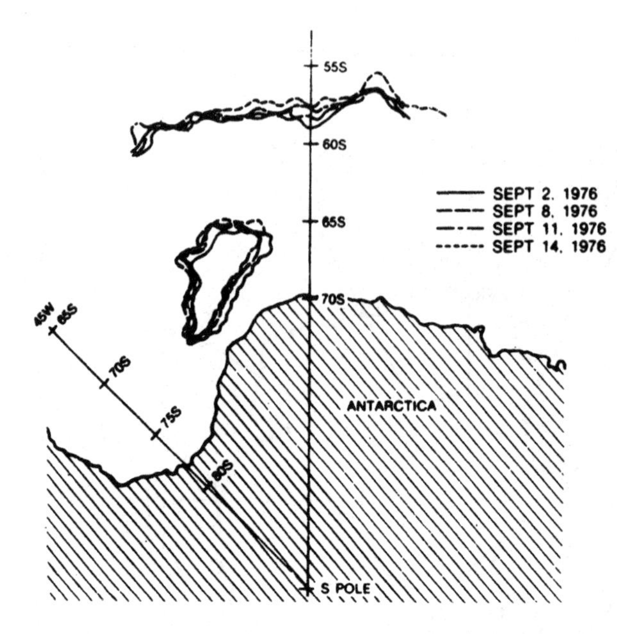

Figure 3.35 The Weddell polynya and its associated ice edge for 1975 and 1976. Lines are 150 K T_B contours taken from three-day average ESMR data. From Carsey (1980).

and Gloersen, 1977). This had an area of about $(2–3)\times10^5$ km^2 and remained in roughly the same location for several months during each winter. Carsey (1980) carried out a detailed examination of the polynya using satellite imagery and meteorological and oceanographic data. He found that the polynya was a complex feature with upwelling taking place at its centre, thermohaline convection and a cycle of ice formation at its edge and transport of this ice away from the area. He concluded from oceanographic models that it had formed because an eddy with specific temperature–salinity profiles had entered the Weddell Gyre and provided the conditions under which convection could take place. More recently, Enomoto and Ohmura (1990) suggested that the polynya formed as a result of its location beneath the circumpolar trough where there was a high spatial variability of the surface wind field. The absence of a Weddell Polynya since the 1970s has highlighted the fact that the conditions which enabled the feature to form are quite rare and suggests that the ocean plays a major role in its formation, with a number of oceanographic factors being important, such as the salinity of the mixed layer (Motoi *et al.*, 1987).

Persistent coastal polynyas that form during the winter months are areas of high sea ice formation and are responsible for a large proportion of sea ice production in certain sectors of the Antarctic. For example, the coastal area of Terra Nova Bay experiences frequent, strong offshore winds associated with katabatic flow down valley glaciers inland of this area that keep the area ice-free throughout much of the winter (Bromwich, 1989a). Because of the very low air temperatures, the open water soon freezes but as it is advected away so it creates new open areas. This area has been estimated to produce 10% of the ice that forms in the Ross Sea and is discussed further in Sections 4.2 (atmospheric energy balance) and 6.1 (katabatic wind effects).

Sea ice motion

Sea ice moves as a result of forcing by the surface wind and ocean currents. Response to the wind field is relatively slow and it has been shown by Budd (1991) that the ice is only advected at about 2% of the geostrophic wind speed. However, sustained flow in the meridional direction for periods of a week or more can create large positive or negative anomalies in the ice extent that can persist for periods of many months. Strong surface winds and ocean currents can also contribute to changes in the thickness of the ice through rafting of floes. Because the sea ice around the Antarctic is not constrained by land, the motion of the ice is generally divergent. Divergence results in the opening up of areas of open water that can initially increase fluxes of water vapour and heat into the atmosphere, but also results in rapid formation of new ice during the winter. Convergent conditions give rise to an increase in the sea ice thickness through rafting and ridge building.

Sea ice motion can be determined by satellite tracking of buoys embedded in the ice (Allison, 1989; Wadhams *et al.*, 1989; Kottmeier *et al.*, 1992). Buoy deployments around the Antarctic have been relatively limited to date, although there is now an Antarctic buoy programme under the auspices of the World Climate Research Programme. The buoys are deployed on suitable ice floes by ships and their position tracked using the ARGOS data collection system on the NOAA spacecraft. Buoys in Antarctic sea ice tend to provide data for periods of only a few seasons because the generally divergent motion takes them northwards and out of the sea ice area.

Ice motion can also be determined by tracking individual floes in satellite imagery for periods of several days or, in the case of long-lived floes, a week or more (Viehoff, 1991). The most suitable data for this purpose are those from the NOAA AVHRR since they are readily available and the horizontal resolution of 1 km allows most floes to be observed. However, the large amounts of cloud in the sea ice zone make it difficult at times to find cloud-free images. More recently, Synthetic Aperture Radar (SAR) imagery from the ERS-1 satellite has been used to observe ice motion in certain parts of the Antarctic. Data from this instrument remove the need for cloud-free conditions and also show fine-scale structure with a resolution of 20 m. However, the instrument only collects data over a 100 km×100 km area so that it is more appropriate for regional studies than it is for broad-scale monitoring.

Sea ice motion in the Weddell Sea has been investigated by a number of workers because this is one of the main areas for sea ice production. Since the early days of Antarctic exploration, when a number of ships were beset in Weddell Sea sea ice, it has been known that there was a clockwise circulation of

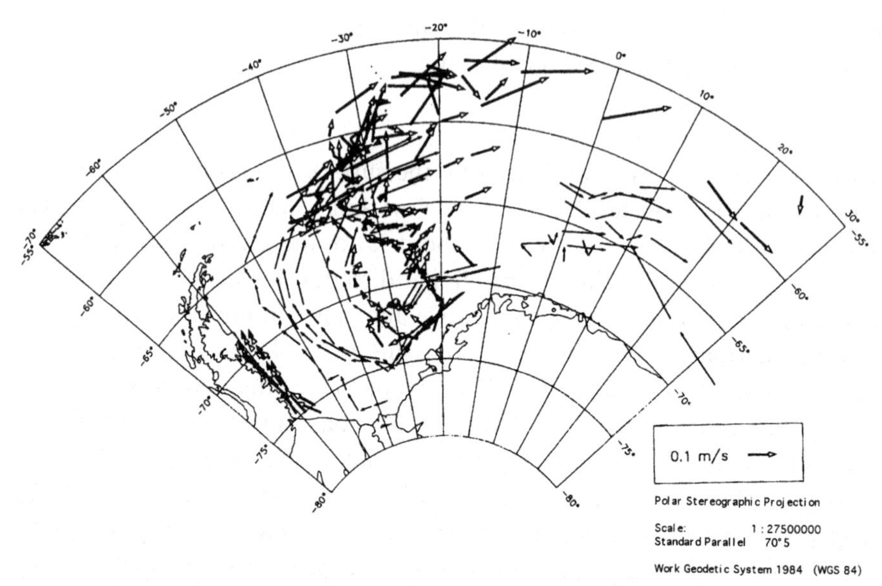

Figure 3.36 The mean monthly ice drift vectors from drifting buoys in the Weddell Sea between 1986 and 1992. Thin arrows are from observations around the 'Winter Weddell Sea Project' 1986, thicker arrows relate to buoys of the 'Winter Weddell Gyre Studies' in 1989 and 1992 and 'AnZone' 1992. From C. Kottmeier *et al.*, *J. Geophys. Res.* **97**, 20374, 1992.

the ice around the area. This can be seen in Figure 3.36, which shows the mean monthly ice drift vectors produced by tracking buoys embedded in the sea ice (Kottmeier *et al.*, 1992). This circulation comes about because of the forcing both from the oceanographic circulation, in the form of the Weddell Gyre, and from the cyclonic wind flow that is a climatological feature of the area.

Sea ice–weather relationships

The extent and concentration of sea ice depend on many atmospheric and oceanographic parameters, including the ocean currents, the temperature and salinity of the upper layer, the atmospheric circulation and the air temperature. However, the role of each of these elements and the interaction among the different physical processes is not well understood and the response of the sea ice may differ in different regions of the Antarctic because of local geographical conditions. Although satellite data can now routinely provide information on the skin temperature of the ocean we have few data on other oceanographic parameters, except those collected by occasional research cruises. However, the meteorological conditions have been fairly well analysed over the Southern Ocean and there have been a number of studies into links between the circulation and sea ice extent that will be reviewed here.

Individual depressions can cause changes in the position of the northern limit of the ice and also affect the ice concentration within the pack (Cavalieri and Zwally, 1985). This can be observed in animated sequences of daily sea ice extent maps produced from passive microwave imagery as a wave at the ice edge that slowly progresses eastwards. The net effect of these disturbances on the ice position appears to be very small but, because the resolution of the current passive microwave satellite data is about 30 km, their overall effect is difficult to determine. For the establishment of sea ice anomalies it appears that the meridional component of the atmospheric circulation is most important (Ackley and Keliher, 1976; Parkinson and Cavalieri, 1982; Jacobs, 1993). Harangozo (1994) has shown that anomalies in the sea ice extent once established can persist for several seasons and has suggested that it is the persistence of strong meridional flow that is important in the establishment of the anomalies. Such conditions occur when the atmospheric long waves are amplified and slow-moving.

Examination of the sea ice record in association with data on the atmospheric circulation has suggested that the motions of the ice edge on the seasonal and synoptic time scales are strongly influenced by the quasi-stationary atmospheric systems (Cavalieri and Parkinson, 1981). They suggested that the Weddell and Ross Seas are particularly sensitive to the interannual variations in the atmospheric circulation and especially to variations in the circumpolar trough. The most important fluctuation in the circumpolar trough is the semi-annual oscillation that results in the trough being farthest north in the summer and winter. These are the times of the sea ice minimum and maximum so that there is no direct connection between the sea ice position and the trough. However, links between sea ice and cyclonic activity have been proposed and it is interesting to note that the latitude of greatest cyclogenesis is furthest north in September and at its most southerly position in March; a cycle that is in phase with the movement of the sea ice limit.

Future research needs

Although satellite systems and *in situ* observations have, over the last two decades, provided us with a great deal of data on the broad-scale sea ice extent and information on its variability on scales from the daily to the decadal, there are a number of parameters and processes of interest for change studies that cannot currently be determined with a high degree of accuracy. These include the following.

(i) The thicknesses of the sea ice and overlying snow cover which are required in order to determine the total mass of sea ice around the continent. This information will probably have to be determined by ship transects and upwards-looking sonar arrays.

(ii) More accurate information on the sea ice concentration, because it is important for computing the heat fluxes from the ocean.

(iii) The mechanisms behind the formation and persistence of polynyas and their role in sea ice production.

(iv) Quantification of ice–ocean–atmosphere interactions and representation of processes in climate models.

3.5.2 The Southern Ocean environment

The Southern Ocean plays an important role in determining the meteorological and climatological conditions of the high southern latitudes through the fluxes of heat and moisture that take place between the upper surface of the ocean and the lowest layers of the atmosphere. The ocean holds a vast amount of heat and has a moderating effect on the climate of the Antarctic coastal region, especially when there is a limited amount of sea ice present. This is evident in the high correlation between the mean temperatures at coastal stations on the western side of the Antarctic Peninsula and the extent of sea ice in the Bellingshausen Sea. During years with little sea ice coverage, temperatures are significantly warmer, as a result of the heat contributed by the ocean, than in years when more extensive ice is present.

Large fluxes of heat into the atmosphere can occur when the ocean is significantly warmer than the air and the effects of this can be seen graphically in satellite imagery during outbreaks of cold air northwards over the ice-free ocean. In these situations extensive convective cloud can develop in the southerly flow with, on some occasions, the large fluxes of heat leading to the development of mesocyclones and even, occasionally, small synoptic-scale depressions (see Sections 6.5 and 5.3.2). During periods when mild, mid-latitude air is being advected polewards it is progressively cooled during its passage south and extensive low cloud can develop. As there is a net flux of heat polewards across the Southern Ocean into the Antarctic (see Section 4.2) so this region is one of the cloudiest areas on Earth. The cooling of the near-surface air by the relatively cold ocean surface also results in extensive fog in certain areas, such as near the major oceanic frontal zones where the sea surface temperature gradient is large.

In this section we will examine the characteristics of the ocean around the Antarctic. Emphasis is placed on the factors that affect the atmosphere of the high southern latitudes.

Characteristics of the Southern Ocean

The Southern Ocean covers an area of 77×10^6 km^2 or 22% of the total area of the world's oceans (Tchernia, 1980). It is the most zonally uniform of the world's oceans since its flow is largely unaffected by land masses. It extends around the Antarctic, its southern boundary being the coast of the continent and the northern limit being generally determined by the characteristics of the water masses. Across the ocean the isopleths (lines of constant water property)

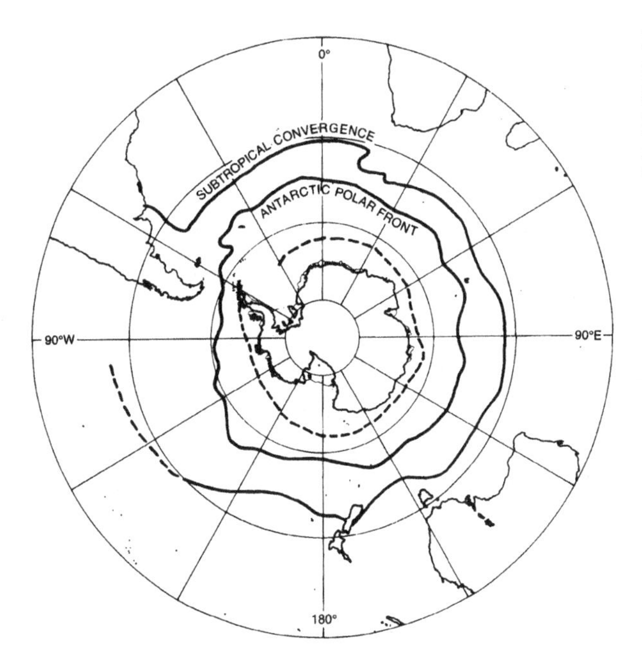

Figure 3.37 Mean positions of the Southern Ocean convergences (solid lines) and divergences (dotted line). After Tchernia (1980).

are approximately parallel to the lines of latitude, although the gradient is not uniform but rather concentrated into narrow frontal zones. In these areas of convergence denser water from the south pushes under less dense water, creating oceanic fronts that can be identified by a strong gradient in properties, such as temperature and salinity. The mean positions of the regions of convergence where the surface fronts are located are shown in Figure 3.37. The two major fronts are the following.

(i) The Antarctic Polar Front (APF), which is sometimes called the Antarctic Front or Antarctic Convergence, is the zone where the cold, dense northwards-moving Antarctic Surface Water sinks below the warm sub-Antarctic water before continuing its progress northwards under the surface. This transition gives rise to a temperature change in the meridional direction of 3–8°C in summer and 1–5°C in winter, although the temperature change experienced by a ship crossing the front can be a single, sharp jump or a transition in two or three stages. The APF is located in the zone 47–61°S, although there are considerable variations in its position around the globe. It has its most southerly location in the Bellingshausen Sea and through the Drake Passage, although the water emerging from the Weddell Sea causes the front to be much further north across the South Atlantic. The structure of the front is not simple and there can be a broader zone of transition rather than the narrow band implied by Figure 3.37. The front is also characterised by meanders and eddies that can break away as cold core vortices containing Antarctic

Surface Water. In the Weddell Sea–Drake Passage region the APF has a seasonal variability of 1–2° of longitude and a long-term variation of up to 4° in latitude.

(ii) The Subtropical Convergence (STC) or Subantarctic Front lies north of the APF and is generally regarded as the boundary between the Southern Ocean and the Atlantic, Pacific and Indian Oceans to the north. It is marked by an increase in salinity from 34.5 ppt (parts per thousand) to the south to 35 ppt north of the front. The STC is less well delineated than is the APF and does not constitute a clear line of convergence. Around the Antarctic it varies in meridional position from close to 37° S in the South Atlantic, 40° S in the Indian Ocean, near 43° S south of Australia and then to its most northerly position at almost 25° S off the coast of Chile. Here the northern limit of the Southern Ocean is usually taken as 40° S.

Water types

Several different water types are found around the Antarctic as a result of the ocean currents experienced in the Southern Ocean and the effects of the large seasonal cycle in sea ice extent. A diagram based on the work of Sverdrup that shows the water masses found in the Antarctic and their movement is shown in Figure 3.38. The main water types found in the Southern Ocean are the following.

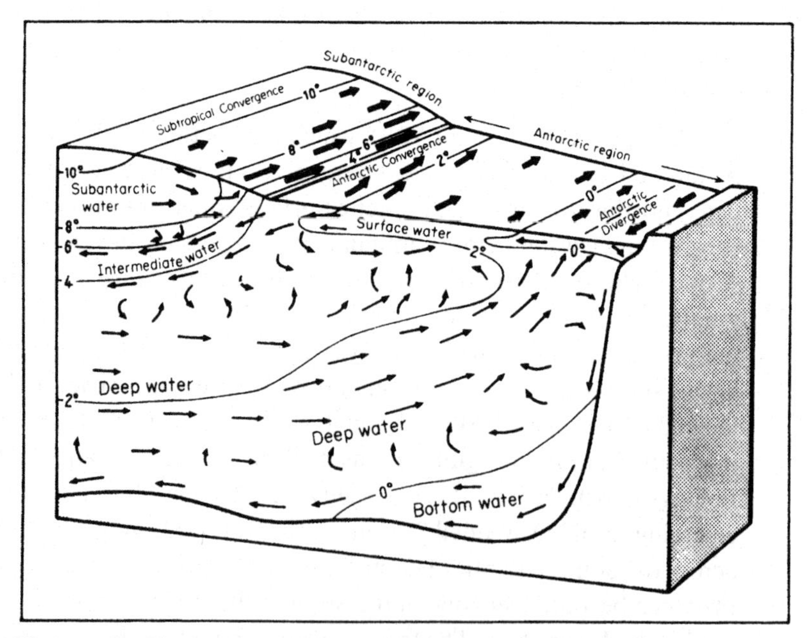

Figure 3.38 The movement of water masses in the Southern Ocean based on the work of Sverdrup. From Tchernia (1980).

(i) Antarctic Surface Water occurs south of the APF and is very cold and quite fresh. Because of its density it is prone to vertical exchange with water at lower levels and there is a marked temperature gradient at its lower boundary. However, its properties are highly dependent on the freezing and melting of sea ice during the year, which has a marked effect on its salinity. When sea ice forms there is rejection of salt, which increases the salinity and therefore the density of the surface water. With the melting of the sea ice the reverse occurs and both the salinity and the density decrease. In the immediate coastal region persistent katabatic winds in certain parts of the continent move the sea ice away from the coast, opening up areas of ice-free water, even in the middle of winter. These are areas of extensive sea ice production during the winter season, resulting in the generation of large amounts of dense, saline water. Antarctic Surface Water extends to a depth of about 100–250 m and has a salinity of less than 34.5 ppt

(ii) Antarctic Circumpolar Water or Warm Deep Water is found below Antarctic Surface Water and is both warmer and more saline. It arrives in the Southern Ocean from the north at a depth of as much as several thousand metres and rises at the APF. Below the surface the temperature steadily increases, has a maximum of around 2°C at about 500 m and then decreases again to close to 0°C near the ocean floor. The mean salinity of Antarctic Circumpolar Water is about 34.70–34.76 ppt and reaches a maximum between 700 and 1300 m depth.

(iii) Bottom Water, as its name implies, is found at the lowest levels around the Antarctic below about 3000 m. It is slightly less saline than is Antarctic Circumpolar Water, having a salinity of 34.66 ppt and a temperature throughout the year of close to −0.5°C. However, its most remarkable characteristic is its very high density and it is, in fact, the most dense water found in any of the world's free oceans. Because it is very stable it flows out from the Antarctic into the lowest areas of the great ocean basins, reaching as far north as 50° N in the Pacific. Eighty per cent of Bottom Water is thought to form in the Weddell Sea (Foldvik and Gammelsrød, 1988) with the Ross Sea being the other main contributor. Its production is closely linked to the particular oceanographic and sea ice conditions in these regions and especially the annual cycle of sea ice formation and melting. The processes behind the production of Bottom Water have become clearer since the mid-1970s and there are now thought to be two main mechanisms responsible. Both rely on the conditions that allow the formation of cold, high-salinity water (shelf water) over the continental shelf by offshore winds exposing open water during the winter. The two theories can be summarised as follows.

(a) Warm Deep Water, modified by winter freezing, mixes with High-Salinity Shelf Water (HSSW) at the shelf break to form Weddell Sea Bottom Water, which is denser than either of its components. This comes about because of the non-linearity of the equation of state for sea water which relates temperature, density, salinity and pressure. Weddell Sea Bottom Water eventually mixes with further Warm Deep Water to give Antarctic Bottom Water. Further details can be found in Foster and Carmack (1976).

(b) A more recently proposed mechanism (Foldvik and Gammelsrød, 1988) has stressed the importance of the modification of the HSSW by passage under the deep ice shelves to form very cold Ice Shelf Water with a temperature of less than −1.95°C, which is below the surface freezing point. This descends the continental slope and mixes with Warm Deep Water to give, ultimately, Antarctic Bottom Water.

Ocean currents

The main feature of the Southern Ocean circulation is the eastwards-flowing* Antarctic Circumpolar Current (ACC) that extends over the latitude band 40–65° S. Figure 3.39 shows the ACC and the other surface currents of the Southern Ocean. The ACC is driven primarily by the prevailing westerly winds of the mid-latitude zone, although when this effect is combined with the effects of the Coriolis force, there is a northwards component to the surface flow. The ACC is characterised by temporal and spatial variations in direction and speed that can disrupt the mean eastwards flow. The main obstacles are the southern part of South America and the Antarctic Peninsula, resulting in convergent flow through the Drake Passage. Here the current is more rapid and speeds of up to 1 m s^{-1} have been recorded. The surface speed of the ACC varies as a function of latitude and is about 0.04 m s^{-1} south of the APF, increasing to about 0.15 m s^{-1} north of the front and then decreasing again further north. Much higher speeds of up to 1 m s^{-1} have been recorded in association with jets near the major fronts, but these are of very limited horizontal extent. A semi-annual oscillation has been noted in the latitudinal position of the speed maximum of the APF from an analysis of data from the First GARP Global Experiment in 1979 (Large and van Loon, 1989). This comes about because of forcing by the atmospheric zonal wind that itself experiences a semi-annual oscillation as a result of the migration throughout the year of the circumpolar trough (see Section 3.3). The ACC extends throughout a deep layer of the ocean and the west to east flow is still found at depths of 3000 m. Such a flow transports large volumes of water and estimates put this at about 110×10^6 m^3 s^{-1} (110 sverdrup)

* The direction of flow of ocean currents is referred to in terms of towards where the water is moving rather than in the atmospheric sense in which a wind direction indicates whence the air has come.

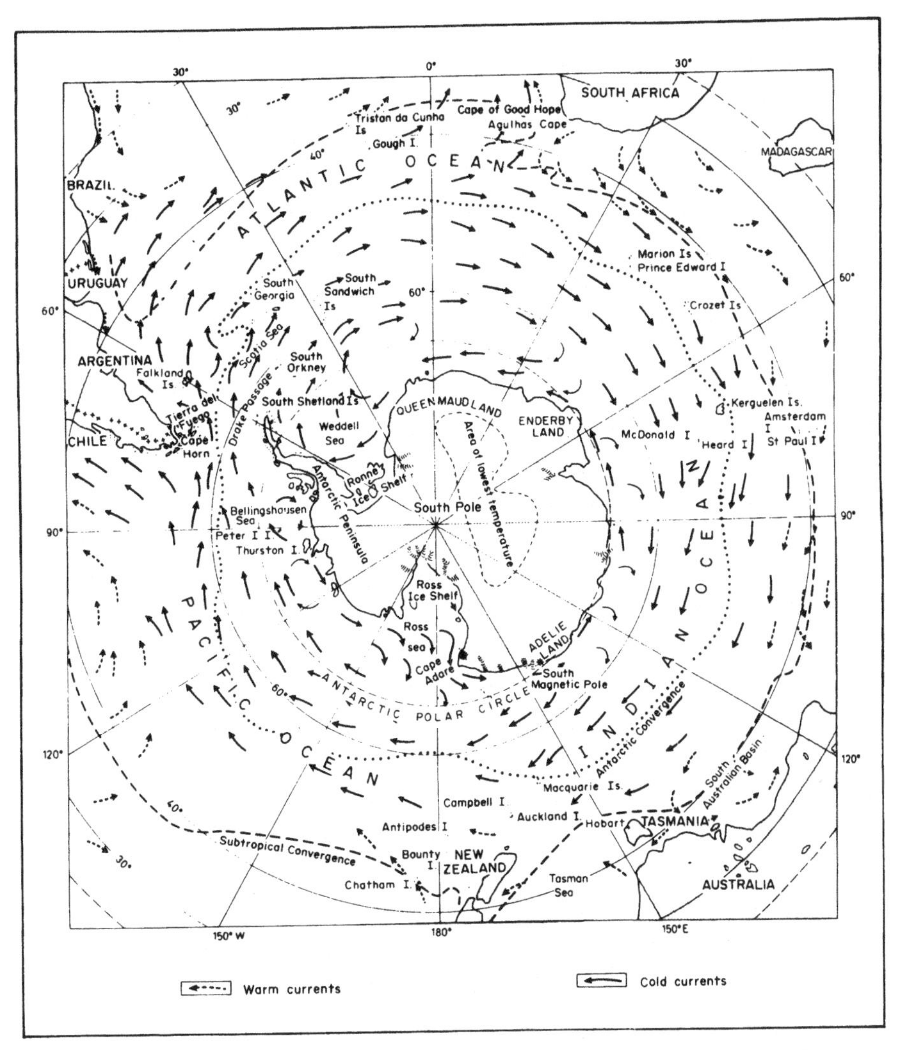

Figure 3.39 The surface currents of the Southern Ocean. From Tchernia (1980).

although large variations from this value have been reported. The constricted flow in the Drake Passage gives the greatest movement of water found anywhere in the world and can reach values of 150 sverdrup. The ACC has been discussed in detail by Nowlin and Klinck (1986).

Close to the coast of the Antarctic there is a westwards-flowing coastal current that is driven by the prevailing easterly winds that are found in this region. This flow is normally limited to the area south of 65° S and only has a flow of around 10 sverdrup. The combined effects of wind and current in the coastal zone can have a significant effect on the sea ice around the coast and are responsible for opening up the coastal leads that are a feature of some areas, such as the eastern Weddell Sea. Similarly, at the edge of the Ross Ice Shelf the circulation is away from the shelf, opening up large areas of open water within the pack ice. In the Weddell Sea the westward coastal current is diverted northwards by the Antarctic Peninsula and then joins the ACC east of the Drake Passage. This results in a cyclonic circulation in the Weddell Sea that is known as the Weddell Gyre. With the climatological cyclonic atmospheric circulation that is found over the Weddell Sea the effects of the wind and current forcing are apparent in the cyclonic rotation of the sea ice in this area. A similar situation occurs in the Ross Sea, although here the extension of the land northwards is not so pronounced as it is with the Antarctic Peninsula. In fact, one branch of the coastal flow westwards passes across the northern boundary of the Ross Sea whereas another turns south and passes under the Ross Ice Shelf, emerging off the coast of Victoria Land and rejoining the main flow.

As the Coriolis force imparts a component to the left on the ocean currents so the eastwards-moving ACC is deflected to the north, whereas the westwards-moving coastal current gains a southerly component. Such a regime gives a divergent flow whereby deep water is drawn to the surface. This upwelling is a feature of much of the region between the ACC and the Antarctic continental margin, although the divergence is not found everywhere around the Antarctic and is absent to the east of the Drake Passage.

The ocean circulations of the Arctic and Antarctic are very different because of the arrangement of the land masses in the two hemispheres. The Southern Hemisphere consists mostly of ocean and there are no land masses extending continuously from mid-latitudes to the polar regions. This results in a more zonal ocean circulation and less transport of warm water to high latitudes than occurs in the Arctic. When this is combined with the more zonal atmospheric flow it results in smaller air–sea temperature differences than there are in the north. The deep convection in the atmosphere and formation of cumulonimbus cloud that occur in parts of the North Atlantic and North Pacific are not found in the Antarctic, although outbreaks of cold continental air can occasionally extend well to the north, giving such conditions outside the Antarctic.

Sea surface temperatures

The sea surface temperature (SST) is a very important quantity because it indicates the amount of energy in the uppermost layer of the ocean which is in contact with the atmosphere. The SST is therefore an important factor in determining the flux of energy between the ocean and the atmosphere or in the reverse direction. South of the Antarctic Polar Front the SSTs are generally between about −1.9 and 1°C in winter and between −1 and 4°C in summer. Figure 3.40 shows the mean January and July SSTs for the Southern Ocean from the GISST (Global Ice and Sea Surface Temperature) data set compiled by the UK Meteorological Office.

The temperatures in summer are relatively low because of the large amount

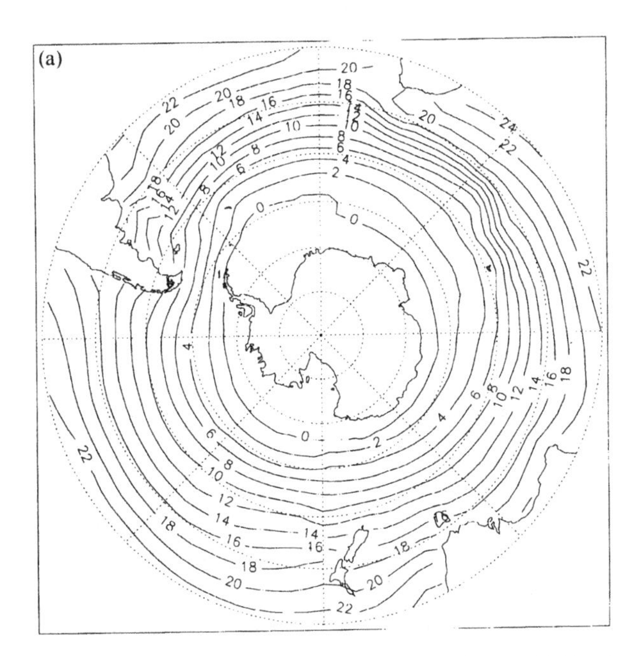

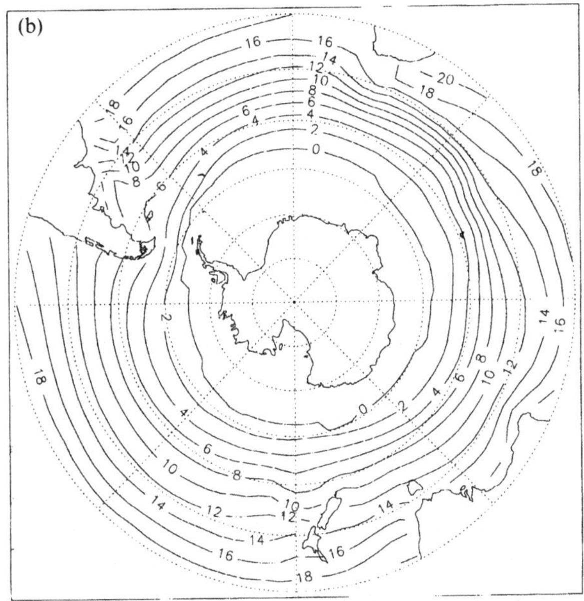

Figure 3.40 Mean sea surface temperatures across the Southern Ocean for (a) January and (b) July.

of heat that is required to melt the sea ice which formed during the winter. The gradients north of the Antarctic coast are relatively small but there is an increase of up to 8°C across the APF. Between the two main frontal zones of the APF and the STC, SSTs are in the range 4–10°C but can be as high as 14°C in summer. The SST gradient here is small but across the STC there is a rapid rise of around 4°C and this feature is usually apparent from an increase in the gradient of the 12–14°C isotherms.

Chapter 4

The large-scale circulation of the Antarctic atmosphere

4.1 Introduction

In this chapter, we shall discuss the mechanisms which maintain the large-scale atmospheric circulation over the Antarctic continent and its immediate surroundings. Observations of the gross features of the circulation were described in Chapter 3; here we shall attempt to explain these observations. The methodology that we have adopted is to examine the budgets of fundamental dynamical and thermodynamic quantities within the Antarctic atmosphere.

First, we examine the heat (or enthalpy) budget. In many ways, this is the fundamental budget, since the Antarctic atmosphere may be regarded as the 'cold' end of the global atmospheric heat engine and the requirements of energy balance place strong constraints on the atmospheric circulation. Furthermore, we shall see that the circulation over the Antarctic continent is strongly controlled by the low-level drainage flow that results from persistent cooling over the sloping surfaces of the continental ice sheets. This is discussed in Section 4.3, in which we consider the atmospheric vorticity budget and the resulting circulation. Finally, in Section 4.4, we study the atmospheric water vapour budget. Water plays a rather passive role in the dynamics of the Antarctic atmosphere but is vital to the maintenance of the Antarctic ice sheets.

Studies of the dynamics of the Antarctic atmospheric circulation have been hindered by the lack of suitable data (particularly upper-air observations) from many parts of the continent. Observations from AWSs and satellite sounders have been of value in diagnostic studies, but many recent advances in our understanding of the large-scale dynamics of the Antarctic atmosphere have come from modelling studies using General Circulation Models (GCMs). In the final

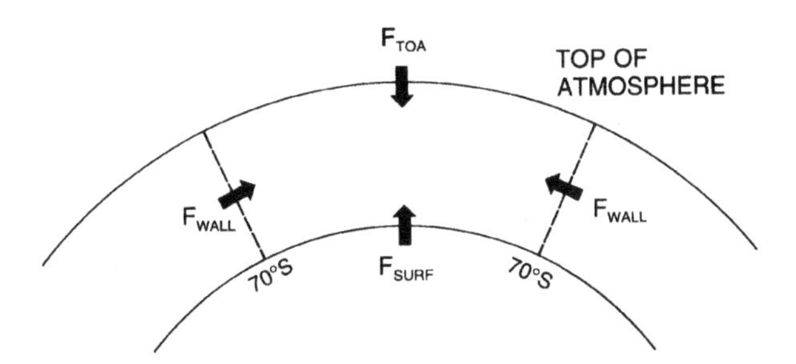

Figure 4.1 A schematic view of the South Polar atmospheric cap and the fluxes of energy into it.

section of this chapter, we consider the representation of the Antarctic circulation in a number of models and consider how effective modern GCMs are at reproducing the essential dynamics of the Antarctic atmosphere.

4.2 The heat budget

4.2.1 Components of the budget

For the purpose of calculating the heat budget of the Antarctic atmosphere, it is convenient to define a 'South Polar atmospheric cap', bounded by the Earth's surface, by the top of the atmosphere and by an imaginary wall running round a latitude circle. The heat budget of the enclosed atmosphere is then determined by the energy fluxes through the bounding surfaces. Such a polar cap is shown schematically in Figure 4.1. A bounding latitude circle at 70° S encircles the major part of the Antarctic continent and was used in the budget studies of Nakamura and Oort (1988) and Masuda (1990). To the south of this latitude, 78% of the Earth's surface is covered by land and only 22% is covered by ocean. Interestingly, these proportions are almost exactly reversed for the Arctic north of 70° N.

The principal energy flux components at each boundary are shown in Figure 4.1. Note the sign convention – fluxes *into* the polar cap are considered positive. At the top of the atmosphere, the energy flux is F_{TOA}, which is the difference between incoming solar radiation incident on the top of the atmosphere and the terrestrial long-wave radiation plus reflected short-wave radiation. These fluxes may be estimated from satellite measurements (Section 3.1). At the Earth's surface, the net flux F_{SURF} is the difference between the net radiative flux from the surface and the atmospheric fluxes of sensible and latent heat to the surface. Finally, the flux F_{WALL} through the bounding wall at 70° S is made up of transports of sensible and latent heat by the mean atmospheric circulation and by transient eddies. The energy budget of the polar atmospheric cap is then given by

$$\frac{\partial \Pi}{\partial t} = F_{TOA} + F_{WALL} + F_{SURF} \tag{4.1}$$

Table 4.1 *Seasonal means of net top-of-atmosphere radiative flux over the region 70° S to 90° S. After Nakamura and Oort (1988).*

Season	F_{TOA} (W m^{-2})
Summer (DJF)	−32
Autumn (MAM)	−130
Winter (JJA)	−131
Spring (SON)	−67
Year	−90

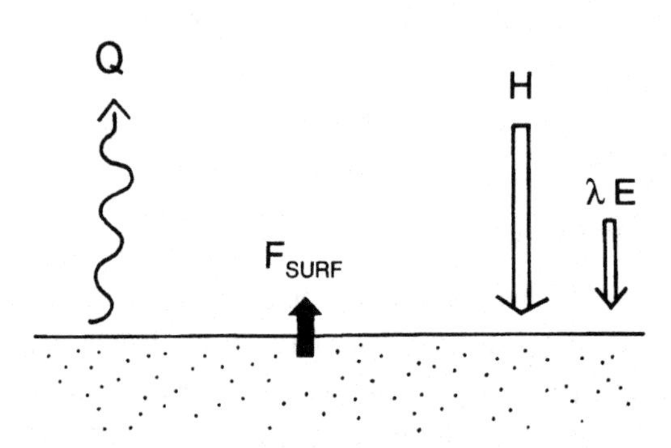

Figure 4.2 A schematic depiction of the components of the surface energy balance. The fluxes are shown as they are normally directed during the Antarctic winter, i.e. H, λE and F_{SURF} positive, Q negative.

where Π is the total thermal energy content of the polar cap. Averaged over an annual cycle, the rate of change term on the left-hand side of Eq. (4.1) will be zero if there is no long-term climate change. In the following sections, we shall consider these energy fluxes in greater detail and then combine them to produce annual and seasonal energy budgets for the South Polar atmospheric cap.

4.2.2 Top of atmosphere radiation

Estimates of the top-of-atmosphere radiation balance are available from polar-orbiting satellites and are discussed in Section 3.1. Nakamura and Oort (1988) gave monthly values of F_{TOA} for the period 1966–77; these have been summarised as seasonal values in Table 4.1. The polar atmospheric cap experiences net radiative cooling in all seasons. Masuda (1990) calculated a net annual cooling of 98 W m^{-2} over the same region for the FGGE year (1978–79).

4.2.3 The surface energy balance

The principal fluxes of energy at the Earth's surface are the net radiative flux, Q and the atmospheric fluxes of sensible heat, H, and latent heat, λE (see Figure 4.2).

F_{SURF}, the net energy exchange between the Earth and the atmosphere, is the residual of these energy fluxes. If fluxes directed towards the surface are defined as positive, then

$$F_{SURF}+Q+H+\lambda E=0 \tag{4.2}$$

In a steady-state climate (i.e. if there were no net heating or cooling of the ice sheet) F_{SURF} must average to zero (or near zero) over the annual cycle over the majority of the continent since horizontal transport of heat within the ice sheets is negligible. The exception is in the limited low-lying coastal regions, where significant run-off of melt water occurs, but this is insignificant on a continental scale. Over the oceans, the annual average of F_{SURF} is not necessarily zero, since heat exchanged between the upper layers of the ocean and the atmosphere can be balanced by horizontal transport of sensible heat in the ocean or by transport of latent heat in the form of sea ice.

The net radiative flux at the surface, Q, is the balance between incoming solar radiation, reflected solar radiation, long-wave radiation from the surface and downwards long-wave radiation from the atmosphere. The climatology of these radiation components was discussed in Section 3.1. For our control region, south of 70° S, the net short-wave flux is zero for the part of the year during which the sun remains below the horizon. Even when the sun is above the horizon, the low solar elevation angles and high surface albedo of the mostly snow-covered surface contrive to keep this flux small. The net long-wave flux was also seen to be quite small in the measurements presented in Chapter 3, as a result of the near balance between upwelling and downwelling long-wave radiation at the surface. Averaged over the whole year, net long wave-cooling of the surface exceeds net solar heating by 10–20 W m^{-2} over most of the Antarctic continent. The only exception is in the few 'oasis' areas (Solopov, 1969), such as the Dry Valleys of Victoria Land, where bare rock remains exposed throughout much of the year, keeping the surface albedo low. In such areas the enhanced solar heating over the summer can be sufficient to maintain a positive annual average radiation balance. However, these regions account for around 1% of the surface area of the continent and are thus not significant for the broad-scale energy balance.

The other components of the surface heat balance are the fluxes of sensible and latent heat which are carried by turbulent transport in the atmospheric boundary layer. We postpone discussion of such transport processes until Chapter 6; here we shall concentrate on the climatology of these fluxes. Accurate measurement of surface fluxes under Antarctic conditions presents a real challenge to the micrometeorologist. Sensitive and well-calibrated equipment is required and long-term measurements have been carried out at only a few Antarctic stations. Seasonal and annual mean values of sensible heat flux for

Table 4.2 *Seasonal values of atmospheric sensible heat flux, H, net radiation Q, and their residual,* $F_{SURF}=-(H+Q)$ *at four stations. Data are from Liljequist (1957), Carroll (1982), Ohata et al. (1985a) and King et al. (1995). The sign convention is that H and Q are positive when directed towards the surface.*

Station	Season	H (W m^{-2})	F_{RAD} (W m^{-2})	F_{SURF} (W m^{-2})
Maudheim	DJF	+5.6	+3.2	−8.8
	MAM	+10.5	−18.9	+8.4
	JJA	+12.8	−22.5	+9.7
	SON	+9.4	+9.1	−0.3
	Year	+9.6	−11.8	+2.2
Halley	DJF		+8.7	
	MAM	+8.4	−18.0	+9.6
	JJA	+11.6	−22.0	+10.4
	SON		−7.9	
	Year		−9.8	
Mizuho	July	+36.7	−37.6	+2.5
	December	25.1	+13.8	−6.4
South Pole	DJF	+22.1	−17.5	−4.6
	MAM	+13.3	−16.2	+2.9
	JJA	+18.7	−20.6	+1.9
	SON	+23.4	−20.3	−3.1
	Year	+19.4	−18.7	−0.7

Halley, Maudheim, Mizuho and South Pole Stations are shown in Table 4.2, together with the corresponding mean net radiation values. At South Pole Station, seasonal means of net radiation and sensible heat flux are balanced to within 5 W m^{-2}. The sign of the residual, F_{SURF}, indicates cooling of the snowpack during winter and warming during summer. Mizuho is situated on the steep coastal slope and is thus subject to strong katabatic winds that maintain a large sensible heat flux during winter. The snow surface thus remains relatively warm and the outgoing long-wave radiation remains large, leading to a high net radiative cooling. At the coastal stations, the seasonal imbalances are larger, of the order of 10 W m^{-2}. The rather large annual imbalance at Maudheim is probably a result of the larger uncertainties in these early pioneering measurements.

Latent heat flux has been measured at even fewer locations but, because the humidity mixing ratio of air at typical Antarctic temperatures is small even when the air is close to saturation, latent heat flux generally makes an insignificant contribution to the surface energy budget over the Antarctic continent. Carroll (1982) estimated that latent heat flux can be no larger than 1–10% of the sensible heat flux at the South Pole. Measurements at Mizuho using an atmometer pan (Fuji, 1979) show effectively zero net evaporation during the winter months but significant evaporation, equivalent to an upward latent heat flux of about 16 W m^{-2} during summer. Profile-based measurements performed at Mizuho during December (Ohata *et al.*, 1985a) indicated that the sensible and latent heat fluxes were both directed upwards and were both around 5 W m^{-2}. King and Anderson (1994) found that latent heat fluxes were no greater than 10% of the sensible heat flux during the winter months at Halley. Given that the uncertainties in most measurements of net radiation and sensible heat flux are at least 10%, it seems reasonable to neglect the small contribution of latent heat flux to the surface energy budget over much of the continental ice sheet.

At present, measurements are far too limited to describe the geographical and seasonal variation of surface heat flux adequately. There is some prospect of this situation improving with the deployment of automatic weather stations equipped for flux measurement. Stearns and Weidner (1993) have reported measurements of heat fluxes over the Ross Ice Shelf from a network of AWSs. Their measurements indicate substantial spatial variation, with monthly mean downward heat fluxes in winter ranging from greater than 50 W m^{-2} at sites close to the Transantarctic Mountains, to less than 10 W m^{-2} in the centre of the ice shelf. They attribute the large values to strong katabatic winds blowing down the glaciers which flow through the Transantarctic Mountains. Studies such as this point to the requirement for an improved network of surface energy balance measurements if we are to estimate the surface energy budget over the whole continent with any certainty.

So far we have only considered the surface energy balance over the Antarctic continent. Over the surrounding sea ice zone and ocean, which account for 22% of the surface area of our South Polar cap, the situation can be very different. Unfortunately, measurements of surface energy balance in this region are even more limited than are those over the continent because of difficulties of access. The inner pack ice zone is inaccessible to shipping for much of the winter, limiting major measurement programmes to the spring and summer. During the winter, fluxes must be estimated from remotely sensed data together with the limited information available from platforms such as drifting buoys (see Section 2.4).

Over sea ice of 100% concentration in winter, the surface energy balance will be similar to that over the continental ice sheets since snow-covered sea ice has a high albedo, comparable to that of the continental ice sheets, and ice of typical thickness 1–2 m effectively insulates the atmosphere from the warmer underlying ocean. However, the presence of even a small fraction of open water in the form of leads or polynyas can radically change this balance (Smith et al., 1990). This is because the overlying air is typically some 10–20°C colder than the water in the leads, which is close to its freezing point (−1.9°C for surface water of typical salinity). Thus, a large upwards heat flux is generated over areas of open water and, even when areally weighted, this dominates the smaller fluxes over the surrounding ice areas. Weller (1980) estimated that the mean upwards fluxes of sensible and latent heat in mid-winter are 89 and 28 W m^{-2} respectively in the inner sea ice zone (ice concentration greater than 85%) and 152 and 37 W m^{-2} respectively in the outer sea ice zone (ice concentration in the range 15–85%). Energy balance measurements made during the Weddell Polynya Expedition in the Antarctic spring (October and November) of 1981 (Andreas and Makshtas, 1985) suggest that these figures may be overestimates. During this expedition, fluxes were measured over a variety of ice and water surfaces between 56° S and 66° S. On average, both the sensible and the latent heat fluxes were upwards, with mean values of about 19 and 29 W m^{-2} respectively. The surface was gaining about 105 W m^{-2} from net radiation, leaving a balance of 57 W m^{-2} available for melting the sea ice. The fluxes showed considerable dependence on synoptic conditions. The largest fluxes upwards were associated with advection of cold air from the south and small heat fluxes downwards were observed when northerly winds advected warm air from the open ocean. Because of such temporal and spatial variability, flux measurements at a single location over a limited period may not be representative of mean values over the whole sea ice zone.

Very large heat fluxes are observed over polynyas around the coasts of Antarctica (Section 3.3). As very cold continental air flows over open water, exceptionally large heat fluxes upwards are generated, cooling the surface water and leading to the formation of sea ice. This ice is driven away from the coast by the action of wind stress and freezing continues. The heat fluxes over several coastal polynyas have been estimated by Kurtz and Bromwich (1985) and by Cavalieri and Martin (1985) using a combination of remotely sensed measurements and meteorological data from nearby land stations. The estimates by Kurtz and Bromwich of the monthly mean energy balance of the Terra Nova Bay polynya are shown in Table 4.3. Note that F_{SURF} does not average to zero over the year since the production and export of sea ice provides an additional source of energy, from the latent heat released during freezing, which balances the local surface energy budget. Cavalieri and Martin estimate slightly smaller heat fluxes – up to 450 W m^{-2} – for polynyas along the Wilkes Land coast. Kottmeier and Engelbart (1992) made direct observations of heat fluxes over a Weddell Sea coastal polynya during October and November 1986. They measured a mean heat flux upwards of 143 W m^{-2} and an extreme maximum of over 400 W m^{-2}.

The total area covered by coastal polynyas is only a small fraction of the sea ice zone, so these anomalously large fluxes are probably not important in the context of the regional surface energy balance. Their primary importance is that

Table 4.3 *Surface energy balance components over a coastal polynya in Terra Nova Bay, from Kurtz and Bromwich (1985). Q is the net radiation, H the sensible heat flux, λE the latent heat flux and $F_{SURF}=-(Q+H+\lambda E)$ the residual. The sign convention is that Q, H and λE are positive when directed towards the surface.*

Month	Q (W m^{-2})	H (W m^{-2})	λE (W m^{-2})	F_{SURF} (W m^{-2})
January	217	−37	−45	−135
February	128	−124	−111	107
March	26	−379	−176	529
April	−25	−403	−179	607
May	−47	−575	−192	814
June	−46	−601	−188	835
July	−50	−574	−192	816
August	−42	−625	−194	861
September	−6	−615	−193	814
October	76	−391	−173	488
November	203	−218	−128	143
December	239	−39	−55	−145

they are the main 'factories' for the production of sea ice during the winter. The net cooling of the Terra Nova Bay polynya of about 800 W m^{-2} during the winter months, for example, would translate to a sea ice growth of about 0.2 m per day were the cooling balanced only by latent heat released during ice formation. In reality, ice growth rates will be slightly smaller since the atmospheric heat flux will be partially balanced by a heat flux from the deep ocean. Kurtz and Bromwich (1985) estimated that approximately 10% of all of the winter sea ice over the Ross Sea continental shelf is formed in the Terra Nova Bay polynya, which has a mean area of only 1300 km^2.

In summary, measurements of the surface energy balance over the Antarctic region are very limited and it is difficult to form representative regional averages of F_{SURF} because of the lack of information on its spatial and temporal variability. Over the continental ice sheets, F_{SURF} is small (perhaps 5–10 W m^{-2}) and upwards during the winter, when the radiative cooling of the surface exceeds the sensible heat flux downwards, and small and downwards during the summer, averaging to zero over the whole year. Over the sea ice zone, F_{SURF} is much larger and more variable. Certain locations, such as ice-free areas on the continent and coastal polynyas, act as anomalously large heat sources but are probably unimportant in the regional heat budget because they are of limited area.

4.2.4 Atmospheric transport of heat

In order to close the heat budget for the South Polar atmospheric cap, we need to estimate F_{WALL}, the flux of energy across the imaginary bounding wall at 70° S. F_{WALL} (expressed as energy flux per unit length of the bounding latitude circle) may be calculated as

$$F_{WALL} = \{[C_p \overline{vT}]\} + \{[g\overline{vz}]\} + \{[\lambda \overline{vq}]\} \quad (4.3)$$

where v is the southwards component of the wind, T is the temperature, z is the geopotential height, q water vapour mixing ratio, C_p is the specific heat of air at constant pressure and λ is the latent heat of vapourisation of water. Time means have been denoted by overbars, zonal means by square brackets and mass-weighted vertical means by curly braces. The first term on the right-hand side of Eq. (4.3) represents the flux of sensible heat into the polar cap. The second term is the flux of potential energy and the final term the flux of latent heat. Each of these terms may be further decomposed into mean flow and eddy contributions. Denoting departures from the time mean by primes, departures from the zonal mean by asterisks and departures from the vertical mean by double primes, we can write

$$\{[\overline{vT}]\} = \{[\overline{v'T'}]\} + \{[\overline{v^*T^*}]\} + \{[\bar{v}]''[\bar{T}]''\} + \{[\bar{v}]\}\{[\bar{T}]\} \quad (4.4)$$

The first term on the right-hand side of Eq. (4.4) is the contribution of transient eddies, generally synoptic-scale weather systems, to the heat flux. The second term represents the contribution of standing eddies, which may be identified with the quasi-stationary planetary waves. The third term is the flux carried by the mean meridional circulation and the fourth term is the contribution of the net mass flux into or out of the polar cap. Since the mass of the polar atmosphere remains approximately constant on timescales of a month or longer, we can neglect this term when considering seasonal and annual energy budgets.

There are two possible strategies for estimating the fluxes which constitute F_{WALL}. First, fluxes may be calculated at station locations from radiosonde observations of winds and temperatures and then interpolated from the irregular station network onto a regular grid in order to estimate the total flux across 70° S. Secondly, the objective atmospheric analyses which are routinely produced as part of numerical weather prediction schemes may be used to calculate fluxes on a regular grid directly. This second approach would appear to have several advantages over the first, both in the ease of calculation and in ensuring dynamical consistency of the results. Furthermore, analysis schemes can incorporate satellite temperature soundings and other

Table 4.4 *Estimates of the components of F_{WALL} ($W\,m^{-2}$).*

	NO88-SPC	NO88-NPC	GCM-SPC	M90-SPC
Total	180	99	95	57
Mean meridional circulation	144	67		
Transient eddies	30	52		
Standing eddies	6	15		

Notes:
Key to columns: NO88-SPC – Nakamura and Oort (1988), South Polar Cap. NO88-NPC – Nakamura and Oort (1988), North Polar Cap. GCM-SPC – estimates for the South Polar Cap from the GCM of Manabe and Hahn (1981), quoted by Nakamura and Oort. M90-SPC – estimates for the South Polar Cap from FGGE analyses by Masuda (1990).

novel sources of data as well as conventional radiosonde observations. However, in regions such as Antarctica where the observing network is sparse, analyses tend to be strongly biased towards the model first guess and different analysis schemes may give very different results (e.g. Trenberth and Olson, 1988), so diagnostics derived from analysed fields must be treated with some caution.

Nakamura and Oort (1988) estimated F_{WALL} using radiosonde data and obtained an annual mean value of 180 W m^{-2}.* Clearly this is too large to balance the observed F_{TOA}, which is 90–100 W m^{-2}. Nakamura and Oort suggested that the mean meridional circulation was poorly represented in their analysis as a result of the sparse radiosonde network in Antarctica and, as a consequence, that they were overestimating its contribution to F_{WALL}. Results from a GCM simulation (Manabe and Hahn, 1981) gave a value of 95 W m^{-2} for F_{WALL}, confirming this suspicion. The breakdown of F_{WALL} into its components is shown in Table 4.4. Taking the GCM estimate of F_{WALL} as 'correct', eddy fluxes account for about 38% of the total energy transport, the remainder being accomplished by the mean meridional circulation. It is interesting to compare these figures with the Nakamura and Oort estimates for the North Polar cap. F_{WALL} at 70° N is similar to that at 70° S, but eddy fluxes account for 68% of the total. In particular, the standing eddy flux is some 2.5 times larger at 70° N than at 70° S, reflecting the greater strength of the planetary waves forced by the Northern Hemisphere's orography.

* For ease of comparison with the surface and top-of-atmosphere fluxes, F_{WALL} is conventionally also expressed as a flux density (W m^{-2}). It is the total flux through the bounding wall, measured in watts, divided by the surface area of the enclosed polar cap (1.5×10^{13} m^2).

Masuda (1990) estimated F_{WALL} using global analyses for the FGGE year (1978–79). His estimates of the eddy contributions to the total flux are very close to those obtained by Nakamura and Oort using radiosonde data. Although this may give us some confidence in both sets of estimates, it is somewhat surprising, since the FGGE year is known to have been anomalously stormy in the Southern Hemisphere (van Loon and Rogers, 1981). In striking contrast to the radiosonde analysis of Nakamura and Oort, the FGGE analyses appear to underestimate the contribution of the mean meridional circulation to the total energy flux. Masuda believes that this is due to the FGGE analyses not exactly conserving mass. The required adjustment to the wind field to restore mass conservation is not unique and changing the adjustment scheme can have a large impact on the energy flux carried by the mean meridional circulation. Clearly, improved observations and analyses are needed before we can estimate the mean meridional circulation component of F_{WALL} with any degree of confidence. In the meantime, diagnostics from GCMs can provide useful estimates, but should be interpreted with care.

4.2.5 The net energy budget and its annual cycle

From the above discussion, it is clear that the only component of the energy balance of the South Polar atmospheric cap known with any great degree of confidence is F_{TOA}, the top-of-atmosphere radiative balance. There are large uncertainties both in the surface fluxes and in the heat transport across 70° S, F_{WALL}, so it is difficult to close the budget. Nakamura and Oort (1988) used the values of F_{WALL} derived from the GCM and then calculated F_{SURF} as a residual term in the budget. Averaged over the year, there is a close balance between top-of-atmosphere radiative cooling (90 W m^{-2}) and transport of energy across 70° S (95 W m^{-2}). The small residual is well within the degree of uncertainty of F_{WALL}.

Although the absolute values of some terms in the energy budget may be uncertain, their seasonal variation, as deduced from observations, may be reasonably realistic. On timescales shorter than a year, energy storage within the atmosphere may be significant and the storage term in Eq. (4.1) must be taken into account. Figure 4.3 shows the annual cycle of the energy budget components as calculated by Nakamura and Oort (1988). All components were computed from radiosonde observations, except for the mean meridional circulation contribution to F_{WALL}, which was taken from a GCM run, and F_{SURF}, which was calculated as a residual of the other components. The South Polar atmosphere is in a state of positive energy balance only from September to December, when mean and eddy transports of heat are sufficient to outweigh the radiative cooling. The ratio of eddy transport to mean transport is relatively large at this time of year, whereas during the winter months the mean transport predominates. All of the atmospheric transport terms show a significant semi-annual variation superposed on the annual

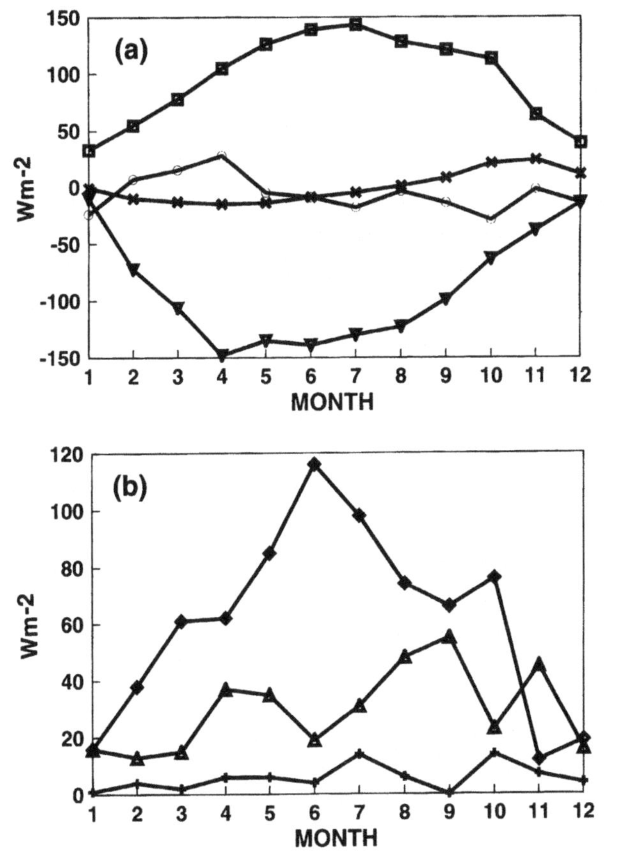

Figure 4.3 (a) The annual cycle of the components of the energy budget of the South Polar atmospheric cap. ∇, F_{TOA}; $\square$, F_{WALL}, $\bigcirc$, F_{SURF}; and $\times$, storage term. (b) The annual cycle of the components which make up F_{WALL}: $\triangle$, heat transport by transient eddies; $+$, heat transport by standing eddies; and $\diamond$, heat transport by the mean meridional circulation. After Nakamura and Oort (1988).

cycle, demonstrating the importance of the semi-annual oscillation (described in Section 3.3) in heat transport processes. The net surface flux, F_{SURF}, remains small throughout the year and becomes positive only during March, April and May, possibly reflecting the rapid production of sea-ice and consequent release of latent heat during this period.

4.2.6 Maintenance of the vertical temperature structure

So far we have only considered the gross thermal budget of the Antarctic atmosphere with little regard to variations within the polar atmospheric cap. The most striking feature of the temperature structure of the Antarctic atmosphere is the strong and persistent surface inversion observed over the continent for much of the year (Section 3.2). In the following section of this chapter we shall see that the surface inversion plays a central role in controlling the dynamics of the Antarctic atmosphere, so it is important to consider how it is maintained.

During the winter months, the surface of the continent is cooling radiatively. The cooling is partially balanced by a turbulent flux of heat downwards through the atmospheric boundary layer. At the surface, the heat flux approximately balances the radiative cooling (see Table 4.2) while, at the top of the boundary layer, it is zero by definition. The boundary layer thus cools relative to the free atmosphere in response to this heat flux divergence and a stably stratified (i.e. inversion) temperature profile develops within the boundary layer. However, heat flux divergence alone cannot account for the observed structure of the surface inversion, since the level of maximum temperature is often more than 200 m above the surface (see Section 3.2) while, under such strong inversion conditions, the turbulent boundary layer may be no more than 50 m deep (King, 1990). Radiative flux divergence within the inversion layer provides an additional mechanism for maintaining the observed temperature profile.

Cerni and Parish (1984) used a simple long-wave radiation model that included some parameterisation of turbulent heat fluxes to simulate the development of an Antarctic surface inversion. Starting from an isothermal atmosphere, they obtained a realistic surface inversion of about 30 K total strength after only 24 h of model integration. By this stage the model had reached a near steady state. The net radiation at the surface had fallen to about 20 W m^{-2} and this was balanced by sensible and sub-surface heat fluxes, both around 10 W m^{-2} (c.f. Table 4.2). Radiative exchange within the atmosphere is thus clearly important in maintaining the surface inversion and allowing it to respond rapidly to changes in surface forcing. It is also responsible for maintaining the smaller, but significant, stable temperature gradient throughout the depth of the Antarctic troposphere.

Vertical profiles of radiative cooling rate in the Antarctic atmosphere can be obtained directly using radiometersondes, or indirectly using radiosonde profiles of temperature and humidity together with a radiation model. Antarctic radiometersonde observations show typical radiative cooling rates of 1–2°C per day in the troposphere during the winter months (White and Bryson, 1967) while observed cooling rates are almost an order of magnitude smaller, implying that radiative cooling must be partially balanced by horizontal advection and by adiabatic warming associated with subsidence. White and Bryson (1967) demonstrated that the latter effect dominates and showed that it is possible to calculate subsidence rates from the difference between calculated radiative cooling profiles and observed cooling rates. These results will be discussed further in the next section of this chapter; for the present we note that the temperature structure of the Antarctic atmosphere is determined both by dynamical and by radiative processes.

4.3 Atmospheric circulation and the vorticity budget

4.3.1 Introduction

In the preceding section, we saw that the Antarctic atmosphere loses more heat by radiative cooling than it gains by surface energy exchanges. This energy deficit has to be balanced by atmospheric transport of heat from mid-latitudes and observations show that this is largely accomplished by the mean meridional circulation of the atmosphere. The large-scale circulation is thus strongly constrained by the requirements of energy balance.

The thermal structure and large-scale circulation are further coupled by the existence of a persistent surface temperature inversion over the interior of the continent. It is no exaggeration to say that this feature couples the circulation to the underlying topography to an extent not seen anywhere else on Earth. In this section we start by examining the low-level flow which results from this combination of topography and thermal structure and then consider the role of this flow in the large-scale vorticity budget of the Antarctic atmosphere.

4.3.2 Katabatic winds and the low-level circulation

A simple model of the katabatic wind

In the previous section of this chapter we saw that, as the surface of the Antarctic continent cools radiatively, the air close to the surface also cools relative to the air aloft. The near-surface air is thus negatively buoyant and, over a sloping surface, will accelerate down the slope in response to the buoyancy force. The resulting flow, known as a 'katabatic wind'†, will be turned by the Coriolis force and retarded by surface friction; it will also respond to free-atmosphere pressure gradients. Over much of Antarctica, however, the stability of the near-surface air is so great that, even over the very modest slopes which characterise the interior of the continent, the surface wind is primarily determined by katabatic forcing. Early explorers on the continent remarked on the persistence and directional constancy of winds blowing down from the high interior plateau but it was only with the advent of radiosonde measurements that the mechanisms maintaining the winds were fully understood.

Katabatic winds are observed in many mid-latitude locations, but are generally considered to be a mesoscale or boundary-layer phenomenon since they only occur at night over sloping terrain. In Antarctica and in Greenland, the combination of extensive sloping surfaces and uninterrupted surface cooling for much of the year allows a large-scale katabatic circulation to develop. This flow dominates the low-level circulation over the Antarctic continent.

One of the earliest theoretical studies of Antarctic katabatic winds was made by Ball (1956, 1960), using the two-layer model shown schematically in Figure 4.4. In this model, the stratification of the lower atmosphere is crudely represented by a layer of potential temperature $\theta - \Delta\theta$ and depth h lying under an infi-

† We choose to use the most general definition of a katabatic wind, that is any wind that receives significant forcing from the pressure gradient resulting from downslope buoyancy forces. Some authors restrict the term to winds for which this is the *only* significant forcing term.

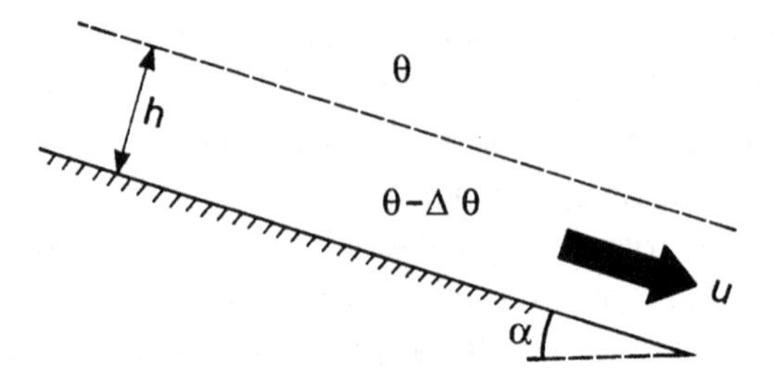

Figure 4.4 The geometry of the two-layer katabatic wind model.

nitely deep layer of potential temperature θ. The underlying surface slopes uniformly downwards in the positive x direction at an angle α and is assumed to be infinite in extent. We assume that there is no frictional coupling between the two layers so that flow in the upper layer is geostrophic and is driven only by large-scale pressure gradients. We further assume that the wind in the lower layer, $V=(u,v)$, where u is the downslope component of wind, is independent of height.

With these simplifications, we can write down equations of motion for flow, $V=(u,v)$ in the lower layer. The downslope buoyancy force per unit volume in this layer is given by

$$F_B=\rho g\alpha\frac{\Delta\theta}{\theta} \tag{4.5}$$

where ρ is the air density. The air in the lower layer will accelerate in response to this force and a steady state will be obtained when the downslope buoyancy force is balanced by Coriolis force, large-scale 'external' pressure gradients and surface friction. In Ball's model, the latter process is parameterised using a drag coefficient formulation for the surface stress, τ_s

$$\tau_s=\rho k(Vu, Vv) \tag{4.6}$$

where $V=|V|$. The stress is assumed to decrease linearly with height, becoming zero at height h, so the frictional force acting per unit volume of the lower layer is τ_s/h. The equations for steady state flow in the lower layer are then

$$hg\alpha\frac{\Delta\theta}{\theta}+hf(v-v_g)-kVu=0 \tag{4.7}$$

$$-hf(u-u_g)-kVv=0 \tag{4.8}$$

where f is the Coriolis parameter and $V_g=(u_g,v_g)$ is the geostrophic wind driven in the upper layer by large-scale pressure gradients. Putting $u=V\cos\phi$ and $v=V\sin\phi$ (where ϕ is the angle between the resultant wind and the fall line of the slope) leads to solutions

$$V^2=\frac{\{f^4+4k'^2[(F-fv_g)^2+f^2u_g^2]\}^{0.5}-f^2}{2k'^2} \tag{4.9}$$

$$\cos\phi=\frac{u_g f^2\pm k'V[k'^2V^4+f^2(V^2-u_g^2)]^{0.5}}{f^2V+k'^2V^3} \tag{4.10}$$

where $F=g\alpha\Delta\theta/\theta$ and $k'=k/h$. With no imposed geostrophic flow in the upper layer, the steady-state katabatic wind will be deflected to the *left* of the fall line in the Southern Hemisphere, hence the positive root in Eq. (4.10) is appropriate.

Thus, given the local topographic slope, a measure of the static stability of the near-surface air and the geostrophic wind velocity above the katabatic layer, it is possible to calculate the steady-state katabatic wind. However, the two-layer potential temperature structure is a rather crude approximation to typical Antarctic surface inversion profiles and it is not entirely straightforward to choose representative values for $\Delta\theta$ and for h. Some authors arbitrarily identify these parameters with the strength and depth respectively of the surface temperature inversion. However, this can be misleading since the proper measure of the stratification is the *potential* temperature gradient. At stations such as Mizuho, which experience strong katabatic winds, the potential temperature profile often shows a two-layer structure, with a well-mixed layer adjacent to the ground (see Figure 6.18 later) and the identification of $\Delta\theta$ is straightforward. Over the interior of the continent, an almost exponential temperature profile is often observed. Parish (1980) has shown that the two-layer model can reproduce the winds observed at such interior stations if the potential temperature of the lower layer is identified with the mean potential temperature of the inversion layer, rather than with the surface temperature. Assigning a value to h when the observed profiles do not possess an obvious two-layer structure is possibly more problematical. h is sometimes identified with the depth of the surface inversion but it is not obvious that this is the correct scale for the stress divergence. However, in Eqs. (4.7) and (4.8), h always appears as a ratio with the drag coefficient, k, which is itself not known with any great certainty so the problem can be subsumed into that of choosing an appropriate drag coefficient.

Figure 4.5 shows the steady-state katabatic wind speed, V, and the angle between the resultant wind and the fall line of the slope as a function of the slope angle, α, and the inter-layer potential temperature difference, $\Delta\theta$ for the case of zero upper-layer geostrophic wind. The range of α shown runs from values typical of the interior plateau (≤ 0.001) to values characteristic of the coastal slopes (≥ 0.01) and the chosen values of $\Delta\theta$ are typical of the seasonal and geographic variation of the near-surface stability. The value used for the drag coefficient, $k=0.005$, has been found to give a reasonable simulation of Antarctic katabatic winds (Ball, 1960). When the katabatic forcing term, $F=g\alpha\Delta\theta/\theta$, is small, there is near geostrophic balance between this term and the Coriolis force.

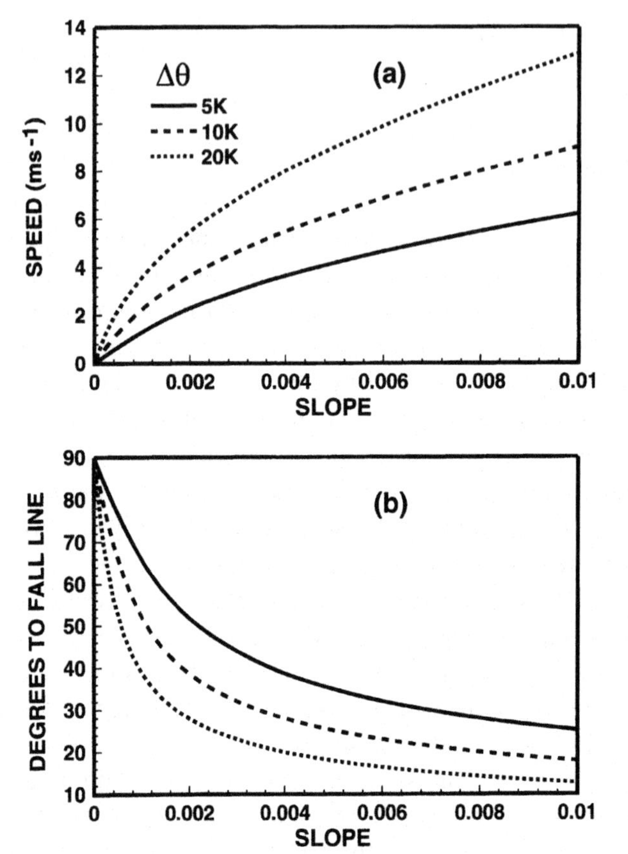

Figure 4.5 Wind speeds (a) and directions (b) calculated from Ball's two-layer model as a function of slope gradient, α, and inter-layer potential temperature difference, $\Delta\theta$. The frictional parameter, k, was set to 0.005 and the lower layer depth, h, to 100 m. The upper layer geostrophic wind speed was set to zero.

The resultant wind is light, of order $F/|f|$, and has a large cross-slope component, blowing with the high ground on its left in the Southern Hemisphere. The winds resulting from this near-geostrophic balance are often referred to as 'inversion' winds (e.g. Parish, 1980) to differentiate them from the 'true katabatic' winds, for which surface friction plays a major role. As the katabatic forcing is increased, the Coriolis term increases linearly with increasing wind speed while the frictional term, kV^2/h, increases quadratically. Eventually, the latter term dominates and the principal balance will be between katabatic forcing and friction. The resultant wind is now stronger, of order $(hF/k)^{0.5}$, and directed more nearly downslope.

Large-scale pressure gradients, associated with synoptic-scale systems or climatological features, will drive a geostrophic wind in the upper layer and will also affect the balance of forces in the lower layer. From Eqs. (4.7) and (4.8), it can be seen that the ratio of the downslope buoyancy force to the pressure gradient driving the upper layer geostrophic flow is

$$R=\frac{F}{|f|(u_g^2+v_g^2)^{0.5}} \tag{4.11}$$

If $R \gg 1$, flow in the lower layer will be essentially independent of the synoptically forced flow above and will depend only on the local katabatic forcing. Surface winds over the continental interior exhibit high directional constancy (see Section 3.3), suggesting that katabatic forcing dominates over synoptic forcing. It is difficult to derive reliable estimates of winds above the katabatic layer because of the sparsity of upper-air observations over the continent. The 500 hPa height climatologies presented in Figure 3.18 show generally light 500 hPa winds over Antarctica and winds recorded by automatic weather stations at sites where the topographic slope is small, such as Dome C (see Table 3.6), are light and variable, suggesting that synoptic forcing is indeed small over the interior of the continent. Simulations of the near-surface wind using the two-layer model with no imposed geostrophic wind are often quite realistic (see the work of Parish and Bromwich (1987), discussed below). However, in the next section of this chapter, we shall see that the upper flow is itself largely determined by the nature of the broad-scale katabatic flow, so regarding it as a free parameter in simulations of the katabatic flow may be an oversimplification.

Of course, the two-layer model is a simplification of the situation pertaining over the Antarctic ice sheets. The temperature and velocity structure in the model is not representative of that observed over much of the interior of Antarctica and the parameterisation of friction is very crude. More sophisticated models are discussed in Chapter 6; although these differ in detail from the two-layer model, they are driven by essentially the same dynamics and produce similar predictions of the continent-wide katabatic flow. A more serious shortcoming of all diagnostic models is the assumption of uniformity in the downslope direction. This simplification allows neglect of the non-linear terms in the equations of motion and will generally be justified if the Rossby number of the flow,

$$\mathrm{Ro} = \frac{Vf}{L_\alpha} \tag{4.12}$$

is small, where L_α is a typical horizontal length scale for variations in surface slope. If we are considering flow variations on the scale of the Antarctic continent itself then Ro is of order 0.1 and the neglect of non-linear terms is justified. In Chapter 6, we shall examine some cases in which large variations in *local* topography violate this assumption. A further weakness of the two-layer model is the neglect of any entrainment at the interface between the layers. Such processes can be included by means of a suitable parameterisation, as discussed in Chapter 6.

Despite its shortcomings, this simple diagnostic model has been used with some success to simulate the near-surface wind field over the Antarctic continent (Parish, 1982; Parish and Bromwich, 1986, 1987). In order to apply the models it is necessary to know i) the surface slope, ii) the near-surface temperature

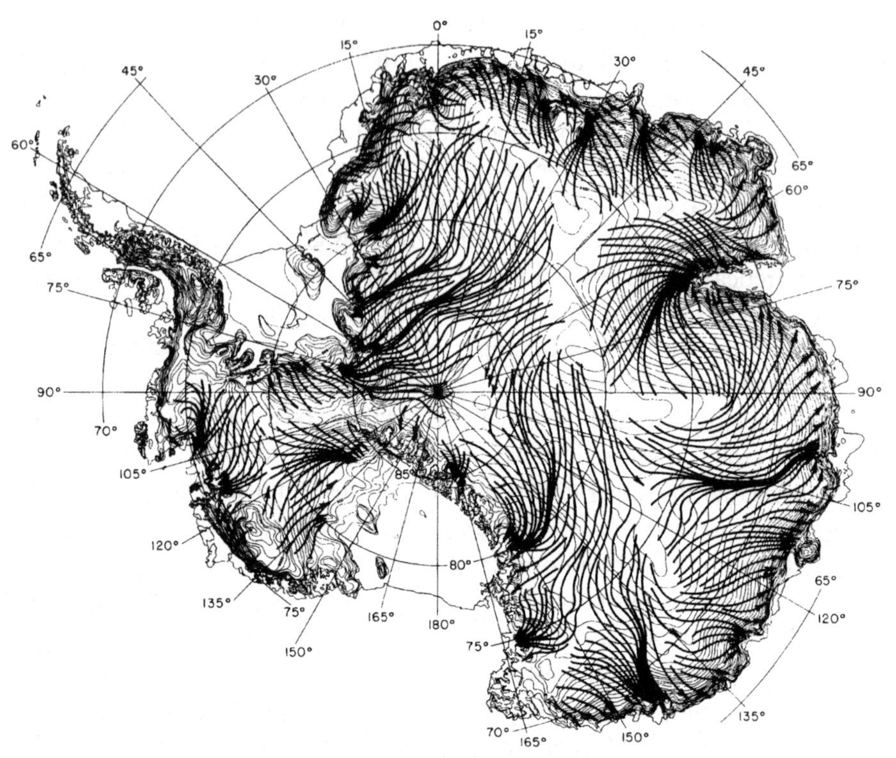

Figure 4.6 Low-level wind streamlines over Antarctica for average winter conditions. From Parish and Bromwich (1987). Reprinted with permission from *Nature*. Copyright 1987 Macmillan Magazines Limited.

structure and iii) the geostrophic wind above the katabatically forced layer at all points over the continent. Parish and Bromwich (1987) estimated terrain slope on a 50 km grid from a digitized version of the Drewry (1983) topographic map of Antarctica and used climatological estimates of surface inversion strength derived from radiosonde observations (Phillpot and Zillman (1970), see Figure 3.12) as a measure of stability. Following arguments given earlier, they assumed that free-air geostrophic wind speeds are low over the continental interior and, to a good approximation, the near-surface flow would be driven only by katabatic forcing. Using these data and assumptions to drive a two-level model, they produced a streamline map of the near-surface flow in winter; this is reproduced in Figure 4.6.

Examination of the streamline map clearly shows the nature of the near-surface flow: cold, stable air drains outwards from the high topography of the continental interior and the flow acquires an easterly component as it is deflected to the left by the Coriolis force. On the broad scale we thus have a low-level anticyclonic vortex, with associated outflow, centred over the highest part of the continent. This broad-scale picture is confirmed by streamline maps produced by compositing surface wind observations and indirect evidence on wind direction

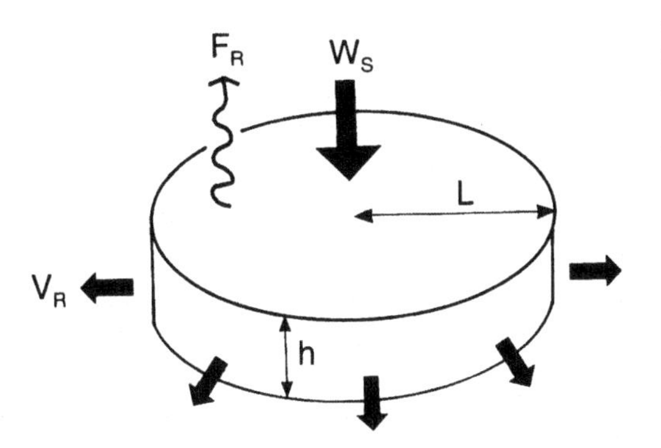

Figure 4.7 The thermal balance of a simplified katabatic layer. The Antarctic continent is represented by a circle of radius L. The katabatic layer is assumed to have constant depth, h, and potential temperature $\theta - \Delta\theta$. The atmosphere above the katabatic layer has constant potential temperature θ.

from observations of sastrugi orientation (Mather and Miller (1967), see Figure 3.20). Looking at the model streamline map in greater detail, we can see that there are localised regions of significant streamline confluence associated with regions of concave topography, most notably the Lambert Glacier drainage basin, Adélie Land and some of the glacial valleys which drain into the Ross and Filchner–Ronne ice shelves. Such confluent regions are likely to be associated with a strengthened katabatic wind; we discuss such regional control of the katabatic wind in Chapter 6.

Over the interior of the continent, winds calculated using the simple model are generally in reasonable agreement with mean wind speeds and directions measured at stations but the model predicts unrealistically high wind speeds over the steep coastal slopes as a result of neglecting the inertial terms in the equations of motion. At coastal stations, well removed from any steep topography, the surface wind will primarily be determined by synoptic forcing and the simple katabatic wind models will no longer be valid. Simulations of the continent-wide surface wind field using a three-dimensional primitive equation model driven by radiative cooling of the boundary layer (Parish and Bromwich, 1991) are in excellent agreement with the simpler models over the interior, suggesting that these diagnostic models capture the essential dynamics of the interior katabatic flow.

Katabatic winds and the atmospheric heat budget

In Section 4.2 we saw that most of the atmospheric heat transport to Antarctica was accomplished by the mean meridional circulation of the atmosphere. Clearly, the katabatic drainage flow contributes to this transport by exporting cold air at low levels and the requirements of heat transport place some constraints on the flow. Dalu *et al.* (1993) considered the thermal balance of the katabatic layer over Antarctica, using a simple model in which the Antarctic continent was represented by a circle of radius L, overlain by a katabatic layer of depth h and potential temperature θ–$\Delta\theta$, where θ is the potential temperature of the overlying atmosphere. Heat fluxes into and out of this layer are shown in Figure 4.7. We have seen (Section 4.2) that the net surface flux, F_{SURF}, is generally quite small – less than 10 W m^{-2} – and the cooling of the katabatic layer will be dominated by the radiative flux, F_R, at the top of the layer. F_R is typically 50 W m^{-2} during the Antarctic winter (Parish and Waight, 1987). This cooling must be balanced by export of cold air by the katabatic outflow, V_R, subsidence of warm air into the katabatic layer at rate w_S and entrainment of warm air into the katabatic layer, parameterised in terms of an entrainment velocity w_e. The thermal balance of the katabatic layer can then be written

$$\oint h V_R(\theta - \Delta\theta)\mathrm{d}s = \iint \frac{-F_R}{\rho C_p} + w_S\theta + w_e \Delta\theta \mathrm{d}a \tag{4.13}$$

where the line integral is around the 'coastline' and the area integral is over the 'continent'. Conservation of mass demands that the katabatic outflow be balanced by subsidence of air into the katabatic layer:

$$\oint h V_R \mathrm{d}s = \iint w_S \mathrm{d}a \tag{4.14}$$

whence

$$-\oint h V_R \Delta\theta \mathrm{d}s = \iint \frac{-F_R}{\rho C_p} + w_e \Delta\theta \mathrm{d}a \tag{4.15}$$

If we neglect entrainment and substitute typical values; $h=200$ m, $\Delta\theta=10$ K, $F_R=50$ W m^{-2} and $L=2000$ km, we obtain $V_R \approx 20$ m s^{-1}. Climatological meridional wind speeds observed at coastal stations are generally somewhat smaller than this, perhaps 5–10 m s^{-1}, so it appears that entrainment may be significant. Dalu *et al.* (1993) followed Manins and Sawford (1979) and set the entrainment velocity equal to the katabatic wind speed multiplied by an entrainment efficiency (which is assumed to be a constant). In order to obtain a realistic value for V_R, an entrainment efficiency of only 10^{-3}–10^{-4} is required. This is significantly smaller than the values suggested by Manins and Sawford (1979), but, even with this low entrainment efficiency, the entrainment heat flux balances about half of the radiative cooling of the layer.

4.3.3 The upper-level circulation forced by the katabatic flow

So far we have largely considered the katabatic flow in isolation and have not been greatly concerned with its interaction with the circulation at higher levels. Indeed, the success of the simple diagnostic models which predict the near-surface flow using only surface layer parameters suggests that such coupling may be relatively weak. However, we have seen that the existence of a persistent katabatic outflow at low levels implies a mass divergence that must be compensated for by subsidence into the katabatic layer and horizontal mass convergence above this layer.

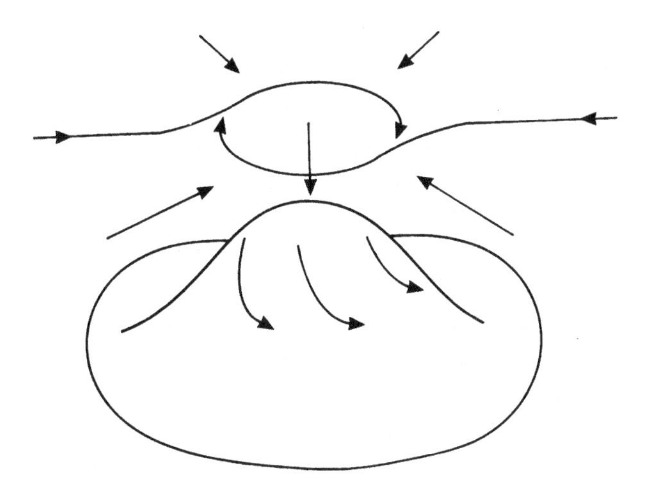

Figure 4.8 A simple conceptual model of the Antarctic tropospheric circulation, after James (1989). Low-level katabatic outflow is balanced by upper-level convergence and subsidence over the continent, which maintains a cyclonic circulation in the middle troposphere.

This conceptual model of the Antarctic atmospheric circulation is shown in Figure 4.8. Air in the katabatic layer cools radiatively and accelerates downslope, with the flow turning to the left under the action of the Coriolis force. The resultant flow is thus an easterly vortex at low levels with associated outflow of cold air. The net mass flux northwards in the katabatic layer is compensated for by subsidence in the middle troposphere that, in turn, implies horizontal convergence at these levels. This completes the thermally direct circulation cell which maintains the heat balance of the Antactic atmosphere: cold air is exported at low levels while warmer air from mid-latitudes is advected towards Antarctica at higher levels, subsiding into the katabatic layer over the continent. Subsidence in the lower troposphere will generate cyclonic vorticity through the action of vortex stretching, driving a westerly upper vortex above the katabatic layer. The complete picture is thus one of divergent, anticyclonic flow in a thin katabatic layer with convergent, cyclonic flow at higher levels. Since the low-level vortex is driven by katabatic drainage, it will be centred over the highest part of the Antarctic ice sheet rather than over the geographic pole and we might expect that the upper cyclonic vortex, which, in this model, is strongly coupled to the katabatic flow, will be similarly displaced.

Studies by Egger (1985) and James (1989) revealed a fundamental flaw in this conceptual model. Cyclonic vorticity is generated in the middle and upper troposphere as a result of vortex stretching by the vertical velocity field induced by divergence in the katabatic layer. Since the rate of production of cyclonic vorticity is proportional to the divergence (and hence the meridional wind speed) in the katabatic layer, the cyclonic vortex would grow without bound if the katabatic flow remained constant. However, westerly flow aloft implies the existence of a low-pressure region in the centre of the vortex and hence a pressure gradient that *opposes* the downslope flow in the katabatic layer. As the upper cyclonic vortex spins up, it thus acts to 'choke off' the katabatic flow until a steady state is reached in which the katabatic flow is just strong enough to balance the local dissipation of vorticity. Integrations of simple two-dimensional models of the flow (Egger, 1985; James, 1989) show that, after about 30 days, an equilibrium state is attained with a strong upper westerly vortex and a very weak katabatic circulation. Downslope velocities in the katabatic layer are reduced to less than 1 m s^{-1}. Parish (1992) carried out a similar study using a more sophisticated axisymmetric model, which includes a more realistic treatment of radiation and boundary-layer friction. This model also exhibits a decay in the strength of the katabatic outflow and upper vortex as it is integrated forward in time and, although the decay is somewhat slower than those in the simpler models of Egger and of James, this confirms that the failure of axisymmetric models to reproduce the observed circulation is dynamical in origin rather than a failing of model physics.

Clearly, such a weak meridional circulation cannot accomplish the required heat transport and we know, from observations, that there is a strong and persistent drainage flow at low levels over the continent. James (1989) argued that what must be missing from the simple conceptual model of the flow is some mechanism for exporting cyclonic vorticity from the upper vortex, thus reducing the associated meridional pressure gradient which would otherwise 'choke off' the drainage flow at low levels. A number of mechanisms could accomplish this. The displacement of the highest part of Antarctica from the geographic pole will tend to generate low-frequency Rossby waves (James, 1988). However, the momentum flux assocated with the propagation of such waves towards mid-latitudes can be shown to be both small and in a sense that would *accelerate* the upper westerly vortex (James, 1989). Another possible mechanism is transport of vorticity by local baroclinic instabilities growing around the periphery of the continent. Vigorous mesocyclone development is seen at many locations around Antarctica (see Section 6.5) but the systems observed are generally quite shallow and a local baroclinic instability calculation (James, 1989) indicates that they are unlikely to have the structure necessary for weakening the upper westerly flow. Drag induced by gravity wave generation and propagation (see Section 6.2) may play some role, but diagnostic studies using GCMs indicate relatively little gravity wave drag over Antarctica. The most plausible mechanism for weakening the upper westerly vortex, and hence maintaining the katabatic flow, would appear to be the interaction of the circulation with decaying high- and mid-latitude weather systems moving towards Antarctica.

Studies of the Southern Hemisphere storm tracks (see Section 5.1) show that many synoptic-scale systems originating in mid-latitudes spiral in towards Antarctica and finally decay in a number of prefered areas for cyclolysis. Egger (1992) investigated the effects of such systems on the Antarctic circulation using a simple two-layer model. The surface of the continent was represented by a uniformly sloping katabatic layer, above which a barotropic troposphere carried the return flow. A prescribed wave-like motion was imposed at the northern boundary of the model to represent the effects of cyclones propagating inwards from mid-latitudes. With no wave forcing, the model could only exhibit one steady state, with a strong upper westerly vortex and very weak katabatic flow. As the strength of the wave forcing was increased, the southwards-propagating waves, modified by the sloping topography of Antarctica, were able to remove progressively more cyclonic vorticity from the upper vortex, weakening the westerly flow and permitting a stronger downslope flow and easterly vortex to develop in the lower layer.

Egger (1992) carried out further experiments in which a stochastic distribution of wave forcings was imposed and the resulting probability distribution of model steady states observed. When the mean wave forcing was large, the distribution of model states was unimodal and corresponded to the observed mean circulation, with a strong katabatic flow and a weak upper vortex. For somewhat smaller mean forcing, a bimodal distribution of model states developed, with the second mode associated with a strong upper westerly vortex and weak katabatic flow at low levels. There is some evidence that the Antarctic atmosphere may exhibit such behaviour. The strength of the katabatic flow at Mizuho Station shows marked intraseasonal variability on a 30- to 50-day period, with weak katabatic flow being associated with anomalously strong upper westerlies and *vice versa* (Yasunari and Kodama, 1993). Such variations in the Antarctic circulation may be connected with changes in the forcing by mid-latitude eddies on similar timescales.

Isentropic circulation diagnostics from the UGAMP GCM (Juckes *et al.*, 1994) confirm the importance of mid-latitude systems in maintaining the vorticity balance of the Antarctic circulation, but suggest that *vertical* eddy transports may be more important than the purely horizontal transports of Egger's (1992) barotropic model. Furthermore, vertical eddy transports play a major role in controlling the momentum balance in the katabatic layer. The balance between katabatic forcing, friction and Coriolis force assumed in the models described in Section 4.3.2 appears to be an oversimplification and it is perhaps surprising that such models have been applied with some success (e.g. Parish and Bromwich, 1987) to simulate the continental-scale near-surface flow. It is noteworthy that these studies have assumed, *a priori,* that the large-scale pressure gradients associated with the upper westerly vortex are small and can be neglected when calculating the low-level flow. This assumption has proved valid, but for rather more complex reasons than those put forward in some early studies.

The upper westerly vortex is almost certainly controlled by factors in addition to the katabatic flow. Even in the absence of an Antarctic continent, radiative cooling of the atmosphere would drive a thermally direct circulation with outflow and easterly winds at the surface and a return circulation with westerly winds at upper levels. Such a circulation is seen over the flat topography of the Arctic regions. In Antarctica, the drainage flow associated with the katabatic winds acts to enhance this circulation. Simmonds and Law (1995) have carried out experiments with a GCM in which the Antarctic katabatic flow was reduced by removing the Antarctic orography or increasing the surface drag coefficient. The response of the upper vortex was complex and indicated that changes in thermal forcing were as important as the katabatic circulation in determining the structure of the upper vortex. Further diagnostic studies are needed to clarify the relative importance of the two effects.

The conceptual model for the Antarctic atmospheric circulation discussed above is clearly in fair agreement with the observations of wind and geopotential height fields above the continent presented in Chapter 3. Both the low-level katabatic outflow and the upper-level cyclonic inflow are apparent in the annual mean wind vectors shown in Figure 3.22. The rather weak westerly flow at 500 hPa, seen on the climatological geopotential height fields shown in Figure 3.18, fits the conceptual model well but the vortex is not centred over the highest part of the continent; rather it is displaced towards the Ross Ice Shelf. The reasons for this are not entirely clear but the low-level katabatic circulation (Figure 4.6) exhibits considerable complexity, which may be reflected in the upper flow. James (1988) has modelled the barotropic disturbance generated by the topography of Antarctica with the observed zonal mean 300 hPa flow imposed. Anticyclonic vorticity developed over the high plateau of East Antarctica and a maximum of cyclonic vorticity was observed over the Ross Sea. Although this technique is inherently linear (since it requires the specification of the zonally averaged flow) the results do show how interaction of the mean flow with the non-axisymmetric topography of Antarctica can lead to the observed displacement of the upper vortex. The effects of transient eddies (which, we have seen, are essential to maintaining the circulation) are unlikely to be symmetric about the pole. Cyclonic disturbances penetrate into West Antarctica quite frequently, but affect the high plateau of East Antarctica more rarely (Schwerdtfeger, 1984, p.129). This assymmetry in eddy activity, together with associated transports of heat and vorticity, may help to explain the displacement of the 500 hPa vortex. It must be remembered that the vortex shown in Figure 3.18 is a long-term average picture. On any particular day, the mean flow at 500 hPa will be strongly perturbed by weather systems and the instantaneous centre of the vortex may be some distance removed from its average position.

The subsidence of air from the mid-troposphere into the katabatic layer is too slow to be observed directly but, as was discussed in Section 2 of this chapter, can be diagnosed from observed radiative cooling rates. Using this technique, White and Bryson (1967) estimated maximum subsidence velocities of about 3×10^{-3} m s^{-1} in the lower troposphere. This is very close to the subsidence rate required to balance the net katabatic outflow from the continent indicating, once again, the close coupling between dynamic and thermodynamic processes in the Antarctic atmosphere.

4.3.4 The circulation to the north of the Antarctic continent

Direct katabatic forcing of the circulation rapidly becomes insignificant as one moves northwards from the Antarctic coast. In this region, the circulation is dominated by the circumpolar low-pressure trough (Section 3.3) which lies at an annual average latitude of about 66° S. The pressure gradient to the south of the trough generates easterly winds around Antarctica. At coastal stations, the surface wind results from a combination of katabatic effects and forcing by this large-scale pressure gradient. Both effects act to force an easterly flow at low levels and the resultant wind often has a high directional constancy, although not as high as that which it has in the continental interior, where katabatic effects dominate (King, 1989).

The circumpolar trough results from the pattern of cyclonic activity in high southern latitudes but its location and structure appear to be controlled to some extent by the circulation over the Antarctic continent. The most obvious departure from zonal symmetry of the circumpolar trough is the presence of three climatological low-pressure centres situated at (approximately) 20° E, 90° E and 150° W (Figure 3.15). The distribution of cyclones and frequency of cyclogenesis follow a similar wavenumber three pattern (Jones and Simmonds, 1993) so the structure of the trough may, to some extent, reflect these zonal variations in cyclonic activity. However, the stationary nature of the climatological lows and their observed barotropic structure suggest that topographic forcing may be involved in their formation. Baines and Fraedrich (1989) investigated this possibility using a topographic model of Antarctica in a rotating tank of fluid. The topographic model was rotated relative to the tank in order to simulate a westerly upper-level flow over the model and neutrally buoyant beads were used to visualise the resulting circulation. After about ten tank rotation periods a clear wavenumber three flow pattern emerged, with cyclonic eddies developing in the three major embayments – the Weddell Sea, Prydz Bay and the Ross Sea. The circulation generated in this experiment exhibits significant similarity to the flow round the continent and the experiment indicates that topographic effects may exert significant control on the structure of the circumpolar trough.

Further evidence for the influence of the Antarctic circulation on the circumpolar trough has come from GCM experiments conducted by Parish *et al.* (1994). In a model experiment with the Antarctic orography completely removed, the circumpolar trough moved five degrees southwards of its position in the control run and was positioned along the strong thermal contrast at the Antarctic coastline. This suggests that the low-level katabatic flow contrives to keep the circumpolar trough well north of the continent in the control run. The topography of Antarctica thus exerts an important influence on the wider scale Southern Hemisphere circulation.

4.4 The water vapour budget

4.4.1 Introduction

Unlike in the tropics or mid-latitudes, water vapour plays a rather passive role in the dynamics of the Antarctic atmosphere. This is because, at the low temperatures which prevail, the air can hold very little water vapour, even when close to saturation. For example, air saturated with respect to ice at 0°C holds only 26% of the water contained in saturated air at 20°C. At −20°C, the water content is reduced to 4% of that at 20°C. There are some processes in which water vapour does become important. Surface latent heat fluxes over the Antarctic oceans appear to play a major role in the energetics of some mesoscale weather systems (see Section 6.5) and water vapour exerts a major (if indirect) control on the regional energy budget through cloud formation and other radiative effects.

In general, however, water vapour may be treated as a passive scalar and the main reason for studying it is that atmospheric water vapour is the source of all of the ice in the continental ice sheets. With increasing concern over the state of balance of these ice sheets and its consequent effect on global sea level, the study of the transport of water vapour into and within the Antarctic atmosphere has assumed a new importance. In this section, we shall study the water vapour budget of the Antarctic atmosphere and briefly consider how this relates to the wider problem of the mass balance of the Antarctic ice sheets.

4.4.2 The water vapour content of the Antarctic atmosphere

Humidity measurements made by radiosondes provide a means of calculating the total column moisture (TCM), that is the water vapour content of an atmospheric column. These measurements must be interpreted with some caution, since radiosonde humidity sensors perform poorly at the low temperatures experienced in Antarctica. Connolley and King (1993) estimate that there is an uncertainty of about 20% in Antarctic TCM values.

Figure 4.9 shows annual average TCM values at 16 Antarctic radiosonde stations, together with an indication of the interannual variation in TCM over a six-year period. The stations fall into three groups. Bellingshausen, in the South Shetland Islands, has a relatively high TCM, as a result of its situation in the relatively warm and moist circumpolar westerlies. The 13 stations around the coast

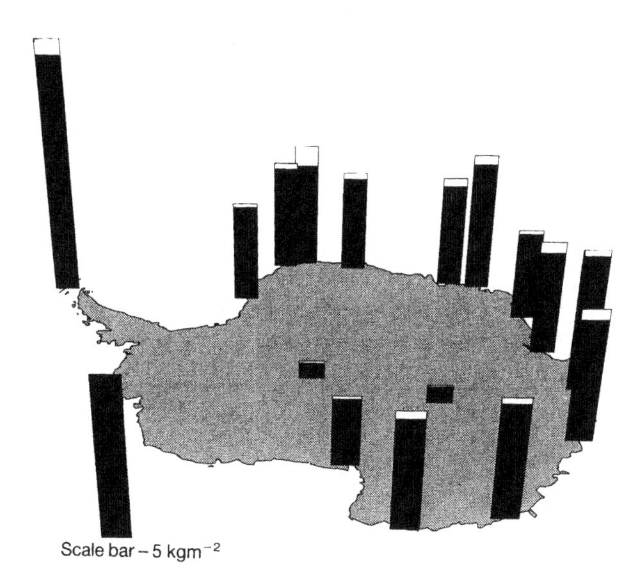

Figure 4.9. The total column moisture at 16 Antarctic radiosonde stations. The dark bar indicates the annual mean TCM; the light bar on top shows the interannual variability of TCM. From Connolley and King (1993).

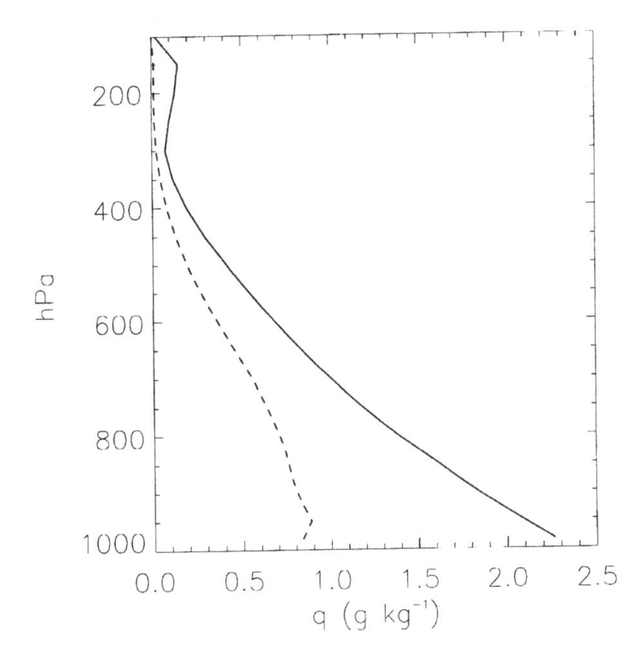

Figure 4.10. Profiles of the water vapour mixing ratio at Molodeznaya. The solid line is for summer (DJF); the broken line is for winter (JJA).

of East Antarctica have comparable values of TCM, with an average of 3.3 kg m^{-2} and an interannual variability of about 10% of the mean. Finally, the two stations representative of the high plateau of East Antarctica have much smaller TCM values as a result of the lower temperatures in the atmospheric column.

Since saturation vapour pressure decreases very rapidly with decreasing temperature, we would expect the greatest contribution to TCM to come from the warmest part of the atmospheric column, that is the region directly above the surface inversion. This is demonstrated in Figure 4.10, which shows seasonal

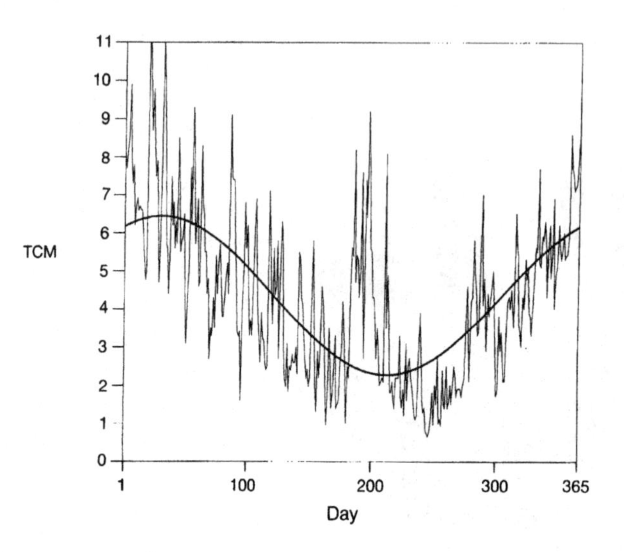

Figure 4.11 One year's daily values of TCM at Molodeznaja. The smooth curve is a best-fit annual sine wave. From Connolley and King (1993).

profiles of water vapour mixing ratio at Molodeznaya. The atmospheric column above 500 hPa contributes very little to the TCM.

The seasonal variation of TCM roughly follows that of temperature. However, as can be seen from Figure 4.11, East Antarctic coastal stations experience large day-to-day variations in TCM at all times of year, reflecting the advection of relatively moist air into the coastal region by synoptic-scale weather systems. At Molodeznaya, only 24% of the standard deviation of daily TCM values is accounted for by the first annual harmonic (Connolley and King, 1993).

4.4.3 Components of the water vapour budget

A water vapour budget for the Antarctic atmosphere may be constructed in an analogous fashion to the heat budget discussed in Section 4.2. The problem is somewhat simplified because there is no exchange of water vapour at the top of the atmosphere, thus only fluxes at the surface and through the 'walls' of the polar atmospheric cap need be considered. The budget equation for the total water content of the Antarctic atmosphere, $\langle Q \rangle$, is

$$\frac{\partial \langle Q \rangle}{\partial t} = \oint_{\text{BOUND}} \boldsymbol{F}_q \cdot \hat{\boldsymbol{n}} \, \mathrm{d}s - \int_{\text{SURF}} (P - E) \, \mathrm{d}a \qquad (4.16)$$

where $\boldsymbol{F}_q$ is the vertically integrated horizontal flux of water vapour, $\hat{\boldsymbol{n}}$ is the unit vector normal to the boundary, P is precipitation and E evaporation. The line integral is taken around the boundary of the polar atmospheric cap and the area integral over the enclosed surface. For timescales of a year or longer, the 'storage' term on the left-hand side of Eq. (4.16) should be negligible and the horizontal transport of water vapour into the polar cap should be balanced by precipitation less evaporation over the enclosed surface. The quantity $P - E$ is conventionally

Table 4.5 *Annual average accumulation estimates for Antarctica.*

Region	Accumulation (mm/year)	Source
Continent (coterminous ice)	124	Giovinetto and Bentley (1985)
70° S to 90° S	186±35	Giovinetto *et al.* (1992)

refered to as the 'accumulation', although the true accumulation, as measured by snow stakes or snow pit stratigraphy, also includes the contribution of blowing snow to the surface mass balance (see Section 3.4.2).

In some studies (e.g. Masuda, 1990; Giovinetto *et al.*, 1992), a latitude circle is chosen as a convenient boundary for the polar atmospheric cap in an analogous manner to the heat budget studies discussed in Section 2 of this chapter. However, since the atmospheric water budget is of interest principally because of its importance in maintaining the continental ice sheets, it is more useful to define a polar cap that just covers the continent. Obviously this makes determining the horizontal transport term in Eq. (4.16) more difficult, since one has to estimate the flux across an irregular boundary.

4.4.4 Precipitation and evaporation

The distribution of precipitation over the Antarctic continent has been discussed in Chapter 3. Because of the difficulties in measuring precipitation directly, it is easier to make measurements of accumulation, using snow stakes or snow pit stratigraphy. In Table 4.5, we give annual average accumulation values for the whole continent and for the region south of 70° S. The latter region includes a significant area of ocean for which no reliable precipitation estimates are available. Giovinetto *et al.* (1992) estimated precipitation in this area by extrapolating accumulation values observed over the adjacent ice sheet.

In order to turn these accumulation estimates into true P–E values, it is necessary to consider the role of blowing snow in the surface mass balance. Apart from enhancing evaporation from the surface (see Section 6.4), the main effect of wind-borne snow is to redistribute precipitation over the ice sheets and to remove mass by transporting some surface snow across the coastline. It is this cross-coastal mass flux that needs to be added to the accumulation values in Table 4.5 to obtain P–E values.

Estimation of the cross-coastal blowing snow mass flux is not a straightforward matter. Microphysical models of blowing snow (see Section 6.4) are at an early stage of development and different parameterisations can give very different results. Estimates of mean wind velocity and its variability are required in order to drive the models. The blowing snow mass flux is very sensitive to wind speed and hence small errors in the wind speeds used, which may be taken from actual observations or from models, can cause large variations in the calculated mass flux. In particular, small regions of concentrated katabatic outflow, which may not be resolved by models or observations, may contribute disproportionately to the total mass flux. Loewe (1970; quoted by Giovinetto *et al.*, 1992) calculated a cross-coastal blowing snow mass flux equivalent to an annual loss of 8 mm water equivalent when averaged over the whole of the Antarctic ice sheets. Giovinetto *et al.* (1992) quote a comparable figure for the flux across 70° S between 0° E and 160° E. Given the problems both with the models used and with the wind data required to drive them, the uncertainties in these estimates are large, probably at least 50%. However, even at the upper end of the range of uncertainty, the cross-coastal blowing snow mass flux is 10% or less of the measured accumulation and is thus less than typical uncertainties in the accumulation itself.

Considerable interannual variability is observed in point measurements of accumulation over Antarctica. The coefficient of variation, or the ratio of the interannual standard deviation to the long-term mean, is about 24% for individual stations (Giovinetto, 1964), with coefficients of variation as large as 50% observed in the low-precipitation region of the high polar plateau (Satow, 1985). Bromwich (1988) suggests that, if coefficients of variation decrease with increasing averaging area, as they do in other regions of the world, then the coefficient of variation for the continent as a whole is likely to be somewhat smaller than the values quoted above. However, accumulation variability estimates from water vapour budget studies (see below) suggest that precipitation averaged over the whole continent exhibits a high level of interannual variability.

Although evaporation and condensation do not make a significant contribution to the surface energy balance (see Section 4.2), they are of importance in determining the surface mass balance at the few locations where measurements have been made. At Mizuho, which is typical of those parts of the high plateau subject to strong katabatic winds, evaporation during the summer removes about 30% of the annual precipitation (Fuji, 1979). The fraction of precipitation evaporated reaches 45% at some automatic weather station sites on the Ross Ice Shelf (Stearns and Weidner, 1993) and is about 30% at Halley (Limbert, 1963). During the winter months, there is a net condensation of water vapour on to the ice sheet surface at most locations but the quantities of water involved are small and do not contribute significantly to the surface mass balance (King *et al.*, 1995).

4.4.5 Horizontal transport of water vapour

Masuda (1990) estimated the total horizontal transport of water vapour across 70° S from FGGE data, in parallel with his energy budget calculation discussed in Section 4.2. His annual mean calculated transport corresponds to an accumulation rate of 148 mm per year for the region bounded by 70° S. This is at the lower end of the range of values estimated by Giovinetto *et al.* (1992) – see Table 4.5 – using glaciological data. Masuda's analysis shows that transient eddies make the greatest contribution to the transport of water vapour southwards across 70° S, with the mean meridional circulation making a small but negative contribution to the water vapour budget of the South Polar Cap. This is in contrast to the heat budget, in which the mean circulation and eddies both contributed to a positive energy transport. Given the problems already noted with the resolution of the mean meridional circulation in this analysis, this observation should be treated with some caution. The FGGE data reveal a marked semi-annual variation in southwards water vapour transport, with maximum transport occurring in spring and autumn. This is not entirely consistent with observations (Bromwich, 1988), which show a winter maximum of precipitation over much of Antarctica.

Estimates of the water vapour transport across the coast of the continent may be obtained from radiosonde measurements made at coastal stations. From the locations of such stations, shown in Figure 4.12, it is clear that a reasonable estimate may be made of the flux across the coast of East Antarctica, whereas West Antarctica is almost devoid of observations. Bromwich (1979, 1990) has used radiosonde data for 1972 to produce water vapour budgets, and hence accumulation estimates, for various sectors of East Antarctica. In carrying out these analyses, he attempted to avoid biases caused by different humidity sensors or reporting practices and those due to balloon launches being missed during periods of strong winds. He also noted the importance of using the highest vertical resolution data available for calculating the water vapour transport. Because water vapour content decreases rapidly above the surface inversion, in the region where the wind itself is changing most rapidly with height, using data from the WMO standard pressure levels alone can give a very biased estimate of $\boldsymbol{F}_q$, even to the extent of reversing the sign of its meridional component.

Bromwich's analysis has been extended to a multi-year study by Connolley and King (1993), who analysed data from 16 Antarctic radiosonde stations for the periods 1988–90 and 1980–82. The total horizontal water vapour fluxes obtained from this study are shown in Figure 4.12, together with their decomposition into contributions from the mean circulation and from transient eddies. The mean fluxes reflect the essentially zonal easterly circulation at low levels whereas approximately 70% of the meridional water vapour transport is accomplished by the transient eddies (Connolley and King, 1993). Also shown in Figure 4.12 are ellipses representing the interannual variablity of the fluxes. Standard deviations of the fluxes are around 18% of their mean value, whereas the interannual standard deviation of TCM at coastal stations (shown in Figure 4.9) is generally less than 8% of the annual mean. This suggests that circulation

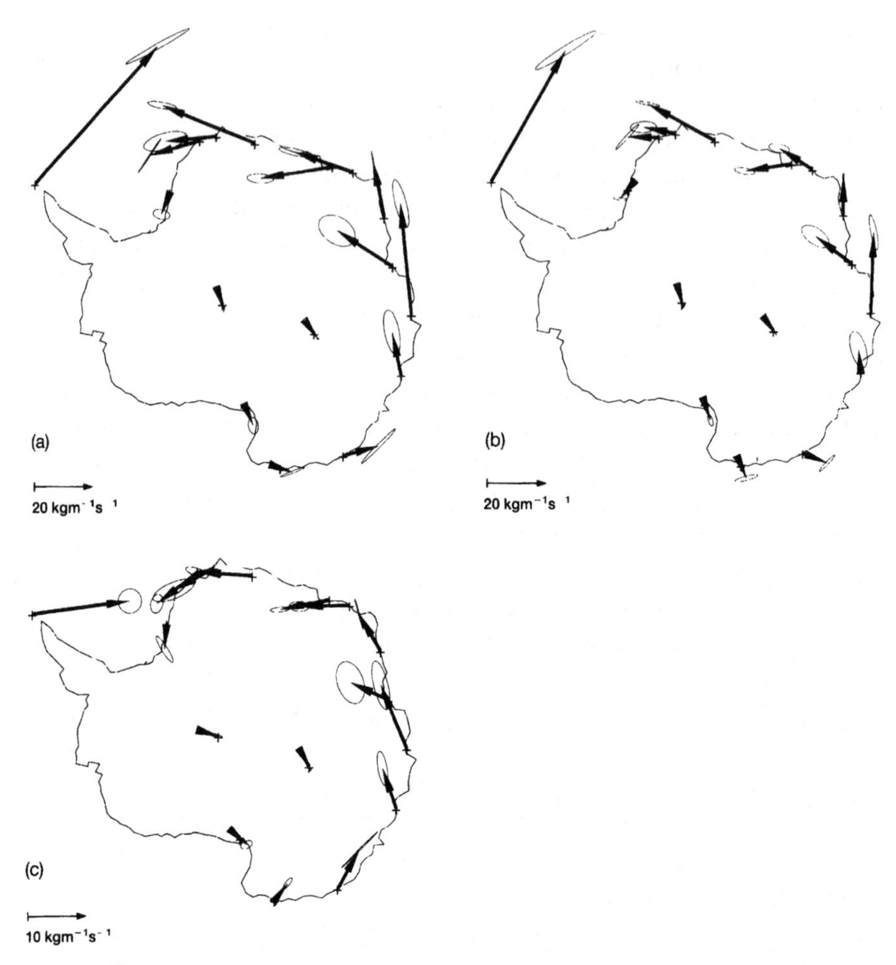

Figure 4.12 The vertically averaged water vapour transport at Antarctic radiosonde stations: (a) total transport, (b) transport by the mean circulation and (c) eddy transport (note different scale). Ellipses give an indication of interannual variability. From Connolley and King (1993).

variations, rather than variations in water content, are the principal cause of variations in water vapour transport and hence in areally averaged precipitation.

Accumulation estimates for the areas considered by Bromwich and by Connolley and King (indicated in Figure 4.13) are given in Table 4.6. In order to derive these estimates, assumptions had to be made concerning the fluxes across the inland boundaries, where there are no radiosonde stations to provide useful flux estimates. Bromwich made the assumption that all water vapour transported southwards across the coastline is deposited as precipitation in the region between the coast and the crest line of the plateau, which formed his southern boundary. At the longitudinal boundaries of his sectors, he either assumed that

Table 4.6 *Regional estimates of atmospheric water vapour flux convergence, expressed as equivalent accumulation rates, together with corresponding glaciological estimates.*

Sector	Atmospheric estimate (mm/year)	Glaciological estimate (mm/year)
A 0° to 110° E, 68.4° to 78.2° S	105 ± 23 (Bromwich, 1990) 170 (Bromwich, 1979)	108
B 0° to 55° E, 69.3° to 76.8° S	131 ± 43 (Bromwich, 1990)	109
C 55° to 110° E, 67.2° to 79.8° S	105 ± 23 (Bromwich, 1990)	108
D 0° to 110° E, 68.4° to 90° S	168 ± 32 (Connolley and King, 1993) 75 (Bromwich, 1990)	
E 'East Antarctic Polygon'	71 ± 29 (Connolley and King, 1993)	
Coterminous ice		124

the zonal fluxes cancelled or attempted to estimate these fluxes, assuming that a certain proportion of the flux as measured at the coastal stations would be 'cut off' by the rising topography of the plateau. The estimate by Connolley and King for sector D differs from that of Bromwich as a result of different assumptions being made concerning fluxes along unsampled boundaries. The variability in areally averaged accumulation deduced from the multi-year atmospheric analysis is of the same magnitude as that observed at individual stations, suggesting that there are very large year-to-year variations in the total mass input to the Antarctic ice sheets.

4.4.6 Water vapour source regions for Antarctic precipitation

Hemispheric water vapour budget studies (e.g. Howarth, 1983) show an excess of evaporation over precipitation in the tropics and subtropics, balanced by a sink of atmospheric water vapour in high latitudes. This simple picture might

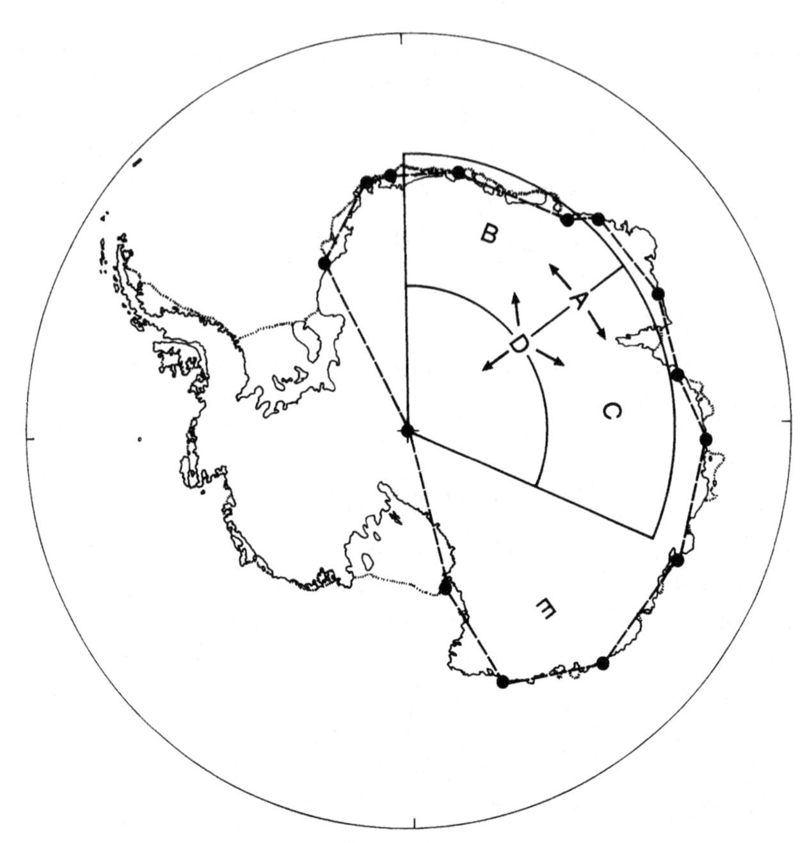

Figure 4.13 A map showing the boundaries (slightly simplified) of the sectors in Table 4.6. The filled circles are the radiosonde stations which form the vertices of the 'East Antarctic Polygon', sector E.

suggest that water precipitated over Antarctica has originated in the tropics or subtropics. However, analysis of the isotopic composition of snow at Syowa Station indicates that the primary moisture source for this precipitation is just to the north of the pack ice, in the vicinity of the 0°C or 1°C sea surface temperature isotherm (Bromwich and Weaver, 1983). This indicates that, as moist air moves southwards from the subtropics, some of the water vapour lost by precipitation is replaced by local evaporation. By the time the air reaches the Antarctic, the original moisture content has been replaced by evaporation from much cooler seas. Bromwich (1988) noted that there is sufficient evaporation between 62° S and the northern edge of the pack ice (about 65° S) to supply all of the precipitation falling on Antarctica.

In contrast, model studies carried out by Ciais *et al.* (1995) indicate a water source between 20° S and 40° S for precipitation falling over the interior of Antarctica. This suggests that the source regions for coastal and interior precipitation are different, reflecting the different circulation regimes and precipitation processes at work in these two regions (see Section 3.4).

Table 4.7. *Estimates of the terms in the mass budget of the Antarctic ice sheets (Jacobs et al., 1992).*

Term	Mass rate Gtonnes/year	Uncertainty (%)
Accumulation		
Grounded ice	1528	20
Ice shelves	616	
Total	2144	
Iceberg calving	−2016	33
Sub-ice-shelf melting	−544	50
Run-off	−53	50

4.4.7 The mass balance of the Antarctic ice sheets

A detailed discussion of the dynamics of the Antarctic ice sheets is beyond the scope of this book. However, it is of interest to see how atmospheric sources and sinks compare with the other terms in the ice mass budget. The subject has received considerable attention in recent years because of interest in the contribution that Antarctic ice mass changes could make to variations in sea level. Here, we review our knowledge of the present state of balance of the ice sheets. Possible future changes are considered in Chapter 7.

We have already seen that some of the snow falling over the Antarctic ice sheets is re-evaporated while a rather small fraction is transported across the coastline by the wind. The remainder, the 'accumulation', feeds the growth of the ice sheets. The rate of growth, or decay, of the ice sheets is then determined by the difference between this mass input and the mass losses associated with iceberg calving, run-off of melt-water and sub-ice-shelf melting‡. Recent estimates of these mass balance terms (Jacobs *et al.*, 1992) are shown in Table 4.7.

Mass loss by iceberg calving is clearly the largest negative term in the budget, but sub-ice-shelf melting cannot be neglected. Run-off of melt-water from the grounded ice makes only a small contribution to the mass budget. It is largely controlled by the small amount of melting which takes place in coastal regions during the summer, with a small additional contribution from subglacial melting. If the terms in Table 4.7 are summed, they indicate a small negative net mass balance. However, given the large uncertainties in some of the terms,

‡ Only changes in the mass of *grounded* ice will directly affect global sea level. The mass loss processes for the grounded ice are the flux of ice across the grounding line and melt-water run-off.

particularly mass loss due to calving, the significance of this imbalance is not very high. A study of the mass balance of the grounded ice by Budd and Smith (1985) indicated a small positive mass balance. It seems likely that the Antarctic ice sheets are currently in a state of near equilibrium, with the input of water from the atmosphere being in approximate balance with mass loss due to calving and ice shelf melting.

4.5 Representation of the Antarctic atmosphere in general circulation models

4.5.1 Introduction

General circulation models (GCMs) can provide useful insight into the processes which control atmospheric circulation, particularly in regions such as Antarctica where observations are sparse and analyses based on these observations may be unreliable. However, before using a GCM as a diagnostic tool, it is important to ensure that it is capable of producing a realistic simulation of the atmospheric circulation in the region of interest. One can then have confidence that the model correctly represents the major physical and dynamical processes which are controlling the climate.

In this section, we shall limit our attention to attempts to model the present-day Antarctic climate using GCMs. Such simulations can usefully be carried out using an atmosphere-only model, with climatological values of sea surface temperatures and sea ice distributions specified as boundary conditions. In order to predict future climates resulting from, for example, an increase in atmospheric carbon dioxide, it is necessary to couple the atmospheric model to an ocean and sea ice model. On timescales of a century or more, the polar ice sheets may change significantly in response to changed atmospheric and oceanic forcing so the ice sheet topography may no longer be regarded as a fixed boundary condition. Thus, if predictions on very long timescales are required, it may be necessary to couple the atmosphere–ocean model to an ice sheet model. Climate predictions using coupled models are considered in Chapter 7.

In the earlier sections of this chapter, we identified some of the important processes which shape the circulation of the Antarctic atmosphere. With a knowledge of these processes, it is possible to state some basic requirements that a GCM must fulfil in order for it to be capable of producing a reasonable simulation of the Antarctic circulation.

(i) The model must include realistic parameterisations of surface energy exchange and long-wave radiative exchange in the lower atmosphere, since these processes determine the structure of the surface inversion which, in turn, controls the katabatic flow.

(ii) The horizontal resolution of the model must be adequate to represent mid-latitude cyclones moving towards Antarctica and the transports associated with these transient eddies. The topography of Antarctica must also be represented at a reasonable resolution (300 km or better) in order to simulate the low-level katabatic flow accurately.
(iii) Very high vertical resolution is required in the boundary layer, since katabatic flows are typically only 100 m deep.
(iv) The model must be capable of a good simulation of the position, strength and seasonal variation of the circumpolar trough.

Simmonds (1990) has reviewed the development of GCM simulations of the Antarctic climate. Most of the early GCM studies, published during the 1970s, produced a very poor simulation of the Southern Hemisphere circulation in general, pointing to the need for improved resolution and parameterisation of physical processes. Efforts at this stage were concentrated on validating (and improving) model performance in the tropics and mid-latitudes. As models improved, some attention was paid to the simulated climate at higher latitudes. Herman and Johnson (1980) described the Antarctic climate simulated by the GLAS GCM. The 500 hPa flow around Antarctica was reasonably realistic in this simulation but the representation of the surface pressure field was poor, with the circumpolar trough being too weak and the stationary low-pressure centres within the trough almost entirely absent. Surface air temperatures over the Antarctic plateau were some 10°C too warm in summer and up to 30°C too warm in winter.

Xu *et al.* (1990) compared the performance of four spectral GCMs with moderate resolution in simulating Southern Hemisphere climate. The models studied were the Canadian Climate Centre model (CCC), the NCAR Community Climate Model, version 0 (CCM0), the Geophysical Fluid Dynamics Laboratory (GFDL) model and the ECMWF model. Only the CCC model was able to produce a circumpolar trough of realistic depth and none of the models simulated the 500 hPa standing wave pattern accurately. An important failing of all four models was the inability to reproduce the semi-annual oscillation of surface pressure and other variables. The authors concluded that this pointed to inadequacies in the parameterisation of surface–atmosphere energy exchanges in all of the models considered.

Mitchell and Senior (1989) studied the Antarctic winter climate simulated by the UK Meteorological Office 11-level GCM. This finite-difference model was run on a fairly high-resolution (2.5° latitude by 3.75° longitude) grid, although the vertical resolution was rather poor, with the lowest grid point being about 200 m above the surface. The depth and latitude of the circumpolar trough were well represented by this model, including the wavenumber three pattern of climatological lows, although one low centre was somewhat misplaced.

Simulated surface temperatures over Antarctica were in good agreement with observations but the upper troposphere was more than 5°C too cold over the continent. The pattern of surface winds produced by the model was in accord with our understanding of the katabatic flow but the modelled wind speeds were somewhat weaker than those observed. This almost certainly reflects the poor boundary layer resolution in this model. Given the artificially large depth of the katabatic flow imposed by the height of the lowest grid point, the total katabatic *transport* (i.e. the product of speed and layer depth) may actually have been simulated correctly and it is this quantity that is of greatest dynamical interest.

Genthon (1994) compared Antarctic climate simulations using the GISS and French Arpège GCMs. Poor resolution was again identified as contributing to model failings. The snow surface albedo parameterisation in the GISS model, based on measurements over seasonal snowpacks, was shown to be inappropriate for Antarctica, leading to errors in simulated temperatures. Although the two model simulations differed in significant respects, the differences were no greater than those between some of the data sets available for validation, pointing to the need to develop better climatologies against which model performance can be judged.

4.5.2 A detailed comparison of two models

From the above review, it is clear that GCM simulations of Antarctic climate have improved considerably in recent years. It is instructive to compare results from one of the latest generation of models with those from an earlier simulation to see how this improvement has come about. Here, we compare results from the UK Meteorological Office Unified Climate Model (UKMOUCM), presented by Connolley and Cattle (1994), with the Antarctic climatology of version 1 of the NCAR Community Climate Model (NCARCCM1) described by Tzeng *et al.* (1993). These two simulations differ greatly in the resolution used and there are also important differences between the physical parameterisation schemes incorporated into the two models. NCARCCM1 is a spectral model, run at R15 truncation in the study under discussion, giving an effective resolution of about 4.5° latitude by 7.5° longitude. Twelve vertical levels were used, with the lowest about 65 m above the ground over Antarctica. In contrast, UKMOUCM is a finite-difference model with a 2.5° by 3.75° latitude–longitude grid. The lowest of the 19 vertical levels is about 30 m above the ground.

Figure 4.14 shows the average winter (JJA) sea-level pressure pattern simulated by both models (values over the high plateau should be ignored because of the problems involved in reducing surface pressures here to sea level). The UKMOUCM simulation is in quite good agreement with observed climatology (see Figure 3.15), whereas, in the NCARCCM1 simulation, the trough is about 6° too far north and the lowest pressures are 15 hPa too high. Furthermore, the

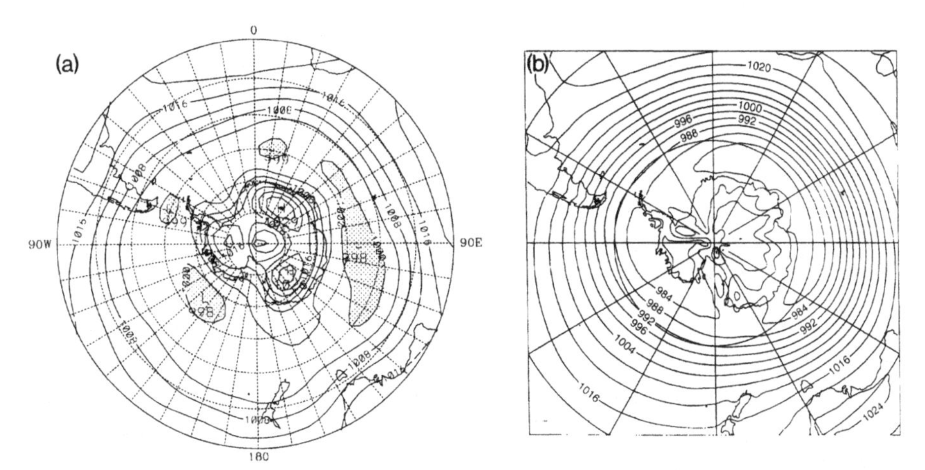

Figure 4.14 Winter (JJA) mean sea level pressure fields from (a) NCARCCM1 (Tzeng *et al.*, 1993) and (b) UKMOUCM (Connolley and Cattle, 1994). The contour interval in (a) is 2 hPa, with values less than 1002 hPa stippled. In (b), the contour interval is 4 hPa.

wavenumber three pattern is not well represented by NCARCCM1. Tzeng *et al.* (1993) suggest that the failure of NCARCCM1 to produce a realistic circumpolar trough is due to the highly smoothed version of the Antarctic orography used in this model. It is also likely that individual mid-latitude cyclones within the trough are inadequately resolved. Both NCARCCM1 and UKMOUCM simulations exhibit a realistic semi-annual oscillation in the strength and position of the circumpolar trough.

Surface temperatures in the NCARCCM1 simulation are in reasonable agreement with observations around the coast of Antarctica, but are about 15°C too warm over the high plateau. The strength of the surface inversion is well simulated by this model, so lower tropospheric temperatures must also be too warm. Excessive cloudiness, resulting from the moisture adjustment scheme used to prevent humidity becoming negative, may be responsible for this anomalous warmth. UKMOUCM surface temperatures are in better agreement with observations over much of Antarctica but even this model produces winter surface temperatures that are about 5°C too cold in the vicinity of Vostok. Free tropospheric temperatures in UKMOUCM are also in good agreement with the observed climatology over much of Antarctica but, like surface temperatures, are somewhat cold at Vostok. Both models simulate the annual variation of temperature well and exhibit a realistic coreless winter over the high plateau.

Near-surface winds in both models (Figure 4.15) follow the pattern expected of katabatic flow. However, on closer examination, the NCARCCM1 winds exhibit much less structure in the low-level easterly vortex than is apparent from observations and the strongest model winds are over the interior of the continent, rather than over the steep coastal slopes as is observed. The topography

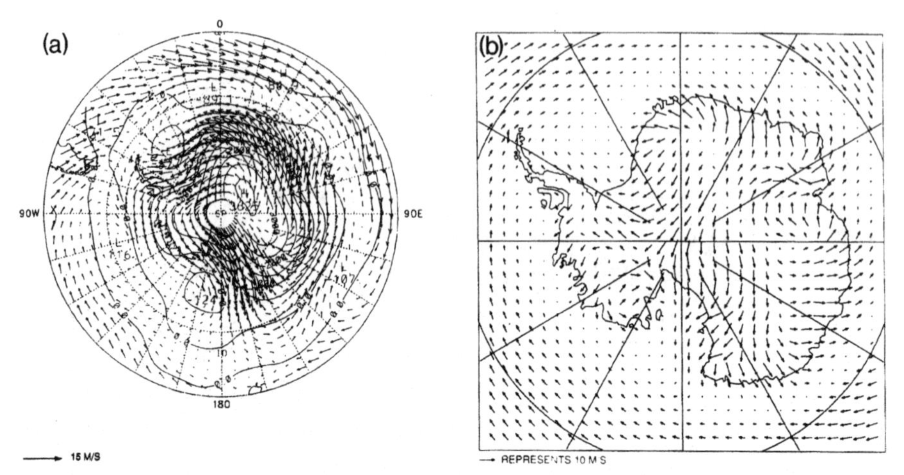

Figure 4.15 Near-surface average wind vectors for winter from (a) NCARCCM1 (Tzeng *et al.*, 1993) and (b) UKMOUCM (Connolley and Cattle, 1994). Contours of the NCARCCM1 model orography are also shown in (a).

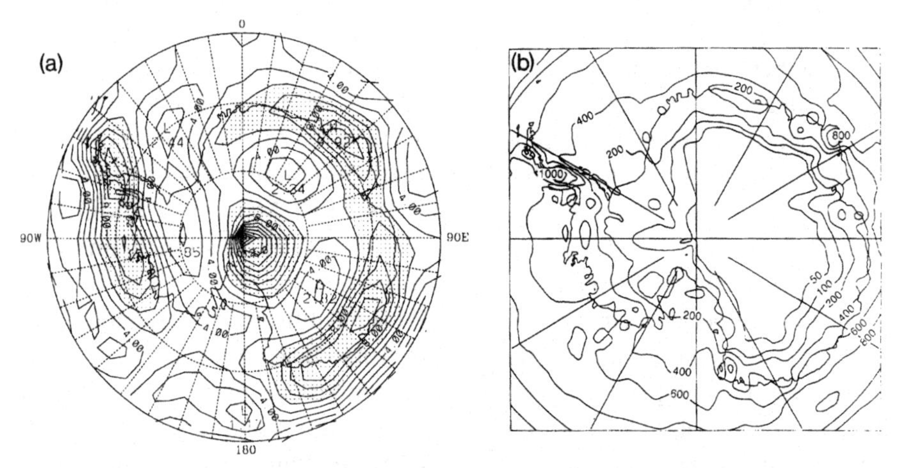

Figure 4.16 Annual average accumulation as simulated by (a) NCARCCM1 (Tzeng *et al.*, 1993) and (b) UKMOUCM (Connolley and Cattle, 1994). The contour interval in (a) is 100 mm per year, with values less than 400 mm per year stippled and values greater than 600 mm per year hatched. In (b), contours are drawn at 50, 100, 200, 400, 600 and 1000 mm per year.

used in this model is highly smoothed and lacks the pronounced embayments which complicate the observed katabatic wind. Furthermore, the smoothing renders the topography concave, so the steepest slopes are found in the interior. The model winds thus simply reflect the structure of the model topography. UKMOUCM winds are more realistic, reflecting the higher resolution topography used in this model.

Annual accumulation (precipitation minus evaporation) distributions for the two models are shown in Figure 4.16. Given that NCARCCM1 has an anomalously warm lower troposphere and that there are problems with the moisture adjustment scheme used in this model, it is not surprising that precipitation rates are too high. Accumulation in the region south of 70° S is about 3.5 times that observed. The accumulation pattern simulated by UKMOUCM is much more realistic, with the central 'desert' of less than 50 mm per year accumulation occupying approximately the right area. The average modelled accumulation rate over the whole of Antarctica, excluding the Peninsula, is 170 mm per year, which compares well with an estimate from observations (Giovinetto and Bentley, 1985) of 143 mm per year.

Neither model is perfect, but the somewhat better overall performance of UKMOUCM can largely be attributed to the higher resolution and better representation of the Antarctic topography used in that model, although differences in physical parameterisations are clearly important. A version of NCARCCM1 run at T42 resolution produced a greatly improved simulation of the pressure field around Antarctica (D. Bromwich, personal communication 1995) whereas version 2 of the NCAR Community Climate Model, which incorporates improved resolution both in the vertical and in the horizontal dimensions, produces a more realistic representation of many aspects of the Antarctic climate (Tzeng *et al.*, 1994).

4.5.3 **Limited-area models**

From the above discussion, it is clear that adequate resolution is the key to producing good simulations of the Antarctic climate. Of course one pays the price for increased resolution in additional computational effort and running a global model at the high resolution necessary is very computationally expensive. If interest is centred on Antarctic regions, then an attractive option is to 'nest' a high-resolution limited-area model, restricted to the environs of Antarctica, within a global model of lower resolution .

Nested limited-area models have been used for regional climate simulations in a number of areas but the technique has only recently been applied to Antarctica. Walsh and McGregor (1993) reported results from a 250 km grid limited-area model covering the whole of Antarctica, which was nested within a low-resolution GCM. The limited-area model produced a significantly better simulation of the circulation over the continent and in the circumpolar trough than did the GCM alone. In the future, limited-area models could usefully be applied to regions such as the Antarctic Peninsula, where the steep topography is poorly represented even in the highest resolution GCMs.

Poor vertical resolution, with consequent failure to represent the katabatic wind adequately, was identified as a failure of some of the GCM simulations discussed above. Pettré *et al.* (1990) attempted to remedy this problem by coupling a GCM to a two-layer simulation of the katabatic flow using Ball's model, which was discussed in Section 4.3.2. The katabatic flow was coupled to the flow at the lowest GCM level by specifying a drag coefficent at this level. Introduction of this parameterised katabatic flow had a significant impact on some aspects of the GCM simulation, including a 50 m increase in the 500 hPa geopotential height and a 2°C increase in surface temperatures at the South Pole. The katabatic wind model, however, required considerable tuning and this study is probably more significant as a demonstration of the need for adequate vertical resolution than as a description of a practical technique for improving GCM performance.

Chapter 5

Synoptic-scale weather systems and fronts

5.1 Introduction

Synoptic-scale weather systems, comprising the extra-tropical cyclones (depressions) and anticyclones (highs), are the main atmospheric systems found in the Antarctic coastal region and over the Southern Ocean. They typically have a horizontal length scale in the range 1000–6000 km and a lifetime of between one day and a week. Within the spectrum of atmospheric disturbances they lie between the mesoscale (less than 1000 km diameter) phenomena, such as cloud clusters, mesocyclones and squall lines, and the planetary-scale long waves that have wavelengths of many thousands of kilometres. Satellite imagery of the high southern latitudes, such as that reproduced in Figure 5.1, shows clearly the locations of the major synoptic-scale low-pressure systems through their frontal cloud bands that spiral into the centre of the systems. Such imagery is an important tool in the production of operational analyses over the Southern Hemisphere and allows the depressions to be tracked in areas sparse in data. The infra-red image shown here is of the Bellingshausen Sea on 28 April 1993 when the area was dominated by an active depression. The high, cold cloud of the front and the shower clouds in the cold air to the west of the system are the most significant features of the image.

Case studies and climatological investigations have shown that depressions form mainly on horizontal temperature gradients (baroclinic zones) in the troposphere and grow through baroclinic instability. In the Southern Hemisphere the major meridional temperature change in the troposphere occurs at the polar front separating the temperate mid-latitude air to the north and the cold polar

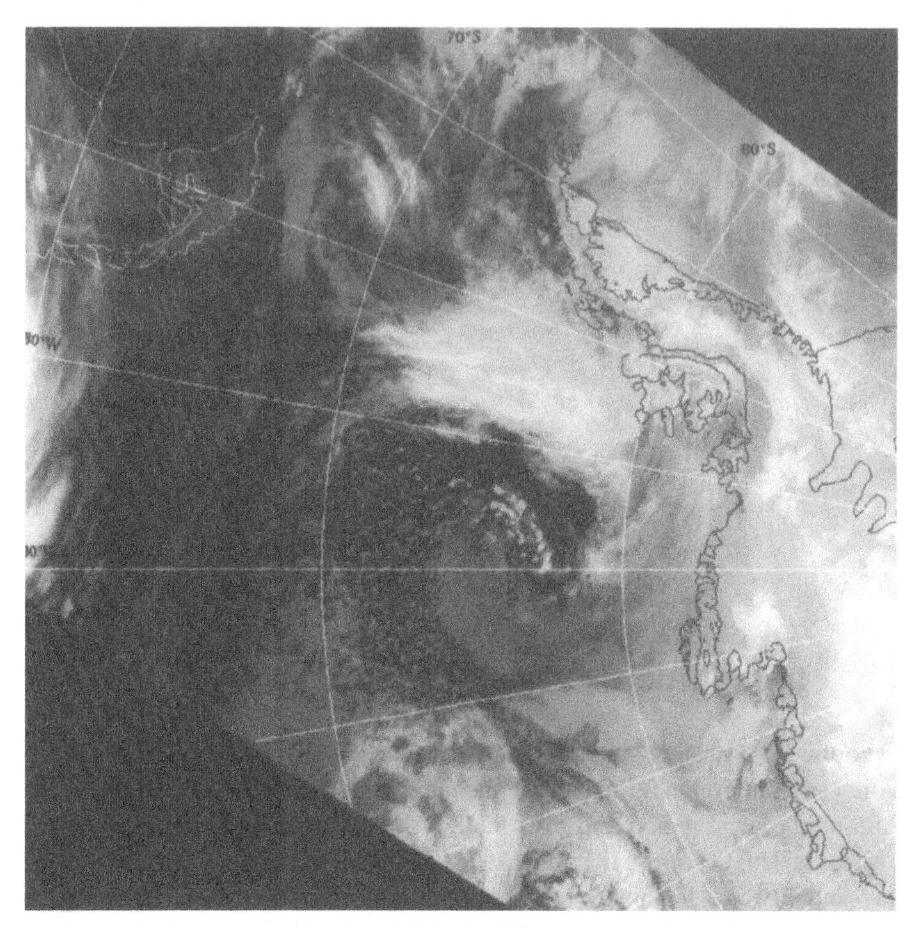

Figure 5.1. An infra-red satellite image of an active depression over the Bellingshausen Sea at 1122 GMT 28 April 1993.

air masses to the south. Although the polar atmospheric front is not a continuous feature extending around the hemisphere and can have breaks or consist of two or more minor frontal bands, it is usually found between the latitudes of 40° S and 50° S. However, it extends much further south in the vicinity of the Tasman Sea and southwest Pacific Ocean in winter. The polar front is therefore one of the primary regions for the development of depressions (cyclogenesis) in the Southern Hemisphere. Another area of significant thermal contrast is in the Antarctic coastal region where the cold, continental air from the interior meets the relatively temperate, maritime air masses. Here, the thermal contrast at low levels results in the development of mesoscale or small synoptic-scale disturbances that can track eastwards parallel to the coast or occasionally move northwards over the Southern Ocean.

As in the Northern Hemisphere, most mid-latitude depressions move polewards from the polar front to higher latitudes. However, because of the distribution of the land masses south of the equator, depressions over the Southern Ocean tend to follow more zonal tracks before arriving at the main area of cyclone dissipation (cyclolysis) in the circumpolar trough, just north of the Antarctic coast. The Antarctic is therefore generally characterised by the frequent occurrence of mature and declining depressions, although, as will be discussed later, recent studies have shown that developments on the meso- and synoptic scales occur more often than had previously been thought in this area. Many of the declining depressions around the coast of the Antarctic still have low central surface pressures and, because they are slow moving and often exist in this state for many days, have a major influence on the climatology of the region. This can be seen in the summer season mean surface pressure chart shown in Figure 3.15 that has a ring of centres of low pressure located around the Antarctic in the latitude band 60 to 70° S caused by the many depressions in this zone. This region is known as the circumpolar trough and it affects many aspects of the climate of the coastal area.

There are major differences in the synoptic-scale activity between the Arctic and the Antarctic as a result of the markedly different topographic conditions in the two polar regions. Whereas, in the Arctic, depressions are often found quite close to the North Pole, the high Antarctic plateau presents a major barrier to the movement of weather systems southwards over the continent itself and the majority of lows do not penetrate far into the interior and usually stagnate and dissipate close to the coast. A further difference is the greater vigour of the depressions over the Southern Ocean as a result of the larger equator–pole temperature difference found in the Southern Hemisphere. This results in stronger winds and more frontal precipitation in the mid-latitude areas of the Southern Hemisphere than is found at corresponding latitudes in the north.

Anticyclones bring relatively quiet weather conditions with low wind speeds and little precipitation, although they can be associated with extensive and persistent cloud. Over the Southern Ocean, anticyclones are usually relatively short-lived features, existing for only one or two days. However, more persistent anticyclones do sometimes occur, which may divert depressions to the north or south. Over the continent, surface pressure is usually high and mean sea level charts normally show a large, persistent anticyclone. However, care must be taken when using such charts because much of the surface of the plateau is at a level above 700 hPa and the reduction of data at the topographic surface to mean sea level is fraught with problems, as is discussed in Sections 5.5.1 and 2.2.1.

The individual fronts observed around the Antarctic have a cross-frontal horizontal length scale of only a few tens or hundreds of kilometres, making them mesoscale features according to the definition of Orlanski (1975). However, because of their close association with synoptic-scale depressions, they will be dealt with in this chapter and we will consider them both in terms of the effect of individual fronts on the meteorology of the continent and in the establishment of climatological frontal zones.

In this chapter we will also examine the question of the production of operational analyses and forecasts for the Antarctic since these are mainly concerned with weather systems on the synoptic scale. It is probably true to say that there is no area on Earth where there are so few meteorological observations with which to attempt to produce routine analyses and forecasts, making this a significant challenge. Nevertheless, the complex logistics of modern Antarctic operations, which involve air-, sea- and land-based activities, require accurate forecasts of the main meteorological elements for up to 24 h ahead and preferably longer. We will therefore describe the current state of analysis and forecasting activities that take place on the research stations using manual techniques and the output of numerical models.

5.2 The role of depressions

The extra-tropical cyclones found over the Southern Ocean and around the Antarctic continent play a major role both in the general circulation of the atmosphere and in the coupled atmosphere–ocean–ice system of the high-latitude areas. In the Southern Hemisphere the standing waves (Rossby waves) are much more barotropic (the surfaces of pressure and density tend to be parallel) than are those in the Northern Hemisphere and hence they transport much less heat and momentum polewards (van Loon, 1979) (see also Section 4.2). Therefore the role of the transient eddies (depressions) is greater and they are responsible for much of the flux of moisture and energy polewards through the transport of sensible and latent heat (Radok *et al.*, 1975; Physick, 1981). These eddy heat transports act to reduce the equator–pole temperature gradient which is maintained by radiative forcing. The heat transports are maintained by flow equatorwards of cold, continental air on the western flank of depressions and the transport polewards of more temperate air masses on the eastern side. When the southwards-moving depressions reach the coast of the Antarctic, they usually turn towards the east and move parallel to the coast because they are unable to ascend the steep topographic gradient at the edge of the plateau. With the advection of warm air towards the coast, the meridional temperature gradient in the coastal region is increased, resulting in the strengthening of the coastal easterlies and the establishment of a jet in the upper troposphere.

Depressions are extremely important in determining the distribution and quantity of precipitation that falls in the Antarctic. Not only is significant precipitation associated with frontal systems, but also advection of microparticles polewards (Thompson and Mosley-Thompson, 1982) plays a major role in precipitation formation. As discussed in Section 3.4, depressions are responsible for most of the precipitation in the coastal region and their limited penetration into the interior is rendered apparent by the rapid decrease in accumulation away from the immediate coastal area. This can also be deduced by inspection of the composite satellite images of the whole continent, which show extensive cloud over the ocean and in the immediate coastal strip, but only isolated, disorganised areas of cloud in the interior. Regions of greater accumulation in the coastal region, such as south of the Bellingshausen Sea, also indicate areas of greater cyclonic activity or where depressions become slow-moving.

Depressions around the Antarctic bring strong surface winds to the coastal area when the surface pressure gradient increases as deep lows over the ocean move south and encounter high pressure, associated with the semi-permanent anticyclone over the continent. These strong winds then affect the surface conditions over the ocean and sea ice. Over the ocean, when strong winds are combined with advection of cold air over the relatively warm sea, such as occurs to the west of low centres, there are increased fluxes of heat and moisture into the atmosphere. This gives a cooling in the uppermost layers of the ocean, resulting in a destabilization of the water column and downward convection. Over the Southern Ocean, depressions also have a major effect on sea ice. During the winter months the ice can extend northwards beyond 60° S into the zone of extensive cyclonic activity, where many lows are moving parallel to the edge of the pack as they spiral in towards the circumpolar trough. Ahead of these systems there is a movement polewards of the ice edge to the east of the centre and a corresponding transport equatorwards on the western side. Most depressions at these latitudes are moving fairly rapidly eastwards and on daily maps of ice extent their effect can be observed as a wave of motion in the meridional direction. With depressions that have a symmetric wind field around their centre, there may be negligible net effect on the position of the sea ice edge, although, insofar as most depressions have stronger winds on their western side, because of the greater instability of the air from the south, there may be some small resultant shift equatorwards of the ice edge after the passage of the disturbance. However, when lows become slow-moving or are blocked for several days then there can be significant movement of ice that can have climatological consequences. Such a situation is created during the winter months when wind forcing on the sea ice over several days or more can move it away from the edge of the continent, creating an area of open water called a coastal polynya. With the low air temperatures and humidity levels at this time of year, there are large fluxes of heat and moisture into the atmosphere and the creation of large quantities of new ice (see Section 4.2).

Over the continent itself the winds in the coastal area are strongly influenced by topography and the downslope, katabatic winds are a climatological feature of many parts of the Antarctic. As depressions move eastwards parallel to the coast the southerly katabatic flow can be enhanced considerably by the synoptic-scale circulation (Murphy and Simmonds, 1993), giving a transport of very cold air far out over the ocean. The interaction between the flow over the ocean and the katabatic winds has also been observed to produce areas of convergence near the coast that can give rise to the formation of fronts or result in cyclogenesis. Interactions between synoptic-scale weather systems and katabatic winds are discussed further in Section 6.1.3. A further consequence of the strong winds associated with depressions is the horizontal transport of snow. These blowing snow episodes (see Section 6.3) can move considerable quantities of snow and can also be a major problem in logistical operations. They also severely hamper meteorological observing programmes and make the measurement of precipitation by stakes or snow gauges very difficult.

5.3 Depressions in the Antarctic and over the Southern Ocean

Our current knowledge of the development and structure of extra-tropical cyclones has been described in a number of recent text books on dynamical and synoptic meteorology, e.g. Carlson (1991) and Newton and Holopainen (1990); these should be consulted by those interested in the mechanisms behind cyclogenesis and the structure of weather systems in the mid-latitude areas of the Southern Hemisphere. Here we will provide only a brief account of developments in the temperate regions and instead concentrate on their effects on the meteorology of the Antarctic and synoptic developments in the coastal areas of the continent.

5.3.1 Cloud signatures, central pressures and pressure perturbations

With the limited amount of *in situ* data over the largely oceanic Southern Hemisphere, satellite imagery has played a more important role in the investigation of extra-tropical cyclones around the Antarctic than it has in other parts of the world. From the 1960s, when the first imagery became available from the polar orbiting satellites, meteorologists have studied the form of the cloud signatures in relation to the major weather systems. One of the first investigations to consider the relationship between cloud patterns in satellite imagery and the synoptic environment was carried out by Streten and Troup (1973), who examined visible wavelength satellite imagery to determine the climatology of developments and to relate the cloud features observed to surface pressure observations from nearby ships. From an examination of 12 months of data from the summer, spring and autumn seasons of 1966–69 they developed a classification scheme for cloud vortices based on the six categories shown in Figure 5.2 and described in Table 5.1. These forms of cloud signature have been

Table 5.1 *The categories of cloud vortex identified by Streten and Troup (1973).*

Vortex type	Characteristics
W	A wave developing on a frontal band
A	Early vortex development, either in isolation or merged with a cloud band
B	Late vortex development – a hook-shaped cloud with a marked development of a slot of clear air
C	Full maturity. A cloud vortex with alternating bands of cloud and clear air spiralling into a well-defined centre
D	Decay, either with considerable cloud near the centre (Dx) or with fragmentary cloud near the centre (Dy)
F/G	Frontless vortices. Corresponding to Dx and Dy types but with no cloud band
E/K	Tropical cyclones south of 20° S

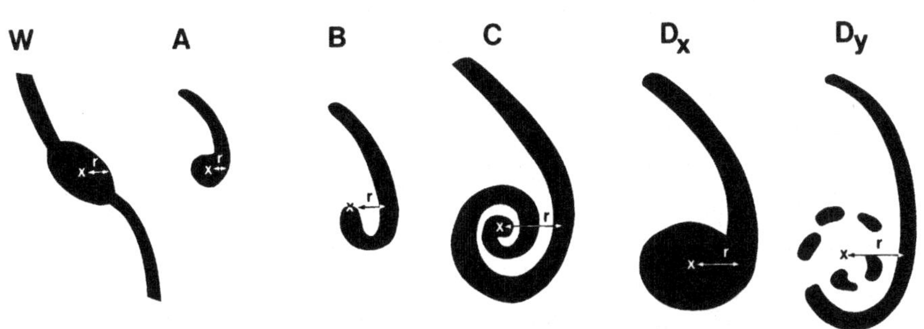

Figure 5.2 A schematic diagram showing the primary classification of extratropical vortex evolutionary patterns. The distance taken as the vortex 'radius' in each case is shown by 'r'. After Streten and Troup (1973).

used in a number of subsequent investigations and will be referred to here frequently.

The scheme they adopted assumed the development of a vortex either as a wave on a main frontal band (type W) or as a comma of cloud in isolation or merged with a major cloud band (type A). The normal stages through which a vortex passes are from type W or A to the late vortex development stage (type B), to full maturity (type C) and dissipation with a disintegrating cloud signature (type D). The vortex shown in Figure 5.1 is typical of the late development, type B stage when there is a shield of high cloud to the east of the centre, a pronounced dry slot behind the front and cold, showery air drawn northwards to the west of the centre. The decay category is split into types Dx and Dy that indicate considerable or fragmentary cloud near the centre of the vortex respectively. A further type of vortex was identified in the study as the frontless system (type F/G) that corresponds to the Dx and Dy classes, except that the vortex was not associated with a cloud band. Subsequent investigations into synoptic-scale cloud vortices in satellite imagery have suggested that the signatures observed over the Southern Hemisphere are much larger than those found in the Northern Hemisphere (Carleton, 1987), with the vortices possibly having a longer lifespan. This may be the result of the extra-tropical cyclones playing a far greater role in the transport polewards of sensible heat in the Southern Hemisphere because of the more barotropic nature of the long waves, although further research is needed to confirm this.

Streten and Troup related the available surface and upper-air data over the Southern Hemisphere to the synoptic environment as implied by the satellite imagery and computed mean anomalies for each class of cloud vortex defined above. Figure 5.3 shows the departures of the surface pressure and 300 and 500 hPa heights from the long-term means for a simplified development cycle of systems passing through the stages A, B, C and D in that order. It can be seen

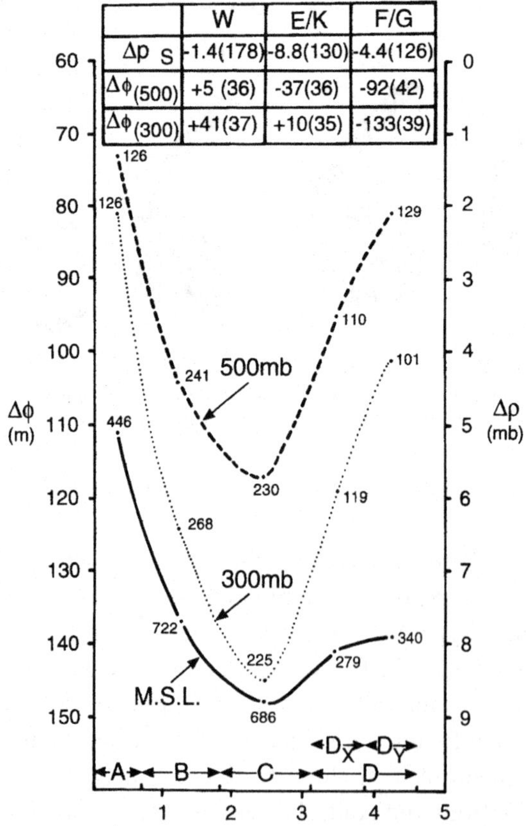

Figure 5.3. Mean departures from long-term monthly means of mean sea level pressure and 500 and 300 hPa geopotentials within 6° of latitude radius of vortex centres, shown as a function of time for a regular extra-tropical sequence. Small numbers on plotted points indicate the number of observations upon which each anomaly is based. The inset table shows equivalent values for less frequent vortex types and the number of observations. After Streten and Troup (1973).

that the surface pressure anomaly in the early development stage (vortex type A) is around −5 hPa, but increases by the time of maximum development (type C) to almost −9 hPa, before decreasing to −8 hPa in the dissipating phase. This small anomaly decrease from stages C to D is somewhat surprising since the cloud signatures in the satellite imagery often show quite disorganised cloud elements. However, the anomalies suggest that the vortices have not declined as much as their cloud signatures would imply and that they still have relatively low central pressures and therefore quite strong winds around their centres. This is one of the reasons for the low surface pressure values within the circumpolar trough around the coast of the Antarctic, even though satellite imagery shows that many of the depressions over this area have only disorganised cloud and are clearly in their declining phase. At upper levels the shape of the height anomaly curve is very similar to that for the surface pressure and there are maximum anomalies of −115 m and −145 m at the 500 and 300 hPa levels respectively during stage C. In the study no attempt was made to use the imagery or *in situ* data to determine the nature of the fronts observed, but this will be considered below in Section 5.3.5.

The central pressures and pressure tendencies of depressions around the Antarctic were also investigated by Jones and Simmonds (1993), who used an automatic depression identification scheme to find lows in a 15-year (1975–89) series of surface analyses produced by the Australian Bureau of Meteorology. The scheme has a number of advantages over manual analysis, including its being consistent, very much less time-consuming and able to detect systems that have a weak cloud signature, which may not be identifiable in low-resolution satellite imagery. Their results are summarised in Figure 5.4, which shows the zonally averaged mean central pressures and pressure tendencies for lows during the four seasons of the year. The lowest central pressures were found in the Antarctic, with a minimum between 65° S and 75° S of close to 967 hPa during the spring. In the Antarctic the seasonal variation in central pressure is greater than it is at lower latitudes, there being a difference of about 10 hPa between the spring minimum and the summer maximum. The pressure tendencies show the deepening of lows as they move southwards in the 30–60° S band, with the greatest deepening rate of 2.5 hPa/day close to 45° S. Lows in the Antarctic are generally filling. As will be discussed later, there are many cyclogenesis events taking place at the latitude of the circumpolar trough, but the pressure tendencies for this zone are such that these systems cannot be deepening very rapidly and the tendencies must be dominated by the filling of the mature depressions. The seasonal cycle of the pressure tendencies shows a large variability in the Antarctic, but with filling of lows predominating. Over the Antarctic continent, care needs to be exercised in interpreting the figures because of the reduction of pressures to mean sea level.

The speed and direction of travel of depressions existing for two days or more

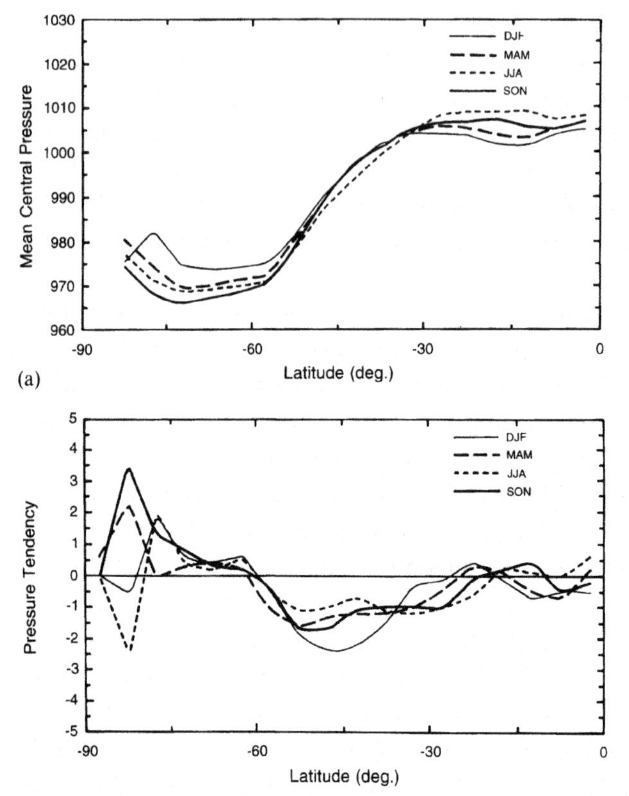

Figure 5.4 Zonally averaged central pressures and pressure tendencies for extra-tropical cyclones in the four seasons as determined from the Australian surface analyses for 1975–89. After Jones and Simmonds (1993).

was also considered by Jones and Simmonds (1993) and their figures showing mean velocity vectors are reproduced in Figure 5.5. It can be seen that the depressions generally move in an east–southeasterly direction towards the Antarctic continent with peak speeds being found in mid-latitudes. The area with the highest depression speeds is a band south of Africa extending from the Atlantic Ocean into the southern Indian Ocean. In this area, which is marked on Figure 5.5, the systems are found to move on average with a speed in excess of 10 m s^{-1}. Their study found little seasonality in the speed of the depressions over the Southern Ocean.

Depressions north of 60° S tend to have fairly well-defined centres at the 500 hPa level that can easily be tracked on the operational charts for this level. However, as they move southwards into the Antarctic coastal region the upper centres become more ill-defined and are sometimes difficult to relate to surface pressure features and the cloud signatures seen in the satellite imagery.

5.3.2 Cyclogenesis

Examination of early satellite imagery and *in situ* data suggested that the main area for cyclogenesis in the Southern Hemisphere was close to the latitude of the

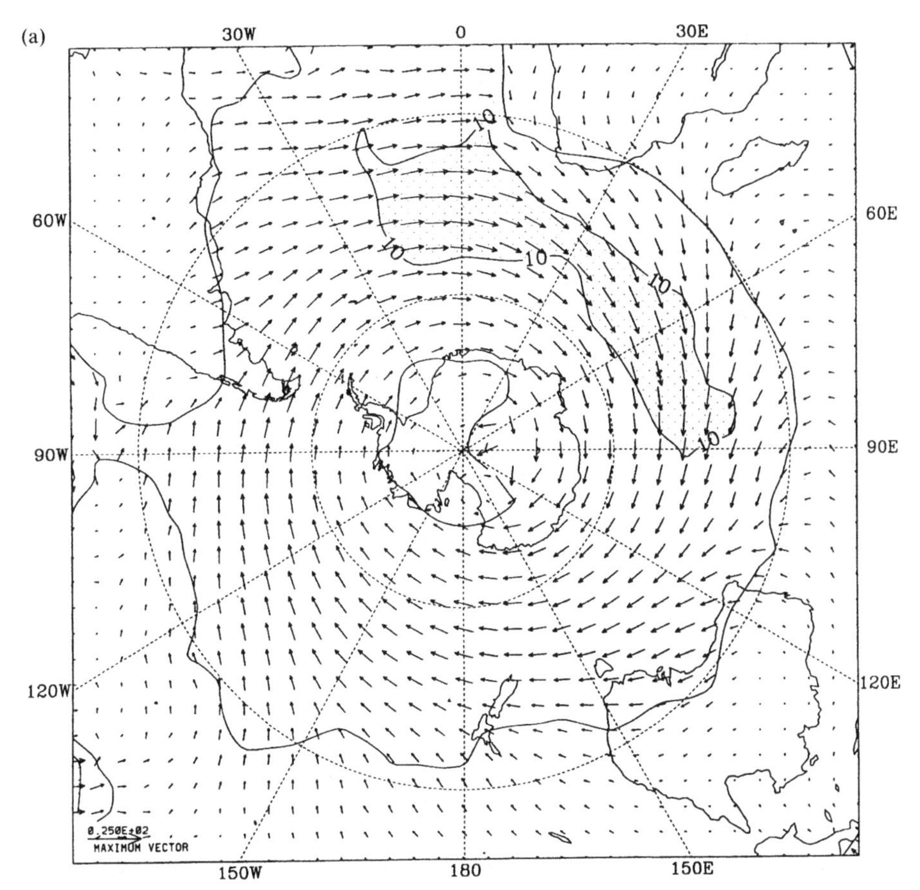

Figure 5.5 Velocities and overlaid speed isotachs of cyclones for (a) summer and (b) winter for the 15 year period 1975–89. From D. A. Jones and I. Simmonds, A climatology of Southern Hemisphere extratropical cyclones, *Climate Dynamics*, **9**, Figure 8, 140, 1993. Copyright (1993) Springer-Verlag.

polar front near 45° S. Here systems were observed to form either as waves on the main front or as isolated lows with a comma-shaped cloud signature forming well away from the frontal bands. During the 1980s satellite imagery with a higher spatial and radiometric resolution became available that showed that cyclogenesis was much more common around the coast of the Antarctic than had previously been thought. The reasons behind these developments are not fully understood, but the vortices are observed to form in a number of ways, including as frontal developments, as systems evolving from fairly featureless areas of low to moderate cloud and as minor vortices developing in isolation. It is known that, on average, the lows do not usually deepen rapidly, in that the figures of Jones and Simmonds show a general filling of depressions in this zone.

Cyclogenesis on the polar front takes place as a result of baroclinic instability and, because it is not significantly different from the mid-latitude developments

(b)

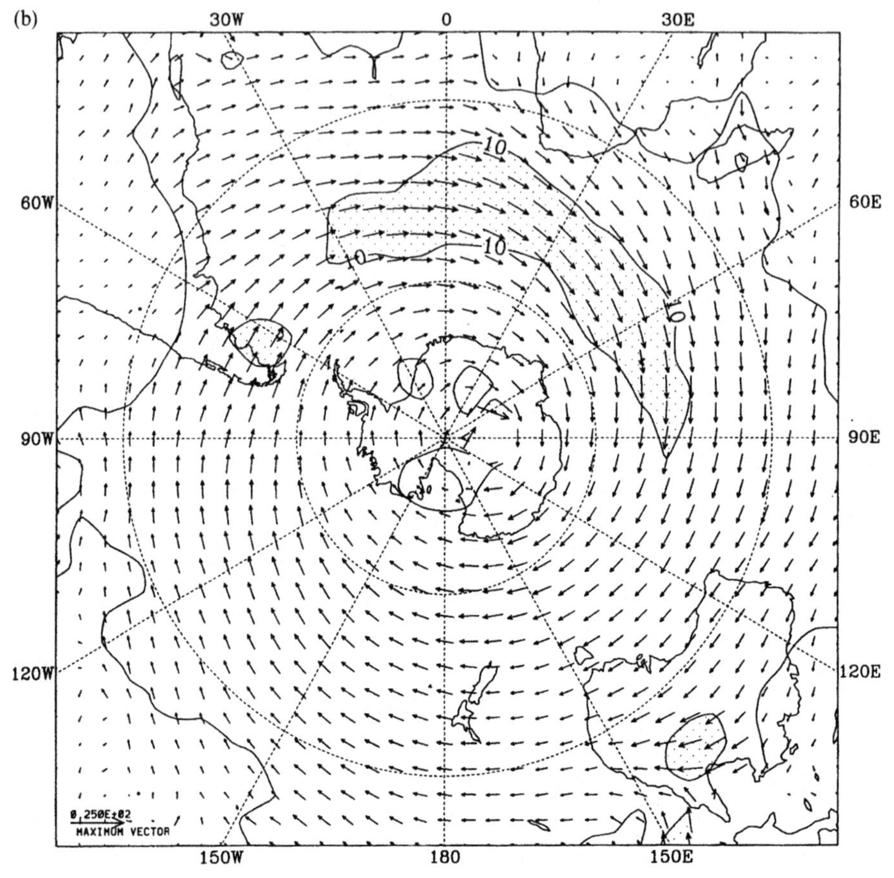

Figure 5.5 (*cont.*)

that take place in the Northern Hemisphere, it will not be considered further here. Other forms of cyclogenesis that are common in mid-latitudes, such as the formation of breakaway lows on fronts ahead of major depressions, are also observed to occur over the high Southern Ocean and the mechanisms behind such developments seem to be similar to those in lower latitudes.

There are three main mechanisms behind cyclonic developments in the Antarctic – baroclinic instability on frontal cloud bands, thermal instability resulting in cold air cyclogenesis and topographic forcing. These types of development will be considered in the following sections and examples of typical vortices will be shown.

Frontal developments

Some vortices develop at high latitudes as a result of the moderate baroclinicity that is found in certain areas, including at the limit of the sea ice, next to the Antarctic coast and at minor thermal boundaries in cold air masses. Satellite imagery shows that many of the vortices which form are rather small, but some do develop into synoptic-scale disturbances as a result of forcing, such as that which comes from upper-level short-wave troughs.

The horizontal temperature gradient in the immediate coastal area develops as a result of the meeting of continental and maritime air masses and is often characterised by extensive low-level cloud that can be observed on satellite imagery. Mechoso (1980) suggested that this thermal gradient is maintained by the flow of cold air down the steep topography in the coastal region and the flux of warm air polewards from the declining depressions. The depression shown in Figure 5.1 is an example of a system that formed over the southern Bellingshausen Sea in a coastal baroclinic zone and developed into a vigorous depression.

The baroclinicity at the edge of the sea ice is usually weaker than that close to the edge of the continent, but is nevertheless sufficient to allow the development of many mesoscale and synoptic-scale disturbances. When first observed on satellite imagery, vortices developing here usually have quite weak cloud signatures with low cloud, and their development into major disturbances is again dependent on the upper level flow. Just like near the coast, serial development of lows is often observed. Figure 5.6 shows a typical example of a small depression that formed close to the northern limit of the sea ice in the Weddell Sea during the winter months.

Strong gradients of sea surface temperature are also responsible for the establishment or re-inforcement of atmospheric baroclinic zones within which developments can take place. Many atmospheric thermal gradients near ocean fronts will be relatively shallow features and most of the vortices developing there will be mesocyclones rather than major synoptic systems. However, when the sea surface temperature gradient is substantial and there are other factors, such as topographic forcing, encouraging cyclogenesis, synoptic-scale developments can be observed. One such area where frequent cyclogenesis has been noted is to the east of South America, where convergence of the Falklands and Brazilian ocean currents takes place, resulting in a large sea surface temperature difference.

Cold air cyclogenesis

Streten and Troup (1973) recognised, in their study of satellite imagery, that non-frontal developments (cyclogenesis taking place well away from pre-existing frontal bands) were more common at southerly latitudes and subsequent investigations have shown that many of the lows developing around the coast of the Antarctic are of this type. Non-frontal developments usually occur in outbreaks of cold, polar air and are often associated with upper air cold core anomalies, especially at higher latitudes. Such regions have also been linked with the formation of polar lows in the Arctic (Businger, 1987) and Antarctic mesocyclones (Turner and Thomas, 1994), and the destabilization of the atmosphere by

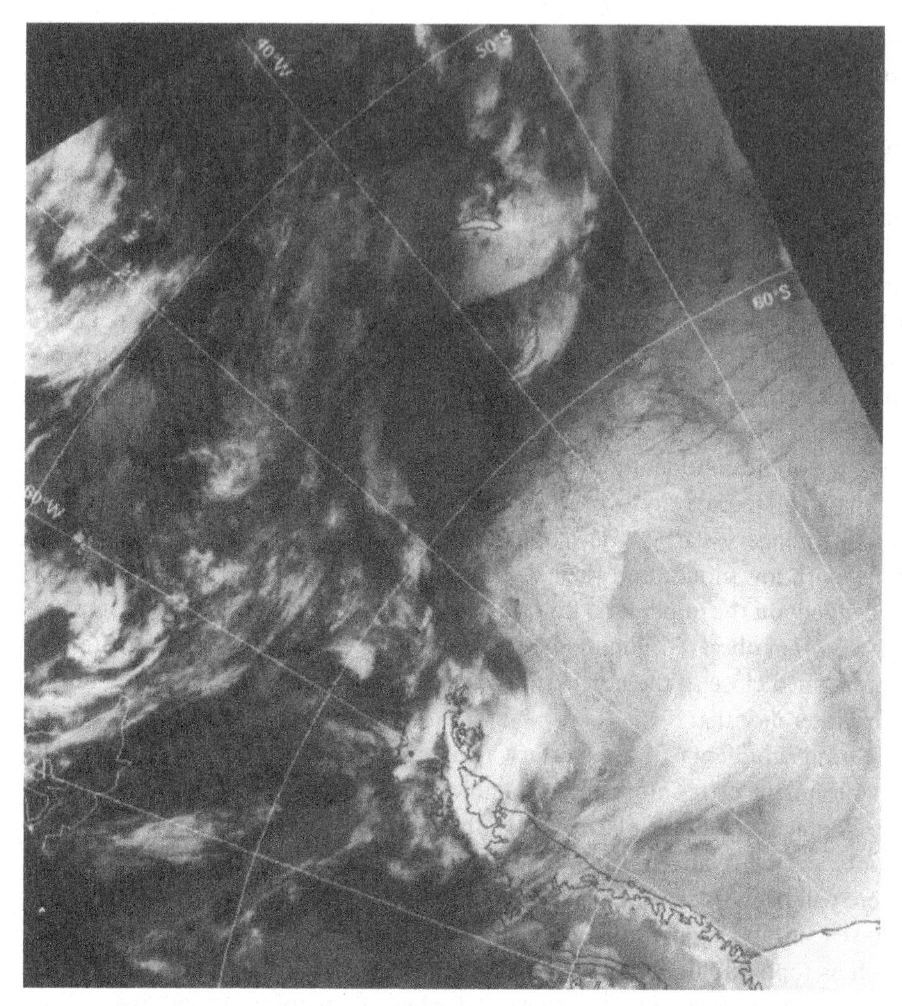

Figure 5.6 An infra-red satellite image of a weak, synoptic-scale depression that formed just north of the edge of the sea ice in the Weddell Sea at 6:21 GMT 18 September 1993.

the cold conditions aloft is thought to play a role in many developments. Such vortices are usually apparent on satellite imagery as a hook of cloud with a comma-shaped appearance and can develop in a field of cumulus cloud. Comma-shaped cloud signatures have been observed on a range of scales from the small, non-developing mesocyclones, with a diameter of 100–200 km, to major synoptic-scale cyclones with a diameter of 2000 km or more. However, there is no fixed length scale boundary between the different types of system and there is undoubtedly a spectrum of atmospheric disturbances around the Antarctic. Comma-shaped cloud vortices often develop in families of several individual centres that may include mesocyclones and larger circulations. Some mesocyclones in polar airstreams have been observed to grow into major extra-tropical low-pressure

Figure 5.7 An infra-red satellite image of a synoptic-scale low-pressure system developing in a southerly cold air outbreak over the Bellingshausen Sea at 19:41 GMT 5 June 1993.

systems (Sinclair, 1993) and eventually acquire all the characteristics of a synoptic-scale disturbance. Mesocyclones can also move northwards and come into contact with the polar front, forming an 'instant occlusion'. Here the combined system has the appearance of a large frontal cyclone with the occluded front being provided by the mesocyclone and the warm and cold fronts coming from the polar front. The form and development mechanisms of mesocyclones are discussed fully in Section 6.5 and will not be considered further here. Instead we will illustrate a case of a synoptic-scale cold air vortex that developed in an outbreak of continental air over the Bellingshausen Sea. Figure 5.7 shows the system at 19:41 GMT 5 June 1993 when the cloud signature of the low had a diameter of about 1200 km. The low formed at a time when there was a large, deep low in the Weddell Sea and a blocking anticyclone in the Bellingshausen Sea, resulting in major outbreak of cold air to the west of the Antarctic Peninsula.

Topographically induced lows

Topographic forcing is important in the development of some depressions around the Antarctic by virtue of the increase in vorticity that occurs when an airstream passes over a topographic barrier. This lee cyclogenesis is apparent in climatological maps of cyclogenesis, in which a greater number of depressions is found to occur just downstream of major mountain ranges. In the Antarctic there is only one high topographic barrier that spans the belt of strong westerly flow and that is the Antarctic Peninsula. Satellite imagery shows frequent cyclogenesis in the western Weddell Sea as a result of air passing over this barrier, with some of these systems growing into major depressions. An example of a depression that developed in the lee of the Antarctic Peninsula is shown in Figure 5.8. A further cyclogenetic region, located some distance outside the Antarctic, is over the South Atlantic, which is under the influence of a semi-permanent trough to the east of the Andes. Depressions forming in this area usually track into the eastern Weddell Sea or move parallel to the coast of the Antarctic to the east of the Greenwich Meridian.

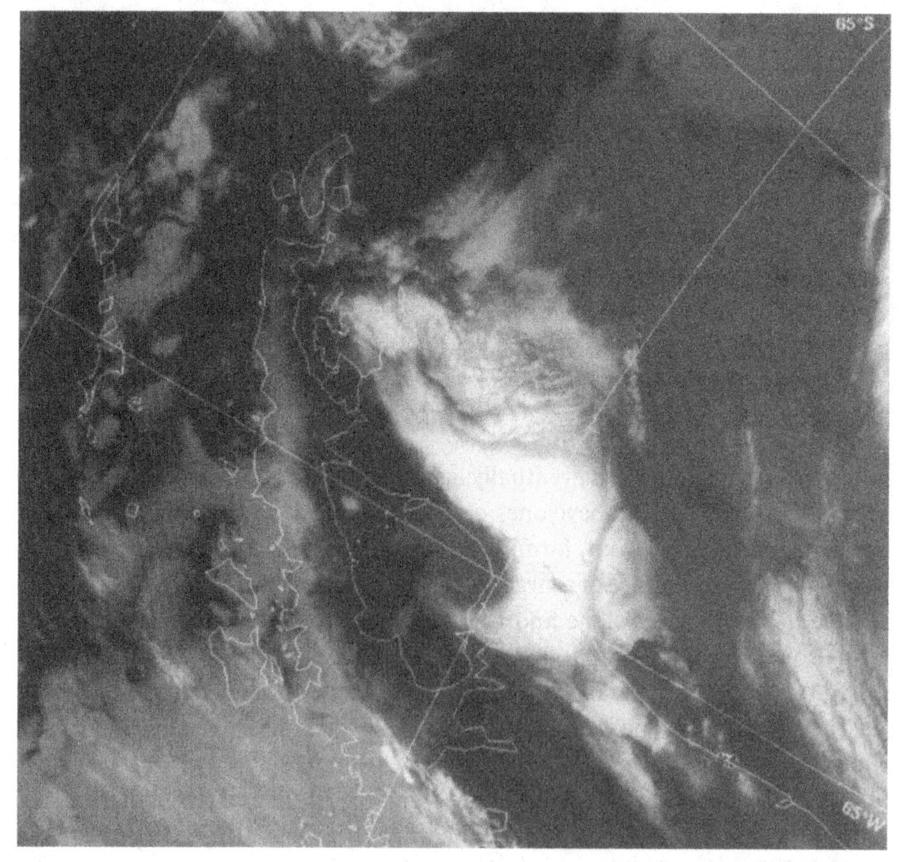

Figure 5.8 An infra-red satellite image of a depression that formed in the lee of the Antarctic Peninsula at 21:14 GMT 4 December 1993.

5.3.3 Deep depressions in the Antarctic

The circumpolar trough that rings the Antarctic continent between 60° S and 70° S is the result of frequent cyclonic activity in southerly latitudes where slow-moving depressions that have moved southwards from mid-latitudes stagnate and decline and where new vortices form. Even though the surface pressure in this zone is generally low and climatologically in the range 980–990 hPa, the depressions here are not especially active, as is shown by the high frequency of slight precipitation that is reported from the coastal stations. However, on occasion, very deep depressions can be found close to the coast of the Antarctic, where they give rise to very strong winds as the pressure gradient increases between these lows and the perennial high pressure over the continent itself. The reasons for depressions attaining very low central pressures in the Antarctic have not been investigated in detail, but the processes involved are probably essentially the same as those operating in the mid-latitude areas of the Northern Hemisphere, although the high topography of the continent will affect the systems as they move southwards. It is hoped that greater insight into these systems will come from case studies that use satellite imagery, *in situ* data and model output, but to date there have been few such studies. One low that attained a central pressure of close to 950 hPa and that gave mean winds of up to 80 kt was examined by Pendlebury and Reader (1993). This storm gave the highest ever wind gust for March (66.3 m s^{-1}, 130 kt) at the Australian Casey Station when it was close to the Antarctic coast, resulting in major disruption to logistical operations. The low formed well outside the Antarctic in mid-latitudes but deepened rapidly and moved quickly into the Antarctic coastal area. Because the low was a significant synoptic event affecting operations in the coastal region its development will be described in the following section. It is not claimed that this low is typical of deep lows in the Antarctic coastal area, but it will serve to illustrate the type of vigorous low that can be found.

A case study of a deep low in the coastal region, 17–22 March 1992

The low (identified as low A on the accompanying charts) that eventually gave the strong winds in the Antarctic coastal region, developed well to the north of the continent, deepened rapidly in mid-latitudes and became slow-moving near to the coast. The low formed as a wave on a cold front and was first analysed on the UK Meteorological Office surface chart at 00 GMT 17 March 1992, close to 38° S, 83° E (Figure 5.9 shows the track of the low over its complete lifetime). Over the next 72 h the low moved slowly southeastwards as a minor feature that, towards the end of the period, had no clear centre analysed on the operational surface charts, but was still apparent on satellite imagery through the associated cloud.

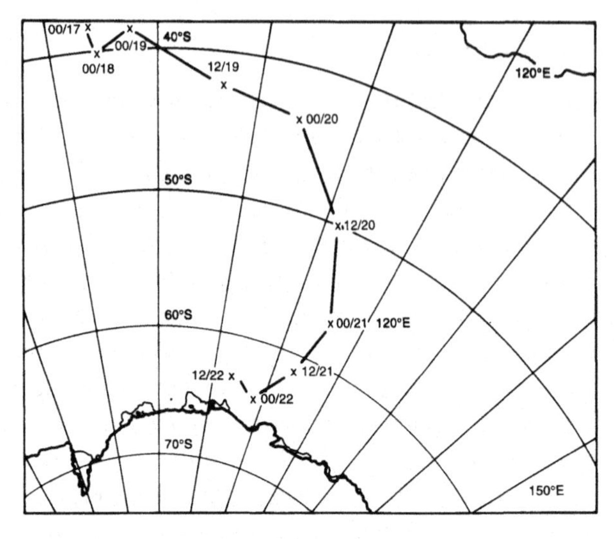

Figure 5.9 The track of the low that gave very strong winds at Casey Station during the period 21–22 March 1992.

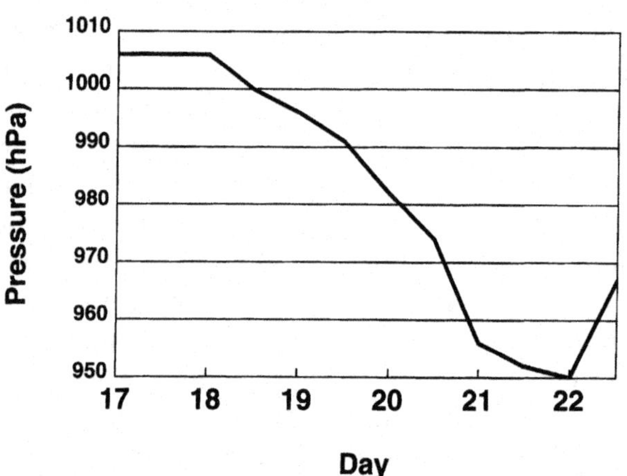

Figure 5.10 The central surface pressure of the low that affected Casey as a function of time.

The marked deepening of the low and its rapid movement towards the Antarctic took place on 20 March, as can be seen in the depression track in Figure 5.9 and in Figure 5.10, which shows the central pressure of the low as a function of time. The UK Meteorological Office surface pressure chart for 00 GMT 20 March 1992, shown in Figure 5.11, indicates that low A at this time was centred near 45° S, 103° E, some 800 km to the north of a barotropic low in the coastal region. The 500 hPa height field for the time (Figure 5.12) had a deep upper low, with a central height of 499 dm, above the low near the coast and a marked upper trough extending north along the 100° E line of longitude. The rapid deepening of low A took place ahead of this upper trough close to the right exit area of a jet in the southwesterly flow behind the upper, cold barotropic low. An analysis of the model output also showed that the deepening coincided with the arrival above the minor surface low of a tropospheric fold that was characterised by high values of isentropic potential vorticity (IPV).

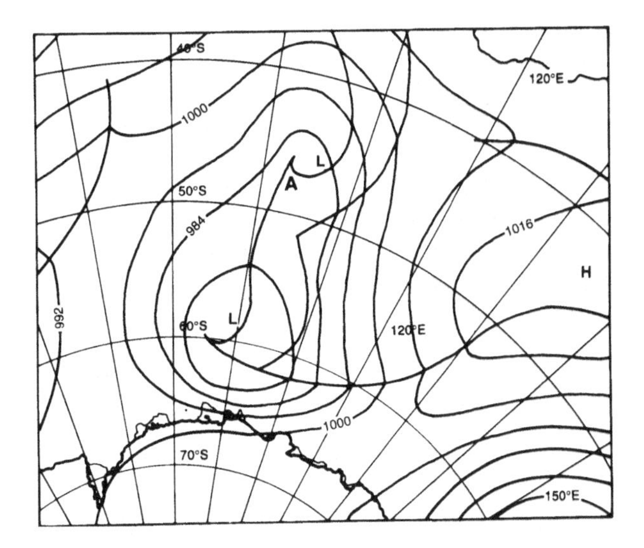

Figure 5.11 The UK Meteorological Office surface pressure analysis for 00 GMT 20 March 1992.

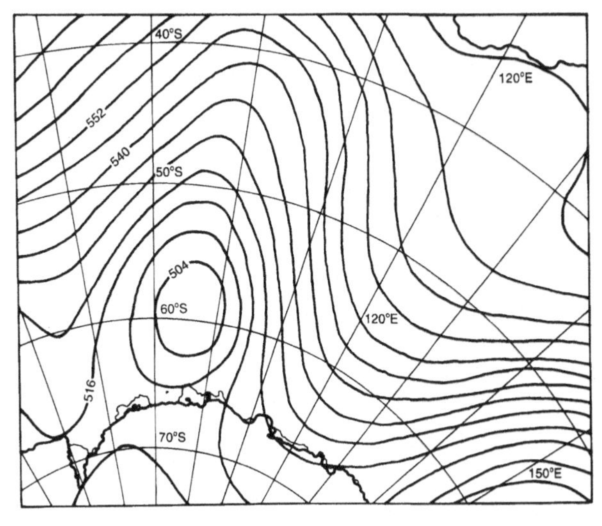

Figure 5.12 The UK Meteorological Office 500 hPa analysis for 00 GMT 20 March 1992.

Case studies of major storms in the Northern Hemisphere have also shown that intrusions of high-IPV air into the upper troposphere can be important in the rapid spin-up and deepening of depressions (Young *et al.*, 1987; Hoskins and Berrisford, 1988).

By 12 GMT 20 March 1992 low A had a surface pressure of less than 976 hPa and had moved southeastwards to 52° S, 110° E with the low under a strong upper level northerly flow between the old, barotropic low at 60° S, 100° E and a pronounced upper ridge extending from the north along 130° E. As can be seen in Figure 5.13, which shows the 500 hPa height and 1000–500 hPa thickness fields for 12 GMT 20 March 1992, there was advection southwards of very warm air associated with the upper ridge and the creation of a very strong baroclinic zone close to the 110° E meridian.

The central pressure of low A had dropped to around 957 hPa by 00 GMT 21 March 1992 and had moved rapidly southwards to 59° S, 113° E in the strong upper northerly flow. The slow moving barotropic low near the coast was still present at this time, but low A had become dominant in the area. With the surface pressure over the continent greater than 1000 hPa and a deep low off the coast, the surface pressure gradient had increased in the coastal region and Casey Station was reporting an easterly wind of 85 kt at this time. The thermal structure of the low was most pronounced at 00 GMT 21 March 1992 and the 1000–500 hPa thicknesses field for this time is shown in Figure 5.14. This indicates the extent to which warm air had reached high latitudes, thickness values of 528 dm being recorded as far south as 70° S, while values of 546 dm were analysed just off the coast at 63° S.

(a)
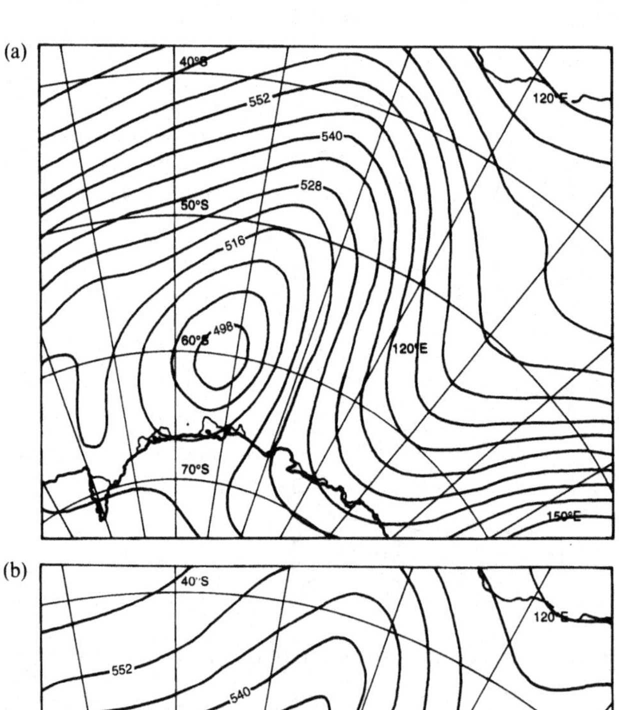

(b)
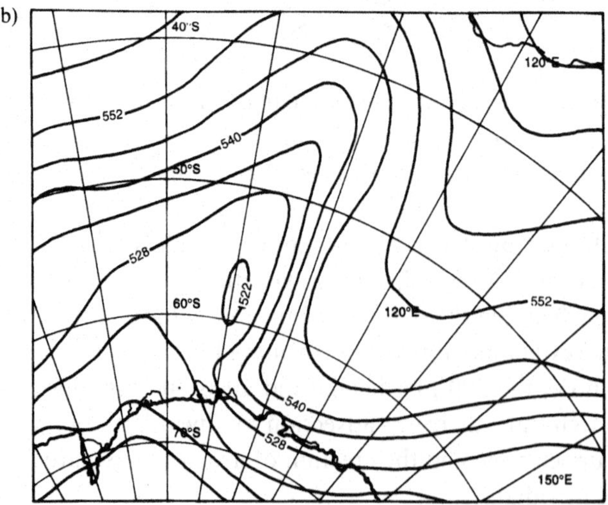

Figure 5.13 The UK Meteorological Office analyses for 12 GMT 20 March 1992: (a) 500 hPa height and (b) 1000–500 hPa thickness.

By 12 GMT 21 March the UK Meteorological Office surface analysis had only one low in the coastal region at 62° S, 112° E and the central pressure had decreased to 953 hPa. The wind at Casey Station was still 85 kt from the east as the pressure gradient was maintained between the deep low and the high pressure over the interior of the continent. At this time the 528 dm 1000–500 hPa thickness line was analysed at its most southerly position close to 74° S, while cold, continental air was carried well to the north. The 85 kt winds persisted until 00 GMT 22 March when the central pressure of the low close to 950 hPa. However, the low filled rapidly after this time and had a central pressure of 967 hPa by 12 GMT 22 March, when the warm air had receded to the north.

The low considered here developed in a strong baroclinic zone between cold air associated with an upper trough and warm air advected northwards to the west of a pronounced ridge. The tropospheric fold also played an important part in its development as high-IPV air descended well into the troposphere. As the system moved southwards the system deepened rapidly and the strong easterly winds experienced at Casey Station were the result of the very strong pressure gradient between the low centre and the high pressure over the continent. Other cases examined of very deep lows in the coastal regions and systems penetrating

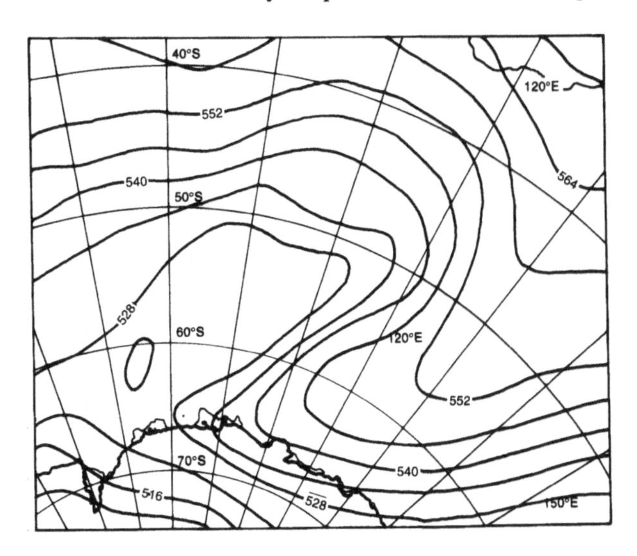

Figure 5.14 The UK Meteorological Office 1000–500 hPa thickness analysis for 00 GMT 21 March 1992.

into the interior suggest that an amplification of the long waves is very important in steering the lows southwards and in creating the strong baroclinicity that is required for the deepening of the systems.

5.3.4 Depressions over the continent

Composite satellite images of the whole Antarctic continent, such as that for 4 July 1994 shown in Figure 5.15, indicate that most of the cyclonic disturbances occur over the Southern Ocean and in the Antarctic coastal region. Inland, far less cloud is apparent and that which can be observed is usually disorganised and slow-moving. Some well-organised depressions do penetrate into the interior and these can bring mild, maritime air over the continent and give precipitation over the plateau. Alvarez and Lieske (1960) found that one event that brought precipitation to the South Pole also caused the surface temperature to rise by nearly 40°C over four days as the oceanic air reached the interior.

The high topography of East Antarctica and the very steep slope in the coastal area around that part of the continent are a formidable barrier to the movement southwards of depressions and few systems manage to reach the plateau here. However, over West Antarctica the topography is lower and there is more cyclonic activity in the interior than there is in the eastern hemisphere. Using IGY data, Astapenko (1964) examined the occurrence of depressions over the interior and found that a significant number tracked from the Ross Sea to the Weddell Sea, affecting stations such as Amundsen–Scott. These events were found to occur in all seasons and are in accord with the general picture of west to east movement of depressions over the continent found by Jones and Simmonds (1993) and shown in Figure 5.5. Astapenko also found that many depressions and low-pressure troughs that affected South Pole Station were accompanied by tropospheric jets, i.e. winds of greater than 50 kt, suggesting considerable baroclinicity in the lower layers.

Phillpot (1968) examined the penetration of depressions into the interior through a study of three years of observations from Vostok station on the high plateau. He found evidence for 42 depressions having been in the vicinity of the station and detected these systems by precipitation falling and relatively high surface temperatures. Of the depressions, 75% were found in the winter months of June to October, suggesting significant cyclonic activity over the continent at that time of year. However, because no satellite imagery was available for this study it is possible neither to determine whether the low-pressure systems were frontal nor to obtain further details about their origin.

A further study of weather systems in the interior was carried out by Neal (1972) using data for the period November 1969 and June 1970. He found no developments at all over the continent in November 1969, whereas several cyclogenesis events were found near the South Pole and over Coats Land in June 1970. Clearly

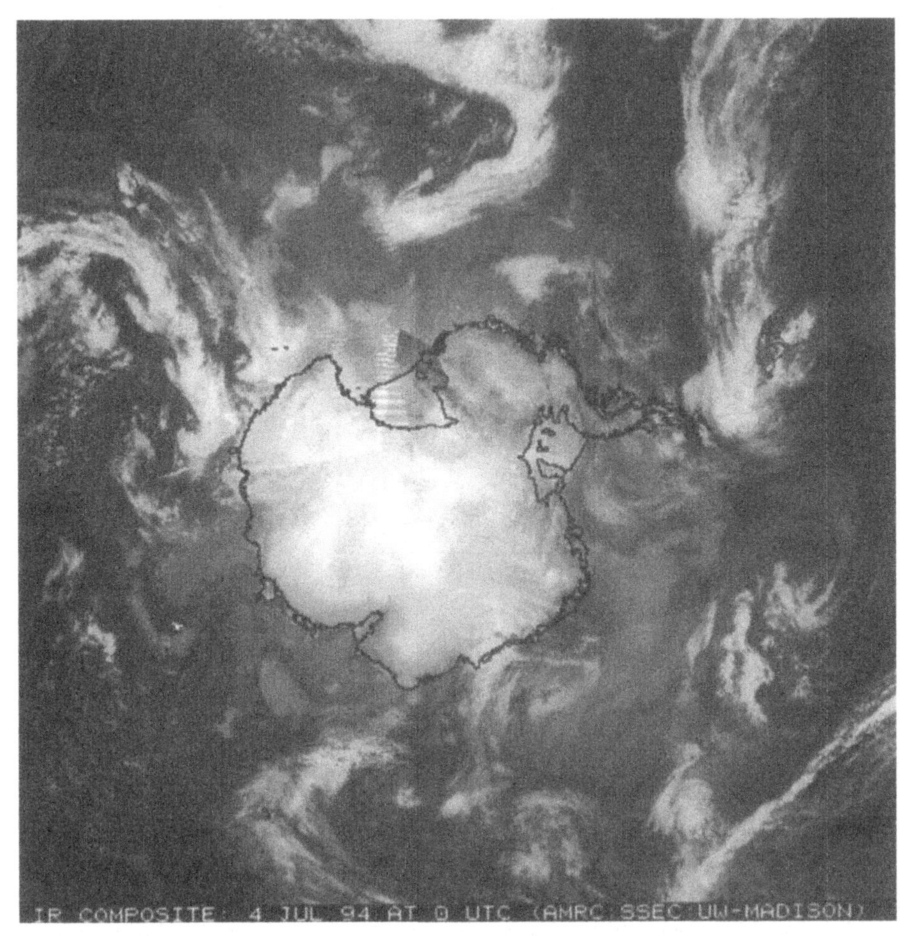

Figure 5.15 A composite infra-red satellite image of the Antarctic and Southern Ocean for 00 GMT 4 July 1994 created from NOAA AVHRR passes. (Courtesy of the University of Wisconsin-Madison.)

the short-period studies carried out so far are not able to give any guidance on the climatological occurrence of weather systems over the interior and major, long-term investigations are required before any firm conclusions can be drawn.

Case studies have suggested that it is not only conventional low-pressure systems that pass into the interior of the Antarctic since other synoptic events had been observed over the continent and found to give severe weather conditions at stations on the plateau. One such case was investigated in detail by Sinclair (1981) who became aware of it by virtue of the record high temperatures that were reported at South Pole (−13.6°C), McMurdo (+9.6°C) and Vostok Stations (−15.7°C). The event took place during the period 25–29 December 1978 when there were two intrusions of warm, humid mid-latitude air from the

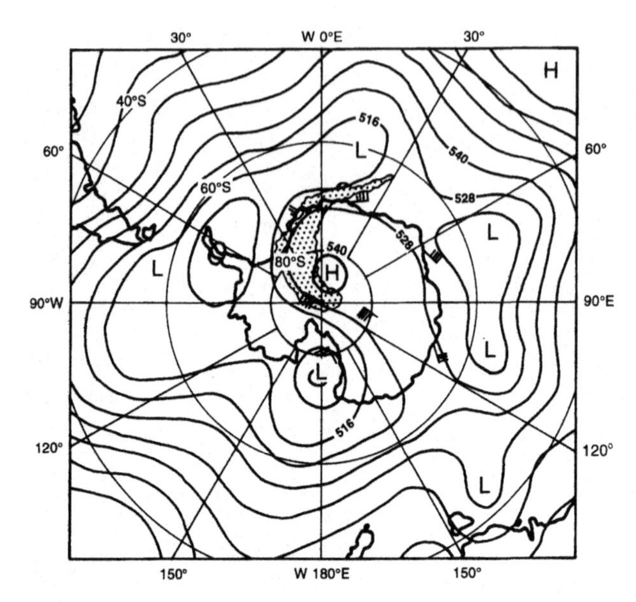

Figure 5.16 The 500 hPa height field and superimposed details of cloud cover at the time of a warming at the South Pole at 00 GMT 26 December 1978. After Sinclair (1981).

Atlantic and Indian Ocean sectors of the Antarctic. An analysis of observations from the research stations, satellite imagery and synoptic charts showed that the warming at the South Pole coincided with the passage of a major cloud system with a jet-like feature and associated baroclinic zone. Figure 5.16 shows the 500 hPa height field for 00 GMT 26 December 1978 with a superimposed outline of the cloud apparent on satellite imagery for that time. The cloud that affected the pole was multi-layered, had a sharp, anticyclonically curved edge on its polewards side and had moved onto the continent via Dronning Maud Land. The highest temperature recorded at the pole occurred under a layer of stratus after the main mass of middle and high cloud had passed. The 500 hPa height field also showed a high to the east of the main cloud band and strong anticyclonic flow between this and a low in the Weddell Sea. It was in this strong gradient that the wind maximum was located. This had the characteristics of a polar jet stream, through the presence of anticyclonically curved cirrus on its polewards side and a low-level baroclinic zone.

As discussed in the previous section, weather systems move southwards over the continent when the upper air long-wave pattern has a large amplitude and there is a strong meridional component to the upper flow. Under these conditions, depressions can be steered into the interior while retaining their frontal structure. Such a situation occurred during the period 24–26 May 1988 when a major low tracked southwards from the Weddell Sea, over Marie Byrd Land and onto the Ross Ice Shelf. This movement westwards of a major depression over the conti-

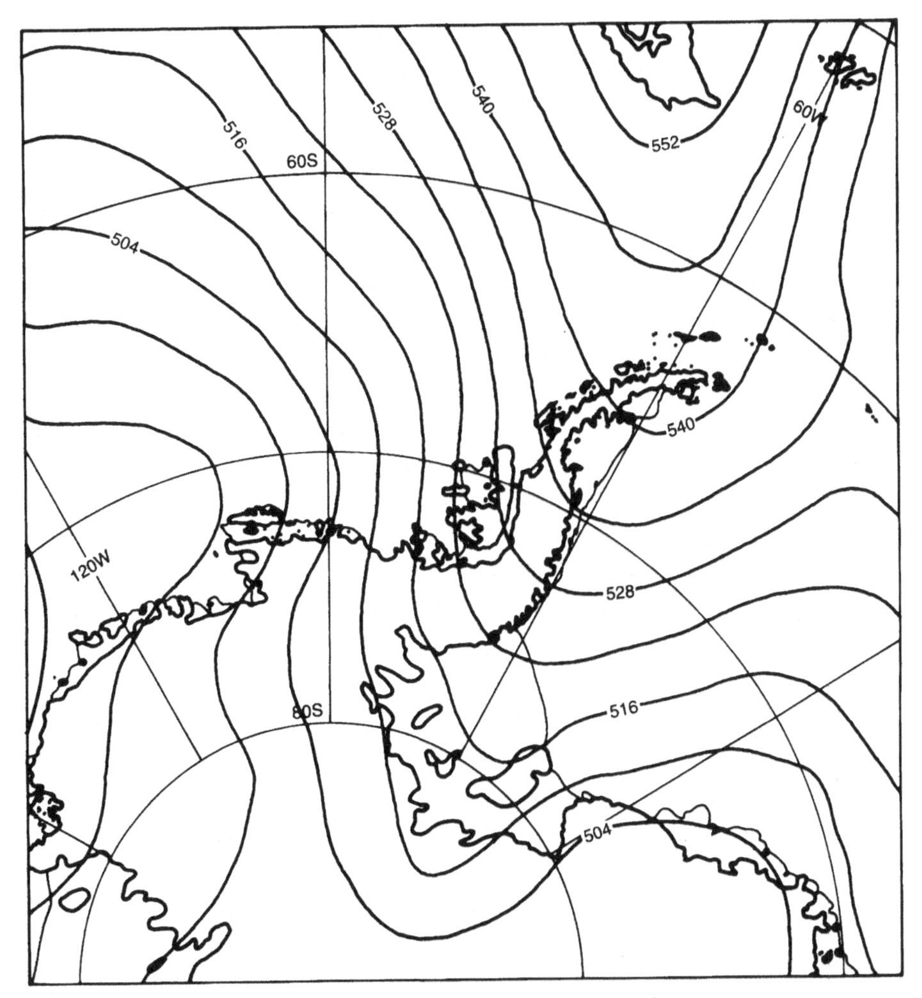

Figure 5.17 The UK Meteorological Office 500 hPa height analysis for 12 GMT 24 May 1988.

nent is rare, but because this is one of the few well-documented cases of a depression over the interior it will be examined in detail in the following section.

A major weather system crossing West Antarctica, 24–26 May 1988

The synoptic situation around the Bellingshausen Sea/Antarctic Peninsula area prior to the development of the low was blocked, with the progression eastwards of depressions impeded. This came about because of an anticyclone over the Peninsula and a barotropic, quasi-stationary, cold-core low-pressure centre over the Amundsen Sea. This high/low pair helped to maintain a strong west–east surface pressure gradient across the Bellingshausen Sea and northerly flow at the mid- and lower tropospheric levels. The upper level flow can be seen from Figure 5.17, which shows the UK Meteorological Office 500 hPa height field for 12 GMT 24 May 1988. At this time there was a ridge over the Antarctic Peninsula that was feeding warm air down the Bellingshausen Sea on its western flank and into the interior of the continent. It was in this strong northerly flow that the low developed and moved into the Antarctic.

The low that tracked across West Antarctica formed in a baroclinic zone over the Bellingshausen Sea that lay north–south close to 90° W at the boundary between cold air to the west and milder air that had arrived from South America to the east. A number of waves developed on this front, but the low of interest here was analysed near 67° S, 104° W on the surface pressure chart for 00 GMT 24 May, which is shown in Figure 5.18. By 00 GMT 25 May the low had moved southwards to 74° S, 100° W and the satellite imagery for that time shows an active front associated with the low that extended across the base of the Antarctic Peninsula. TOVS temperature profiles retrieved around the low support the analysis of the cloud band as a cold front, although, because of the problems in processing TOVS data over high ground, it was not possible to determine any additional information on the structure of the front.

Observations from the 'Siple' AWS located at 76° S, 84° W allowed the

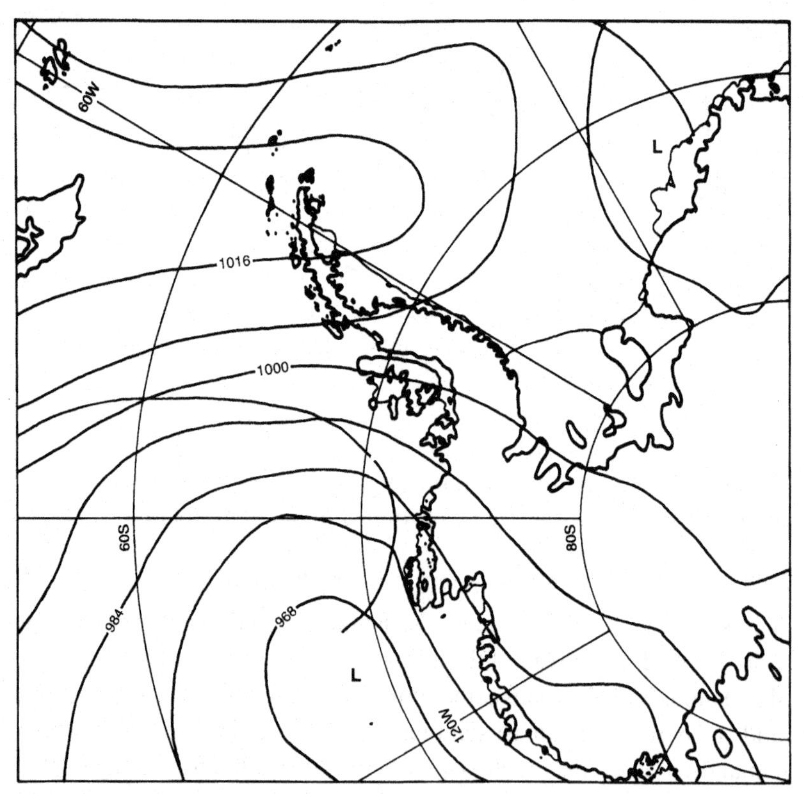

Figure 5.18 The UK Meteorological Office surface pressure analysis for 00 GMT 24 May 1988.

progress of the low to be followed into the interior of the continent. These data, which are summarised in Figure 5.19, show that at this location the surface pressure dropped by about 20 hPa from 15 GMT 23 May to 3 GMT 25 May as the system approached from the north. Before 00 GMT 24 May the wind was around 5 kt from the southwest and the surface temperature close to −38°C. Upon the passing of the cold front between 00 and 03 GMT 24 May the wind backed to southeasterly and increased to 10 kt. Although the front passed over the AWS at about this time, the surface pressure continued to drop for a further 24 h as the low-pressure centre, which was located behind the front, moved over the continent. As the front approached Siple the cloud cover increased and the surface air temperature rose by almost 10°C.

Satellite imagery for 7 GMT 25 May 1988 (Figure 5.20) shows the front as a hook of cloud located over Marie Byrd Land with the cloud band extending from the eastern Ross Ice Shelf, across the Ronne Ice Shelf and northwards along the eastern side of the Antarctic Peninsula. A diagram showing the location of the cold front at various times on 25 and 26 May is shown in Figure 5.21. As can be seen in Figure 5.20, the cloud band associated with the low extended over a great distance and the tail crossed Rothera Station on the western side of the

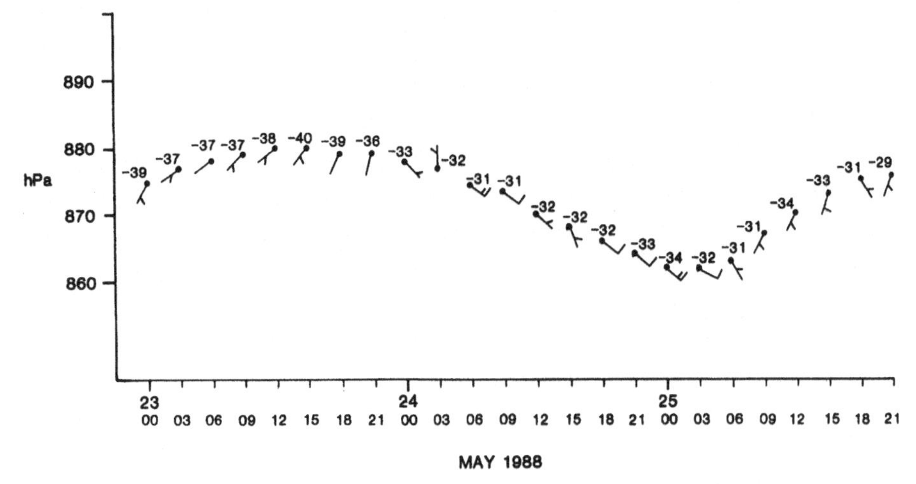

Figure 5.19 Observations of pressure, temperature and wind speed/direction from the 'Siple' AWS during the passage of a depression.

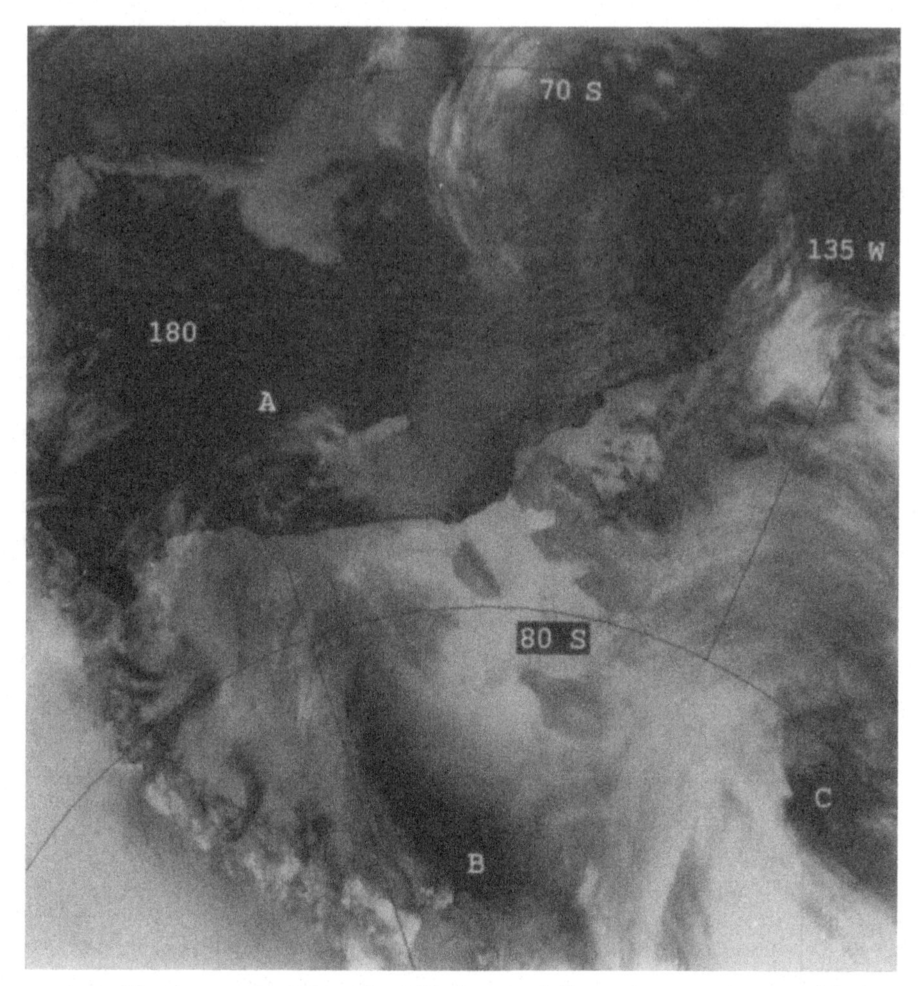

Figure 5.20 An infra-red satellite image of a low-pressure system over Marie Byrd Land, West Antarctica, at 0711 GMT 25 May 1988.

Antarctic Peninsula some 24 h after the forward section crossed Siple. The pressure minimum at Rothera occurred at 00 GMT 25 May when the station experienced continuous moderate snow and a wind speed maximum of 24 m s^{-1} at 06 GMT.

Subsequent satellite images showed that the cold front moved along the Transantarctic Mountains towards Victoria Land, becoming stationary over Terra Nova Bay by the early part of 26 May with the cloud band dissipating during the 27 May. Even in its declining phase the low-pressure centre was playing an important role in the synoptic environment of West Antarctica as it merged with a pre-existing mesoscale low on the Ross Ice Shelf, which became more vigorous and developed a significant comma-shaped band of cloud. The low which had crossed Marie Byrd Land also played a further part in the development of the mesocyclone through the presence of the baroclinic zone

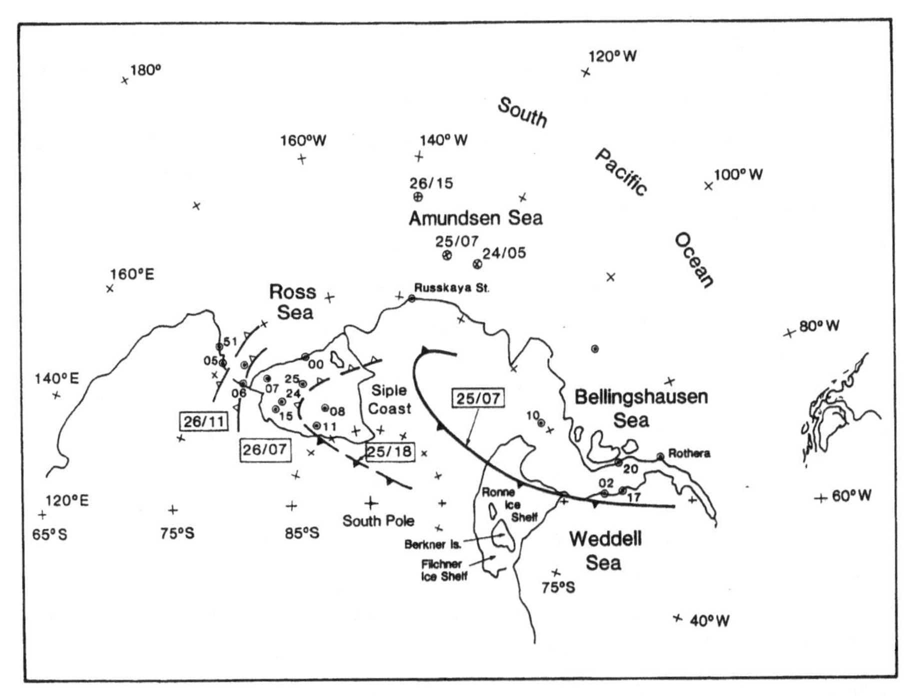

Figure 5.21 The location of the cold front as it crossed Marie Byrd Land, West Antarctica, at various times on the 25 and 26 May 1988.

associated with the system that aided the spin-up of the sub-synoptic-scale vortex.

The warm air associated with the low was detected on the Ross Ice Shelf via the automatic weather station network and in the radiosonde ascents from McMurdo Station. One interesting aspect of this case is that, when the synoptic-scale low arrived on the Ross Ice Shelf and the mesocyclone was intensified by the baroclinic zone, the meridional temperature gradient was reversed, with the warmest air that had traversed Marie Byrd Land being closest to the pole and colder air further to the north.

This case illustrates that, far from being an area of cyclolysis only, the Antarctic has major cyclonic developments taking place that can penetrate well into the interior of the continent. It also shows that interactions between systems on the meso- and synoptic scale take place that will be a challenge for numerical models to represent correctly.

5.3.5 Fronts

Satellite imagery has shown that the major depressions over the mid-latitude areas of the Southern Ocean have a frontal structure that fits the well-known Norwegian model proposed in the early part of the twentieth century. However, by the time these systems have moved polewards and become slow-moving in the coastal area, many are occluded and have much less distinct cloud signatures and lower cloud top temperatures than they had when they were in their most vigorous phase. Although Figure 5.1 shows an active frontal depression, the vast majority of systems in the coastal region have frontal cloud bands far less distinct than those in this image. An example of a deep low with disorganised cloud structure is shown in Figure 5.22(a). The corresponding surface analysis for this time is shown in Figure 5.22(b). The satellite image is typical of many large lows around the Antarctic and highlights the problems facing the operational analyst.

As discussed in Section 5.3.1, the central pressure of many lows is still quite low when they are off the coast, resulting in the climatological circumpolar trough that appears in the mean surface pressure charts. However, the fronts associated with these lows are much less active than those found at mid-latitudes, so most of the precipitation which falls at coastal stations is of slight intensity.

(a)

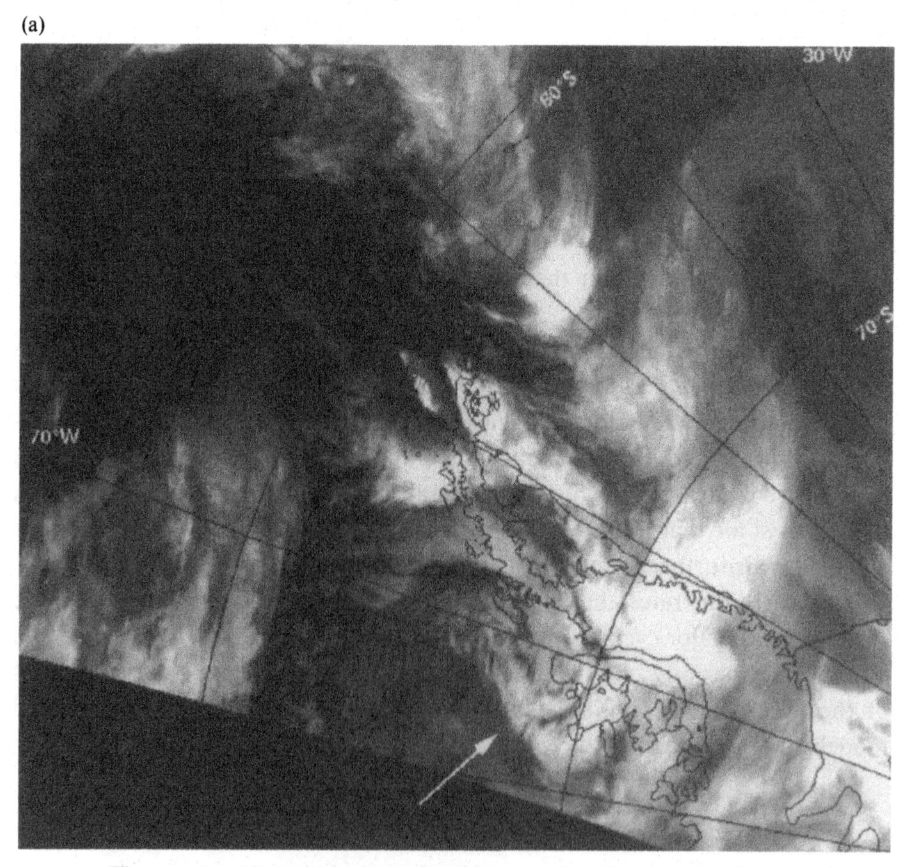

Figure 5.22 (a) An infra-red satellite image of a deep low with disorganised cloud in the southeast Bellingshausen Sea at 21:13 GMT 30 July 1994. (b) The corresponding UK Meteorological Office surface analysis for 00:00 GMT 31 July 1994.

In fact the large amount of low cloud that is found in the coastal area means that it can be difficult to determine the frontal positions from satellite imagery, although the ability to observe the data as a 'movie-loop' on a computer system greatly aids the process of interpretation.

The strength of the fronts near the coast is usually dependent on the upper level flow and the extent to which warm air masses are brought polewards from mid-latitudes. This in turn is dependent on the amplitude of the long waves that are necessary to maintain the strength of the baroclinicity through a substantial depth of the atmosphere. An example of an active warm/cold frontal system is shown in Figure 5.23(a) on satellite imagery and in the corresponding 1000–500 hPa thickness field. This case occurred over the Bellingshausen Sea when the long waves were amplified and there was pronounced warm air advection over the western side of the Antarctic Peninsula. The warm air coincides with the mass of cloud over the eastern Bellingshausen Sea and the western side

(b)

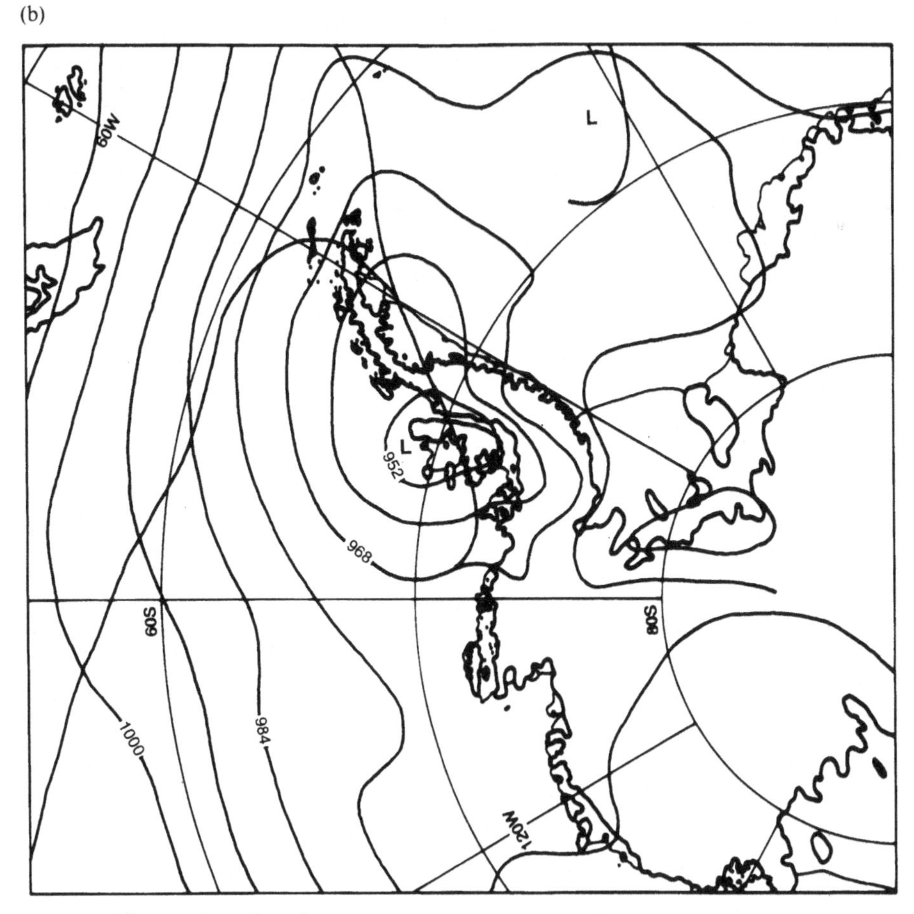

Figure 5.22 (*cont.*)

(a)

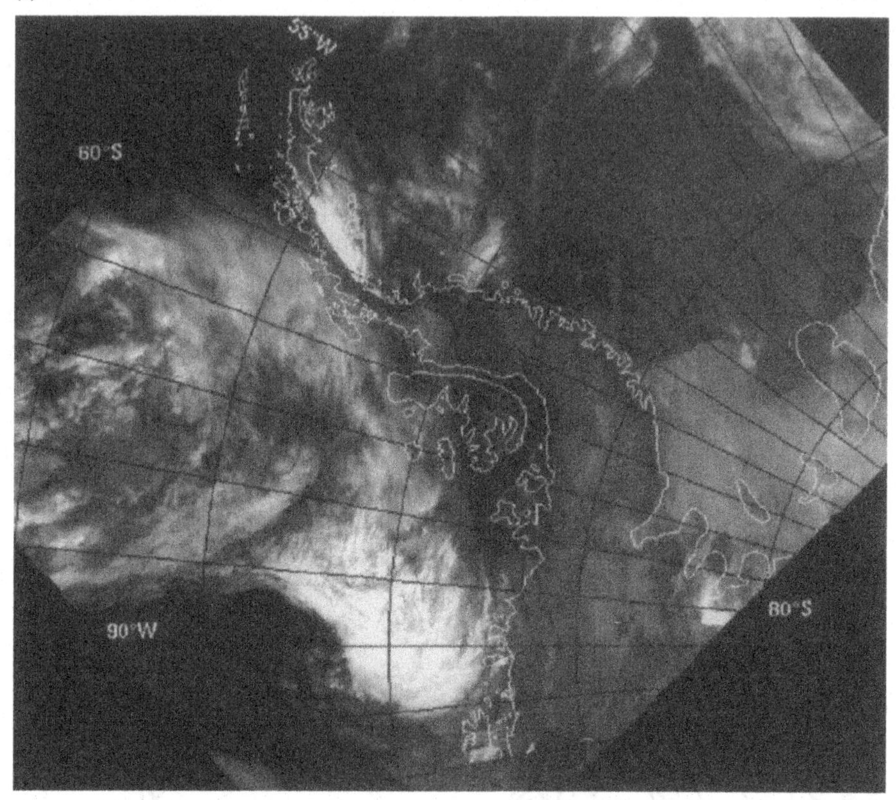

Figure 5.23 A case of an active frontal system over the Bellingshausen Sea on 1 March 1993 illustrated via (a) the infra-red satellite imagery for 10:23 GMT and (b) the 1000–500 hPa thickness field for 12:00 GMT as analysed by the UK Meteorological Office.

of the Peninsula, but the exact position of the surface warm front cannot easily be determined from the imagery alone because of the large amount of high cloud. However, examination of the surface observations from the research stations on the Peninsula suggests that it crossed Rothera Station on Adelaide Island at about 22:00 GMT 1 March 1993.

Many of the depressions developing over the Southern Ocean do so in cold air outbreaks well removed from the major frontal zones and have a single hook of cloud in their circulation and no conventional warm and cold frontal structure. Although this comma of cloud indicates slantwise ascent up a frontal surface, the temperature difference across the cloud band is usually small and has developed as a result of minor temperature variations in the cold air outbreak. Similarly, developments on fronts near the coast or at the northern limit of the ice edge tend to result in vortices with rather weak cloud signatures because the temperature gradient across the front is often not especially strong.

(b)

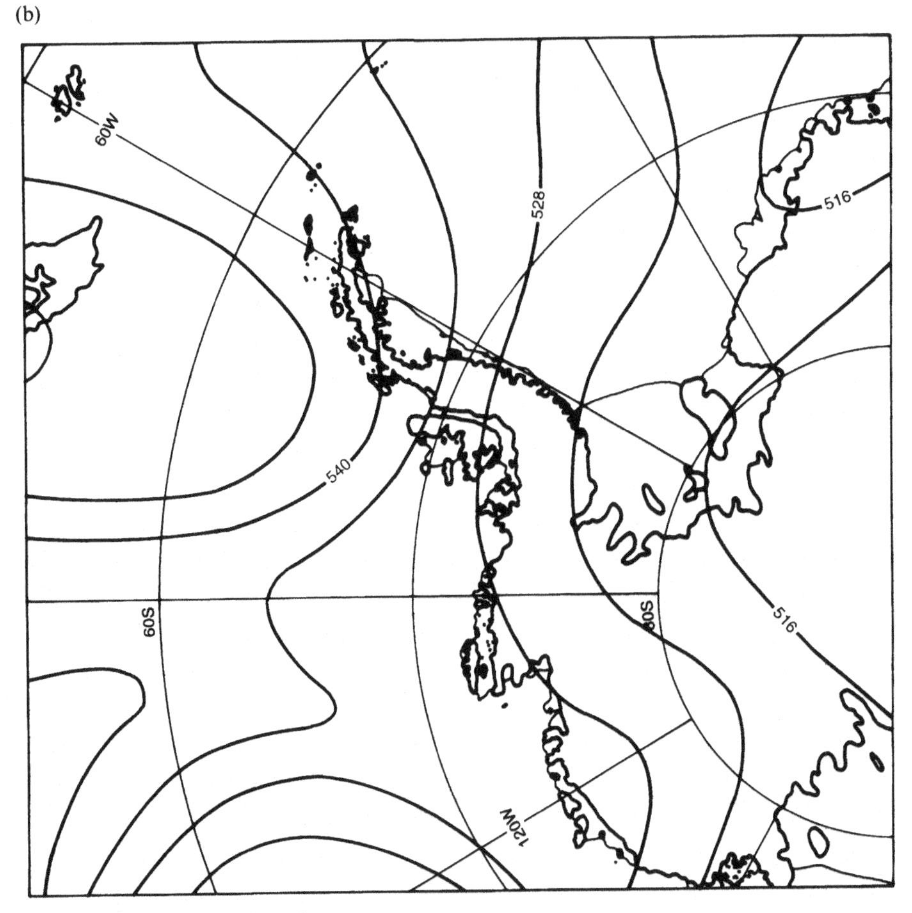

Figure 5.23 (*cont.*)

The role of fronts in cyclogenesis over the western Weddell Sea

The north–south barrier of the Antarctic Peninsula has a major impact on depressions and associated fronts moving from west to east in this area. Inspection of satellite imagery has shown that there are many depressions in the eastern Bellingshausen Sea that have either moved into the area from the South Pacific or developed near the coast or edge of the sea ice. Some of the more vigorous depressions cross the Peninsula with relatively little change to their structure, but many lows become slow-moving in the Bellingshausen Sea. However, fronts on the eastern side of such lows often rotate around the low centre, become orientated in an east–west direction and pass southwards down a considerable length of the Peninsula. Because the low-pressure centres usually remain quasi-stationary in the Bellingshausen Sea, the fronts often become slow-moving in the area of the mid-Peninsula. However, the fronts are an important catalyst in triggering cyclogenesis in the lee of the Peninsula, where the combined effects of vortex stretching, as a result of air flow over the topographic barrier, and the pre-existing convergence in the vicinity of the front results in frequent cyclogenesis events. The outcome is often that the original depression over the Bellingshausen Sea dissipates and the new vortex over the Weddell Sea develops and moves away to the east. An example of a cyclogenesis event taking place on a pre-existing frontal cloud band over the western Weddell Sea is shown in Figure 5.24.

Fronts over the continent

As discussed in earlier sections of this chapter, some depressions do penetrate into the interior of the continent and retain some frontal structure that is apparent in satellite imagery. So far there have been no long-term climatological investigations of the occurrence or structure of such systems but it is very

(a)

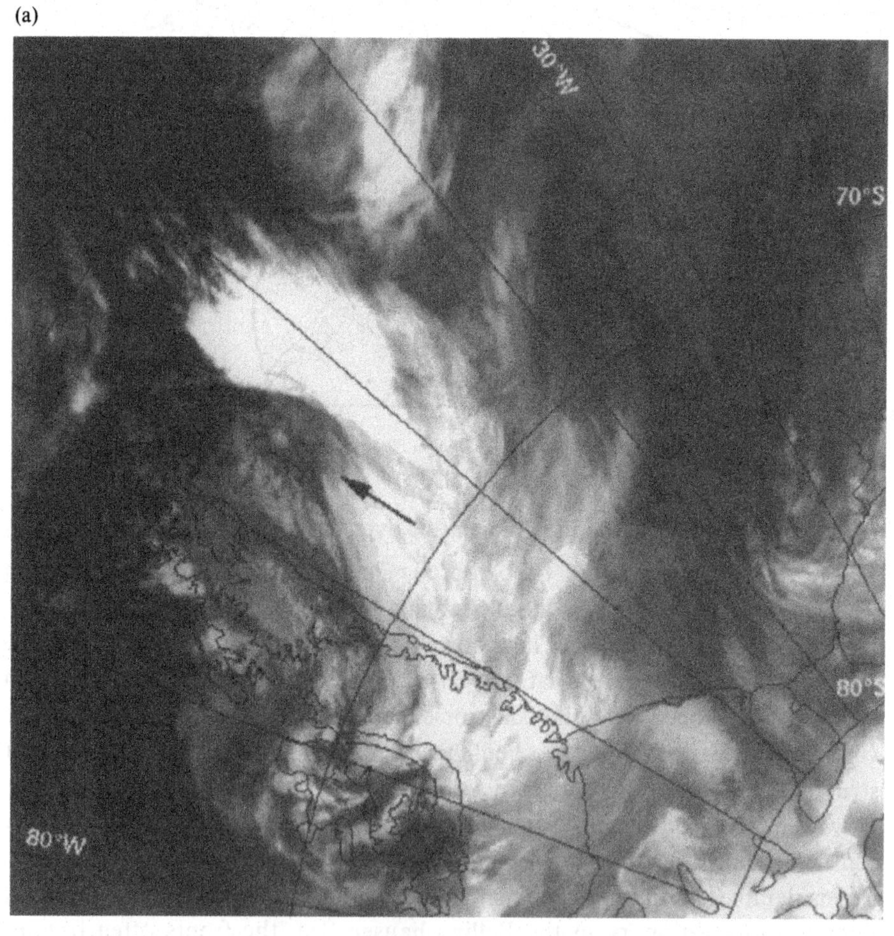

Figure 5.24 Two infra-red satellite images illustrating cyclogenesis in the lee of the Antarctic Peninsula on a pre-existing frontal band. The images were taken at (a) 2:34 GMT and (b) 21:00 GMT 31 July 1994.

unlikely that many have the warm, cold and occluded frontal bands of mid-latitude weather systems. From the studies of sequences of satellite images that have been carried out, it is more likely that the case of late May 1988 presented in Section 5.3.4 is more typical, with the depression having a single band of cloud. However, case studies need to be carried out before generalising about such systems.

Because there are very few radiosonde ascents over the continent itself, and the spacing between stations with upper-air programmes is often several thousand kilometres, it is not possible to produce detailed upper air analyses from *in situ* measurements. However, advances in processing of satellite sounder data have meant that some thickness fields over the more low-lying parts of the continent have been determined in the context of research projects into tropospheric processes over the Antarctic. The case study of a mesocyclone in the

(b)

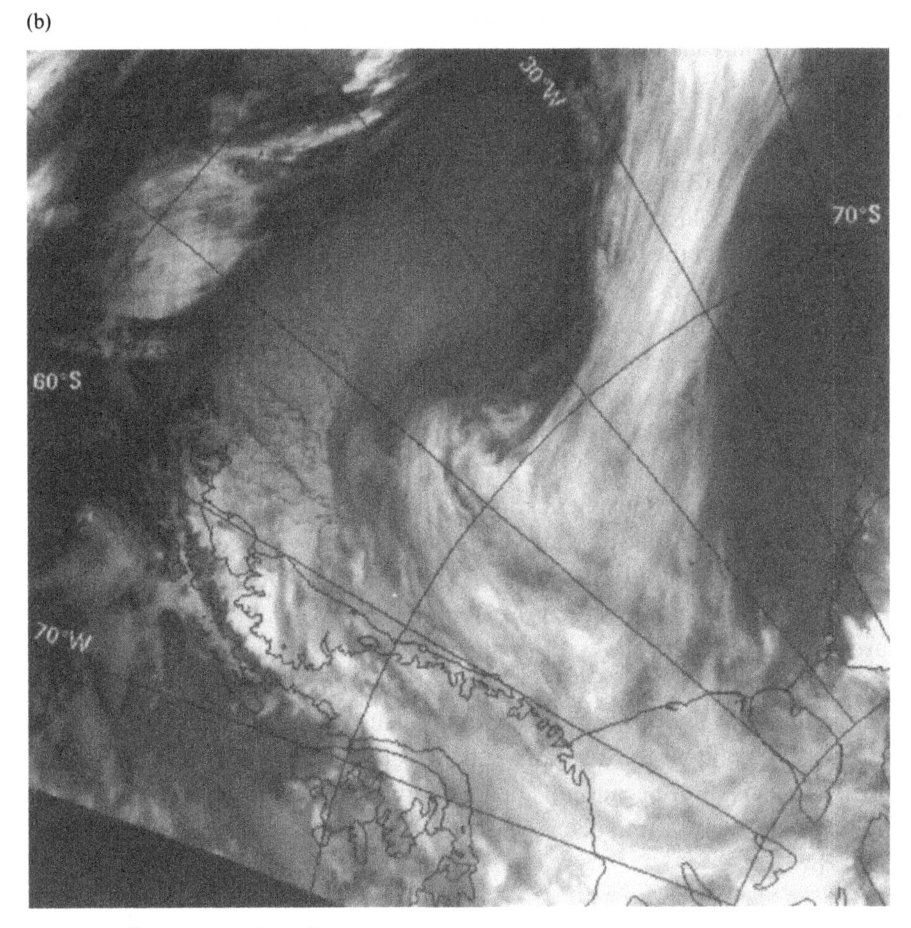

Figure 5.24 (*cont.*)

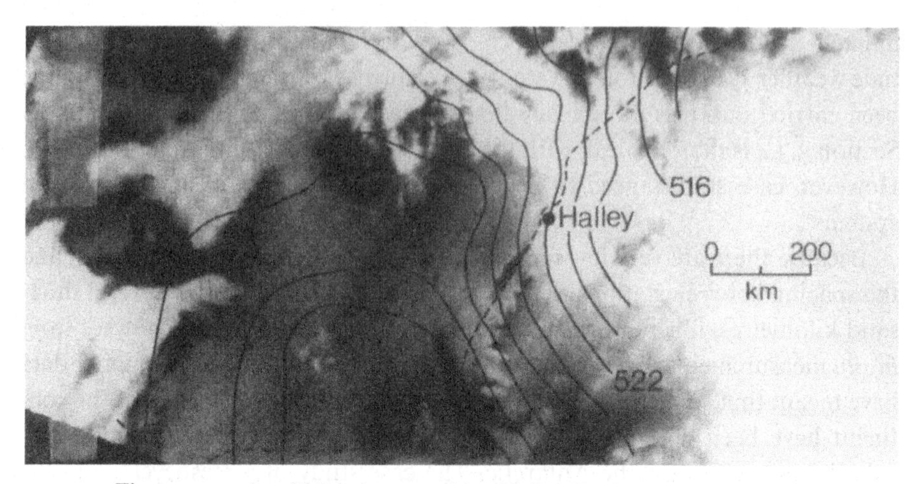

Figure 5.25 A cold front approaching Halley Station from the interior of the continent at 23:49 GMT 3 January 1986. The infra-red satellite imagery is overlaid with contours of 1000–500 hPa thickness determined from TOVS sounder data.

coastal region that is discussed in Section 6.4.3 reveals that fronts with relatively limited horizontal length can develop in the interior of the continent at the leading edge of cold air outbreaks. In this case the very dry conditions over the plateau meant that the front had no cloud associated with it, but it was detected using water vapour imagery and observations from Halley Station. A 1000–500 hPa thickness field produced from TOVS data for this case is shown in Figure 5.25. These data show the cold front approaching Halley Station from the interior of the continent at the leading edge of a cold air outbreak associated with an upper trough extending from a deep, barotropic low to the north of SANAE station. The TOVS 6.7 μm water vapour imagery revealed a 'dark band' of dry upper tropospheric air, such as can be observed in the vicinity of cold fronts in the mid-latitude areas. The satellite data showed this cold front to be only several hundred kilometres in horizontal extent, although it merged with a mesocyclone and existed over the eastern Weddell Sea for about three days. The surface data showed that the front was associated with a weak pressure trough while the imagery revealed an intrusion of very dry upper tropospheric air to lower levels behind the front. Such minor fronts are hard to detect in routinely available surface and satellite data but are probably worthy of further investigation because of their role in coastal cyclogenesis events. This case has been discussed in detail by Turner and Ellrott (1992) and Turner *et al.* (1993a).

Minor fronts in cold air masses

Outbreaks of cold, continental air moving northwards over the ocean can usually be identified on satellite imagery by virtue of the large number of

Figure 5.26 Two minor fronts in a cold airstream over the Drake Passage as seen in infra-red satellite imagery for 5:53 GMT 5 June 1993.

shower clouds that give the imagery a mottled appearance in the visible and infra-red data. This is apparent in Figure 5.1, in which the showery airstream can be seen to the west of the depression centre located over the Bellingshausen Sea. In major outbreaks of cold air that extend a significant distance in the north–south direction, the temperature gradient often is not uniform but rather is concentrated into one or more narrow zones. These are apparent as bands of cloud lying perpendicular to the wind direction and are essentially cold fronts of weak or moderate strength. Such cloud bands in the high-latitude areas of the Northern Hemisphere are often called Arctic Fronts and have been discussed by Rasmussen *et al.* (1992). A comparable pair of fronts over the Drake Passage is shown in Figure 5.26. Here the differences between the air masses on either side of the fronts are apparent from the different types of convective cloud.

5.4 Climatology

Our present knowledge of the formation, movement and dissipation of synoptic-scale systems has developed slowly during the last 40 years as conventional observations, satellite data and accurate numerical analyses of the meteorological environment of the Southern Hemisphere have become available. Before the late 1950s, the only observations available came from the occasional expedition to the Antarctic or from ships operating in the Southern Ocean. The limited amount of data available from these sources meant that only the broadest features of the circulation could be determined. Nevertheless, the available ship reports were used in studies, such as that of Lamb and Britton (1955), who considered the circulation during the summer and autumn months of 1947 and 1948, to produce maps of depression frequency over the whole southern hemisphere. Surprisingly, considering the data available, these charts of depression location agree broadly with the maps produced later when far more comprehensive data sets were available. The International Geophysical Year of 1957/58 resulted in the establishment of a number of new observing stations in the Antarctic and the availability of data caused a revival of interest in the general circulation of the Southern Hemisphere. However, it was the arrival during the 1960s of imagery from polar orbiting satellites that provided the most important tool for monitoring weather systems in areas sparse in data. With their frequent coverage of all parts of the Southern Hemisphere, it at last became possible to assemble accurate, long-term records of the development and movement of all weather systems on scales greater than a few tens of kilometres. By the 1980s the performance of global numerical models had improved to the point that their output could also be used to examine the distribution of cyclogenesis and cyclolysis events and the locations of the main cyclone tracks. With the development of automatic techniques for identifying depression centres in digital model output and tracking the movement of the lows between fields 12 or 24 h apart (Murray and Simmonds, 1991a; 1991b), the extremely time-consuming task of manually analysing long series of charts should no longer be necessary. However, such schemes have mainly been tested on series of operational analyses and, as considered in Section 5.5, the ability of these models to represent correctly the wide range of depressions found in the Antarctic is still in question.

Here we examine our present knowledge of the climatological occurrence of extra-tropical cyclones and anticyclones using observational data. The first results of the automatic detection of depressions and anticyclones in operational analyses are also introduced and compared with the purely observational investigations. Although there are no continuous observational data sets spanning many years with which to examine the inter-annual variability of the systems, the studies that have been carried out for periods of several years are considered to show the degree of variability in the circulation around the Antarctic.

5.4.1 Climatological aspects of cyclogenesis

The first major effort to use multi-year satellite imagery to determine the locations where depressions were forming was carried out by Streten and Troup (1973). Their study was based on visible wavelength data for the period 1966–69 from the ESSA series of satellites that had been combined into hemispheric mosaics. They considered the vortex positions during the summer (December–March), winter (June–September) and the intermediate months (April and May, October and November). Such a division of the year had been used previously by Taljaard (1967) and is in line with the cycle of sea surface temperatures. Their study found many depressions forming in mid-latitudes but few in the circumpolar trough as the satellite imagery at that time had a poor radiometric and spatial resolution. Because of low levels of illumination during the winter they could not carry out an analysis of vortices during that season. However, Carleton (1979) undertook a similar study of cyclone developments for the winter months over the period 1973–77 using infra-red imagery from the USA's ITOS satellite. These two studies remain the only multi-year, satellite-based investigations of cyclogenesis over the extra-tropical Southern Hemisphere carried out to date. Their results were in broad agreement with the analysis of cyclogenesis regions by Taljaard for the IGY period, although the greater incidence of cyclogenesis within the circumpolar trough was not found in the earlier study.

Because of the lack of a year-round analysis of cyclogenesis based on satellite imagery, we will consider developments determined from the assessment of operational analyses carried out by Jones and Simmonds (1993). Their results showing cyclogenesis events in the four seasons for the period 1975–89 are illustrated in Figure 5.27. They found that depressions developed throughout the Southern Hemisphere, with the exception of the most isolated interior parts of the Antarctic continent. Here satellite imagery has shown minor, mesoscale developments taking place but these would not be represented in the model fields. The greatest number of cyclogenesis events were found in the 55° S to 65° S band, which is at the latitude of the circumpolar trough. Earlier studies had placed the latitude of maximum cyclogenesis further to the north, although this was probably the result of a number of factors, including the poor quality of the analyses when the work took place, the problems of using satellite imagery at high latitudes and the limitation of specifying a minimum lifetime for depressions in some investigations. However, during some periods even surface data alone have shown considerable cyclogenesis at high latitudes. This was the case during July 1958 (Taljaard, 1965) when many new lows were observed in the high-latitude areas of the Indian and Pacific Oceans. Within the circumpolar trough there is a maximum of cyclogenesis in the spring with a minimum during the summer months, although this situation reverses to the north of the trough

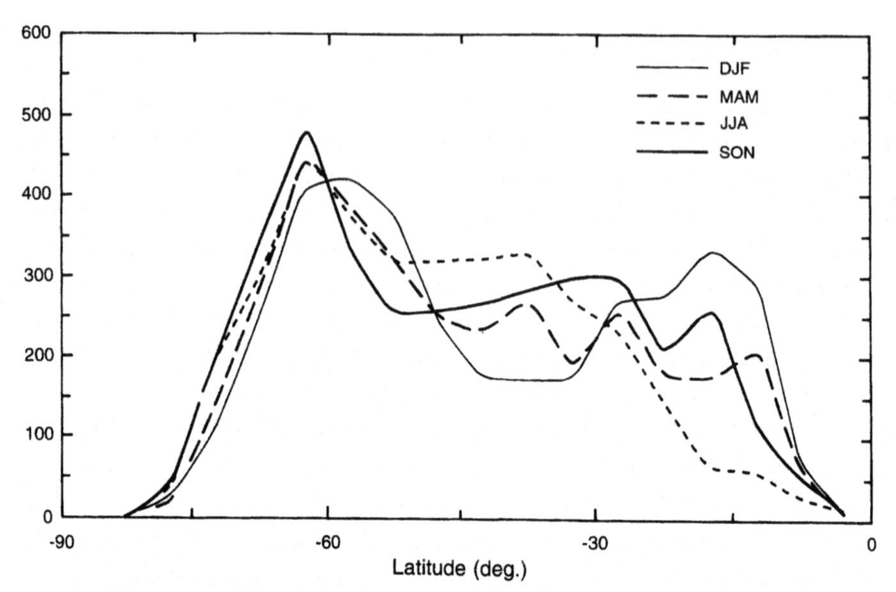

Figure 5.27 The total count of cyclogenesis in 5° latitude bands for summer, autumn, winter and spring, for the period 1975–89. The units are cyclones. After Jones and Simmonds (1993).

in the 60° S to 50° S band. Near to the latitude of the polar front, close to 40° to 45° S, there is a clear maximum of cyclogenesis during the winter.

The studies of cyclogenesis based on satellite imagery allowed the distribution of different types of newly developed vortices to be examined. They showed that comma cloud developments occurred farther south than the frontal waves in all seasons.

The geographical distribution of vortex development events will be considered via the maps of density of cyclogenesis for the summer and winters seasons produced by Jones and Simmonds and shown in Figure 5.28. The corresponding chart produced by Carleton (1979) on the basis of hemispheric mosaics of relatively coarse-resolution satellite imagery is shown for comparison in Figure 5.29. The units on the charts based on the satellite and analysis fields are different, but they allow the intercomparison of areas of high and low cyclogenesis derived from these two very different forms of data.

The main areas of cyclogenesis as determined in the Jones and Simmonds study are the following.

(i) A belt spanning the circumpolar trough and the high-latitude parts of the Southern Ocean. In this zone there are more depressions forming than in all other part of the hemisphere, except the subtropical land areas, where thermal lows are very common. However, as can be seen in hemispheric mosaics of satellite imagery, such as that in Figure 5.15, the systems developing at high latitude are usually not major frontal depressions,

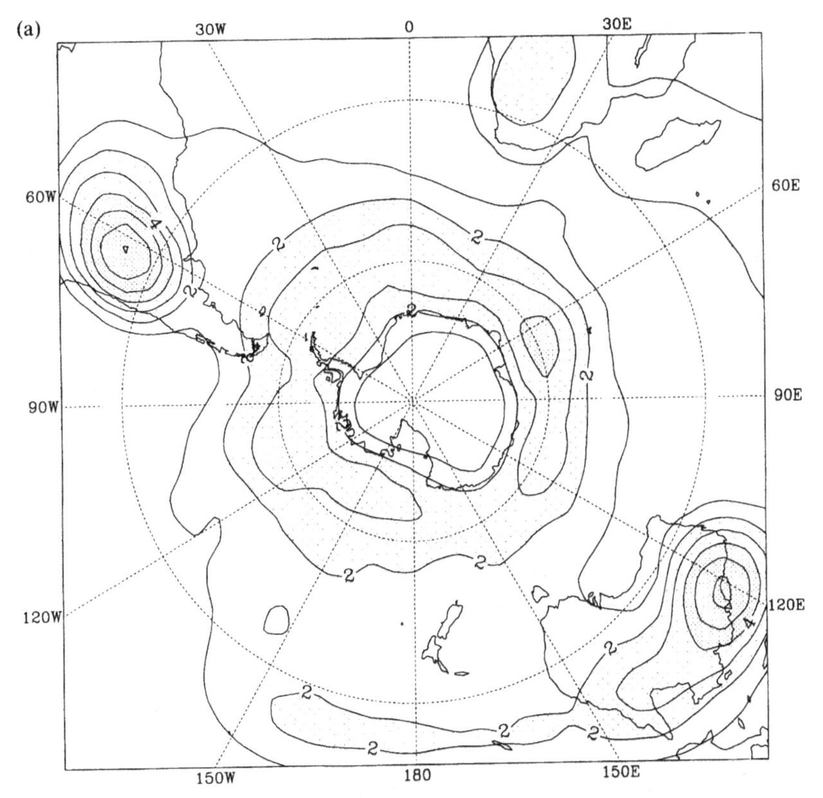

Figure 5.28 The density of cyclogenesis for (a) summer and (b) winter. The contour interval is 1.0×10^{-4} cyclones per day (degree of latitude)$^{-2}$. Light and heavy stippling denotes areas above 2.0 and 4.0 respectively. From D. A. Jones and I. Simmonds, A climatology of Southern Hemisphere extratropical cyclones, *Climate Dynamics*, **9**, Figure 4, 136, 1993. Copyright (1993) Springer-Verlag.

which tend to occur further to the north, but rather systems with more disorganised and lower cloud. Examination of high-resolution satellite imagery of the circumpolar trough shows that many of the large depressions in this area turn into multi-centred lows that contribute to the large number of developments taking place. The positions of the maxima of cyclogenesis within the circumpolar trough are most pronounced in winter and are located in the Ross Sea, near the Greenwich Meridian and south of the Indian Ocean. The winter season satellite data analysed by Carleton show these three centres, but the maxima are displaced a little further to the north.

(ii) Over the southwest Atlantic and the Drake Passage region. The Australian analyses show little change between the summer and winter seasons, there being only a decrease in the number of developments in winter over the southeast Pacific. Carleton found the maximum over the

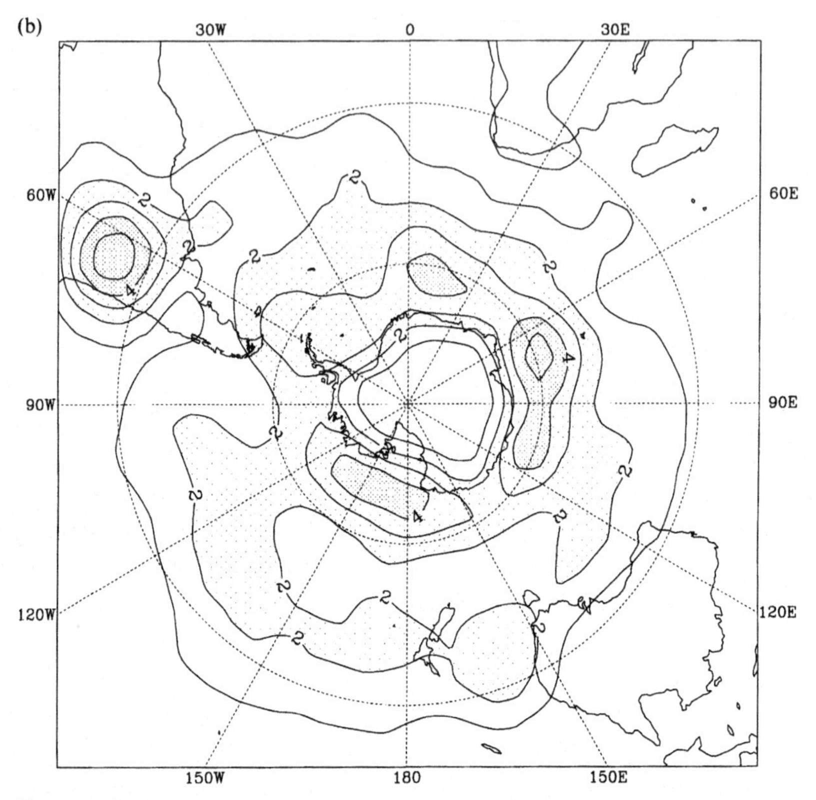

Figure 5.28 (*cont.*)

Drake Passage and noted that developments in this area consisted predominantly of waves on frontal bands with few 'inverted comma' formations. This was attributed to the frequent developments taking place over the land, which produce waves on fronts rather than the commas evolving out of cumulus fields over the ocean areas.

(iii) In the central and northwestern South Pacific. In the northwest Pacific cyclogenesis takes place more frequently than at the same latitude in other parts of the hemisphere. This area is particularly active in winter.

(iv) South and southwest of Australia. The number of vortices developing in this area is greater in winter than at other times of the year. There is good agreement here between the study based on satellite imagery and the analysis of Australian model data.

(v) In the Tasman Sea during the winter season. Again this area was apparent in the satellite and model studies.

A feature brought out by all the studies of regions of cyclogenesis is the basic hemispheric three-wave pattern of the upper-air long waves, which results in cyclonic developments throughout the year in the South Atlantic, the South

Figure 5.29 The mean distribution of the 'early development' (W, A, B (see Section 5.3.1)) vortices for the five winters 1973–77. Isopleth values refer to normalised vortex frequencies in each 5° latitude by 10° longitude tessera. After Carleton (1979).

Pacific and the Indian Ocean. This was the case during the winter of 1979, which has been studied intensively since it was part of the FGGE period. Physick (1981) examined cyclogenesis regions using high-quality analyses from the project and reported significant regions of cyclogenesis in the above three regions in association with troughs at 110° E, 120° W and 40° W.

Although recent studies have highlighted the large number of developments taking place at high latitudes, it is still not clear how important these systems are. In their study of the synoptic origins of precipitation over the western Antarctic Peninsula, Turner and Colwell (1995) found that half the precipitation reported came from depressions that had developed in the Antarctic. In contrast, Neal (1972) found that lows that formed in the circumpolar trough rarely became significant systems, although there were exceptions, such as in the Weddell Sea, where some major lows developed in the lee of the Antarctic Peninsula. Clearly further research is required on these subjects using combined observational and model data. Two specific types of vortices investigated via satellite imagery were the frontless systems that corresponded to the Streten and Troup classes Dx and Dy, but without a cloud band. Carleton (1979) mapped the distribution of these systems for the 1973–77 winters and his results are shown in Figure 5.30. It can be seen that the highest density of these systems was in the Tasman Sea and south-west of Australia, but there were also maxima located in the Antarctic coastal region. These were close to the Amery Ice Shelf, north of Dronning Maud Land, in the Bellingshausen Sea and off Wilkes Land. Fewer vortices of this type appear to form in winter than at other times of the year. Since such systems seem to be associated with upper-air cold anomalies and unstable conditions, this may be the

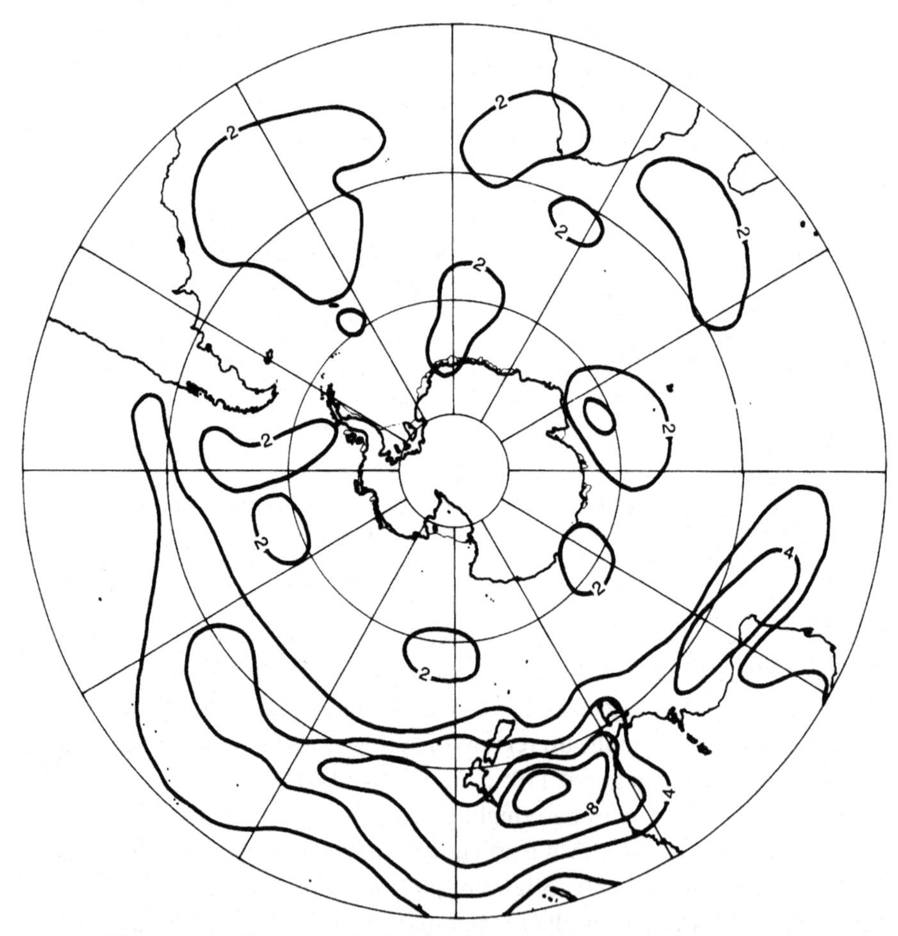

Figure 5.30 The mean distribution of the frontless 'cut-off' (F/G, see Section 5.3.1) vortices for the five winters 1973–77. Isopleth values refer to normalised vortex frequencies in each 5° latitude by 10° longitude tessera. After Carleton (1979).

result of the extensive sea ice cover at high latitudes and the stability of the atmosphere at this time of year.

Another objective study of cyclogenesis and cyclone activity across the Southern Hemisphere based on model fields has recently been carried out (Sinclair, 1994; 1995). The results of this work agreed well with those of Jones and Simmonds and confirmed the general picture of depression development, track and areas of dissipation presented here.

Inter-annual variability

The large inter-annual variability of developments around the Antarctic has been known for many years and was noted during the IGY by Alvarez (1958) and Taljaard (1965). However, until long series of satellite images became available it was possible neither to quantify this variability nor to identify regions where the variability was high. The five-winter study by Carleton examined the large inter-annual variability in the locations of cyclogenesis over the Southern Hemisphere and, although the data used did not cover a long period, some conclusions could be drawn. Figure 5.31 shows this through the difference between the highest and lowest number of winter season early development cloud signatures found in 5° latitude by 10° longitude boxes during the period 1973–77. Across the latitude band 40° S to 60° S there are several centres with ranges greater than 20 units, with further centres of ranges exceeding 15 units extending to lower latitudes. The greatest variability is found in four areas, which are over the southern Indian Ocean to the southwest of Australia, in the South Pacific, near the Drake Passage and near the coast of eastern Antarctica, south of New Zealand. Despite this study there is still a great deal of work to be carried out to relate the cyclogenesis taking place around the Antarctic during each year to the major climatic cycles of the Southern Hemisphere and the tropical and mid-latitude circulation.

5.4.2 Depression tracks

The tracks of depression around the Antarctic have been investigated for over 40 years, with the work splitting into three distinct phases. First, during the 1940s and 1950s there were still many ships operating in the mid-latitude areas of the Southern Hemisphere and their observations were used by workers such as Lamb (1959), van Loon (1960) and Astapenko (1964) to produce the first estimates of track locations from these *in situ* data. Secondly, the advent of meteorological satellites in the 1960s provided a powerful tool with which to track depressions by their cloud signatures in visible and infra-red wavelength imagery and the studies described in previous sections resulted in improved estimates of tracks, especially in the more southerly and sparse in data latitudes. Finally, the numerical models covering the Southern Hemisphere had improved by the 1970s

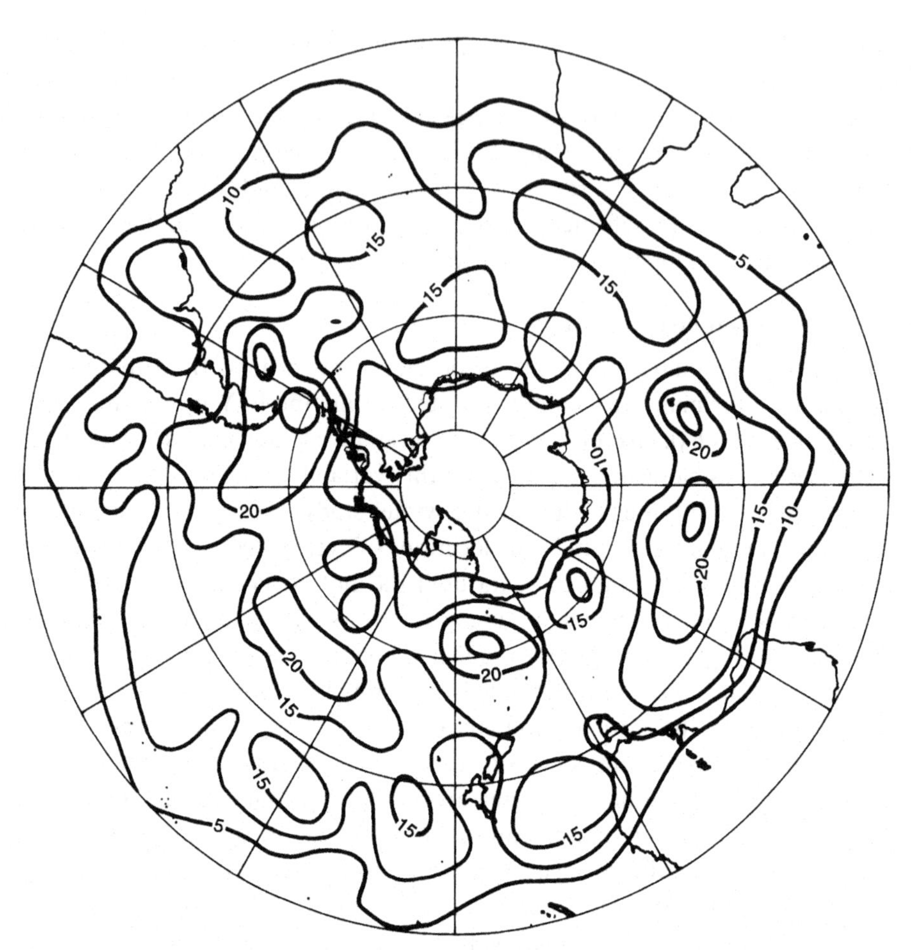

Figure 5.31 The 'range' of 'early development' (W, A, B (see Section 5.3.1)) vortices. Isopleths represent the difference between the lowest and the highest normalised values in each 5° latitude by 10° longitude tessera over the five winters 1973–77. After Carleton (1979).

so that their representation of the general circulation could be used to infer the movement of depressions. All these forms of data will be used in the following section to examine climatological tracks of depressions in the Southern Hemisphere.

The work on depression tracks over the last few years has shown that the paths followed by the synoptic-scale systems around the Antarctic exhibit a high degree of variability, so that simple climatological maps of depression tracks, such as the work of Astapenko (1964), which was based on IGY data, cannot show the very variable pattern of the tracks. We have therefore decided to show the tracks of all winter and summer depressions as determined from the Australian analyses for the period 1975–89 and these are reproduced in Figure 5.32. These tracks link the main areas of cyclogenesis discussed in Section 5.4.1

(a)

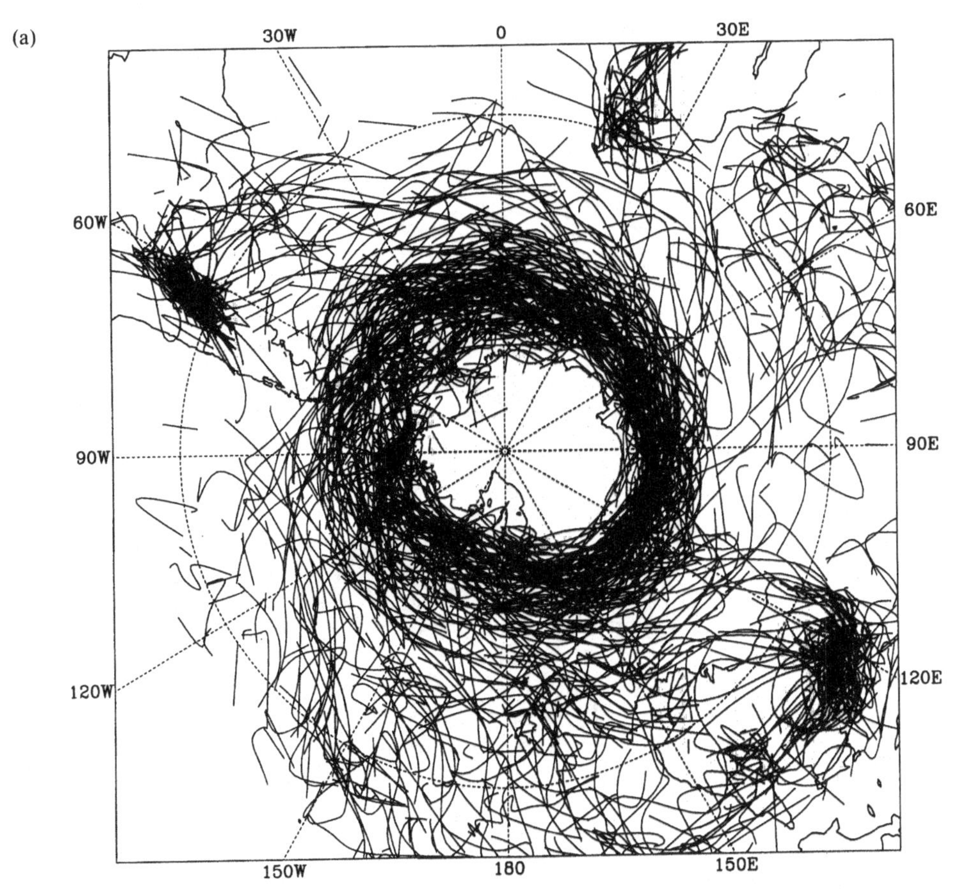

Figure 5.32 Tracks of all depressions for (a) summer and (b) winter for the years 1985–89 inclusive. From D. A. Jones and I. Simmonds, A climatology of Southern Hemisphere extratropical cyclones, *Climate Dynamics*, **9**, Figure 2, 135, 1993. Copyright (1993) Springer-Verlag.

with the regions of dissipation in the circumpolar trough. These tracks result from a combination of mid-latitude systems spiralling into the high Southern Ocean area and systems that have developed in the circumpolar trough and are moving parallel to the coast. This merging of the lows from high and mid-latitudes results in the large number of tracks straddling the circumpolar trough, where the lows move in a more eastward direction than do those further to the north. Many of the lows can be long-lived, lifetimes of up to ten days having been found by Neal (1972), during which period they can cover distances of up to 7000 km.

An association of depressions with embayments in the coast of the Antarctic has been suggested for many years and noted in IGY data (Taljaard, 1967) and in the analyses produced in the GARP Basic Data Set Project (Neal, 1972). However, this does not stand out in the summer and winter season track

(b)

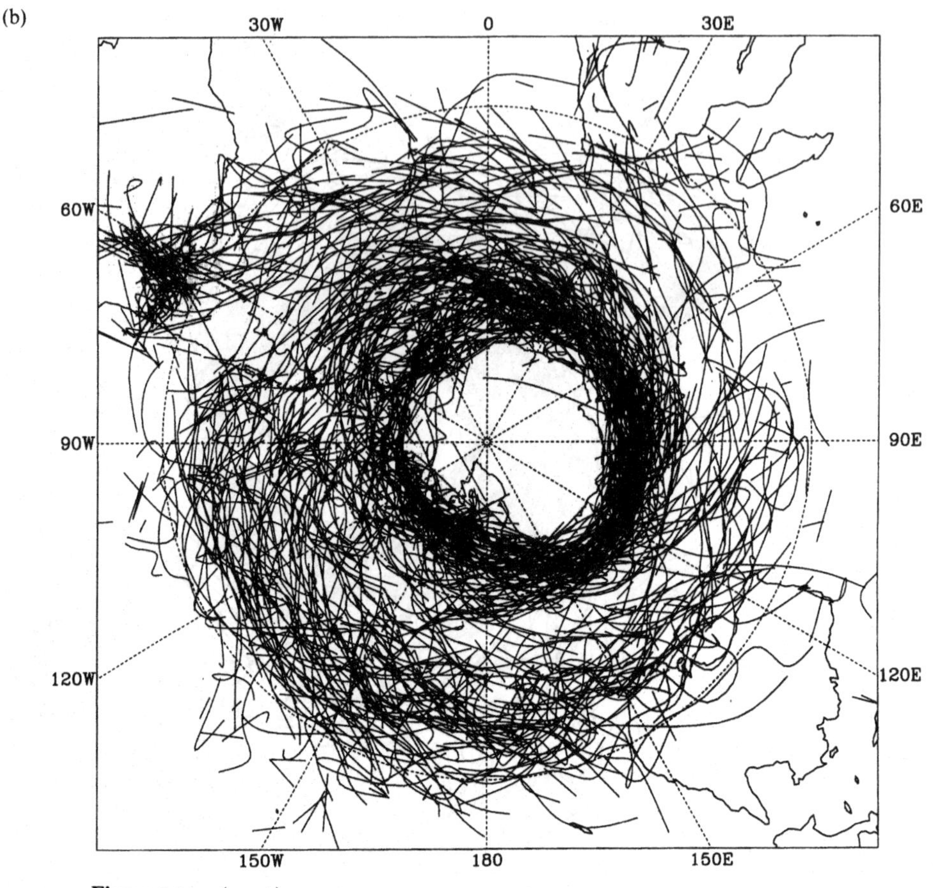

Figure 5.32 (*cont.*)

diagrams from Jones and Simmonds (1993) shown in Figure 5.32. This is probably a result of these diagrams having been produced from the output of numerical analysis schemes which still have a poor representation of the Antarctic topography. For example, the figures show a large number of depressions crossing the Antarctic Peninsula from the Bellingshausen Sea to the Weddell Sea. However, we know from examination of satellite imagery that this 2 km high topographic barrier is a major obstacle to the progression eastwards of depressions and causes many systems to stagnate in the eastern Bellingshausen Sea. Nevertheless, these figures are very effective in illustrating the very chaotic form of depression tracks in the Southern Hemisphere and also in showing that the very generalised maps of tracks produced from IGY data are an oversimplification of the situation.

Although the direction of travel of depressions is generally towards the Antarctic, Sinclair (1993) has suggested that a significant number of systems initially move equatorwards soon after formation and before they then turn polewards. This characteristic has been attributed to the large number of cold air disturbances that are involved in cyclogenesis events either through the evolution of small disturbances into major extra-tropical depressions or through the merging of waves on frontal bands and cold-air vortices through the 'instant occlusion' process. In both cases the formation in a southerly flow imparts an initial equatorwards component to the larger vortices; a feature that has been observed in ECMWF model fields. This can be observed in some of the depression tracks shown in Figure 5.32.

When depressions move south of the mean position of the circumpolar trough and enter the Antarctic coastal region they can re-curve and begin to move westwards or cross the coast onto the continent itself. Lamb (1959) attributed the motion westwards of depressions at southerly latitudes to more vigorous, younger depressions to the north and the easterly flow on their southerly flank. Easterly, or even slightly equatorwards, motion has been observed in some areas, such as in the Drake Passage, or when depressions cross the Antarctic Peninsula. Here they usually have a zonal track to the south of South America, but then follow a curved path to the east–northeast once they have entered the Atlantic Ocean or the Weddell Sea. Taljaard (1967) linked such northwards components with the regions between large-amplitude ridges and troughs, such as are found in the South American and New Zealand areas.

Although the paths of depressions are very variable, it is possible to identify some consistent tracks that have emerged from the studies carried out to date.

(i) The band of eastwards moving depressions close to the circumpolar trough where systems both develop and decline. Knowledge of the large number of relatively minor depressions in this area has only emerged in the last few years as a result of the analysis of model fields and high-resolution satellite data. Consequently it was not mentioned in the early papers on Southern Hemisphere depression tracks.

(ii) East of South America to Enderby Land and the Antarctic coastal region to the east. This track was evident in the IGY data examined by Taljaard (1967), in the FGGE data of Physick and the GARP Basic Data Set Project analyses (Neal, 1972). This was called the 'Falkland' track by Astapenko (1964) and is apparent both in the summer and in the winter depression track charts of Jones and Simmonds.

(iii) From the central and northwest Pacific and through the Drake Passage or southwards to the Bellingshausen Sea. This track was identified by Lamb (1959), who found that large numbers of depressions formed in the area 20° S to 40° S, between New Zealand and Tahiti, and then moved to the Drake Passage and the Bellingshausen Sea. The track was confirmed by Taljaard (1967) using IGY data, by Physick with FGGE observations of 1979 and by Neal (1972) with GARP BDS data, especially from the central Pacific. It incorporates the 'East Pacific' and 'South American' tracks of Astapenko but is broader across the central Pacific. Physick and Taljaard suggested that the positions and definition of this and the previous track are clearest in winter, which would seen to be confirmed by the work of Jones and Simmonds. The strong southerly winds to the west of these depressions are thought to be responsible for the large number of icebergs reaching latitudes as low as 50° S in the central Pacific area.

(iv) Across the southern part of the Indian Ocean from Kerguelen to the south of Australia and declining near Adélie Land and the Ross Sea. To the south and southwest of Australia there is a large area of extensive depression activity that has a high degree of variability. During the FGGE experiment the area south of Australia was very active, many

(a)

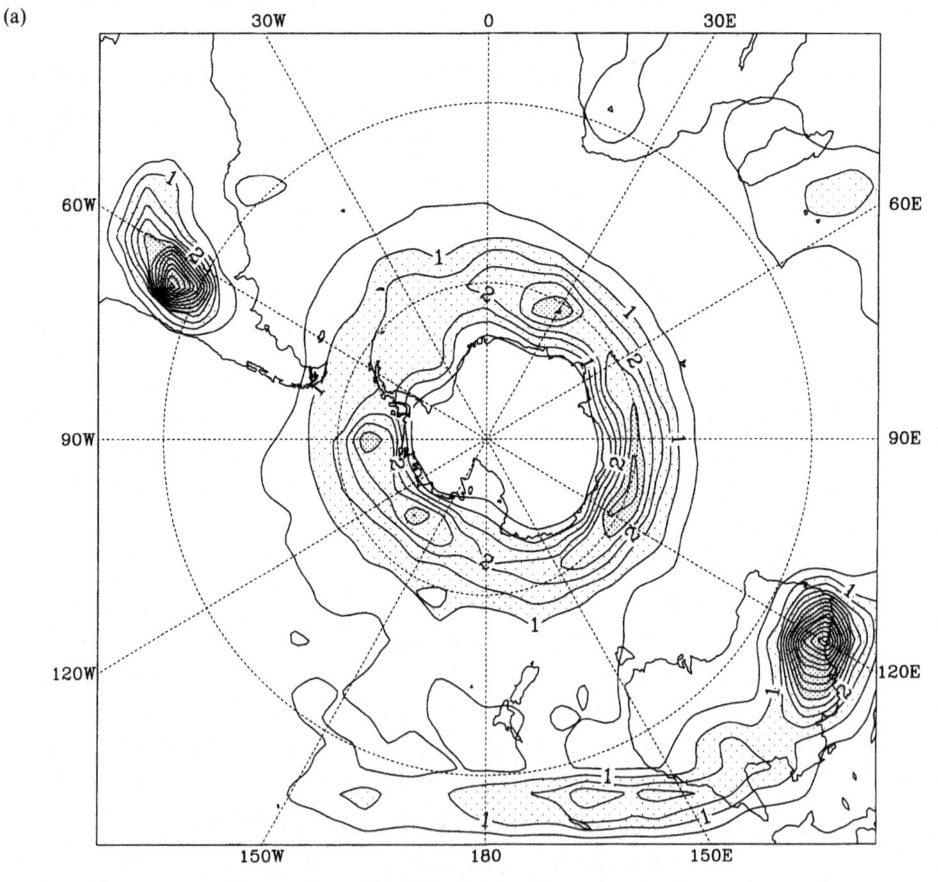

Figure 5.33 The cyclone system density for (a) summer and (b) winter as determined from an analysis of the Australian Bureau of Meteorology operational surface pressure charts for the period 1975–89. The contour interval is 0.5×10^{-3} cyclones (degree of latitude)$^{-2}$. Light and heavy stippling denotes areas above 1.0 and 3.0 respectively. From D. A. Jones and I. Simmonds, A climatology of Southern Hemisphere extratropical cyclones, *Climate Dynamics*, **9**, Figure 3, 136, 1993. Copyright (1993) Springer-Verlag.

depressions being present. Neal (1972) reported a number of cyclone centres southeast of the Ross Sea during June 1970 in his analysis of the charts prepared as part of the GARP Basic Data Set Project.

(v) From the Tasman Sea and south of New Zealand to the Ross Sea and the coastal area of the southern Amundsen Sea. This track, named the 'New Zealand' branch by Astapenko, was very active during FGGE and appears to be a feature of all seasons.

The density of cyclones over the Southern Ocean has been examined by a number of workers including Taljaard (1967) using IGY data and Jones and Simmonds (1993) with their analysis of Australian operational analyses.

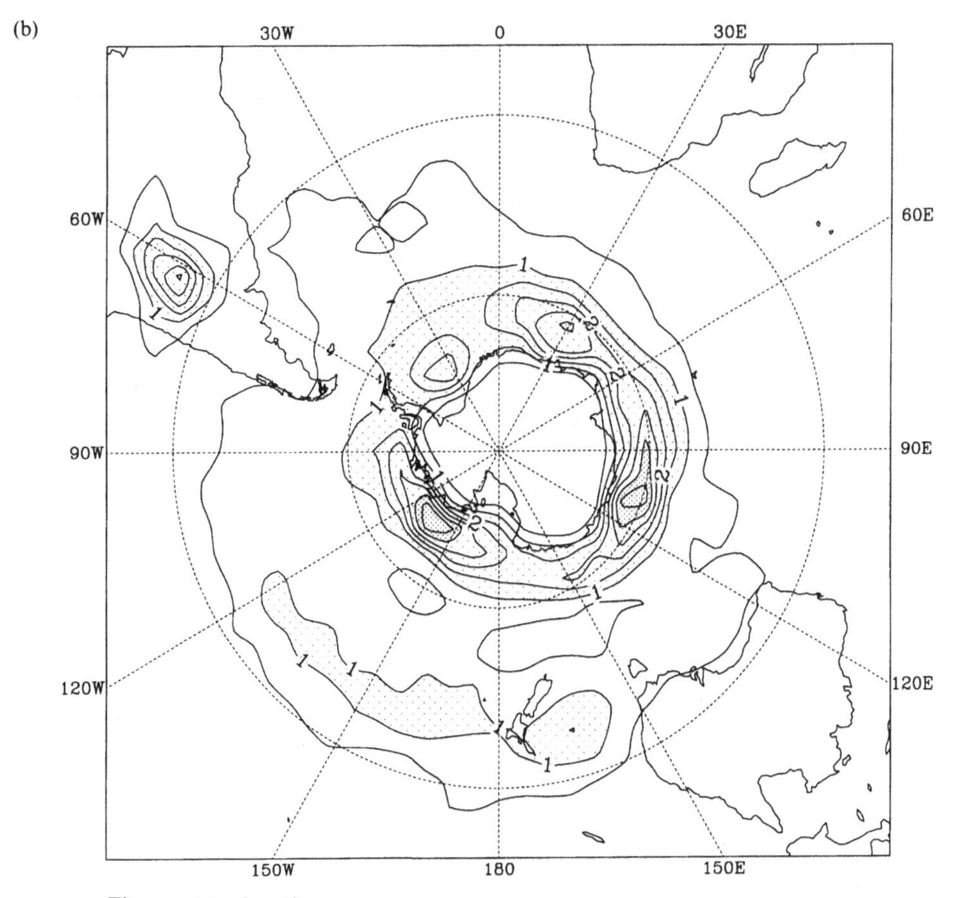

Figure 5.33 *(cont.)*

Taljaard found that cyclone centres were most common between 62° S and 64° S, some 2–6° equatorwards of the mean position of the circumpolar trough. A similar picture was found by Jones and Simmonds who, as discussed, in Section 5.4.1, found that the peak of cyclonic activity was in the 60° S to 65° S band. Their maps of summer and winter season cyclone system density are shown in Figure 5.33 and provide more information on the spatial distribution of depressions. They show that large numbers of depressions occur throughout the year north of the coast of East Antarctica, in the area to the southwest of Australia. A further maximum is found in the Amundsen Sea with a summer-season peak in the Bellingshausen Sea. The winter chart also shows the depressions associated with the Tasman Sea cyclogenetic region and the area of depression activity across the western Pacific to the East of New Zealand.

Depressions are much less common in the interior of the continent than they are over the surrounding ocean and it is very difficult to determine any persistent depression tracks over the continent. Lamb (1959) suggested that a number of depressions pass from the Ross Sea across West Antarctica to the Bellingshausen and Weddell Seas; a result that was confirmed by Astapenko (1964) using IGY data. However, the depression tracks of Jones and Simmonds shown in Figure 5.33 do not show any consistent pattern in depression tracks over the continent, but this is probably a limitation of the numerical analyses from which these fields were produced.

Variability of depression tracks

There is large variability in the atmospheric circulation of the high southern latitudes between years, resulting in large inter-annual differences in precipitation, sea ice extent and temperature recorded at the coastal stations. The problems of studying 'typical' conditions over the high southern latitudes were highlighted by Physick (1981), who found that, during the FGGE winter of 1979, there was significantly more cyclonic activity in the eastern hemisphere than there was in the western one. There being so many depressions in this sector, he considered the whole area from south of Africa to south of New Zealand and over the zone 45° S to 60° S as a single storm track. The tracks of depressions on intra-seasonal and inter-annual timescales are determined, to a large extent, by the atmospheric long waves and in particular by their amplitude and location around the hemisphere. Physick (1981) suggested that the tracks followed by depressions that form to the east of South America are very sensitive to the upper air circulation in the South Atlantic, with the lows moving to Dronning Maud Land under strongly meridional conditions and to Wilkes Land or Victoria Land via the Indian Ocean under predominately zonal conditions. The depression tracks discussed earlier should only be regarded as very general indications of where the systems will move during periods of a few weeks or months, for they can either traverse long distances around the continent before dissipating or move quite rapidly polewards into the coastal region, depending on the zonal and meridional components of the upper flow. However, on average the long waves around the Southern Hemisphere are of limited amplitude compared with those found in the Northern Hemisphere, resulting in a strong zonal component to the track of depressions. Other factors such as the sea ice extent can influence the tracks of depressions. The relationship between sea ice extent and the tracks of depressions is not well understood, but it has been noted that the latitude of maximum mid-latitude cyclonic activity tends to move southwards after July just as the sea ice limit shifts equatorwards.

5.4.3 Cyclolysis

Since the 1950s the high-latitude part of the Southern Ocean has been recognised as an area of cyclolysis where the extra-tropical cyclones become slow moving and decline. Streten and Troup (1973), in their study of satellite imagery for the summer and intermediate months, found that declining depressions were, in general, only found between 50° S and the coast of the Antarctic, with a median latitude of 56° S to 57° S. Carleton (1979) also found that the winter season maximum of cyclolysis occurred between about 55° S and 60° S. Both studies noted that this type of system was more frequent in the imagery since depressions remain in this state for periods longer than those of their formative or mature phases. However, the analysis of Australian model fields by Jones and Simmonds (1993) suggests that the main zone of cyclolysis is slightly further south than had previously been thought. Their diagram of the total number of cyclolysis events for the four seasons is shown in Figure 5.34 and suggests that the peak of cyclolysis is close to 62° S throughout the year, with a rapid drop to the north and south.

The main regions of cyclolysis are obviously at the southernmost ends of the main depression tracks from mid-latitudes and from the cyclogenesis areas in the circumpolar trough. There is good agreement between the areas of cyclolysis determined from studies based on satellite imagery and model analyses and Figure 5.35 shows the summer and winter results of Jones and Simmonds (1993) and, for comparison, the winter season data of Carleton (1979) are presented in Figure 5.36.

All the main areas of cyclolysis are located in the circumpolar trough with the main centres being located in the following areas.

(i) Southwest of Australia between 100° E and 150° E. Here there is frequent cyclolysis throughout the year, but with a maximum of activity during the winter months.

(ii) In the Bellingshausen Sea. This appears to be one of the most consistent regions for cyclolysis at all seasons, although Physick found the centre to be located somewhat to the west during the winter of 1979.

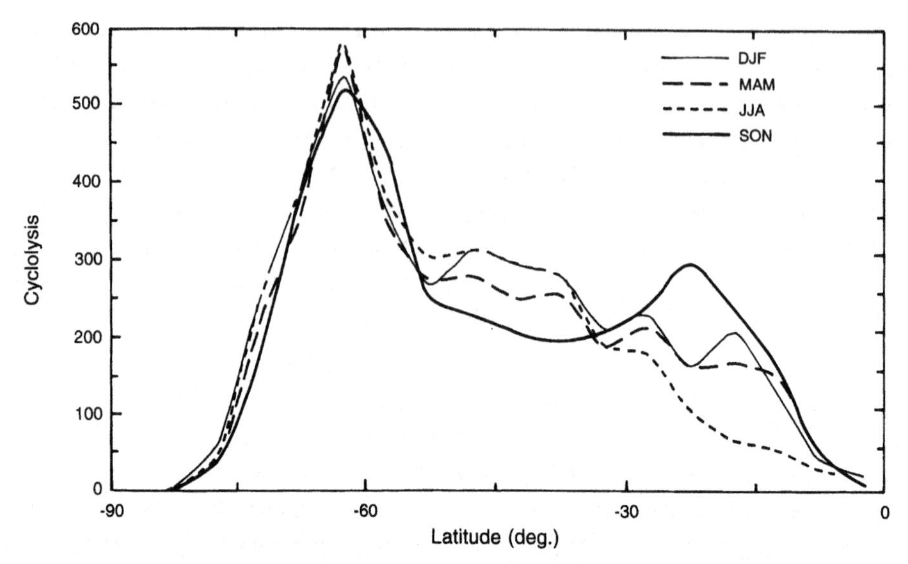

Figure 5.34 The total count of cyclolysis in 5° latitude bands for summer, autumn, winter and spring for the period 1975–89. The units are cyclones. After Jones and Simmonds (1993).

(iii) North of the central and eastern Ross Sea. Taljaard (1967) called this area a 'graveyard' of cyclones, since he found a maximum of depression activity here in the IGY period.

(iv) West of the Enderby Land Peninsula at 40° E to 60° E. Here there is a peak of cyclolysis during the winter months.

(v) In the central and eastern Weddell Sea.

Carleton (1979) found little interannual variability in the areas of cyclolysis around the Antarctic in his study of five years of winter season satellite imagery. However, several areas showed a high degree of variability in the number of dissipating vortices found in different winters and the range of frequencies is shown in Figure 5.37. The greatest variability was found around the Ross, Weddell and Bellingshausen Seas and north of Dronning Maud Land. Along the coast of eastern Antarctica, however, the range was low, indicating the high degree of cyclolysis taking place in each of the winters.

5.4.4 The climatological frontal zones

In the Southern Hemisphere, the equator-to-pole temperature difference is greater than that in the Northern Hemisphere, resulting in more vigorous weather systems, especially over the extensive ocean areas. This temperature gradient is not uniform, but rather is concentrated in certain latitude bands where the major climatic frontal zones exist. These are characterised by a strong

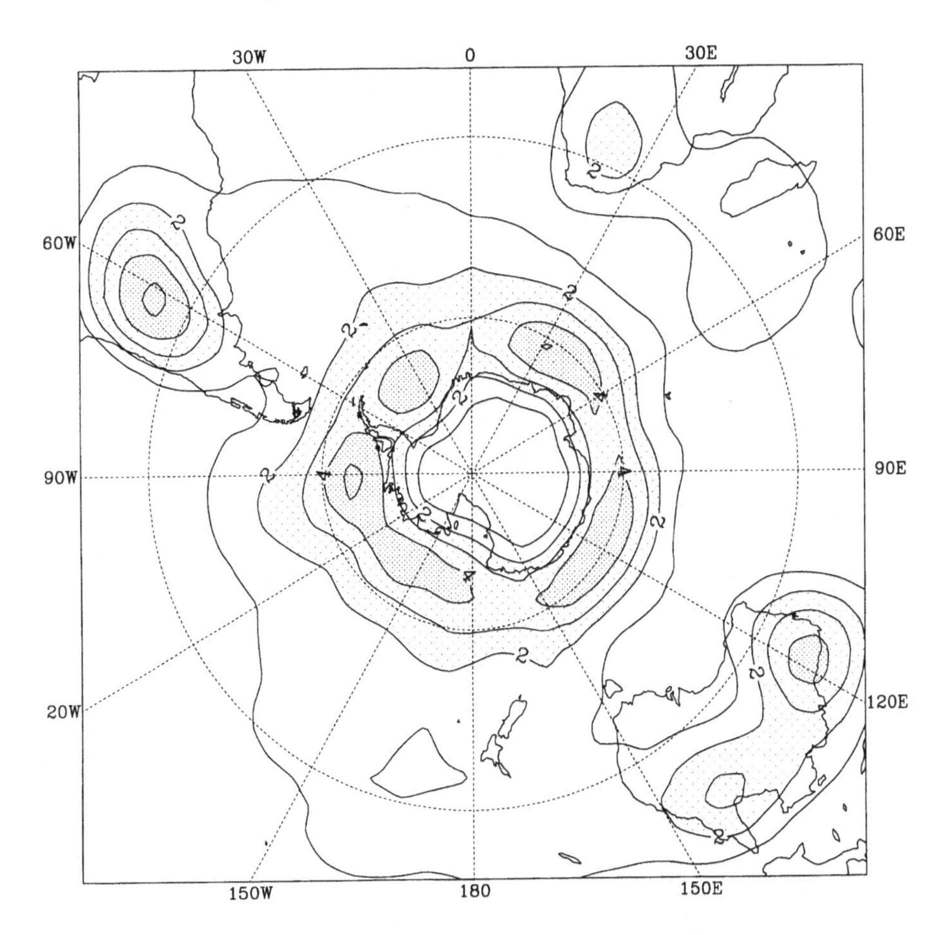

Figure 5.35 The density of cyclolysis for (a) summer and (b) winter as determined from an analysis of the Australian Bureau of Meteorology operational surface pressure charts for the period 1975–89. The contour interval is 1.0×10^{-4} cyclones per day (degree of latitude)$^{-2}$. Light and heavy stippling denotes areas above 2.0 and 4.0 respectively. From D. A. Jones and I. Simmonds, A climatology of Southern Hemisphere extratropical cyclones, *Climate Dynamics*, **9**, Figure 5, 137, 1993. Copyright (1993) Springer-Verlag.

horizontal temperature gradient (baroclinicity), extensive cloud, frequent cyclogenesis and strong winds in the mid-to-upper troposphere. On satellite imagery it can be observed that these zones are often covered by cloud as a result of the frequent passage of depressions and the development of new systems. The two most pronounced frontal zones are the polar front in mid-latitudes and the zone of enhanced thermal gradient in the Antarctic coastal zone. There are regions where less pronounced semi-permanent fronts can be found, such as close to the edge of the sea ice and near strong gradients in the sea surface temperature. However, there is rarely zonal flow across these regions for long periods so that strong, lower tropospheric temperature differences tend not to develop in these areas. The weak frontal zones that do form tend also to be quite

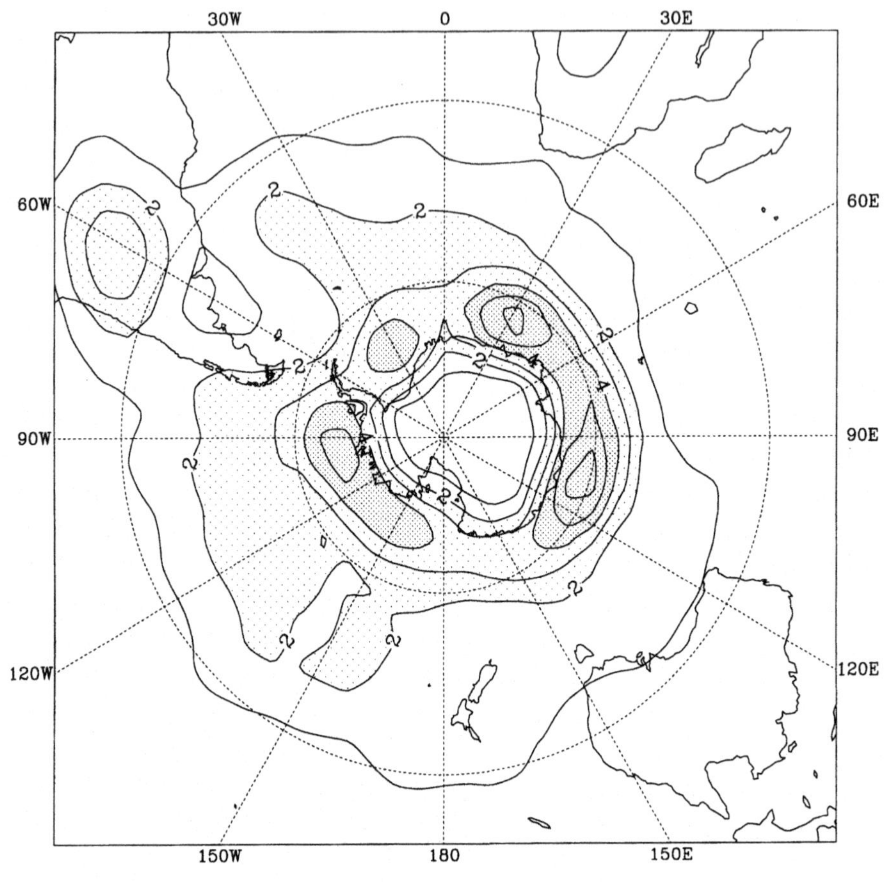

Figure 5.35 *(cont.)*

shallow, as can be seen from the large amounts of low cloud in such areas in satellite images.

The polar front

The early picture of the polar front was that it separated the temperate westerlies, which have their origins in the sub-tropical high-pressure belt, from the colder air masses coming from an easterly direction. However, examination of synoptic charts during experiments, such as the IGY, showed that air on the southern side of the front often came from the west and did not originate in the Antarctic, but was of a heavily modified polar origin. The polar front is an area of frequent cyclogenesis and is often characterised by families of frontal cyclones stretching over many thousands of kilometres. Since these are transient disturbances, the polar front is not always apparent on individual synoptic charts, but it is a climatological feature that can be observed in the mean 1000–500 hPa thickness fields, in the same way that the circumpolar trough is apparent in the mean surface pressure fields. It is certainly not continuous at any one time and

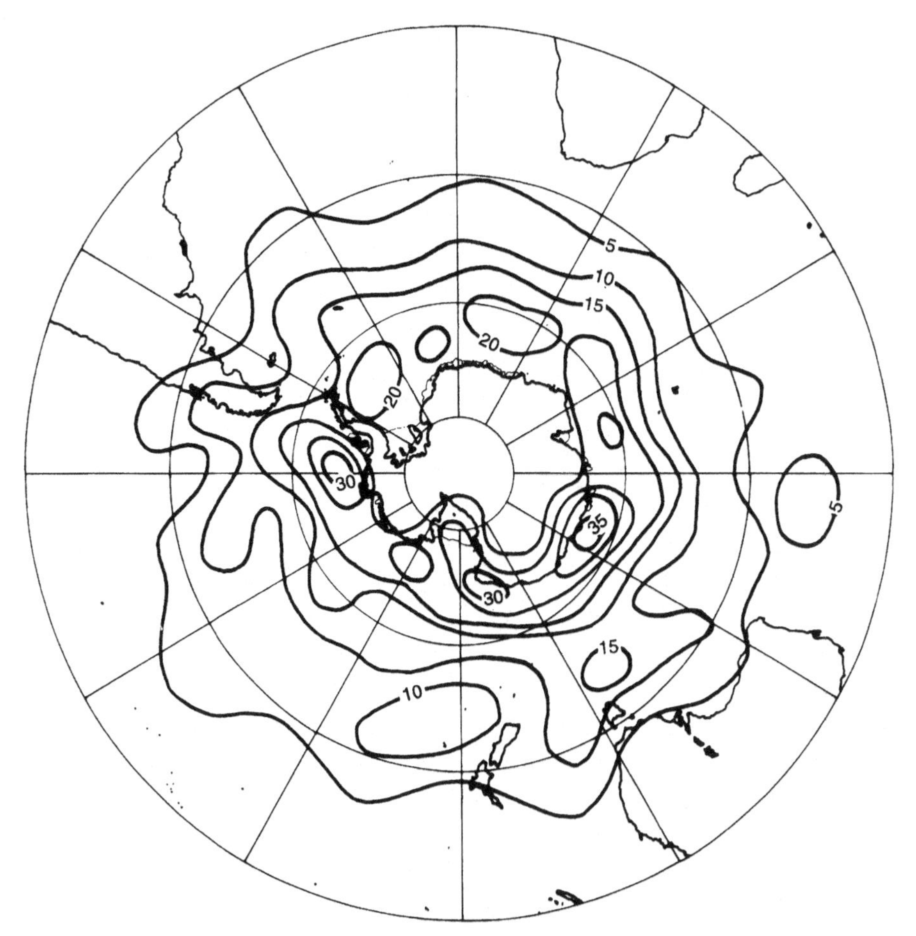

Figure 5.36 Mean distribution of the 'dissipating' (D, see Section 5.3.1) vortices for the five winters 1973–77. Isopleth values refer to normalised vortex frequencies in each 5° latitude by 10° longitude tessera. After Carleton (1979).

there are always breaks in the front and regions where the structure is more complex. The orientation of the polar front is often west–northwest to east–southeast, reflecting the direction of travel of depressions towards the Antarctic.

The polar front has a mean position near 40° S to 50° S, but there are considerable variations around the hemisphere and a change of several degrees of latitude between the summer and the more southerly wintertime position. Over the northwest Pacific Ocean the polar front extends further to the north and is coincident with the region of greater cyclogenesis between New Zealand and Tahiti, apparent in Figure 5.28. Another extension northwards of the front is towards the cyclogenetic region to the east of the tip of South America.

It will be apparent from the above that the polar front is closely related to the tracks of depressions. However, as can be seen in Figure 5.32, the tracks of

Figure 5.37 The 'range' of dissipating (D, see Section 5.3.1) vortices. Isopleths represent the difference between the lowest and the highest values in each 5° by 10° longitude tessera over the five winters 1973–77. After Carleton (1979).

depressions over the Southern Ocean are not well defined, so that the location of the polar front will be very variable.

The frontal zone in the Antarctic coastal region

The Antarctic coastal region is the meeting point of very cold air masses from over the continent and milder, maritime airstreams arriving from the north that have often been brought from lower latitudes by the extra-tropical cyclones that have migrated southwards. This results in a tightening of the thermal gradient, apparent in the mean thickness fields for the lower layers of the atmosphere. In certain areas the cold, katabatic winds that have descended from the plateau enhance the gradient because these strong southerly winds can extend out over the ocean for considerable distances. The strongest gradient is in the latitude band 60° S to 70° S, which is close to the axis of the circumpolar trough and where most high-latitude cyclogenesis takes place. Such conditions imply the presence of a frontal zone, but this is much less distinct that the polar front to the north. Satellite imagery certainly shows a great deal of cloud just to the north of the coast of the Antarctic, but this is generally quite shallow and does not usually have the appearance of mid-latitude fronts, which consist of ribbons of medium- or high-level cloud.

5.4.5 Anticyclones

The distribution of anticyclones over the Southern Hemisphere will be considered through the results of the study of Jones and Simmonds (1994), who developed an automatic tracking scheme for anticyclones and applied it to 15 years of once-daily surface pressure charts produced by the Australian Bureau of Meteorology. The period covered was 1975–90 and the anticyclones were followed using a version of the vortex tracking scheme described by Murray and Simmonds (1991a; 1991b).

The tracks of anticyclones

The tracks of all summer and winter season anticyclones that lasted for one day or more within the period 1985–89 are shown in Figure 5.38. The greatest number of anticyclones occurred in the band 25° S to 42° S, corresponding to the sub-tropical high-pressure belt. This is in agreement with the work of Taljaard (1967) based on synoptic charts for the IGY period of July 1957 to December 1958, who found the greatest number of anticyclones in the band 25° S to 45° S. South of this zone, over the Southern Ocean and the Antarctic coastal region, there were few anticyclones, although the Weddell Sea was an exception to this. Jones and Simmonds found that, south of 55° S, anticyclones were short-lived and generally existed for less than four days, illustrating the predominance of cyclonic activity around the Antarctic. The ridges and anticyclonic centres that do form over the ocean develop both from the north and from the south (Streten, 1980a) whereas the preferred areas for ridging and the development of anticyclones are from 10° W to 10° E, 40° E to 70° E and 120° E to 160° E and to a lesser extent from 120° W to 30° W (Streten and Pike, 1980). The semi-permanent high centred over East Antarctica is apparent in the figures, although, as discussed in Section 5.5.1, care needs to be taken in interpreting the fields of mean sea level pressure over the continent because of the reduction of data from the surface at an elevation of 2–4 km to sea level (Le Marshall and Kelly, 1981). Neal (1972) also found that there were many anticyclones in the vicinity of Vostok, especially during the winter months, when there were more systems there than over West Antarctica. During the winter the East Antarctic high is generally more intense (Jones and Simmonds, 1994).

Figure 5.38 shows that the anticyclones are found at all longitudes in the sub-tropical high-pressure belt, but there are variations around the globe. For example, the greatest density of systems is more polewards in the western parts of the ocean areas and there are fewer anticyclones to the west of the continents. There are also fewer systems found over the continents.

The mean motion of anticyclones over the Southern Ocean is towards the east with a small component towards the equator. Systems tend to be faster moving in mid-latitudes and slower in the Antarctic coastal zone, whereas the semi-permanent high over the continent is quasi-stationary during all seasons.

Anticyclogenesis

The summer and winter season maps of anticyclogenesis density normalized with respect to area for summer and winter from the study by Jones and Simmonds are shown in Figure 5.39. The maxima of anticyclogenesis are mostly in the band 30° S to 40° S and are found slightly further south during the summer

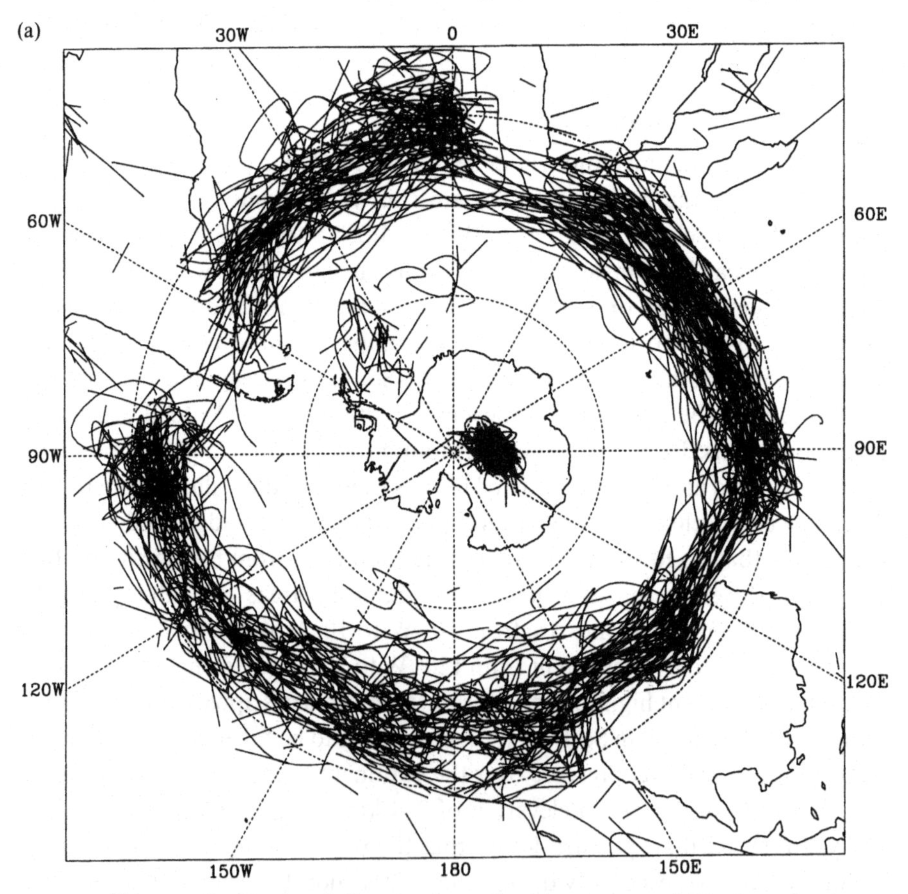

Figure 5.38 Tracks of all anticyclones for (a) summer and (b) winter for the period 1985–89. From D. A. Jones and I. Simmonds, A climatology of Southern Hemisphere anticyclones, *Climate Dynamics*, **10**, Figure 1, 336, 1994. Copyright (1994) Springer-Verlag.

season. The maxima tend to occur close to the southern tips of the sub-tropical continents. A further peak of anticyclogenesis is present during all seasons over East Antarctica, but the maximum of activity occurs during the summer and the relatively low density values indicate the semi-permanent nature of the high here. The only other clear centre within the Antarctic is found close to the tip of the Antarctic Peninsula with an extension over the northwest Weddell Sea during summer. This region is the source of anticyclones that move eastwards and northeastwards into the rest of the Weddell Sea and the Southern Ocean.

Anticyclolysis

Anticyclones that have developed in the sub-tropical high-pressure belt tend to decline in the zone 28° S to 35° S, some 2–4° equatorwards of the maximum in anticyclone density and anticyclogenesis (Jones and Simmonds, 1994). Over the Antarctic continent the peak of anticyclolysis is coincident with the maximum

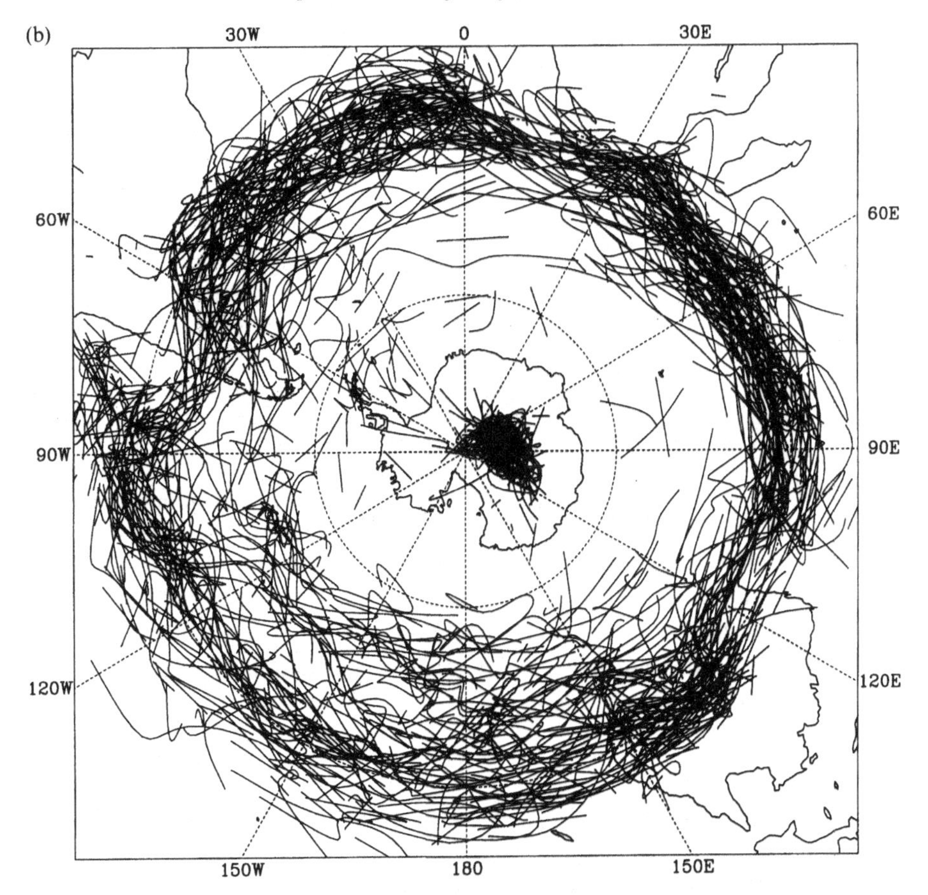

Figure 5.38 *(cont.)*

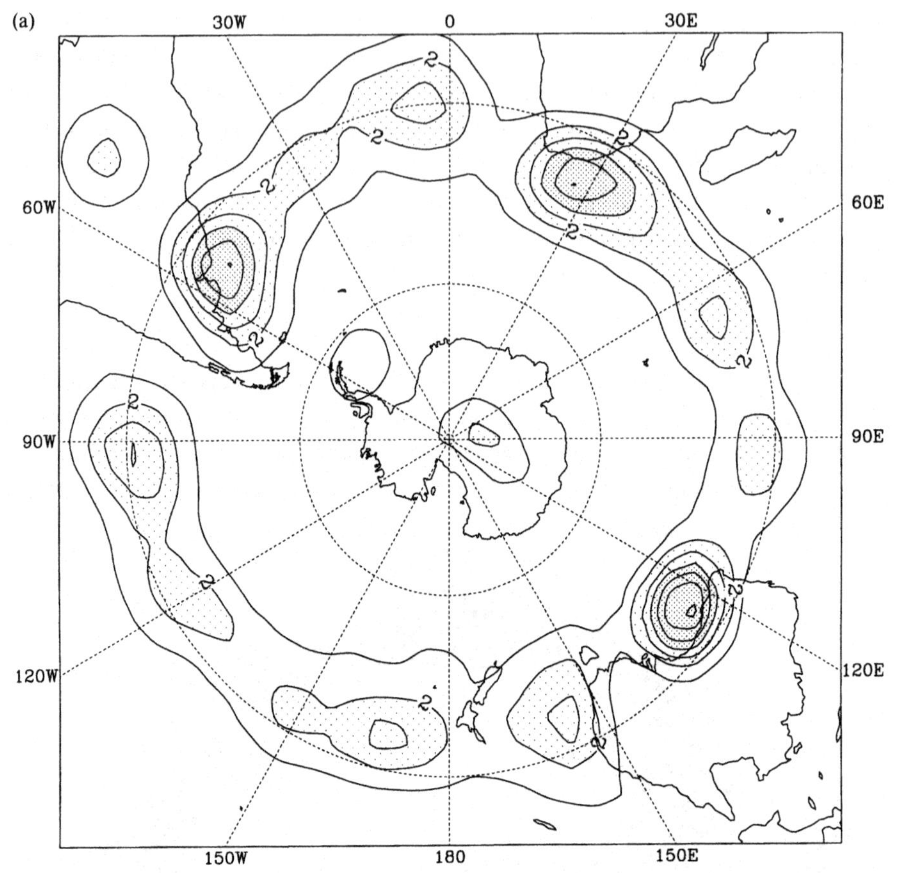

Figure 5.39 The density of anticyclogenesis for (a) summer and (b) winter. The contour interval is 1.0×10^{-4} anticyclones per day (degree of latitude)$^{-2}$. Light and heavy stippling denotes areas above 2.0 and 4.0 respectively. From D. A. Jones and I. Simmonds, A climatology of Southern Hemisphere anticyclones, *Climate Dynamics*, **10**, Figure 3, 337, 1994. Copyright (1994) Springer-Verlag.

in anticyclone development, indicating the quasi-stationary nature of systems over East Antarctica. A further peak in anticyclolysis occurs over the Weddell Sea and this is the sink for the highs that develop close to the tip of the Antarctic Peninsula. This sink is located over the north central Weddell Sea in summer and just to the east of the Peninsula in winter.

5.5 Preparation of operational analyses and forecasts

The harsh climate of the Antarctic was a great danger to those who ventured south on the early expeditions and many explorers perished as a result of the appalling conditions that they encountered. However, with the only meteorological observations available to them being those that they were able to make

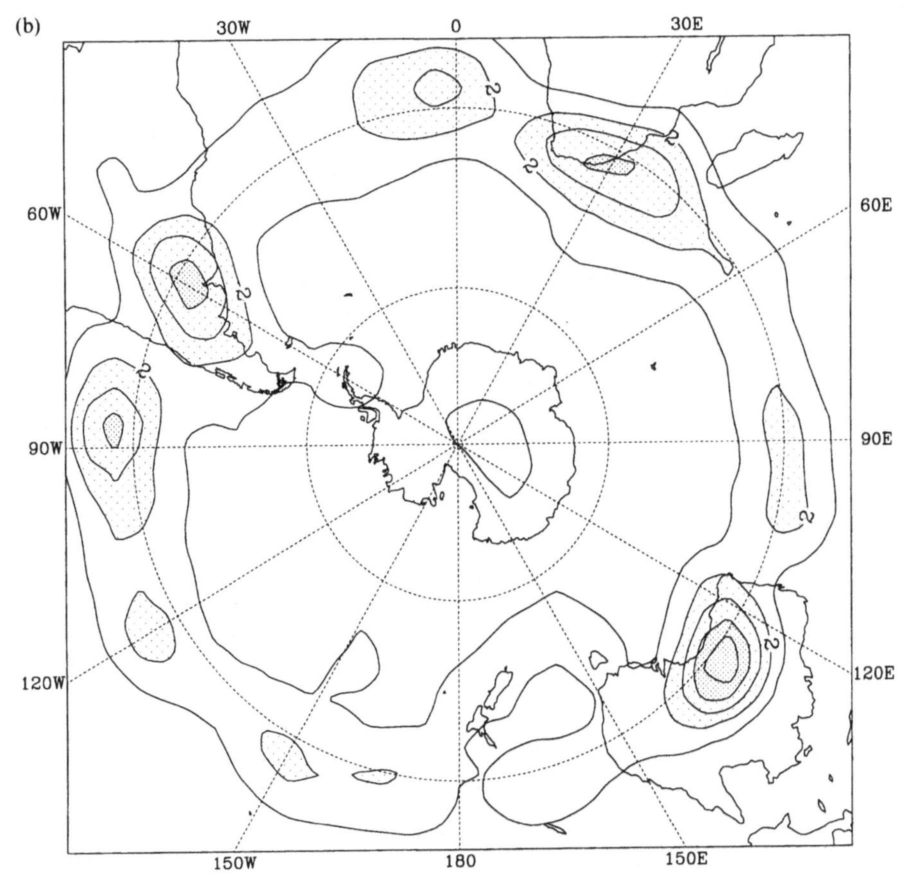

Figure 5.39 *(cont.)*

themselves, there was no way that they could determine the broader scale atmospheric conditions or make any form of forecast for more than a few hours ahead. Instead, information on the meteorological conditions and the sea ice extent at various times of the year was gradually accumulated from the expeditions so that a climatology for the continent was assembled and a knowledge built up of the likely departures at any time of the year. However, by the late 1940s there were enough observations being collected from areas such as the Antarctic Peninsula for crude surface analyses to be prepared, although over the ocean areas there were still very few data and only the broadest indications of the circulation could be determined. Although experiments such as the International Geophysical Year of 1957–58 helped to provide insight into the circulation of the Antarctic atmosphere, it was the routine availability from the mid-1960s of imagery from the polar orbiting satellites that allowed the major weather systems over the data-sparse ocean areas to be monitored routinely. Although the imagery only showed the positions of the frontal cloud bands associated with the depressions, techniques were developed to allow the estimation of the central pressures from the cloud features (Guymer, 1978). The other valuable form of data that could be obtained from satellites was the temperature and humidity sounder data that gradually became available in the 1970s. These profiles to some extent compensated for the lack of radiosonde ascents over the oceans and allowed upper-air analyses to be prepared over the Southern Ocean, although not over the Antarctic continent itself. At about this time the numerical weather prediction models run by the major weather services were being extended to have global coverage and the satellite sounder data ensured that they had at least a reasonable representation of the circulation over the Southern Hemisphere. Today's numerical weather prediction systems produce analyses of very high quality for the tropical and mid-latitude areas of the world and give forecasts that are much improved over those produced even ten years ago. However, in the Antarctic region there are still problems in producing accurate analyses and forecasts using manual and numerical techniques and these topics will be considered in the following sections.

5.5.1 Manual analyses

Over the ocean areas around the Antarctic continent the manual analysis process is very similar to that employed at mid-latitudes, since the main weather systems encountered are frontal depressions and anticyclones. Surface analyses, such as that shown in Figure 5.40 which was prepared at the forecasting office at Rothera Station, are produced using a combination of output from numerical models, satellite imagery and the available *in situ* observations. Over the ocean areas there are very few conventional observations so that the surface pressure values are usually taken from the output of the numerical analysis systems. Satellite imagery is used to adjust the positions of the centres of depressions, to determine the locations of fronts and to check that the model thickness fields are consistent with the frontal locations. The imagery also provides very valuable information on whether the major lows are multi-centred and on features such as the development of breakaway triple-point lows. Small-scale low-pressure systems, such as mesocyclones (see Section 6.5), are usually not represented on the numerical analyses but may have a significant effect on local weather. They are therefore often indicated on the surface charts even though it may not be possible to determine their central pressure and analyse their structure on the chart.

The topographical conditions over the Antarctic continent present particular problems in the production of analyses, especially over the high ground of East Antarctica, where the surface is close to the 700 hPa level, and in the coastal area where the gradient is very steep. Over the interior, the Norwegian frontal model, which is used in most areas outside the tropics, is not usually applicable and relating the clouds observed in satellite imagery to the pressure field poses many

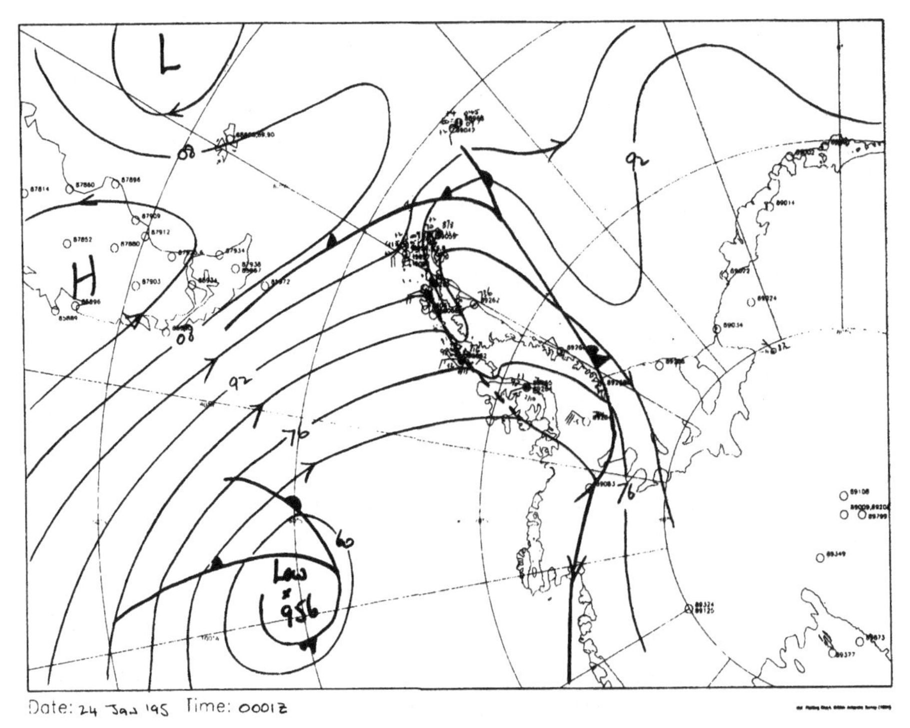

Figure 5.40 The surface analysis for the Antarctic Peninsula region for 00 GMT 24 January 1995. The chart was prepared at the forecasting office at Rothera Station.

problems. Different approaches to the preparation of routine analyses have therefore to be adopted. One of the mainstays of operational meteorology is the mean pressure at sea level (MSLP) charts which are prepared both manually and from numerical model output for all extra-tropical areas. In the Antarctic coastal region, up to levels of about 1000 m, there are no problems in producing such charts in the normal fashion. However, over the high ground of the Antarctic the reduction of surface observations, which can be made at elevations of 2000–3000 m, down to mean sea level can introduce errors so that MSLP charts from different analysis centres can differ considerably depending on the reduction procedure used.

Since there are so many problems in the production of MSLP charts over the Antarctic continent, some alternative procedures for analysis in these areas have been proposed. During the IGY the 700 hPa height field was frequently used as the main analysis level and 600 hPa has also been suggested as a suitable base level (Voskresenskii and Chukanin, 1986). Phillpot (1991) suggested the use of the height of the 500 hPa pressure surface and proposed a technique for the estimation of this quantity from AWS observations made at elevations above 2500 m. Because this technique has now been used by a number of workers it will be described in detail in the following section.

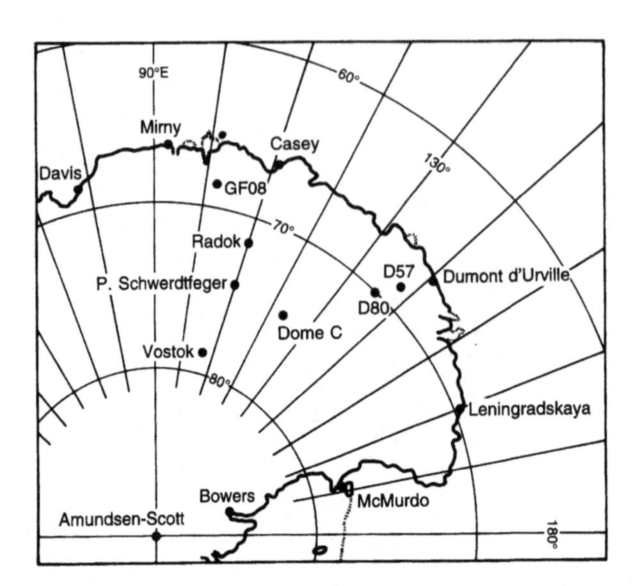

Figure 5.41 The sector of the Antarctic where the technique of Phillpot has been applied to derive a 500 hPa height field. AWS sites and over-wintering stations are shown. After Phillpot (1991).

The Phillpot technique for analysis of the height of the 500 hPa surface over the continent

The 500 hPa surface is at a level of about 5 km above sea level and about 1–2 km above the snow and ice surface of East Antarctica. It is therefore above all the topography of the continent, as well as the strong surface temperature inversion. The technique adopted by Phillpot was to take the operational Australian Bureau of Meteorology 500 hPa height fields and modify them in the light of the observational data, which provide information concerning the circulation features. Figure 5.41 shows the area where the technique was applied and the locations of conventional reporting stations and AWS sites. The AWSs provide observations of surface pressure, temperature and wind, although the wind observations are thought to be of least value because of the limited correlation to the 500 hPa surface. The technique requires the surface pressure and temperature and the estimated vertical profiles of temperature and humidity between the station level and the 500 hPa surface. The depth of the layer over which the profile must be estimated varies over the continent from about 5 km at the coast to about 1.5 km over much of the interior. The strong surface temperature inversion is a major feature of the interior of the continent, but Phillpot showed that the surface temperature could be taken as being representative of a deep layer; a fact that can be used in the estimation of the 500 hPa conditions. This assumption appears to be reasonable for a layer of about 1.5 km depth, although the error will increase as the technique is applied to areas towards the coast. Phillpot has estimated that the error in the derivation of the 500 hPa height via this method is about 40 m over the high ground.

The technique was developed by analysing the many radiosonde ascents that were launched from the Antarctic plateau during the IGY to determine the

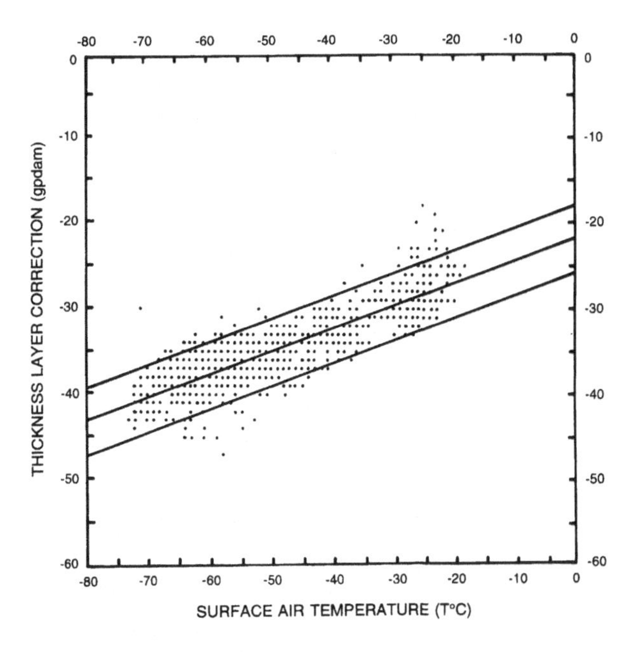

Figure 5.42 The distribution, over the observed range of surface air temperature, of the corrections (gpdam) required to the thickness of the atmospheric layer between 500 hPa and the surface by the departure of the layer-observed mean virtual temperature from 0°C, at Amundsen–Scott Station during the IGY period July 1957–December 1958. The derived linear relationship and the limits representing thickness departures of ± 4 gpdam from the mean are also shown. After Phillpot (1991).

relationship between temperature and height. These ascents were each used to compute

(i) the pressure difference (Δp) between that observed at station level and 500 hPa,
(ii) the thickness of this layer (ΔZ_1) for a mean virtual temperature of 0°C,
(iii) the radiosonde-calculated thickness of the layer between 500 hPa and the surface (ΔZ_2) (this gives the true mean virtual temperature for the layer),
(iv) the difference $\Delta Z_1 - \Delta Z_2$, representing the thickness correction ΔZ_T that needs to be applied to correct for the difference of the true mean virtual temperature from 0°C,
(v) linear regression equations relating the observed surface air temperature T°C (1°C class intervals) and the thickness correction ΔZ_T and
(vi) the standard deviation of the departure of the layer thickness from the mean value for each 1°C surface temperature class interval.

Figure 5.42 shows the regression equation developed for Amundsen–Scott Station from the 918 ascents available. This shows that, over the range of surface air temperatures experienced at the pole, the thickness correction is within the range ±4 gpdam (geopotential decametres) for the vast majority of cases. The regression coefficients for other stations were presented by Phillpot (1991). From an assessment of the standard deviations computed for all the stations in the interior, Phillpot estimated that the technique probably cannot be applied to stations with an elevation of below 2.4 km.

With the large number of AWSs over the interior of the continent it is now possible to use the above technique to produce a 500 hPa height field over extensive parts of the continent. Regression coefficients have been produced for the AWSs over East Antarctica from the information gained by the analysis of the IGY radiosonde data allowing the 500 hPa height to be estimated by

(i) finding the thickness of the layer from 500 hPa to the AWS pressure level for a mean virtual temperature of 0°C,
(ii) using the regression relationship to correct this thickness value as a result of the departure of the layer mean virtual temperature from 0°C and
(iii) adding the corrected layer thickness to the AWS elevation.

An example of the use of the technique is shown in Figure 5.43. Here Figure 5.43(a) shows the operational 500 hPa height analysis for East Antarctic and the adjacent ocean areas for 00 GMT 5 December 1986. At this time there was a slack area of low 500 hPa heights lying along the coastal region with a suggestion of a ridge extending southwards into the interior between 130° E and 160° E. The revised analysis in Figure 5.43(b), which incorporates the 500 hPa height estimates produced from the AWS observations, shows far more detail over the continent, including a closed high over the plateau and more structure around the coastal lows.

The Phillpot technique has great potential for those investigating synoptic-scale weather systems over East Antarctica and the penetration of disturbances inland from the coastal region. The careful analysis of the 500 hPa heights presented in the earlier referenced paper shows that the values can be used with confidence over the higher parts of the plateau. Future study must assess the contribution that the data can make to the numerical analysis procedures because there are now much fewer upper-air data over the plateau than there were during the IGY period of 1957/58.

Other analysis techniques

The problems involved in using conventional meteorological charts over the interior of the Antarctic have resulted in a number of novel analysis techniques being applied. Here we will briefly discuss other approaches that have recently been documented.

(i) Pendlebury and Reader (1993) proposed the use of wind streamlines over the continent. This technique has found application in the preparation of operational analyses and forecasts at Casey Station.
(ii) Wind vectors from the lowest level of a numerical model can be plotted on a chart, giving a good indication of the near-surface flow over the continent. An example of this form of output is shown in Figure 5.44. These are from the Australian GASP model 0.991 sigma level at approximately 75 m above the topographic surface and show the wind field at 23:00 GMT 20 March 1992.

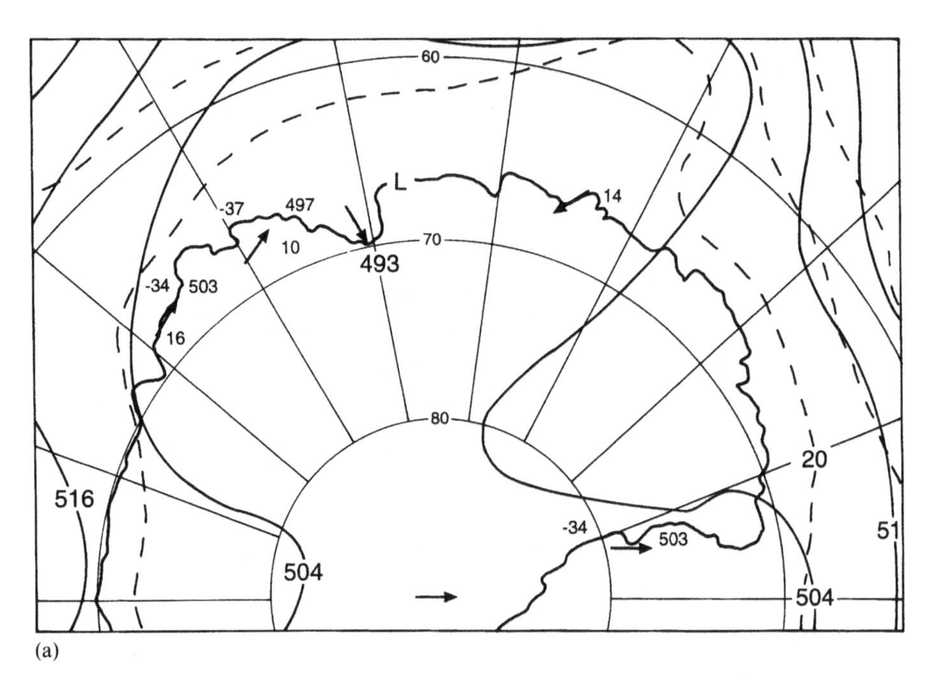

(a)

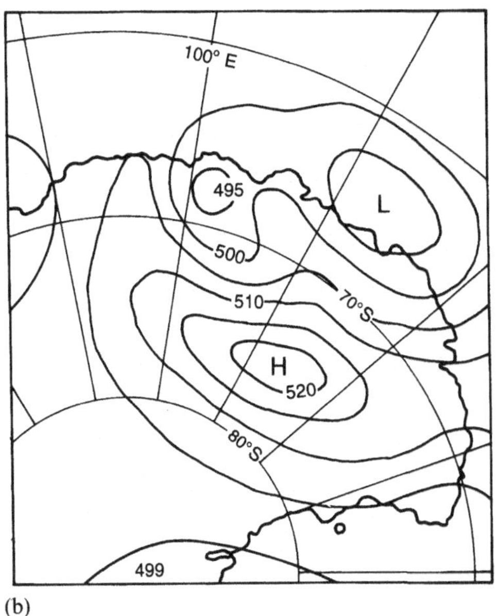

(b)

Figure 5.43 (a) The Melbourne NMC 500 hPa height and isotach analysis over East Antarctica for 00 GMT 5 December 1986. (b) The revised 500 hPa height analysis for the same occasion incorporating the 500 hPa heights estimated from the AWS observations. After Phillpot (1991).

(iii) Carrasco and Bromwich (1993a) have used maps of the difference between monthly mean values of station pressure at AWS sites and instantaneous pressure values in studies of mesoscale cyclogenesis in the Antarctic coastal region. Such a technique has the advantage of compensating for the large differences in elevation among AWS sites yet still providing a synoptic view of atmospheric developments.

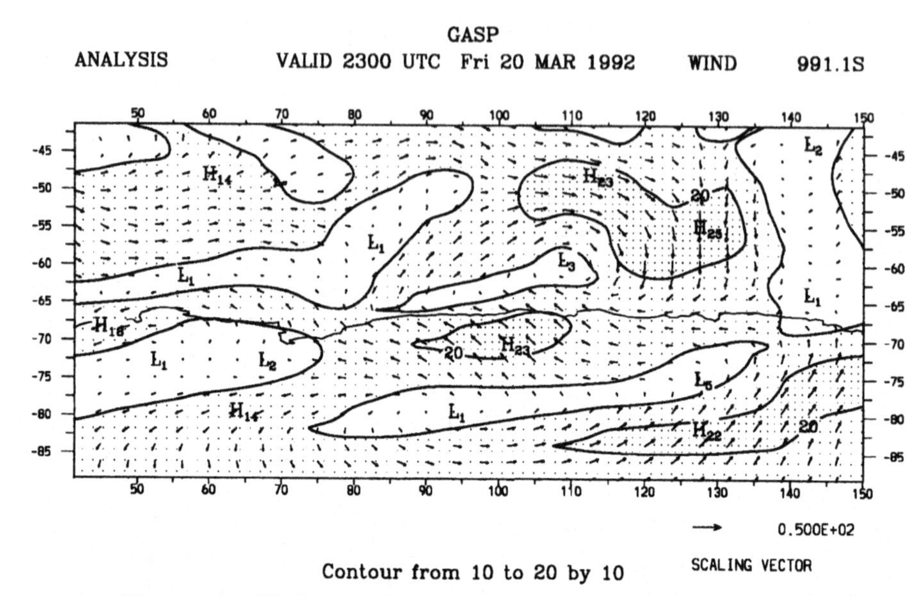

Figure 5.44 Wind vectors from the Australian GASP model 0.991 sigma surface for 23:00 GMT 20 March 1992. From Pendlebury and Reader (1993).

5.5.2 The production of numerical analyses

Today, most of the major weather services produce global analyses using numerical weather prediction systems running on very powerful supercomputers. These systems incorporate data assimilation schemes that merge all the available observations into a model run, so producing surface and upper air analyses that are dynamically balanced. These, it is hoped, make optimal use of the data from different observing systems and can take account of the data that are collected away from the main observing times. These schemes are obviously tuned for the tropics and mid-latitude areas where most of the users of the analyses and forecasts are located. In these areas the analyses are very good and generally provide an excellent representation of synoptic-scale disturbances. As the forecasts run from the analyses have been extended into the medium range of 6–10 days so it has been necessary to ensure that the analyses are realistic over the whole globe, since errors in the initial conditions can propagate a significant distance during the forecast period. Although the Antarctic is well removed from the main population centres, it is still necessary to have a reasonable representation of the surface and upper air conditions over the continent and Southern Ocean if errors are not to affect the forecasts over the Southern Hemisphere in the forecast period. The forecast centres make use of the surface and upper air observations from the coastal stations around the Antarctic since these are all at low level and can be used in the same way as any other land observations. Over the interior of the continent radiosonde data are currently available from one site only – Amundsen–Scott Station. Around the coast of the continent the separation of upper air stations is variable but is not far short of the 1000 km recommended by the WMO for oceanic areas.

The sigma coordinate system (pressure normalised with respect to surface pressure) is now frequently used in numerical analysis and forecast systems and this greatly helps in the representation of air flow over the continent and the incorporation of surface observations over areas of high topography and steep topographic gradient. For example, it allows the UK Meteorological Office to use the observations made by AWS on the high plateau and these are incorporated into the analyses in the same way as are any other surface observations. The only problem is that the data must be reduced to mean sea level for the quality control of the data against the model first guess field and other data.

Over the Southern Ocean, where there are few conventional observations, data from satellites play a more important role. For surface analysis over the ice-free ocean the ERS-1 scatterometer wind vectors (see Section 2.6.4) are very valuable and recent studies have suggested that these data can lead to a lowering of the central pressures of mid-latitude cyclonic depressions by 3 hPa compared with the values from European Centre for Medium Range Weather Forecasting (ECMWF) analyses (Brown and Zing, 1994). The TOVS temperature soundings (see Section 2.6.2) are the main source of data for upper air analysis and are included in the numerical analysis schemes. The SATEM messages, which are available over the GTS at a horizontal resolution of 250 km, can resolve the main details of the thermal field and have been shown in observing system experiments to extent the forecasting capability by about one day (Bengtsson, 1989; Hart *et al.*, 1993). Over the high Antarctic plateau few centres make a great deal of use of the TOVS retrievals and the ECMWF and the UK Meteorological Office only use the satellite data above 100 and 250 hPa respectively so that there are few tropospheric data to help maintain realistic conditions. In fact, comparisons of the ECMWF analysed 200 hPa heights against the radiosonde data from South Pole Station showed a bias of 40 m during 1986 (Lutz *et al.*, 1990).

With the research that has been carried out over the last few years into the derivation of temperature profiles from raw TOVS data and in the assimilation of these data into numerical weather prediction systems, there has been a noticeable improvement in the quality of the analyses. Comparison of the numerical analyses with high-resolution satellite imagery at the research stations shows that, over the ocean areas at least, there are rarely major errors in the representation of the synoptic-scale disturbances. Model surface pressure analyses usually have the positions of the depression centres analysed correctly and troughs around these systems can often be correlated with frontal bands in the imagery. However, mesoscale weather systems that are often of great importance in day-to-day forecasting are rarely resolved in the model fields.

5.5.3 Forecasting in the Antarctic

Producing accurate forecasts over the Antarctic continent and in the immediate coastal area is a major challenge to meteorologists because of the lack of observations and the complexity of the atmospheric flow as a result of the steep topography. However, the need for accurate forecasts has increased in recent years as Antarctic logistical operations have become more complex and the use of aircraft and helicopters has increased. Today there are year-round flights within the Antarctic and frequent long-haul flights from outside the Antarctic, such as the C130 flights from Christchurch, New Zealand to the USA's McMurdo Station. This takes between 6 and 8 h and requires good forecasts of the expected conditions in the Antarctic.

Forecasts for the Antarctic are usually based on a synthesis of output from numerical models and data gleaned from the inspection of satellite imagery and the limited amount of *in situ* data that are available. The lack of data is the greatest handicap to producing accurate forecasts and despite the expansion of the network of automatic weather stations in recent years there are still far fewer *in situ* observations in the Antarctic than there are over other land areas. Satellite imagery can help overcome some of the problems in determining the structure of the major weather systems and identifying the locations of areas of extensive cloud, but its value is mainly in the 'nowcasting' period of up to 6 h ahead. Beyond that time it is necessary to use more dynamical methods and to make use of model output. However, in many areas over the continent it is difficult to give an accurate prediction of most weather elements for more than a few hours ahead.

Forecasting in the Antarctic can be considered from two points of view – that of representing the continent in operational numerical weather prediction systems and in terms of forecasting the various meteorological elements using conventional techniques developed in mid-latitude areas. These two approaches will be considered separately in the two following sections and the successes and limitations of the methods employed examined.

The accuracy of forecasts from numerical models

As discussed above, the quality of analyses over the Southern Ocean has improved significantly in recent years and this is reflected in the parallel improvements noted in the forecast fields. Subjectively, forecasters working at the research stations report that the model guidance over the ocean areas is very good up to about 24 h ahead and that, during this period, the positions and central pressures of the synoptic-scale systems are well forecast. Over this time frame the model upper winds are also generally accurate and suitable for use in aviation forecasting. Between 24 and 72 h the models show skill in handling the large depressions, although there are errors in timing and location. Importantly,

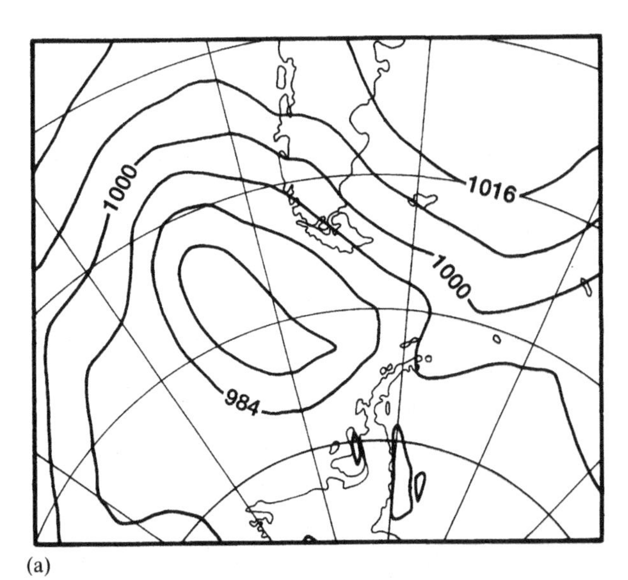

(a)

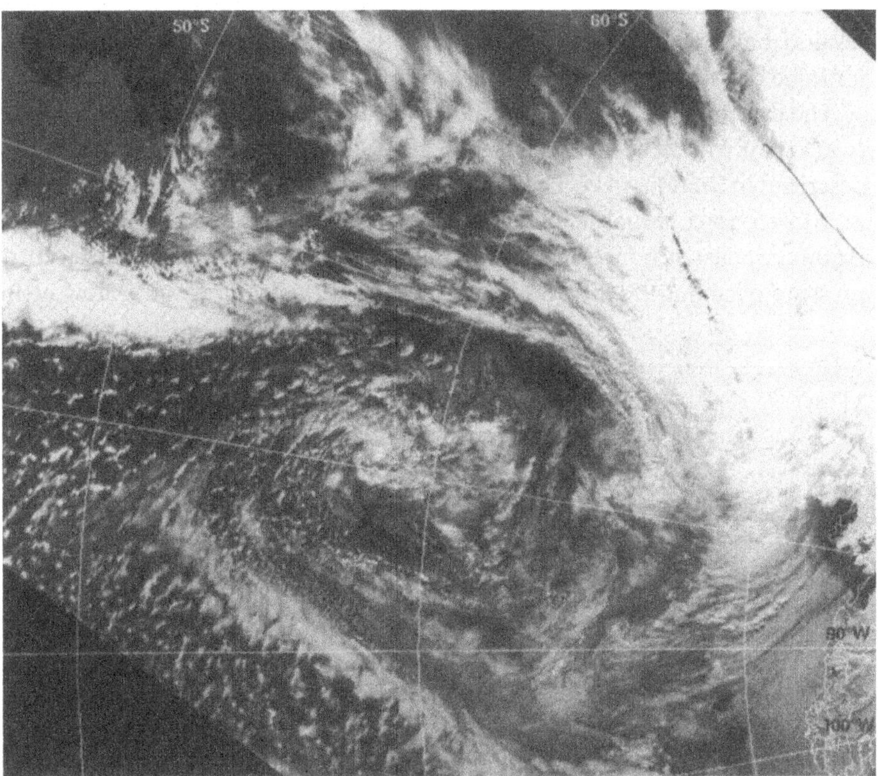

(b)

Figure 5.45 (a) The UK Meteorological Office 36 h forecast, from data at 00 GMT 16 December 1994, valid at 12 GMT 17 December 1994. (b) A visible wavelength satellite image at 12:45 GMT 17 December 1994.

the models seem able to predict changes in 'type', such as a change from a blocked to a mobile situation. An example of a successful 36 h forecast produced by the UK Meteorological Office is shown in Fig 5.45(a), together with the visible satellite image at the verifying time, which is shown in Figure 5.45(b). The forecast shows the surface pressure field at a time when a large depression was predicted to move into the Bellingshausen Sea from the northwest. The satellite image shows that the position of the low was well forecast and that an indication of the position of the front could be determined from the trough of low pressure extending from the low centre into the Drake Passage.

The main problems in running numerical models over the Antarctic are in dealing with the high topography of the continent and with the very stable boundary layer structure and in representing the extensive area of sea ice. Although the forecasts produced by numerical models are quantitatively assessed on a routine basis against numerical analyses and high-quality observations for many areas of the world, none of the major weather services produce validation statistics specifically for the Antarctic. It is therefore necessary to consider forecast accuracy for shorter periods and individual cases. We will therefore examine the predictions issued for specific synoptic events and present results from a recent project that examined the accuracy of operational forecasts for three one-month periods.

The case of a deep synoptic-scale low that gave the highest ever wind gust (130 kt) for March at the Australian Casey Station (66° 17′ S, 110° 48′ E) was described in Section 5.3.3. The ECMWF surface pressure analysis for 12 GMT 21 March 1992, which shows the low when it was about 400 km off the Antarctic coast, is reproduced in Figure 5.46. At this time the low had a central pressure of about 951 hPa and was giving a very strong easterly flow at the

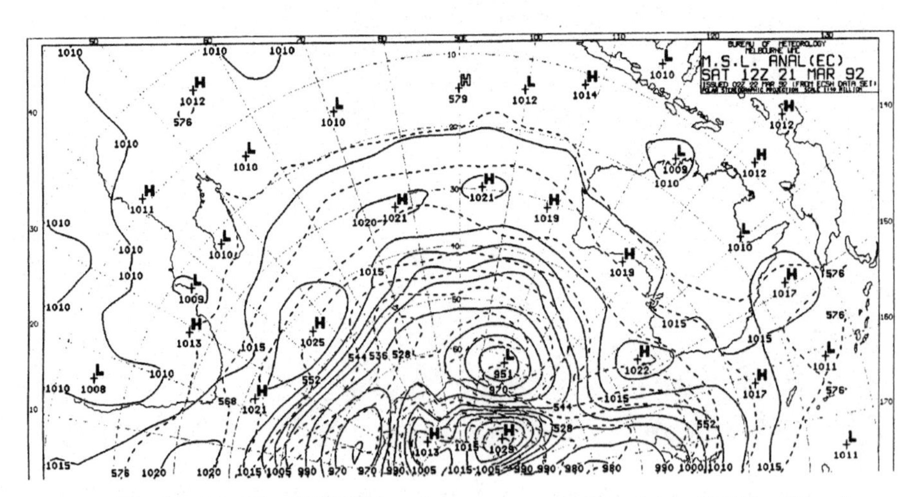

Figure 5.46 The ECMWF surface pressure analysis for 12 GMT 21 March 1992. From Pendlebury and Reader (1993).

station. The ECMWF 96 h forecast that verified at this time was very successful and is shown in Figure 5.47. The position of the low off the coast was quite good and the central pressure was especially well predicted at 950 hPa. Although the central pressure of the high in the interior was too large, the pressure gradient over Casey was essentially correct and the strong winds were well forecast. This case illustrates that the current generation of forecast models can give good guidance in the coastal region, even out into the medium range on occasions.

To try to assess the current accuracy of the operational analyses and forecasts over the Antarctic, the Physics and Chemistry of the Atmosphere group of the Scientific Committee on Antarctic Research recently organised a project to consider three months of operational products in detail. The project was called the Antarctic First Regional Observing Study of the Troposphere (FROST) and was conducted by a number of nations active in Antarctic meteorological research. They considered the forecasts for the area south of 55° S and verified the forecasts from the ECMWF, the UK Meteorological Office, the USA's National Meteorological Center and the Australian National Meteorological Centre against each model's own analyses. The results for July 1994 are shown in Figure 5.48 in terms of the bias and anomaly correlation of the 500 hPa height field. The anomaly correlation figures suggest that the forecasts have

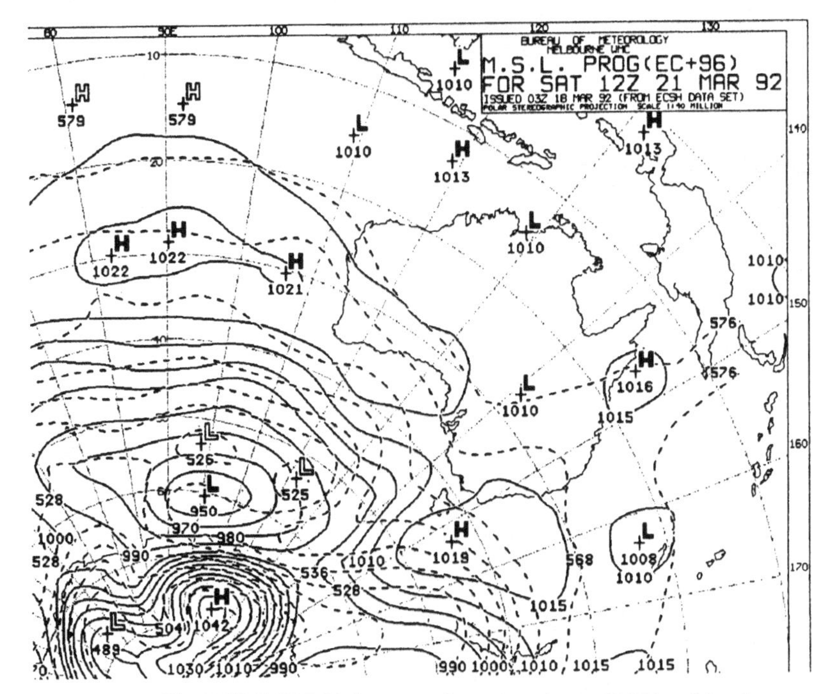

Figure 5.47 The ECMWF 96 h forecast, from data at 12 GMT 17 March 1992, valid at 12 GMT 21 March 1992. From Pendlebury and Reader (1993).

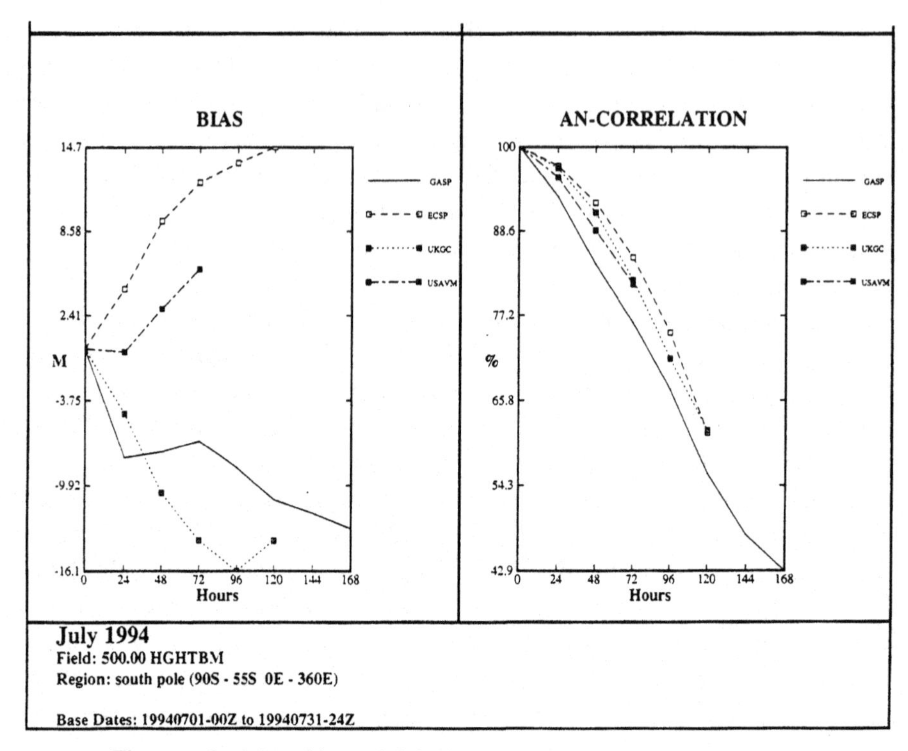

Figure 5.48 Mean biases (left-hand panel) and anomaly correlations (right-hand panel) of 500 hPa height forecasts from the GASP, ECMWF (labelled ECSP), UKMO (UKOC) and USANMC (USAVM) models during July 1994.

value out to about five days if a value of 60% is taken as the limit of useful skill. This indicates that the forecasts are about 20% poorer than in the mid-latitude regions. The 500 hPa height bias figures, shown in Figure 5.48(b), indicate quite large bias errors developing during the forecasts, although the signs for the various models are different. This is probably a result of the forecast centres concentrating their model assessment and development work on the mid-latitude regions.

Forecasting the main meteorological elements

Even in mid-latitude areas numerical models of the atmosphere cannot provide all the information needed to give a comprehensive weather forecast and there is still an important role for the forecaster in assessing the model output and in utilizing other data, such as satellite imagery. This is especially true in the Antarctic where the models cannot represent the smaller scale developments and topographic forcing affects day-to-day weather variations in many areas. In this section we will examine how each of the main meteorological elements is forecast and how the numerical model output and other data are merged to make a prediction.

(i) **Surface pressure.** The surface pressure fields from numerical models are used to predict the movement of the main weather systems and to monitor new developments on the synoptic scale. In forecasting for periods up to 12 h ahead, satellite imagery is also of great value insofar as it shows the detailed structure of depressions and features such as frontal bands and the locations of low-pressure centres. In a situation in which there is a large, quasi-stationary low in the coastal region, the system may be much more complex than analysed by the model and the imagery is very important in allowing the details of multiple centres to be resolved.

Mesoscale low-pressure systems (mesocyclones) are rarely analysed or forecast by the models because they often have a horizontal length scale of only a few grid lengths. Predicting where mesocyclones will form is therefore almost impossible at present, although the model thickness fields can show regions of increased thermal gradient, showing where the satellite imagery should be monitored for possible developments. A number of mesocyclones develop in association with minor thermal troughs that should be carefully monitored for developments on the mesoscale. Once a mesocyclone has formed, satellite imagery can be used to follow its track and estimate its position over the following 6–12 h. The model's upper wind fields are also valuable in determining the steering of a mesocyclone.

(ii) **Temperature.** To predict near-surface temperatures accurate knowledge is needed of the air mass that is expected to be over a location. This can be obtained from model thickness fields with modifications being made for the passage of the air mass over a warm ocean or cold ice surface.

(iii) **Precipitation.** The main means of predicting precipitation in the coastal region is through the output of the numerical models. If such data are not available then satellite imagery can be used to estimate the time of arrival of frontal bands, although there is no way at present to determine how active a particular front is. The precipitation on a front is therefore estimated from the form of the cloud, the cloud top temperatures and the nature of the upper circulation. Decisions on whether precipitation will be in the form of rain or snow are made using the model thickness fields. A good knowledge of local weather characteristics is important in estimating how much enhancement of precipitation will take place via lifting over topographic barriers.

(iv) **Surface wind.** The accuracy with which surface winds can be predicted is very dependent on the area being considered. Over the ocean areas the low-level winds from numerical models can be used with some minor modifications in the light of satellite imagery, which may result in the adjustment of the positions of fronts or vortex centres. In the coastal region, more care needs to be taken because local wind systems affect many areas so that the output of the models, which often have a very crude representation of the topography and cannot reproduce local effects, cannot be trusted. Certain areas are prone to persistent katabatic or barrier winds and the surface winds here will often bear little relationship to the broad-scale flow. Under these conditions good knowledge of the local topography and how the local effects can enhance or reduce the gradient wind is essential for a forecaster trying to predict the conditions likely to be experienced.

Over the interior of the continent the surface winds from AWS data can be used to predict conditions where there is a reasonable distribution of sites, such as on the Ross Ice Shelf and in parts of East Antarctica. Elsewhere, the motion of low cloud in sequences of satellite images can provide some indication of how the wind field may develop, but this must be used with great care. In mountainous regions it may be possible to see waves in low cloud and these can show the wind direction, although no estimate of the speed can be obtained.

(v) **Visibility.** The prediction of visibility is difficult because it is dependent on many factors, including the surface wind speed, air temperature, surface skin temperature, the availability of moisture and the amount of cloud over an area. Although, as discussed elsewhere in this section, many of these elements can be predicted, the formation, maintenance or dispersion of fog or mist can be affected by one crucial parameter in a particular situation. However, a knowledge of the air masses in an area is of great importance in predicting visibility as well as the factors that can change the origins of the air, such as the arrival of a front. Shallow fog can usually be detected in the daytime 3.7 μm satellite imagery with sequences of images showing the movement, increase or dispersion of the fog.

Blowing snow can also affect the visibility quite markedly and, on occasions, bring it down to fog limits. However, blowing snow reduces visibility over a relatively shallow depth of several metres and above this the airborne visibility will remain quite good, provided that there is no falling snow. The predicted surface wind is a good guide to whether blowing snow will occur and as a rough guide, blowing snow, i.e. snow blowing above head height (otherwise classed as drifting snow) can be expected to begin with speeds of 10 m s^{-1} or more.

(vi) **Contrast.** This parameter, which indicates the extent to which the surface and sky can be discriminated, is very important in flying operations insofar as it determines the feasibility of making a landing on a snow or ice surface.

It can be forecast using the predicted low-cloud field and knowledge of the local topography. Prediction of contrast is usually restricted to the next 12 h.

(vii) **Horizontal definition.** This is defined as the ease with which the boundary between the ground and the sky can be determined. It is a parameter most appropriate over ice shelves or areas where there are no mountains or nunataks visible. As with surface contrast, the horizontal definition is affected by the cloud cover, with an opaque layer of water droplet cloud producing the worst conditions. The presence of clear water leads off an ice shelf may be an important element in enhancing the horizontal definition in these areas. Horizontal definition can be predicted once the forecast of cloud has been made.

(vii) **Upper winds.** Forecasts of upper winds are required for aviation purposes, data usually being needed at the standard flight levels of 5000, 10 000 and 18 000 feet. Model winds at various standard levels form the basis of these forecasts, with modifications only being made if the model clearly has a significant error in the location or intensity of some feature.

(viii) **Cloud.** The coastal region of the Antarctic has extensive cloud, a great deal of it being well removed from the major weather systems. The arrival of frontal cloud bands can be forecast using a combination of the model surface pressure fields and satellite imagery. However, the non-frontal cloud is more difficult to predict and satellite imagery is often used to determine the amount of cloud that is associated with a particular air mass and then estimating its arrival time by extrapolation from a sequence of images. Such a technique can obviously only be applied for about 12 h ahead.

In the interior parts of the continent, prediction of where cloud will develop is not possible at present and the only option is to adopt a nowcasting approach and to consider the advection of pre-existing cloud features. With good knowledge of the topography it is sometimes possible to estimate where cloud will be enhanced by forced ascent. Because few major weather systems penetrate inland of the coastal region, clouds here are usually disorganised and slow-moving. Monitoring sequences of satellite images can show the direction of travel and whether the cloud is developing or dissipating.

Determining the height of clouds is relatively simple if infra-red satellite imagery is available because the brightness temperatures from the cloud tops can be converted into a height by the use of a nearby radiosonde ascent or a climatological temperature profile. On the other hand, estimating the cloud base is extremely difficult

Operational forecast centres

In order to improve the forecasts issued within the Antarctic a number of forecasting offices have been established on the research stations in recent years. These use analyses and forecasts produced by the main numerical weather prediction centres located outside the Antarctic as well as observational data collected within the Antarctic. Obviously, good satellite communications links are required to outside the Antarctic because the NWP products consist of many megabytes of data that must be transferred to the stations every 12 or 24 h, depending on the requirements. High-resolution satellite imagery is also an important element in any forecasting system so that most forecast centres have advanced satellite receiving systems. At present there is limited international co-operation concerning meteorological services in the Antarctic and the development and improvement of meteorological facilities is being dealt with at the national level. However, consultation takes place via the SCAR Committee of Managers of National Antarctic Programmes and the WMO Working Group on Antarctic Meteorology.

Some of the main forecast centres in operation at present are the following.

(i) McMurdo Station, where guidance is provided for the C130 flights from New Zealand. Here data from the AWSs on the Ross Ice Shelf are collected via the Data Collection System on the NOAA polar orbiting satellites and used to monitor the surface conditions in the vicinity of the station. With the high-resolution satellite imagery, AWS observations and computer forecasts available it is possible to follow developments on the mesoscale.

(ii) Australia has recently established an Antarctic Meteorological Centre (AMC) at Casey Station to produce analyses (including sea ice analyses) and forecasts for the Australian Sector of the Antarctic. The AMC receives NWP products from NMC Melbourne and has an AVHRR satellite receiver to take high-resolution satellite products. A team of up to three forecasters provides an 18 h per day forecast service during the summer months, with products issued on HF fax as well as via satellite communication methods. A reduced winter service involves NWP products issued from NMC Melbourne being re-transmitted on HF fax from the AMC. The forecasts being produced are particularly important for aviation and shipping. The aviation forecasts serve intra-continental flights between Australian stations and field camps, and serve the recently re-instated tourist overflights of coastal regions of eastern Antarctica which depart and return to Australia the same day. Marine products serve coastal shipping in the Australian sector as well as shipping on the high seas south of 50° S between 80° E and 160° E.

(iii) Rothera, on the Antarctic Peninsula, is the centre for British flying operations within the Antarctic and also receives routine flights from the Falkland Islands. During the Austral summer a forecaster from the UK Meteorological Office is based at the station and provides a full forecasting service for the air operations and the two British research ships. Output from the UK Meteorological Office 19-level global model is sent to the base twice daily via Inmarsat whereas satellite data are available from the AVHRR receiver at the station. Experiences of forecasting for the Antarctic Peninsula region have been described by Wattam and Turner (1996).

(iv) Chile operates an Antarctic Meteorological Centre at its Presidente Frei Station, where analyses are prepared and forecasts issued for aviation and ship operations. Surface and upper air charts are received via fax from Buenos Aires and the station also broadcasts their analyses and forecasts via fax.

(v) Vicecomodoro Marambio (Argentina) has an Aeronautical Meteorological Office, where manually plotted charts are prepared and issued, together with forecasts of upper winds. Area forecasts, alerts, shipping forecasts and gale warnings are also issued for the Antarctic Peninsula sector.

(vi) Molodeznaja is the main Russian station for meteorological analysis and prediction and the station also broadcasts various products by fax. The station is also equipped with a satellite link.

Casey, McMurdo, Marambio, Molodeznaja and Presidente Frei are all stations that have responsibility for carrying out data processing activities and providing meteorological services in the Antarctic. They all prepare analyses and forecasts for the whole Antarctic or for a specific sector and make these available to other stations. More details can be found in the *WMO Manual on the Global Data Processing System*, Volume II (World Meteorological Organisation, 1992).

5.6 Future research needs

In this chapter we have considered our present understanding of the development and structure of synoptic-scale weather systems and their climatological distribution in the Antarctic. Despite almost 40 years research into these systems since the IGY there are still major gaps in our knowledge that can only be filled by intensive *in situ* data gathering campaigns, modelling experiments and the use of satellite data. Some of the most pressing needs are for the following.

- A long-term study of the areas of cyclogenesis, depression movement and decay based on high-resolution satellite imagery. The work of Streten and Troup (1973) and Carleton (1979) provided a valuable first look at this problem but needs to be repeated for a long, multi-year period.
- Information on weather systems over the continent is especially poor and the data from the research stations and AWSs need to be used in conjunction with satellite imagery to produce a reliable climatology of depression activity in the interior and to assess the role of these systems in the precipitation budget.
- The degree to which cyclogenesis is taking place in the coastal area has only become apparent in the last few years and case studies of the mechanisms behind these developments are needed.
- The performance of operational numerical models over and around the continent is still not well understood and more case studies of their ability to simulate cyclogenesis events and the penetration of depressions into the interior are required.

Chapter 6

Mesoscale systems and processes

6.1 Local wind systems

6.1.1 The effect of local topography on katabatic flow

In Section 4.3, we saw how the near-surface wind field over much of Antarctica could be explained using simple diagnostic models of the katabatic wind. Such models provide realistic simulations of the mean wind at stations where the local topographic slope is reasonably uniform and not too great. However, the neglect of the non-linear inertial terms in these simplified models is not justified in regions where the topographic slope varies significantly. In such regions the advection of momentum and heat by the katabatic wind must be taken into account in order to model the local wind system correctly.

A number of coastal regions of Antarctica are subject to exceptionally strong katabatic winds. These include the coast of Adélie Land between Port Martin and Cape Denison. The annual mean wind speed at Cape Denison, measured by Mawson's Australasian Antarctic Expedition during 1912–14, was 19.3 m s^{-1}, with monthly mean speeds never dropping below 12.9 m s^{-1} (Parish, 1981). At nearby Port Martin, occupied by a French expedition during 1950–52, mean wind speeds were only slightly lower, with an annual mean of 16.9 m s^{-1}. However, the zone affected by these intense katabatic winds appears to be limited and does not extend as far as Dumont d'Urville, some 65 km west of Port Martin. Limited information from inland traverses suggests that the region of extreme winds extends at least 250 km inland from Cape Denison (Parish and Wendler, 1991). A second region where an 'extraordinary' katabatic regime prevails is Terra Nova Bay on the western coast of the Ross Sea. Intense winds, comparable to those of Adélie Land, were first observed during the enforced wintering of Scott's Northern Party in 1912 and recent automatic weather station (AWS) measurements (Bromwich, 1989a) indicate monthly mean wind speeds of 14–18 m s^{-1}, with a directional constancy of 0.94–0.99.

It is clear that katabatic winds of such intensity and persistence can only be localised phenomena. Simple heat budget considerations (Section 4.3.2) demonstrate that, if such strong katabatic outflow were to prevail continuously all around the coastline of Antarctica, the supply of cold, negatively buoyant air generated by radiative cooling in the interior would be rapidly exhausted. Topographic slopes in the vicinity of the two areas mentioned above are no greater than those found in other coastal regions of Antarctica, where a much weaker katabatic outflow is observed. The limited geographical extent of regions subject to strong katabatic winds suggests that the wind regime in these areas is controlled by some peculiarity of the local topography rather than by slope alone.

In Section 4.3, we noted that simulations of the continental-scale katabatic flow using simple diagnostic models indicate the existence of regions of confluence and diffluence associated with mesoscale variations in topography. Parish (1982) demonstrated that confluent regions, which will be associated with enhanced katabatic flow, are generally situated just to the east of meridional ridge features or just west of meridional valleys. A significant ridge–valley feature extends inland from the Cape Denison region. Application of the two-layer diagnostic model described in Chapter 4 to this local topography (Parish, 1981) indicates that there is a considerable confluence of streamlines just inland of Cape Denison and Port Martin. This suggests that negatively buoyant air is being drawn from a large area of the interior and then concentrated onto a small section of the coastline, giving rise to strong katabatic winds in this region.

Although diagnostic models can usefully indicate the expected positions of such regions of confluence, they cannot be used to predict the resulting wind speeds since non-linear inertial terms in the equations of motion will be significant. Kikuchi and Ageta (1989) estimated the magnitude of the inertial term in a simple model of katabatic flow and concluded that it is significant when slopes vary on scales of 50–200 km. Parish (1984) used a three-dimensional hydrostatic primitive equation model to study the dynamics of confluence zones. Model integrations were performed over an idealised parabolic ice sheet profile, both with and without a 400 km wavelength topographic 'wave' imposed on the basic profile to generate a meridional ridge–valley system resembling the topography inland of Cape Denison. With a topographic 'wave' amplitude of only 250 m, a strong confluence zone developed to the east of the ridge and coastal wind speeds downstream of the confluence zone were enhanced by about 60% of their value in the control run with no east–west topographic variations, reaching a realistic

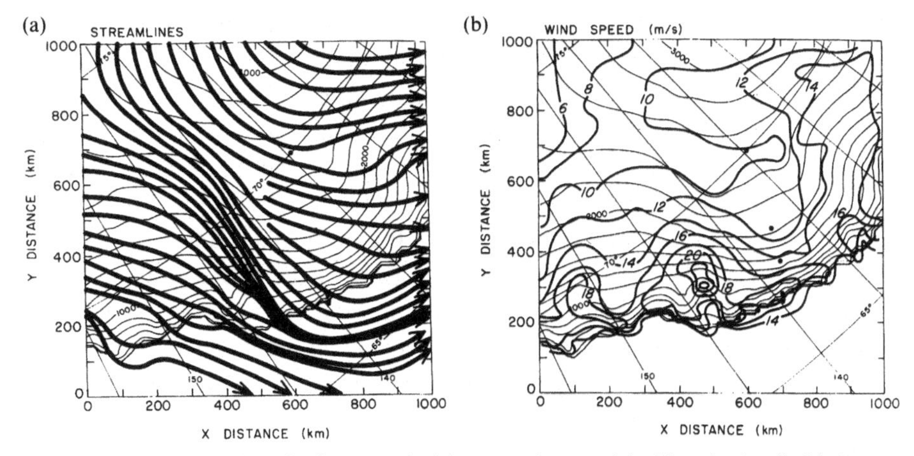

Figure 6.1 Results from a primitive equation model of katabatic wind behaviour in Adélie Land showing near-surface streamlines (left-hand panel) and wind speeds (right-hand panel). Contours of the model topography at 200 m intervals are also shown. From Parish and Wendler (1991), *International Journal of Climatology*. Copyright (1991) Royal Meteorological Society. Reprinted by permission of John Wiley & Sons Ltd.

18 m s^{-1} in the vicinity of Cape Denison. Reducing the amplitude of the topographic 'wave' to 125 m had only a small effect on the flow. Further runs carried out with an imposed synoptic pressure gradient in addition to the katabatic forcing demonstrated that synoptic forcing could change the location of maximum outflow slightly but its effects were of secondary dynamical importance to the katabatic convergence. Parish and Wendler (1991) applied this model to a more realistic representation of the Adélie Land topography, their results are shown in Figure 6.1. The streamline analysis shows a strong confluence region extending about 300 km south-eastwards inland of Cape Denison, with wind speeds in excess of 24 m s^{-1} along the confluence axis. A secondary confluence zone is indicated between the Mertz and Ninnis Glaciers but few observations are available to confirm the existence of extreme katabatic winds in this region.

The katabatic wind regime in the vicinity of Terra Nova Bay has been studied by Bromwich *et al.* (1990). The local topography here is somewhat more complex than in Adélie Land, with a number of glaciers converging on the bay. Katabatic wind streamlines inferred from an aerial survey of sastrugi orientations indicated the existence of a broad-scale confluent region inland of Terra Nova Bay and application of a primitive equation model produced a surface flow pattern that was generally in agreement with these observations, except where local topographic features were not resolved on the 32 km model grid. An instrumented aircraft was used by Parish and Bromwich (1989) to study the structure of the confluence region. Measurements made at a flight level approximately 170 m above ground level showed a clearly marked vertical boundary to the negatively buoyant strong katabatic flow within the confluence zone. The horizontal extent of this confluence zone appears to be substantially smaller than that seen in Adélie Land and the region of strong katabatic winds only extends some 100 km inland from the coast. Despite this difference in scale, it appears that the same dynamical processes are operating in both regions: local topography generates an inland confluence zone, concentrating negatively buoyant air drawn from a wide area into an intense katabatic stream that propagates down to the coast. This process has been studied most intensively in the two regions described above. However, the continent-wide katabatic wind simulations by Parish and Bromwich (1987) indicate the existence of other confluence zones where enhanced katabatic winds might be expected but, as yet, no measurements are available to confirm their presence.

The existence of confluence zones thus appears to explain the concentration of the katabatic outflow around the coasts of Antarctica into a number of narrow regions. Such a distribution of outflow has confounded attempts to calculate water vapour and energy budgets for the Antarctic atmosphere from radiosonde data (see Chapter 4), since a large proportion of the horizontal transport must occur in the regions of concentrated flow, which are not necessarily well sampled by the sparse radiosonde network. The regions of intense katabatic flow also play an important role in the energy budget of the South Polar regions by maintaining coastal polynyas (Kurtz and Bromwich, 1985; Bromwich and Kurtz, 1984). Although limited in geographical extent, heat fluxes and sea ice production rates in these polynyas (see Section 4.2) are of great significance. Such polynyas acted as a magnet for early expeditions, since they provided easy access to the Antarctic coast during early summer. Accesibility, however, comes at a price and as one violent katabatic storm followed another throughout the winter, the members of Mawson's expedition and Scott's Northern Party may have regretted choosing such sites for their bases!

6.1.2 Local thermal effects and the development of katabatic winds

Besides neglecting the inertial terms in the equations of motion, the simple diagnostic models of katabatic winds discussed in Section 4.3 also made the simplifying assumption that there were no variations in the strength or depth of the surface inversion along the slope. Under certain circumstances, such variations can contribute significantly to the net katabatic forcing and can cause what, at first sight, appears to be anomalous behaviour of the katabatic wind. Much of our understanding of these effects has come from the results of the IAGO (Interactions Atmosphère–Glace–Océan) experiment, a joint USA–French programme that established a chain of five AWSs from the Adélie Land coast near Dumont d'Urville to Dome C, 1080 km inland at an elevation of 3280 m (Wendler *et al.*, 1993).

Kodama and Wendler (1986) analysed surface winds measured by these AWSs. During the winter months, the vector mean wind at station D80 (elevation 2450 m) was directed more downslope than was that at station D57 (elevation 2103 m), even though the surface slope at the latter station was nearly four times that at the former. This is at variance with the simple two-layer model, which predicts that the angle between the wind and the fall line decreases monotonically with increasing slope.

If variations in potential temperature and inversion layer depth along the slope are taken into account, the total downslope pressure gradient force in the two-layer katabatic wind model becomes (following Mahrt and Larsen (1982))

$$F=\underbrace{\frac{g}{\theta}\alpha\Delta\theta}_{\text{DB}}-\underbrace{\frac{g}{\theta}h\frac{\partial(\Delta\theta)}{\partial x}}_{\text{IS}}-\underbrace{\frac{g}{\theta}\Delta\theta\frac{\partial h}{\partial x}}_{\text{ID}} \tag{6.1}$$

The notation is as in Section 4.3.2: α is the magnitude of the slope, h the katabatic layer depth, θ the potential temperature of the overlying layer and $\Delta\theta$ the potential temperature deficit of the katabatic layer. The first term on the right-hand side of Eq. (6.1), DB, is the familiar downslope buoyancy force, the second (IS) and third (ID) terms arise as a result of variations in the strength and depth respectively of the katabatic layer (which is often identified with the surface inversion layer). The ratios of these three terms are

$$\begin{array}{ccccc} \text{DB} & : & \text{IS} & : & \text{ID} \\ 1 & : & \dfrac{h}{\Delta Z}\dfrac{\Delta\theta_x}{\Delta\theta} & : & \dfrac{\Delta h}{\Delta z} \end{array} \tag{6.2}$$

where $\Delta\theta_x$ and Δh are scales for changes in $\Delta\theta$ and h respectively over a downslope distance l, and $\Delta z=\alpha l$ is the change in surface elevation over this distance.

Kodama and Wendler (1986) found that the surface potential temperature at Dome C was significantly lower than that at D57, particularly during the winter months, while near-adiabatic conditions prevailed from D57 to the coast. Using these data, it is possible to estimate the relative magnitude of the pressure gradient due to changes in inversion strength (Table 6.1). This term appears to be comparable in magnitude to the downslope buoyancy force during the winter. Since the surface potential temperature decreases (and hence the inversion strength increases) towards Dome C, this term will act to enhance the basic katabatic forcing, thus explaining the anomalous downslope component of the wind at D80. No measurements of inversion depth were available at the AWS sites so it is not possible to assess the importance of the ID term in Eq. (6.1) directly. However, Kodama and Wendler (1986) suggested that it is significant, because the katabatic layer will deepen towards the coast as a result of entrainment of

Table 6.1 *Calculation of the relative magnitude of the inversion strength gradient term (IS) in Eq. (6.1) at AWS D80. After Kodama and Wendler (1986).*

	Summer	Winter
h (m)	200	350
$\Delta\theta$ (K)	3	15
$\Delta\theta_x$ (K)	6	10
Δz (m)	400	400
IS/DB	1.0	0.6

overlying air. It is clear from Eq. (6.1) that such deepening will act to decelerate the katabatic flow.

Horizontal temperature gradients can also exert considerable control over katabatic winds as they approach the coast. During the summer months, katabatic winds from the interior of Adélie Land frequently do not reach the coastal station of Dumont d'Urville . At this time of year, the sea is ice-free and temperatures at coastal stations have only a weak diurnal cycle while, over the continental ice slopes, a large diurnal cycle is observed (Wendler *et al.*, 1988). The temperature at AWS D10, 5 km inland of Dumont d'Urville at an elevation of 240 m, is observed to exceed that at Dumont d'Urville at some time on 88% of all summer days (Pettré *et al.*, 1993). The pressure gradient associated with this reversal of the temperature gradient can drive an anabatic 'sea breeze' circulation, which can prevent the katabatic flow reaching the coast (Figure 6.2(a)). During the winter, the effect is reversed. Katabatic outflow maintains open water in a coastal polynya, where large heat fluxes will warm the atmosphere relative to that over the continental slopes. The horizontal temperature gradient will now act to enhance the katabatic flow (Figure 6.2(b)). This feedback mechanism may be important in the maintenance of coastal polynyas.

6.1.3 The response of katabatic winds to the synoptic environment

In Chapter 4 we saw that, over the interior of the continent, the downslope pressure gradient resulting from katabatic forcing is generally much larger than that associated with synoptic-scale developments. As a result of this, the interior surface winds show little response to changes in synoptic forcing and are largely determined by local topography. In the interior, such synoptic forcing is weak, since few vigorous weather systems either develop on or penetrate into the high plateau (see Chapter 5). In order to assess the effect of synoptic forcing on katabatic winds, Parish (1982) calculated power spectra of surface pressure for a

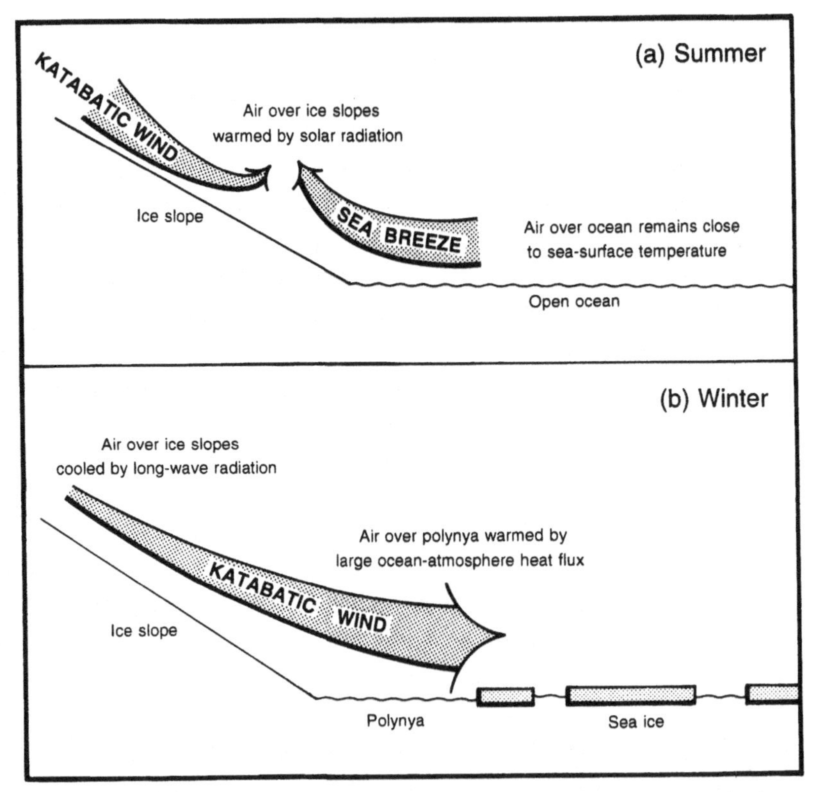

Figure 6.2 Illustrating the effect of land–sea temperature contrasts on the katabatic wind. (a) The summer situation and (b) the winter situation.

number of Antarctic stations. Spectra for coastal stations showed a significant peak at 3–5 days, a typical timescale for cyclonic development, whereas such a peak was much weaker or entirely absent for stations on the interior plateau. We may conclude that synoptic forcing is likely to modify katabatic winds to a significant extent only in coastal regions.

Murphy and Simmonds (1993) used a GCM to investigate the relative importance of katabatic and synoptic forcing in generating strong winds in the vicinity of Casey station. Compositing surface pressure anomalies for 20 strong wind events observed in a 600-day perpetual July simulation revealed a clear pattern: the strong wind events were associated with a blocking anticyclone to the east of the station and a mobile low to the northwest. A similar anomaly pattern was seen at 500 hPa, indicating that the anomalous forcing is largely barotropic. Interestingly, the geostrophic wind forced by the pressure anomaly is not in the same direction as the katabatic wind and some cancellation results; however, both the katabatic and the synoptic forcing were observed to increase during strong wind events. This indicates that the interaction of the two forcing mechanisms is not simply additive and that more complex processes need to be considered. Strong wind events were also observed to be associated with a strengthening of the inversion over the high plateau, which would increase the katabatic forcing. It is not clear whether these changes in the inversion are related to synoptic forcing.

The relative importance of synoptic and katabatic forcing of the low-level winds appears to vary considerably from one coastal location to another. At Dumont d'Urville, katabatic forcing generally appears to be dominant, although synoptic gradients do affect the observed wind speeds to some extent (Parish *et al.*, 1992). At Halley (King, 1989) and Neumayer (Kottmeier, 1986), which are both characterised by somewhat gentler slopes than is Dumont d'Urville, katabatic forcing and synoptic pressure gradients appear to be of comparable importance in determining the surface wind regime.

6.1.4 Katabatic jumps – 'Loewe's phenomenon'

A dramatic feature sometimes observed in regions subject to strong katabatic flow is the katabatic jump or 'Loewe's phenomenon' (Valtat, 1959). This is a region of rapid transition from strong katabatic flow to near-calm conditions, occurring over a horizontal distance of less than 100 m. Common characteristics of katabatic jumps noted by a number of observers (Valtat, 1959; Lied, 1964; Pettré and André, 1991) are shown schematically in Figure 6.3 and include the following.

(i) A shallow layer of dense drift snow is observed in the strong katabatic wind upstream of the jump.

(ii) The jump itself is marked by a narrow region of intense turbulence. Blowing snow in this region is lifted up to 100 m above the ground,

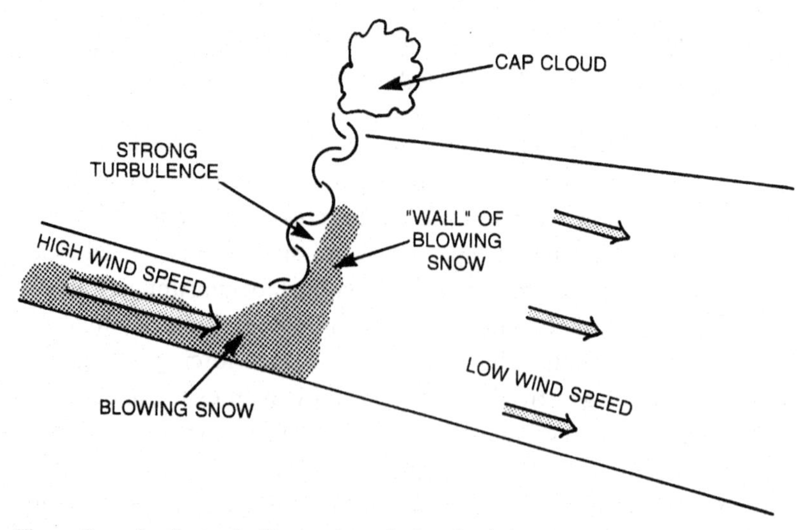

Figure 6.3 A schematic illustration of a katabatic jump.

giving the appearance of a 'wall' of drift snow, which is often capped by small cumuliform clouds. These features indicate large vertical velocities at the jump.

(iii) Downstream of the jump, winds are calm or light and variable, with no drift snow.

(iv) Large temperature and pressure differences are observed across the jump. Lied (1964) measured pressure differences of up to 20 hPa. Schwerdtfeger (1984, p. 71) questioned the reliability of such observations on the grounds that dynamical pressure errors may affect the measurements. However, the observed pressure jumps do appear to be genuine.

(v) The jump may be quasi-stationary or may propagate slowly upstream or downstream.

(vi) Katabatic jumps are generally observed close to the coast, on the steep ice slopes descending from the interior plateau.

Ball (1956) attempted to explain katabatic jumps by analogy with the hydraulic jumps observed in channel flow. This simple model is illustrated in Figure 6.4. Upstream of the jump, a thin katabatic layer of depth h, density ρ and speed u_1 lies under still air of density ρ'. If the Froude number

$$F_1 = u_1\left(\frac{\rho-\rho'}{\rho}gh\right)^{-1/2} \tag{6.3}$$

is greater than unity, the flow is 'shooting' and a hydraulic jump may develop. Downstream of the jump, the katabatic layer deepens to $h+\Delta h$, the wind speed

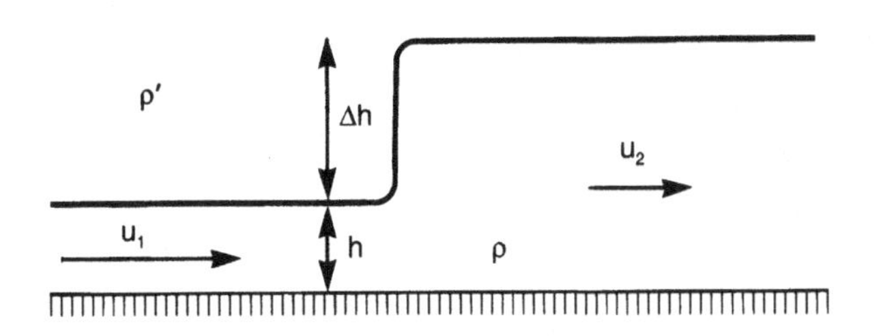

Figure 6.4 The hydraulic model of a katabatic jump

reduces to u_2 and the Froude number, F_2, will be less than unity. In hydraulic terminology, the flow in this downstream region is 'tranquil'.

The pressure difference across the jump is clearly seen to be

$$\Delta p=(\rho-\rho')g\,\Delta h \tag{6.4}$$

Imposing a requirement for conservation of mass and momentum at the jump, this may be written

$$\Delta p=\frac{\rho u_1^2}{2}\frac{(1+F_1^2)^{1/2}-3}{F_1^2} \tag{6.5}$$

The maximum pressure difference thus occurs when $F_1=2$. Pettré and André (1991) showed that this model predicts pressure jumps of 1–2 hPa for conditions typical of katabatic jumps observed during the IAGO experiment in Adélie Land, when pressure changes of up to 6 hPa were measured. Their observations suggest that the simple hydraulic jump model, Figure 6.4, is an oversimplification of the flow within the katabatic jump. High-resolution soundings upstream of two katabatic jumps displayed the following features.

(i) A cold, well-mixed (or slightly stable) layer, 50–80 m deep.

(ii) A capping inversion of about 3 K, 250 m deep. The high katabatic wind speeds extended through both the cold layer and the inversion layer.

(iii) An unstable layer above the inversion, which grew in depth downstream and was bounded by surfaces of approximately equal potential temperature.

(iv) A stable, transition layer in which conditions returned towards those of the undisturbed overlying atmosphere.

The crucial difference between these observations, shown schematically in Figure 6.5, and the simple model, is the unstable layer (iii) above the inversion. This is observed to develop from zero thickness at some point upstream to a depth of several hundred metres at the coast. The layer is bounded by isentropes of equal value, θ_0, which, in an adiabatic model, may be identified with streamlines so the Boussinesq form of Bernoulli's equation may be applied to calculate the pressure variations along these surfaces. The pressure perturbation, p', at point r_1 just upstream of the jump is given by

$$\frac{1}{2}\rho_0 u^2(z_1)+p'(z_1)-\rho_0 ug\left(\frac{T'(z_1)}{\hat{T}}\right)z_1=H_0 \tag{6.6}$$

where H_0 is a constant for the streamline, z_1 is the height of point r_1, T' is the temperature perturbation at this point, u is the wind speed and $\hat{T}$ and ρ_0 are the

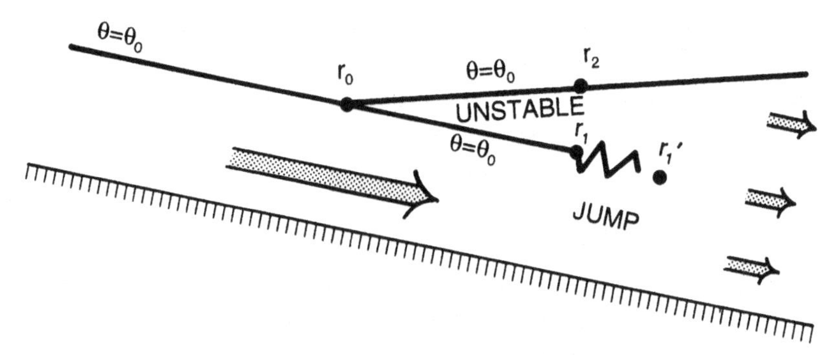

Figure 6.5 The katabatic jump model of Pettré and André (1991).

reference temperature and density. Since the upper bounding streamline also originates at point r_0, the pressure at r_2 is given by

$$\frac{1}{2}\rho_0 u^2(z_2)+p'(z_2)-\rho_0 g\left(\frac{T'(z_2)}{\hat{T}}\right)z_2=H_0 \tag{6.7}$$

By definition, $T'(z_1) = T'(z_2) = T_*'$. The dynamical pressure change along the upper streamline, $p'(z_2)$ will be small, for this is a region of comparatively slow flow. Hence

$$p'(z_1)=-\frac{1}{2}\rho_0(u^2(z_1)-u^2(z_2))-\rho_0 g\frac{T_*'}{\hat{T}}(z_2-z_1) \tag{6.8}$$

At the jump, the lower streamline loses its identity in the strong turbulent mixing. At point r_1', immediately downstream of the jump and at the same height as r_1, the flow is greatly reduced and the dynamical pressure perturbation, p', will be negligible. The background pressure will be the same as that at r_1, so the pressure difference across the jump is simply $p'(z_1)$, given by Eq. (6.8). Pettré and André (1991) showed that the above model can succesfully account for the large observed pressure changes across katabatic jumps in Adélie Land. The first term on the right-hand side of Eq. (6.8), which depends on the wind shear across the unstable layer, appears to contribute most to the pressure jump.

Gallée and Schayes (1992) studied the downslope development of Antarctic katabatic winds using a two-dimensional hydrostatic model. Their modelled wind field exhibits a sharp transition from strong katabatic flow to weak upslope flow about 10–20 km inland from the coast. Because their model was hydrostatic, it could not be expected to reproduce the complex (and clearly non-hydrostatic) flow observed within real katabatic jumps. However, the generation of a jump phenomenon in such a model suggests that rapid adjustment from katabatic flow to calm conditions may occur frequently in coastal regions. King (1993a) showed that, over the Brunt Ice Shelf, katabatic flows do not propagate far beyond the base of the steep coastal slopes and suggested that a katabatic jump may explain this behaviour.

Such behaviour is clearly not universal and we have already seen examples of katabatic flows affecting coastal sites. In the following section we shall examine further instances of katabatic flows that are observed to propagate over large distances from the regions in which they develop.

6.1.5 Propagation of katabatic winds across ice shelves and oceans

Satellite thermal infra-red (TIR) images of the Ross Ice Shelf frequently show narrow regions of relatively warm surface temperatures extending outward from the termini of the glaciers which drain through the Transantarctic Mountains or from Marie Byrd Land. Such features are visible on 98% of wintertime once-daily TIR images of the Ross Ice Shelf (Bromwich, 1989b). An example of these 'warm signatures' is shown in Figure 6.6.

Swithinbank (1973) and d'Aguanno (1986) identified the warm signatures as *Föhn* winds associated with adiabatically warmed air descending from the Transantarctic Mountains. However, Bromwich (1989b) pointed out that a *Föhn* wind posseses positive buoyancy relative to its environment and will not generally propagate far beyond the foot of a mountain range, while the warm signatures are sometimes observed to extend for hundreds of kilometres across the ice shelf. He suggested that the warm signatures result from strong katabatic winds moving across the ice shelf. Turbulence in the katabatic stream will keep the air well-mixed while, in the quiescent region outside the katabatic stream, a strong surface inversion will be present. Hence the *surface* temperature within the stream may be higher than that on the surrounding ice shelf, while the *bulk* temperature of the katabatic stream remains lower than that of the surrounding environment and the stream retains its negative buoyancy. The association of warm signatures with negatively buoyant katabatic winds was confirmed by aircraft measurements, discussed in Section 6.1.1, over the Nansen Ice Sheet in the vicinity of Terra Nova Bay (Parish and Bromwich, 1989). TIR images have proved to be a most useful tool for studying the behaviour of katabatic winds over ice shelves, particularly when other more quantitative data, such as AWS observations, are available to aid in the interpretation of the imagery.

Once a katabatic stream leaves the steep coastal slopes and flows onto an (essentially flat) ice shelf, it loses the forcing associated with the downslope pressure gradient and air within the stream will move subject to the Coriolis force, frictional retardation and synoptic pressure gradients. Strictly, perhaps, one should no longer call the wind 'katabatic', since it is no longer subject to katabatic forcing. However, it is convenient to label it according to its origins. In the absence of frictional and pressure gradient forces, air parcels in the katabatic stream would undergo simple inertial motion, curving to the left along a circular trajectory of radius

Figure 6.6 A thermal infra-red satellite image of the Ross Ice Shelf in mid-winter. Darker shades indicate a warmer surface. Warm katabatic signatures can be seen originating from the glaciers draining the Transantarctic Mountains and a well-developed katabatic surge is seen flowing across the ice shelf from the Siple Coast. Automatic weather station observations of wind and temperature have been overlaid on the image. The image is reproduced by permission of D. H. Bromwich, Byrd Polar Research Centre.

$$R_{\mathrm{I}}=\frac{V}{|f|} \tag{6.9}$$

where f is the Coriolis parameter and V the speed of the katabatic flow as it runs onto the ice shelf. R_{I} will be 50–150 km for typical katabatic wind speeds. Frictional effects will decelerate the flow. King (1993a) estimated that significant deceleration will occur over 20 km, whereas trajectory calculations by Bromwich and Kurtz (1984) suggested that an initial speed of 15 m s^{-1} will only be reduced to 10 m s^{-1} after 40 km of travelling across the ice shelf. Such inertial–frictional behaviour is clearly seen in the curvature of some of the warm signatures visible in Figure 6.6.

However, under certain circumstances, katabatic streams have been observed to propagate for much greater distances without significant deceleration or change in direction. Franklin Island in the Ross Sea occasionally experiences strong winds blowing from the west or northwest, which is in the same direction as the katabatic wind at Terra Nova Bay some 190 km to the northwest. TIR images have shown a warm signature extending from Terra Nova Bay across the Ross Sea pack ice to Franklin Island, confirming the katabatic origins of the winds observed at Franklin Island (Bromwich, 1989b). For the katabatic stream to propagate substantially further than R_I without significant change in direction, some other force must be acting to balance the Coriolis force. It is likely that synoptic or mesoscale pressure gradients provide the required support. No mechanism has been conclusively identified, although Bromwich (1989b) did note that such anomalous propagation of the Terra Nova Bay katabatic stream only occurs when westerly winds are present at mid-tropospheric levels.

An even more dramatic example of long-distance propagation of a katabatic stream is seen over the Ross Ice Shelf. On occasions, a warm signature associated with strong katabatic flow originating in West Antarctica is seen to extend from the Siple Coast in the southeastern corner of the ice shelf for over 1000 km in a northwesterly direction, reaching the Ross Sea in the vicinity of Ross Island. Bromwich *et al.* (1992) examined TIR images of the Ross Ice Shelf for the period April–August 1988. On 92% of days when suitable imagery was available, warm signatures were observed where the steep slopes of Marie Byrd Land run into the southeastern part of the ice shelf. On average, these signatures only extended for 170 km across the ice shelf but, on 21% of all days with imagery available, a katabatic 'surge' was observed and the warm signature propagated all the way to the Ross Sea. Secondary warm signatures were observed to extend from glaciers draining from the Transantarctic Mountains. These appeared to merge into the main 'surge' signature, strengthening it by adding further negatively buoyant air and, on one occasion (described by Bromwich (1992)), a surge developed from the combined outflow from these glaciers alone, with no contribution from Marie Byrd Land. A well-developed katabatic surge can be seen in Figure 6.6.

Clearly some support is required to enable katabatic streams to propagate for such long distances. On 82% of occasions on which katabatic surges occur, a synoptic-scale cyclone is observed over the Amundsen Sea. The cyclone generates a southeasterly cyclonic flow over the Ross Ice Shelf, which will be favourable to the progression northwestwards of the katabatic surge. Composite 500 hPa analyses for katabatic surge days show a cold upper trough extending northeastwards from the Ross Ice Shelf, with a warm ridge to the east extending south into Marie Byrd Land. Bromwich *et al.* (1994) studied the dynamics of katabatic surges using a mesoscale model. With no synoptic pressure gradient imposed, katabatic flows from West Antarctica turn to the left under the influence of the Coriolis force as they move across the flat ice shelf. However, the Transantarctic Mountains block progress in this direction and the cold katabatic air piles up against the northern side of the mountains, leading to the formation of a barrier jet, as discussed in the following section. If the synoptic pressure field associated with an Amundsen Sea cyclone is imposed, a more complex picture results. The layer of cold air becomes shallower adjacent to the Transantarctic Mountains than it is over the ice shelf, but between these two regions there is a narrow zone (about 20 km wide) of very deep cold air. This feature is somewhat reminiscent of the katabatic jumps discussed in the preceding section. The strongest flow is now no longer adjacent to the mountains but displaced northwards onto the ice shelf as a result of the interaction of the pressure field of the cyclone with that associated with the cold air stream.

The observation that, under the right synoptic conditions, strong katabatic flows can cross the entire Ross Ice Shelf has important implications. Katabatic surges are associated with mesocyclone formation (see Section 6.5) and play an important role in keeping open the polynya at the edge of the Ross Ice Shelf (Bromwich, 1992; Bromwich *et al.*, 1993). In addition to contributing significantly to the production of sea ice and to the atmospheric energy balance (see Chapter 4), ice front polynyas are believed to exert important controls over the oceanic circulation beneath ice shelves and, hence, the melting rate of the ice shelf. Any changes in the frequency or intensity of katabatic surges could thus affect the mass balance of the Antarctic ice sheets. It is rare for a mesoscale phenomenon to have such far-reaching consequences.

6.1.6 Barrier winds

If cold, stably stratified air at low levels is forced up against a mountain barrier by the prevailing synoptic flow, it will not have sufficient kinetic energy to cross the barrier if the Froude number

$$\mathrm{Fr} = U_0\left(gH\frac{\Delta\theta}{\theta}\right)^{-1/2} \tag{6.10}$$

is less than unity. Here, U_0 is the speed of the flow approaching the barrier, H is the height of the barrier and $\Delta\theta$ the potential temperature deficit of the cold air. If $\mathrm{Fr}<1$, the cold air stream will be unable to cross the barrier and it will dam up against the upwind side of the mountains. The cold air layer will deepen towards the barrier, generating a pressure gradient perpendicular to the barrier. This pressure gradient

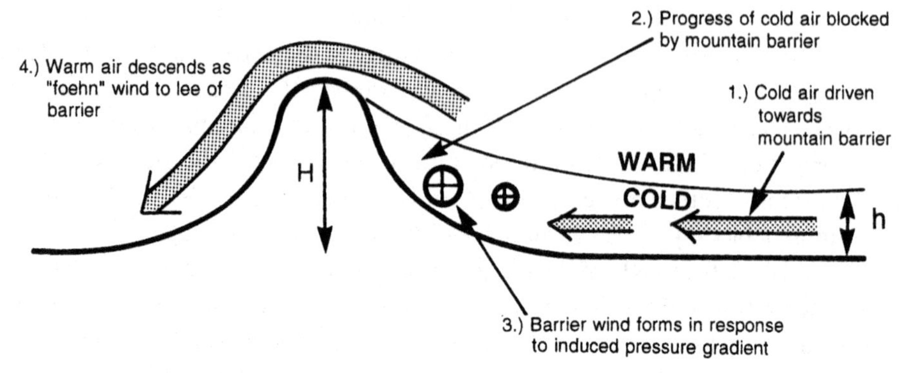

Figure 6.7 Barrier wind formation.

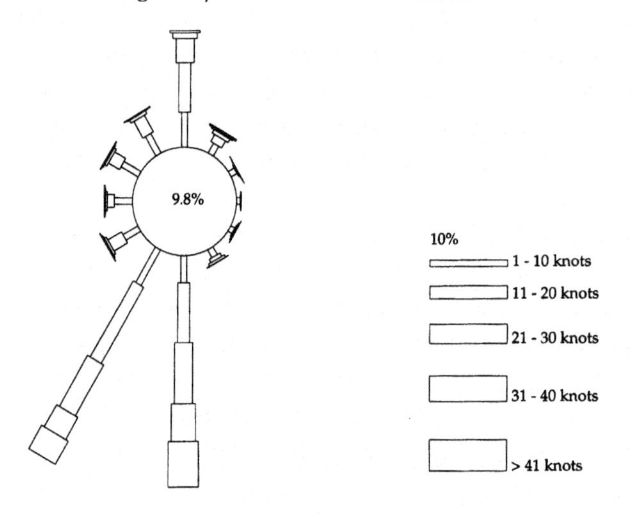

Figure 6.8 A wind rose for Butler Island automatic weather station (72.2° S, 60.3° W). The predominance of winds from the south and south–southwest indicates the influence of the barrier wind.

will support a geostrophically balanced 'barrier wind' in the cold air layer, which blows with the mountain barrier on its left (in the Southern Hemisphere). The process of barrier wind formation is illustrated schematically in Figure 6.7.

The strong static stability seen at low levels makes barrier winds a relatively common phenomenon in Antarctica. The mountains of the Antarctic Peninsula rise to an average height of 1500–2000 m, providing a very effective barrier to the progress westwards of cold, continental air driven towards the Peninsula by the prevailing low-level easterly winds over the Weddell Sea. Few stations have been occupied on the east coast of the Peninsula owing to the extreme inaccessibility of the region but the limited number of observations available show a high frequency of winds blowing from the south and southwest, which is parallel to the axis of the Peninsula. In recent years, measurements from a number of automatic weather stations have confirmed the existence of mountain-parallel winds along the east coast. Figure 6.8 shows a wind rose for the automatic weather station on Butler Island which is only a few kilometres east of the steep coast of the Peninsula and is clearly influenced by the barrier wind.

The influence of the barrier wind is felt at some considerable distance from the foot of the mountains. The observed drift northwards of pack ice in the western Weddell Sea (Limbert *et al.*, 1989; Kottmeier and Hartig, 1990) suggests that barrier winds may extend some way from the coast of the Peninsula and may play an important role in the regional sea ice dynamics (Schwerdtfeger, 1979). Kottmeier and Hartig (1990) produced regional wind roses for the Weddell Sea, based on measurements from drifting buoys and observations made from the ships *Endurance* and *Deutschland*, which became beset in the Weddell Sea pack ice. Frequent southwesterly winds are seen as far east as 45° W in the southern part of the Weddell Sea – this is some 400 km from the east coast of the Peninsula. Dynamical considerations (Overland, 1984) show that the influence of the barrier will be significant within the Rossby radius of deformation

$$R_{\mathrm{R}}=\frac{1}{|f|}\left(gH\frac{\Delta\theta}{\theta}\right)^{1/2} \tag{6.11}$$

where h is the depth of the cold air layer. For conditions typical of the Weddell Sea in winter, $h=1000$ m, $\Delta\theta=10$ K and $\theta=250$ K (Parish, 1983), giving $R_{\mathrm{R}}=140$ km.

Parish (1983) modelled the development of barrier winds along the Antarctic Peninsula using a two-dimensional hydrostatic mesoscale model. The model was initialised using a temperature profile typical of Weddell Sea winter conditions and easterly geostrophic winds of various strengths were imposed below 2000 m. In all model runs, mountain-parallel winds developed on the east side of the Peninsula and extended up to 300 km to the east of the barrier. With an easterly geostrophic wind of 10 m s^{-1}, mountain-parallel winds of over 10 m s^{-1} were generated at the surface and a barrier jet of over 20 m s^{-1} speed developed above the eastern slopes of the Peninsula. An interesting feature of the simulations was the tendency for strong downslope '*Föhn*' winds to develop on the west side of the Peninsula, even when a realistic westerly geostrophic wind was imposed above 2000 m. The association of *Föhn* storms on the west of the Peninsula with barrier winds to the east is well documented and was discussed by Schwerdtfeger (1984, pp. 90–94). Air in the *Föhn* wind originates in the warm layer overlying the blocked cold layer to the east of the Peninsula (see Figure 6.7), and will thus give rise to relatively warm conditions to the west of the Peninsula. This *Föhn* effect, together with the blocking of the cold, low-level easterly flow, contributes to the strong temperature contrast across the Antarctic Peninsula (see Chapter 3).

Barrier winds blowing from the south or southwest are observed along the whole length of the Antarctic Peninsula. Beyond the northern tip of the Peninsula, there is no mountain-generated pressure gradient to support the accelerated flow and the barrier wind must undergo geostrophic adjustment. In the absence of other forces, air parcels in the flow will undergo simple inertial adjustment, moving anticlockwise along a circular trajectory with a radius, R_I, given by Eq. (6.9). In reality, inertial trajectories will be modified by geostrophic forcing and friction. Parish (1977) showed that the cold easterly and northeasterly winds occasionally observed in the South Shetland Islands, situated northwest of the tip of the Peninsula, could have originated as barrier winds along the east of the Peninsula which then moved along pseudo-inertial trajectories beyond the tip of the Peninsula.

Another region favourable to the development of barrier winds is the northwestern part of the Ross Ice Shelf. Here, the progress of cold air moving to the west across the ice shelf is blocked by the Transantarctic Mountains and the steep rise to the plateau of East Antarctica. This low-level blocking generates a southerly barrier wind in the vicinity of Ross Island. O'Connor *et al.* (1994) showed that the presence of a cyclone (either synoptic or mesoscale) over the western Ross Sea or northwestern Ross Ice Shelf favours the development of barrier winds in this region. The easterly flow to the south of the cyclone centre drives cold ice shelf air against the mountain barrier, thus generating the barrier wind. In the two-year study of O'Connor *et al.* (1994), barrier winds were observed over the northwest Ross Ice Shelf for about 9% of the time and about 80% of all barrier wind events were associated with cyclonic forcing. The dense network of automatic weather stations on the Ross Ice Shelf has enabled the generation and propagation of barrier winds to be studied in some detail. In particular, the AWS observations have revealed how the complex terrain in the vicinity of Ross Island often blocks the progress of the barrier wind, with significant stagnation zones forming just upwind of Minna Bluff and Ross Island (O'Connor and Bromwich, 1988).

6.2 Internal gravity waves

6.2.1 Introduction

Internal gravity waves are wave motions in a fluid where buoyancy provides the restoring force. The persistent stable stratification of the lower troposphere over Antarctica enables the atmosphere to support such wave motions and manifestations of internal gravity waves have been observed at a number of Antarctic locations. It is now known that, globally, gravity waves are an important mechanism for vertically transporting momentum in the atmosphere and 'gravity wave drag' parameterisations are incorporated into most GCMs. However, these effects are most important in regions of steep topography, where gravity wave generation is strongest. The topography of much of Antarctica is relatively smooth and gravity wave drag may be of secondary importance. James (1989) has suggested that gravity wave transports may contribute to removal of vorticity from the tropospheric westerly vortex and hence to the maintenance of the katabatic drainage flow (see Section 4.3), but the importance of this process remains to be demonstrated.

A number of mechanisms may be responsible for generating gravity waves in the Antarctic atmosphere. These include air flow over topographic features, instabilities generated in regions of high wind shear and the radiation of energy from features such as mesoscale fronts and atmospheric gravity currents. Gravity waves generated by flow over topography ('lee waves') will be stationary if the flow is steady; instabilities and radiated waves will generally propagate past a ground-based observer.

Gravity waves may be detected with a variety of instruments. Anemograph and thermograph traces frequently reveal wave-like motions with periods ranging from a few minutes up to one hour and waves appear to show up particularly clearly in time series of the temperature difference between two levels within the surface inversion. Anderson *et al.* (1992) showed that, at the surface, the signature of gravity waves is seen most clearly in the pressure signal, although quite sensitive equipment is required for these measurements. Above the surface, gravity waves often manifest themselves as wave-like displacements of the elevated echo layers seen on sodar (acoustic sounder) records. A typical record is shown in Figure 6.9. Although this technique is rather qualitative, it can, with careful interpretation, provide useful information on the vertical structure of the waves. A further technique for obtaining information on vertical structure is to analyse the periodic variations in ascent rate of a radiosonde balloon (Figure 6.10) launched during a period of wave activity (Mobbs and Rees, 1989).

Observations from a single point can only give a measurement of the frequency of a wave. In order to estimate the wavelength and phase velocity, it is necessary to correlate time series measured at three or more points separated by less than one wavelength. An array of microbarographs, as described by Rees *et al.* (1994) is probably the ideal system for gravity wave investigations.

6.2.2 Case studies of Antarctic gravity waves

King *et al.* (1987) observed a large-amplitude gravity wave at Halley using surface instruments and sodar. The wave, which had a well-defined period of about 11 min, caused the temperature difference between 17 and 32 m to fluctuate by over 1°C (Figure 6.11(a)). A radiosonde was launched during the wave event and revealed a low-level jet structure (Figure 6.11(b)) in the wind profile. Application of a linear stability analysis to this profile suggested that the observed wave was generated in the shear layer on the upper side of this jet at around 1500 m elevation. In this layer, the Richardson number

$$\mathrm{Ri} = \frac{g}{\theta}\frac{\partial \theta}{\partial z}\left(\frac{\partial u}{\partial z}\right)^{-2} \tag{6.12}$$

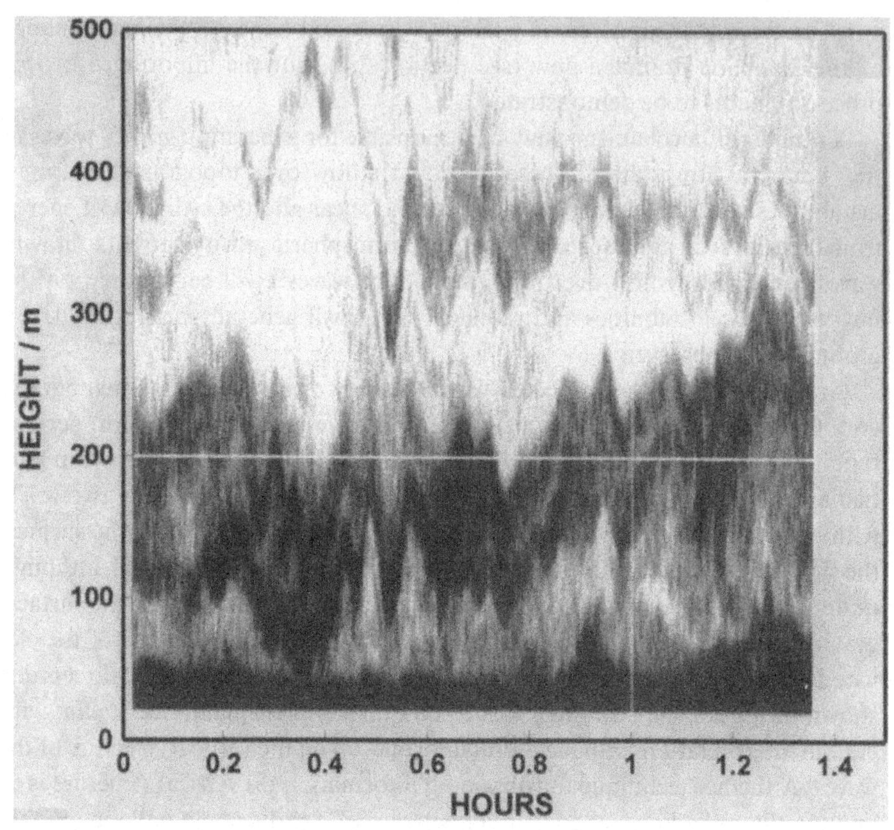

Figure 6.9 A sodar record made at Halley Station during the winter, showing wave-like undulations of elevated echo layers.

fell below the critical value of 0.25 necessary to generate instabilities. The stability analysis suggested that the observed wave had a wavelength of about 1000 m and a phase speed of 1.7 m s^{-1}.

A rather similar wave event was observed by Kikuchi (1988) at the Japanese Advance Camp, at an elevation of 3200 m on the Mizuho Plateau. Oscillations were again apparent in surface instrument records, but wave-like clouds were also present at an elevation of about 400 m above the surface. The period for a complete wavelength to pass the observer was 30 s and the wavelength was estimated to be 600 m from the cloud observations. A radiosonde launched at the time once again showed a low-level jet profile with a layer of low Richardson number and application of deep-water wave theory gave a wavelength and phase speed in good agreement with the observations.

A gravity wave event of short (one hour) duration was observed at Mizuho station by Kobayashi *et al.* (1983) using sodar. Wave-like motions, with a period

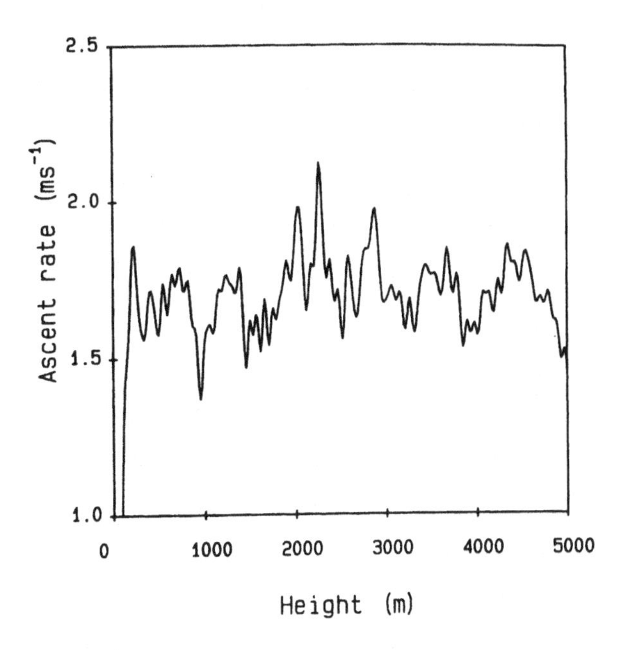

Figure 6.10 The ascent rate of a radiosonde balloon launched from Halley Station on 19 July 1986. From Mobbs and Rees (1989), reprinted with permission from *Antarctic Science.* Periodic variations in ascent rate are caused by vertical velocity variations associated with an internal gravity wave.

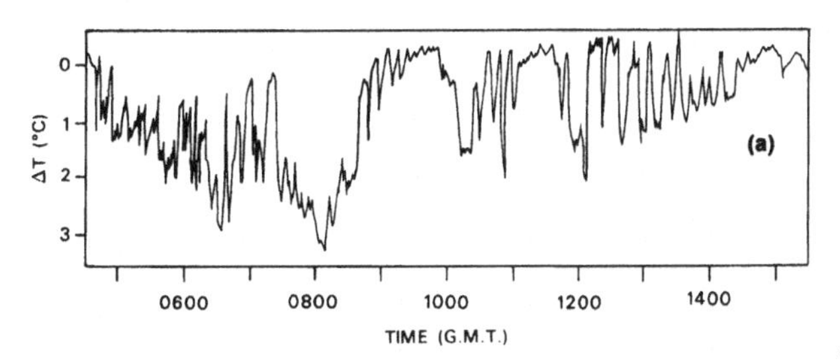

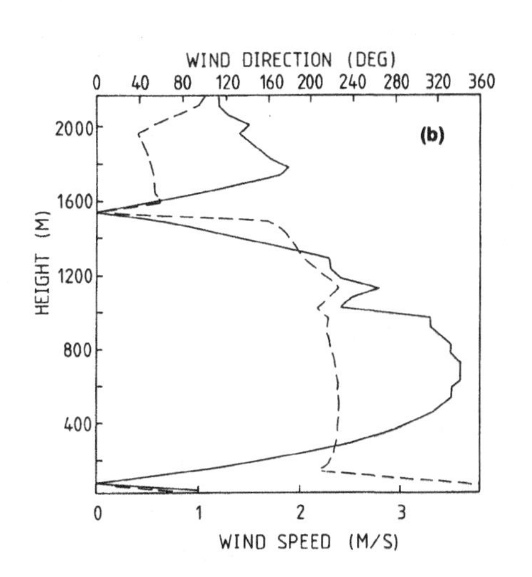

Figure 6.11 Observations of an internal gravity wave at Halley Station on 1 February 1986. (a) The temperature difference between 17 and 32 m, showing well-developed wave-like oscillations. (b) The wind speed profile in the lowest 2 km from a radiosonde launched at 11:30 GMT. From King *et al.* (1987), reproduced by permission of Kluwer Academic Publishers.

of about 2.3 min, were visible in an elevated echo layer between 100 and 300 m. The observed wave period was somewhat less than the Brunt–Vaisala frequency calculated from simultaneous radiosonde observations, indicating that the wave was propagating with the mean flow.

Egger *et al.* (1993) analysed two wave events recorded by a microbarograph array and tower instruments at Neumayer station during 1986. These waves were somewhat longer than the other events described above, with wavelengths of 9.1 and 2.9 km. Linear models provided a good description of the wave behaviour and the authors speculated that the waves could have been generated in the frontal zone separating outflowing cold continental air from warmer Weddell Sea air.

Rees and Mobbs (1988) studied gravity waves at Halley station using wind observations from a triangular array of masts, separated by about 200 m. By obtaining dispersion relationships, relating wavelength to phase speed, they demonstrated that the observed waves did not result from shear instabilities but rather had the characteristics of trapped, neutral modes. They speculated that such waves could be generated by unsteady flow over undulations in the surface of the Brunt Ice Shelf upwind of Halley.

A particularly dramatic gravity wave has occasionally been observed at Halley when the surface layer is strongly stratified. This takes the form of a series of solitary disturbances, clearly visible in near-surface records of temperature, wind speed and vertical velocity (Figure 6.12). Each disturbance, which usually lasts for about 100 s, is characterised by a sudden fall in temperature of several degrees Centigrade, followed by an equally rapid recovery to the undisturbed value. Vertical velocity measurements show that the fall in temperature is associated with upward motion, indicating that cold air is being brought up from the surface. Rees and Rottman (1994) analysed these events using weakly non-linear solitary wave theory, which appears to explain the observed structure and phase speed quite well. The solitary wave is trapped in a shallow 'waveguide' produced by the curvature of the temperature and wind speed profiles near to the ground. The origin of the disturbances remains unclear, although some form of shear instability seems a likely mechanism. Although these waves are associated with very marked changes in the boundary layer, their climatological significance is likely to be small since they only occur under a small range of ambient conditions and are observed rather rarely.

6.2.3 The importance of gravity waves in the Antarctic atmosphere

Few attempts have been made to assess the importance of gravity waves as a mechanism for transporting momentum and heat within the Antarctic atmosphere, although it is frequently speculated that they must play an important role. Mobbs and Rees (1989) estimated the momentum flux carried by gravity waves at Halley by analysing variations in the ascent rate of radiosonde balloons. The

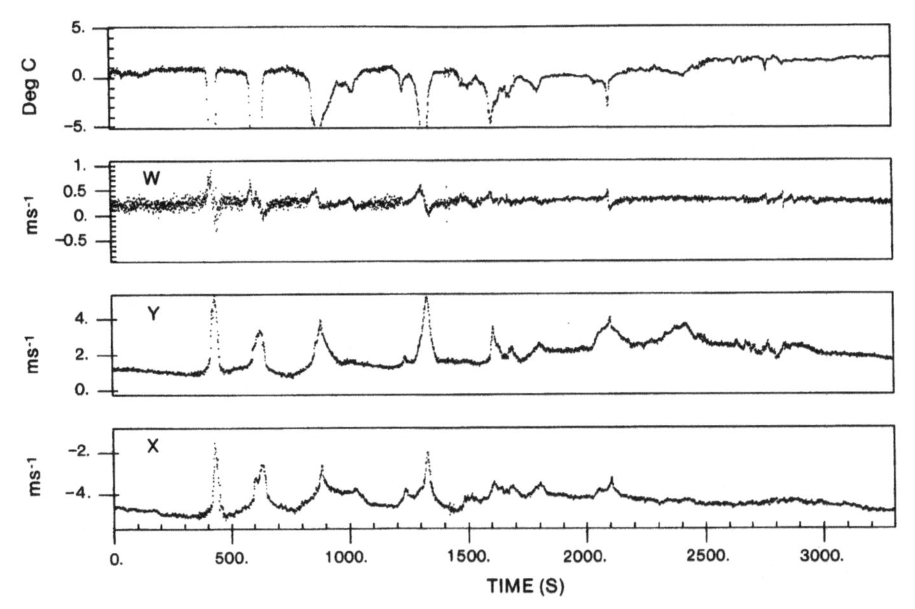

Figure 6.12 A series of solitary waves recorded at Halley Station on 22 June 1986. The waves were observed using a sonic anemometer at a height of 17 m. The four time series show, from the top, temperature (with respect to an arbitrary zero point), vertical velocity and the two orthogonal horizontal velocity components recorded at this level. From King (1993b), reproduced by permission of Oxford University Press.

divergence of gravity wave momentum flux during one event was found to be comparable to that associated with surface friction and equivalent to a flow acceleration of about 10 m s^{-1} per day. This indicates the importance of gravity wave fluxes during certain periods; because of the limited nature of the studies undertaken it is not possible to say whether they are important in a climatological sense.

Gravity waves may also indirectly affect transport through their interaction with turbulence in the atmospheric boundary layer. As discussed in the following section, turbulent transport in the boundary layer may become very small when temperature gradients are large and wind speeds are low. Under these circumstances, the additional wind shear associated with the passage of a gravity wave may generate additional turbulence and thus have a profound effect on boundary layer structure. Edwards (1992) modelled the turbulence generated by typical wave structures observed at Halley and concluded that, although the wave can generate turbulence, there is little feedback from the turbulence on the wave structure. Further studies are needed in this area and the Antarctic atmosphere, with its persistent stable stratification and frequent gravity wave events, provides an ideal natural laboratory for carrying out such investigations.

6.3 The atmospheric boundary layer

6.3.1 Introduction

The boundary layer may be defined as that part of the lower troposphere which is directly influenced by the underlying surface. Many of the phenomena already discussed in this chapter and elsewhere – katabatic winds, hydraulic jumps, gravity waves etc. – may be regarded as boundary layer phenomena. In this section, we shall concentrate on the turbulent processes responsible for the vertical transports of heat, momentum and moisture within the boundary layer and we shall examine the vertical profiles of wind and temperature which result from the action of these processes.

The boundary layer couples the atmosphere to the underlying surface and thus plays a central role in connecting the atmospheric, oceanic and cryospheric components of the Antarctic climate system. Boundary layer transports have to be parameterised in climate models and the parameterisations developed for the mid-latitude boundary layer may not be appropriate for the Antarctic atmosphere. The most notable feature of the Antarctic boundary layer is the strong and persistent stable stratification of the air resulting from the net negative surface radiation balance over the continent (Section 3.1). Stable stratification acts to damp out vertical motions and hence suppresses turbulent transport, so stability effects must be included in any parameterisation of the Antarctic boundary layer. Mid-latitude boundary layer parameterisations have mostly been developed from studies over vegetated surfaces and the snow and ice of Antarctica may require special treatment.

The wide variety of phenomena associated with the stably stratified boundary layer in Antarctica have attracted the attention of meteorologists since the earliest scientific expeditions to the continent but it is only in the last two decades that micrometeorological instrumentation which is sufficiently robust to withstand the rigours of the Antarctic winter has been developed. In particular, ultrasonic anemometers have proved capable of making reliable measurements of turbulent fluxes at a variety of Antarctic locations. However, even this instrument suffers from the perennial problems of icing and interference from blowing snow discussed in Chapter 2. Vertical profiles of wind velocity and temperature are also required for boundary layer studies. In the following section we shall see that the turbulent boundary layer in Antarctica is often quite shallow and a large fraction of the boundary layer can be probed from instruments mounted on 30 m masts that can be installed fairly easily. However, in some studies it is necessary to profile a deeper layer. Conventional radiosondes do not have the vertical resolution required for boundary layer studies but this can sometimes be achieved by using underfilled balloons with a slower ascent rate. Alternatively, sondes mounted beneath small tethered balloons or kites provide a means of obtaining very detailed profiles in the lowest few hundred metres. Ground-based remote sensing techniques, such as sodar (acoustic radar) and RASS (radio acoustic sounding system, see Section 2.5), have also proved of value in Antarctic boundary layer studies. Although the profiles produced are of lower vertical resolution than those obtained from tethersondes, continuous profiles (and hence time–height sections) may be obtained and the instruments can function unattended for long periods.

In the following section we review a number of studies of the atmospheric boundary layer in Antarctica and consider the parameters which control its structure. The most important of these are stability (which is largely determined by the surface energy budget) and the slope of the underlying surface, which will determine the importance of katabatic effects. Since these parameters vary in importance with location we shall divide our review by region.

6.3.2 Boundary layer structure

The boundary layer over ice shelves

The boundary layer that forms over the ice shelves which surround the continent is, in many ways, comparable to the nocturnal boundary layer (NBL) observed at mid-latitudes. The underlying surface is flat or only gently sloping and, in these regions, the stratification of the near-surface layer is generally moderate, even during the Antarctic winter. At Halley, on the Brunt Ice Shelf, the temperature difference between 2 and 32 m is less than 4°C for over 60% of the time. However, there are important differences between the NBL and the Antarctic boundary layer. The NBL is always in a state of evolution as it responds to the diurnal cycle in the surface energy balance. By contrast, the diurnal cycle is weak or absent through most of the year in Antarctica and the boundary layer can evolve towards a steady state, responding only to the slower changes in synoptic-scale forcing. Furthermore, the NBL forms beneath the remnants of the well-mixed daytime boundary layer and is thus generally capped by a layer of neutral stability. In the Antarctic, the whole of the lower atmosphere is stably stratified as a result of radiative cooling (Section 3.1).

Average profiles of wind and temperature measured at Halley during the winter of 1986 (King, 1990) are shown in Figure 6.13. A few observations were made under near-neutral conditions, the remainder have been divided into three broad stability categories based on surface heat flux and wind speed, from near-neutral to very stable. Wind speeds were normalised with respect to that measured at 32 m and temperatures are shown as differences from the temperature at 2.2 m. Even under near-neutral conditions, the wind speed profile above about 10 m departs from the logarithmic form expected in the surface layer and increasing curvature is seen as the stability increases.

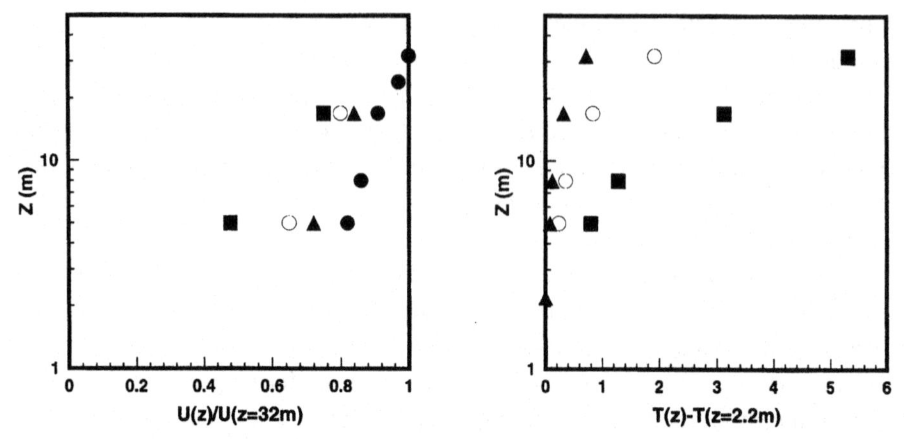

Figure 6.13 Wind and temperature profiles in the surface layer at Halley Station for four different stability categories: near-neutral (●), slightly stable (▲), moderately stable (○) and very stable (■). The wind speed has been normalised with respect to that measured at 32 m; temperatures are differences from that measured at 2.2 m.

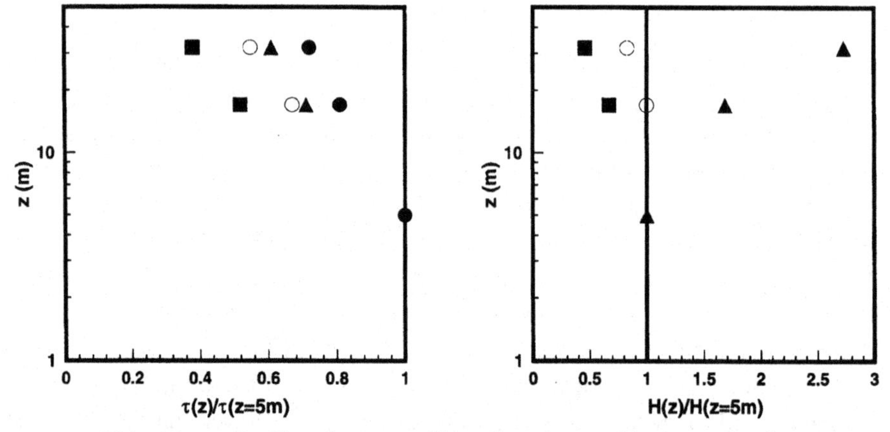

Figure 6.14 Profiles of stress and heat flux measured in the surface layer at Halley for the stability categories shown in Figure 6.13. Both are scaled by values measured at 5 m.

Profiles of stress, τ and heat flux, H, both scaled by values measured at $z=5$ m, are shown in Figure 6.14. τ falls off with height increasingly rapidly with increasing stability, indicating a decrease in the depth of the turbulent boundary layer as stability increases. Even under near-neutral conditions, the stress is observed to decrease by nearly 30% between 5 and 32 m, suggesting that the turbulent boundary layer is perhaps only 150 m deep. This is much shallower than a typical mid-latitude neutral boundary layer, almost certainly as a result of the persistent stability of the overlying atmosphere limiting boundary layer development. With increasing stability, stress falls off even more rapidly with height and, for the most stable data, the turbulent boundary layer depth probably does not exceed 50 m.

Under slightly stable conditions, the heat flux towards the surface actually appears to *increase* with height. Temperature gradients are small close to the ground but strong stratification persists further aloft. The heat flux at low levels is thus quite small while, higher in the boundary layer, there is a large heat flux downwards associated with the entrainment of the overlying stable air into the turbulent boundary layer.

According to Monin–Obukhov similarity theory (see, e.g. Garratt, 1992, p. 49), non-dimensional gradients of wind speed and temperature in the stratified surface layer should be functions of the stability parameter z/L only, where the Monin–Obukhov length L is defined by

$$L=\frac{\rho C_p T u_*^3}{\kappa g H} \tag{6.13}$$

H is the (downwards) heat flux and κ is von Karman's constant (0.4) and the friction velocity u_* is the square root of the surface stress divided by the air density. Surface-layer gradients may then be written as

$$\begin{aligned}\frac{\kappa z}{u_*}\frac{\partial u}{\partial z}&=\phi_{\mathrm{M}}\left(\frac{z}{L}\right)\\ \frac{\kappa z}{T_*}\frac{\partial T}{\partial z}&=\phi_{\mathrm{T}}\left(\frac{z}{L}\right)\end{aligned} \tag{6.14}$$

where $T_*=H/(\rho C_p u_*)$ is a temperature scale. Mid-latitude stable boundary layer studies (e.g. Businger *et al.*, 1971) have demonstrated that the ϕ-functions take the form

$$\begin{aligned}\phi_{\mathrm{M}}&=1+\alpha_{\mathrm{M}}\frac{z}{L}\\ \phi_{\mathrm{T}}&=\mathrm{Pr}+\alpha_{\mathrm{T}}\frac{z}{L}\end{aligned} \tag{6.15}$$

At Halley, wind and temperature profiles below 10 m are well described for $z/L\leq 2$ by Eq. (6.15) with $\alpha_{\mathrm{M}}=5.7$, $\alpha_{\mathrm{T}}=4.6$ and Pr=0.95 (King and Anderson, 1994). However, significant departures from these similarity forms are observed above 10 m, even under near-neutral conditions, as a result of the very limited depth of the turbulent boundary layer. No universal forms have been found to describe the observed profiles at higher levels.

The boundary layer over the interior of the continent

Over the high plateau of East Antarctica the boundary layer is more stably stratified than it is over the coastal ice shelves as a result of intense radiative cooling. Although topographic slopes in this region are generally modest – of the order

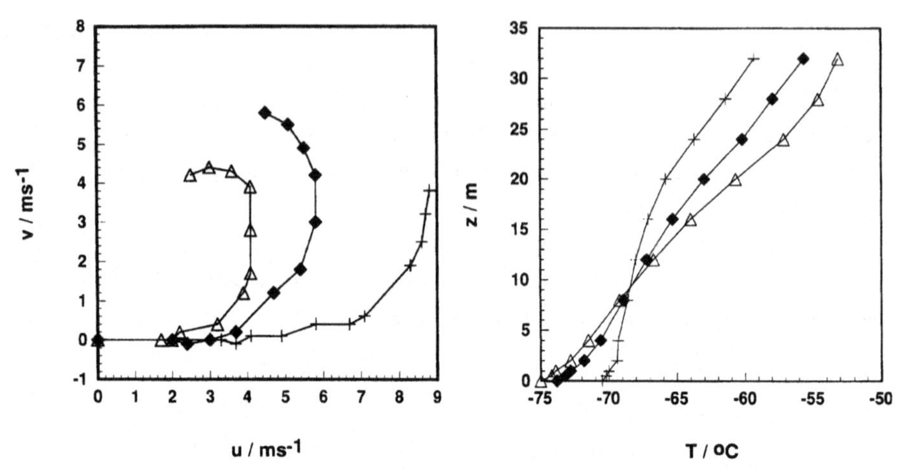

Figure 6.15 Wind hodographs and temperature profiles from Plateau Station for three different stability categories: stable (+), very stable (◆) and extremely stable (△).

of 1 part in 1000 – katabatic effects are important because the near-surface temperature gradient is so strong.

Measurements made during the winter of 1967 at Plateau Station revealed what is probably the most strongly stably stratified boundary layer observed anywhere on Earth. Average wind hodographs and temperature profiles (classified according to stability) collected in the lowest 32 m during this campaign are shown in Figure 6.15. Note that the stability classes used are not the same as those in Figure 6.13. Apart from the extraordinarily large temperature differences between the surface and 32 m, the most striking feature of the profiles is the rapid change in wind direction with height seen in the hodographs. This indicates that the turbulent boundary layer is very shallow indeed and thus the Ekman spiral of the wind profile is apparent even in measurements from a relatively short mast.

Unfortunately, no direct measurements of turbulent fluxes were made at Plateau Station because suitable instrumentation was not available at that time. However, if some reasonable assumptions are made, it is possible to calculate stress profiles from the observed velocity and temperature profiles (Kuhn *et al.*, 1977b). Profiles of the magnitude of the stress deduced from the hodographs of Figure 6.15 are shown in Figure 6.16. In all cases the shallow depth of the boundary layer is clearly apparent and extrapolation of the stress profiles yields boundary layer depths of 44, 48 and 72 m for the three stability categories.

It is clear that the profiles presented above bear little resemblence to the structure of the simple two-layer katabatic wind model presented in Chapter 4. In an attempt to develop a more realistic model for the boundary layer over the Antarctic interior, Mahrt and Schwerdtfeger (1970) obtained a solution for the katabatic wind forced by an exponential temperature profile. Their derivation

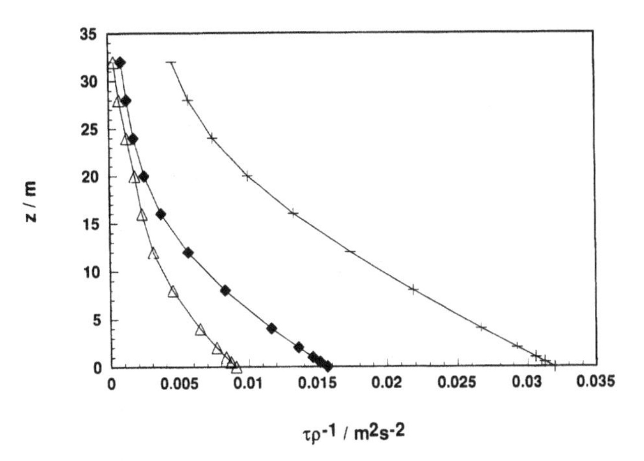

Figure 6.16 Profiles of stress deduced from the hodographs shown in Figure 6.15.

essentially follows that for the classical Ekman boundary layer (see, e.g., Holton, 1979 p. 107) with additional pressure gradient terms due to the katabatic forcing. Friction is parameterised using a constant eddy viscosity, K, and it is straightforward to incorporate the effects of a pressure gradient driving a geostrophic wind, $V_g=(u_g, v_g)$ above the stable layer. The potential temperature profile is assumed to be exponential in form and is characterised by a height scale H and a temperature scale $\Delta\theta$:

$$\theta=\theta_0-\Delta\theta\, e^{-z/H} \tag{6.16}$$

For a slope of gradient α directed downwards in the positive x direction, the steady-state equations of motion are

$$0=-g\frac{\Delta\theta}{\theta_0}\alpha e^{-z/H}+f(v-v_g)+K\frac{\partial^2 u}{\partial z^2} \tag{6.17}$$

$$0=-f(u-u_g)+K\frac{\partial^2 v}{\partial z^2} \tag{6.18}$$

These equations have the solutions

$$u=u_g-(a-u_g)e^{-z/H_e}\cos(z/H_e)-(b+v_g)e^{-z/H_e}\sin(z/H_e)+ae^{-z/H} \tag{6.19}$$

$$\frac{v}{\mathrm{sgn}(f)}=v_g+(a+u_g)e^{-z/H_e}\sin(z/H_e)-(b+v_g)e^{-z/H_e}\cos(z/H_e)+be^{-z/H} \tag{6.20}$$

where

$$a=\frac{H^2KfV_*}{K^2+H^4f^2} \tag{6.21}$$

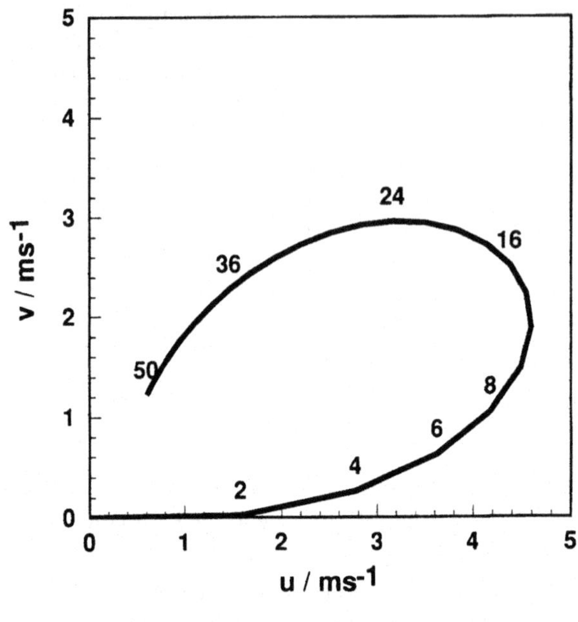

Figure 6.17 A wind hodograph calculated from the exponential inversion model with the following parameter values: $u_g = v = 0$, $K = 0.01$ m^2 s^{-1}, $H = 25$ m, $\Delta\theta = 30$ K, $\theta_0 = 220$ K and $\alpha = 0.001$. The figures by the curve give the height above the surface in metres.

$$b = \frac{H^4 f^2 V_*}{K^2 + H^4 f^2} \tag{6.22}$$

The Ekman depth scale is given by

$$H_e = \left(\frac{2K}{|f|}\right)^{1/2} \tag{6.23}$$

where

$$V_* = \frac{g\alpha}{f}\frac{\Delta\theta}{\theta_0} \tag{6.24}$$

is a measure of the strength of the katabatic forcing.

Even though this system of equations may appear to have fewer arbitrary parameters than does the two-layer model, it is still not straightforward to choose an appropriate value for the eddy viscosity, K. In reality, K is a function both of height and of stability (see Section 6.3.3) and the best compromise value to use for the constant in this simplified model is most easily found by comparing model predictions with observations. Wind profiles calculated from this model (Figure 6.17) show fair agreement with those observed at Plateau Station (Figure 6.15). The restriction of constant K can be relaxed, but it is then no longer possible to obtain an analytical solution and one has to resort to numerical methods (Brost and Wyngaard, 1978; King, 1989).

Boundary layer structure in the strong katabatic wind zone

Between the gentle slopes of the high plateau and the flat ice shelves of the coastal regions lies the zone of more steeply sloping ice which is characterised by

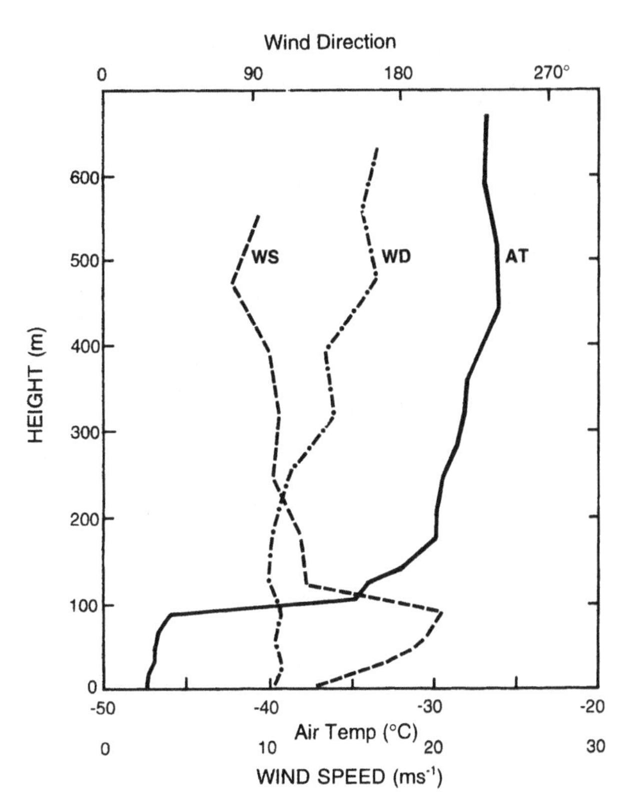

Figure 6.18 Profiles of air temperature (AT), wind speed (WS) and wind direction (WD) through a katabatic flow at Mizuho Station. After Ohata *et al.* (1985b).

strong katabatic winds. Few detailed observations of the boundary layer structure have been made in this region and most measurements have been limited to the summer season, when the katabatic flow is at its weakest. The chain of automatic weather stations between Dome C and Dumont d'Urville, established for the IAGO experiment (see Section 6.1) has provided valuable insight into katabatic wind dynamics throughout the year but boundary layer wind and temperature profiles have only been measured during the summer (Sorbjan *et al.*, 1986).

One of the few locations in the strong katabatic zone for which year-round measurements are available is Mizuho Station, which was occupied for a number of winters. An instrumented mast was used to measure wind and temperature profiles in the lowest 30 m while high-resolution radiosonde ascents provided information at higher levels. An example of a strong katabatic wind profile observed at this station (Ohata *et al.*, 1985b) is shown in Figure 6.18. The temperature profile shows a distinct two-layer structure, with a near-isothermal layer below 100 m capped by a very sharp 15°C inversion with a further near-isothermal layer above. Note how similar this is to the assumed temperature profile in the simple two-layer katabatic wind model discussed in Chapter 4. The wind speed profile exhibits a low-level jet structure below 100 m, with wind speed declining rapidly above the cold layer. On this occasion no sharp changes in wind direction were apparent at the interface between the two layers, suggesting that synoptic and katabatic forcing were in roughly the same direction. A similar structure is seen in wind and temperature profiles measured during strong katabatic wind conditions in Adélie Land (Gosink, 1989).

Ohata *et al.* (1985b) applied the two-layer katabatic model to this temperature profile and obtained a good prediction for the wind speed in the cold layer. In common with the simple two-layer calculations of Chapter 4, their model neglected the stress at the interface between the two layers and the good agreement with observations probably reflects the choice of value for the surface drag coefficient. In reality, fluxes of heat and momentum resulting from turbulent entrainment processes at the sharp interface are likely to be important. Manins and Sawford (1979) demonstrated the importance of interfacial stress in katabatic flows and suggested that the stress, τ_i at the interface may be parameterised as

$$\tau_i = -\rho w_e (V_2 - V_1) \qquad (6.25)$$

where V_1 and V_2 are the lower and upper layer velocities respectively and the entrainment velocity, w_e is given by

$$w_e = \frac{a_1}{a_2 + \mathrm{Ri}} |V_2 - V_1| \qquad (6.26)$$

where a_1=0.002, a_2=0.02 and Ri is the Richardson number of the interfacial layer. Consideration of the heat balance of the Antarctic atmosphere, discussed in Chapter 4, indicates that the turbulent entrainment of heat into the katabatic layer, H_i, must also be important. This may be parameterised as

$$H_i = \rho C_p w_e (T_2 - T_1) \qquad (6.27)$$

where T_1 and T_2 are the layer temperatures.

Although interfacial entrainment processes are likely to be of great importance in strong katabatic winds, they do not appear to have been studied in any Antarctic field experiments. The technical difficulties involved in making such measurements are formidable. The cold katabatic layer is typically 100 m deep, so the interfacial layer is beyond the reach of the sort of masts which can be easily erected in Antarctica. Tethered balloons, the alternative to fixed masts or towers, will not operate satisfactorily in the strong winds associated with katabatic flow, although kites offer an attractive means of supporting instruments for such studies. As yet, ground-based remote sensing techniques, such as sodar, do not have sufficient resolution for these studies and problems with wind noise limit the effectiveness of sodar in the katabatic zone. The climatological importance of the katabatic flow makes studies of entrainment processes an important priority for future research.

6.3.3 Parameterisation of the Antarctic boundary layer

The problem of parameterising turbulent fluxes in atmospheric models falls into two parts. First, fluxes at the surface have to be specified, which requires a knowledge of surface roughness characteristics in order that drag and exchange coefficients can be formulated. Secondly, the variation of fluxes with height within the boundary layer must be calculated.

Surface fluxes are generally calculated by assuming that the flow at the lowest model grid point is in equilibrium with the underlying surface and thus wind speed and temperature profiles below this point will be given by the Monin–Obukhov similarity forms, Eqs. (6.14). For this assumption to be valid and the calculated fluxes accurate, the lowest model level must be at a small fraction of the boundary layer depth. Because the boundary layer in Antarctica is generally very shallow, this places severe constraints on model resolution. In most GCM experiments the vertical resolution near the surface is chosen to provide a reasonable representation of the boundary layer in the tropics and mid-latitudes. This will generally be too large for an accurate simulation of the Antarctic boundary layer and the calculated surface fluxes are likely to be inaccurate.

Integrating the similarity function for wind speed between the surface and the lowest model level at height z_1 gives

$$u(z_1)=\frac{u_*}{\kappa}\left[\ln\left(\frac{z_1}{z_0}\right)+\alpha_M\frac{z_1}{L}\right] \tag{6.28}$$

The surface stress, ρu_*^2, can then be calculated from the wind speed at the lowest model level, provided that the aerodynamic roughness length, z_0, is known. Estimates of z_0 have been made for a number of Antarctic sites where wind profiles and/or surface stress have been measured; these are summarised in Table 6.2. The very small roughness length values obtained – of the order of 0.1 mm – are, at first sight, somewhat surprising given that most Antarctic snow surfaces are not truly flat but rather are composed of sastrugi or snow dunes which are typically 0.1–1 m high. However, the aerodynamic roughness length is mainly related to the microscale roughness of the snow surface rather than to the scale of these streamlined features. Studies over the Mizuho Plateau (Inoue, 1989a; 1989b) have shown that the effective roughness length does vary with wind direction relative to the orientation of the sastrugi, so surface features do exert some influence on the surface drag.

A roughness length of 0.1 mm is equivalent to a neutral 10 m drag coefficient, C_D, of about 1.2×10^{-3}, where C_D is defined in terms of the surface stress, τ, and the 10 m wind speed, u_{10} by

$$\tau=\rho C_D u_{10}^2 \tag{6.29}$$

Table 6.2 *Measurements of aerodynamic roughness length, z_0, at various Antarctic locations.*

Author	Location	z_0 (m)
Liljequist (1957)	Maudheim	1.0×10^{-4}
König (1985)	Neumayer	1.0×10^{-4}
King and Anderson (1994)	Halley	5.0×10^{-5} to 6.1×10^{-5}
Heinemann (1989)	Filchner–Ronne Ice Shelf	1.0×10^{-4}
Inoue (1989a)	Mizuho Plateau	4.0×10^{-6} to 1.5×10^{-4}
Inoue (1989b)	Mizuho Station	4×10^{-4} to 8×10^{-4}
Bintanja and van den Broeke (1995)	Blue ice field	2.8×10^{-6} to 3.7×10^{-6}

In Chapter 4 we saw that a drag coefficient of about 5×10^{-3} gave an optimal simulation of Antarctic katabatic winds with the two-layer model. This is somewhat larger than would be suggested by measurements of z_0 and may result, in part, from the neglect of interfacial stress in the two-layer model. However, some observations suggest that z_0 may increase under strong wind conditions when surface snow starts to drift (see Section 6.4), thus providing an additional means of exchange of momentum between atmosphere and surface. Dimensional analysis suggests that, under such conditions

$$z_0=a\frac{u_*^2}{g} \tag{6.30}$$

where a is a dimensionless constant. Measurements over the Ekström Ice Shelf (König, 1985) indicate that a=0.006 while observations over a frozen lake by Tabler (1980) gave a=0.013.

Surface fluxes of heat and water vapour may be parameterised in an analogous fashion to momentum. In general, the roughness lengths for these scalar quantities will not be equal to z_0, although they are often assumed to be equal in GCM parameterisations. Very few measurements have been made of the scalar roughness lengths of snow and ice surfaces. Andreas (1987) calculated theoretical values by matching a viscous sublayer model to a surface layer profile. He showed that the roughness lengths for heat, z_H, and water vapour, z_Q, were strongly dependent on the surface Reynolds number. At low Reynolds number (i.e. low wind speed) $z_H/z_0=3.49$ and $z_Q/z_0=5.00$ but at high Reynolds number the scalar roughness lengths become much smaller than z_0. King and Anderson (1994) measured z_H at Halley and obtained values that were several orders of magnitude larger than z_0. However, they noted that the value obtained depends critically on how the snow surface temperature is defined. Given the practical difficulties of measuring scalar roughness lengths, the standard practice of setting $z_H = z_Q = z_0$ is to be recommended for the present.

Once surface fluxes have been parameterised, fluxes have to be calculated at all other model levels. In the simplest 'first-order closure' schemes, fluxes are related to mean gradients through an eddy viscosity:

$$\tau = \rho K_M \frac{\partial u}{\partial z} \tag{6.31}$$

The eddy viscosity, K_M, is a function of height above the surface and atmospheric stability. A form suitable for use in the stable Antarctic boundary layer has been described by King (1989). First-order closure schemes are probably adequate for describing much of the dynamics of the Antarctic boundary layer, provided that the model used has sufficient vertical resolution. However, they are unlikely to give good results when applied to strong katabatic flows in which there is a sharp interface between the fast katabatic stream and the overlying air. The simple bulk formulae given in Eqs. (6.25)–(6.27) can provide useful estimates of entrainment fluxes but progress in modelling strong katabatic flows will probably require the use of models with second (or higher) order turbulence closure.

6.4 Blowing snow

6.4.1 Introduction

No one who has spent time in Antarctica can have failed to be impressed by the capacity of the wind to transport large quantities of snow. Deep drifts can form around buildings in a matter of hours, even when no precipitation is falling and wind-borne snow can rapidly reduce visibility to the point at which any outside activity becomes impossible. On a wider scale, blowing snow can re-distribute precipitation falling on a region, radically changing the net accumulation in some areas. It can also play a role in the overall mass balance of the Antarctic ice sheets, both by direct removal of snow across the coastline and by enhancing evaporative loss from the surface.

In strict meteorological terminology, 'blowing snow' refers to wind-borne snow of sufficient depth and density to impair visibility at or above the height of the observer's head. Wind-borne snow not reaching this level is properly referred to as 'drifting snow'. However, the difference between the two is one of degree rather than kind and, in this section, we shall make no distinction between blowing and drifting snow.

Although the effects of blowing snow may be of climatological significance, its origins and structure are inextricably linked to turbulent processes in the boundary layer, whence its inclusion in this chapter. In the following subsection we present some basic theoretical results that are of relevance to blowing snow and we review some of the Antarctic field studies which have been conducted on this important topic. Finally, we consider the significance of blowing snow in the Antarctic climate system.

6.4.2 Basic theory and observations

If the wind stress on a snow surface is sufficiently large, snow grains will become dislodged from the surface and move along shallow trajectories. When they re-impact upon the surface they will, in turn, dislodge further particles and so the process continues. This transport of snow grains along and close to the snow surface is known as saltation.

In order for saltation to occur, the surface stress must be large enough to break the cohesive forces between individual snow grains. This is clearly dependent on the form and history of the snow surface. Observations made at Halley Station (Dover, 1993) showed that the critical friction velocity for the onset of saltation was in the range 0.18–0.38 m s^{-1}, corresponding to 10 m wind speeds of about 5–10 m s^{-1}.

Dimensional arguments suggest that the maximum height reached by a saltating particle will be given by

$$z_{\text{salt}}=\frac{cu_*^2}{2g} \tag{6.32}$$

where u_* is the friction velocity. Observations suggest a value of about 1.6 for the constant c. Thus, even with very strong winds, saltation cannot lift snow to more than about 0.1 m above the surface.

However, once snow grains have been lifted off the surface by saltation, they can be carried to greater heights by turbulent eddies in the boundary layer, a process referred to as suspension. In a steady state, the turbulent diffusion of snow away from the surface will be exactly balanced by gravitational settling of the snow particles, so the snow drift density, η, is given by

$$\frac{\partial}{\partial z}\left(w_{\text{f}}\eta+K_{\text{s}}\frac{\partial\eta}{\partial z}\right)=0 \tag{6.33}$$

where w_{f} is the fall velocity of the snow particles. If the diffusivity of snow particles, K_{s}, is equal to that of momentum then

$$K_{\text{s}}=\kappa z u_* \tag{6.34}$$

where κ is von Karman's constant. Eq. (6.33) then has the solution

$$\eta(z)=\eta(z_R)\left(\frac{z}{z_R}\right)^{-w_f/(\kappa u_*)} \tag{6.35}$$

relating drift density at height z to that at a reference height, z_R. Observed particle fall velocities are in the range 0.3–0.9 m s^{-1} (Takahashi, 1985b), thus Eq. (6.35) predicts a rapid decrease in drift density with height at moderate wind speeds. Most observational studies (e.g. Budd *et al.*, 1966) show that significant drift densities are restricted to the lowest 10 m but, under extreme conditions, the drift layer can be much deeper. Blowing snow of sufficient density to cause white-out conditions at flight level has been observed from an aircraft flying at 183 m over Adélie Land in wind speeds in excess of 45 m s^{-1} (Gosink, 1989).

In deriving the above equations, it has been assumed that all snow particles have the same size and hence the same fall velocity. In reality there will be a distribution of particle sizes and fall velocities. Budd (1966) demonstrated that the analysis could be extended to a distribution of particles quite straightforwardly since, for particle diameters between 0.1 and 1 mm, the fall velocity is directly proportional to particle diameter and most drifting particles lie within this range. For analytical convenience, Budd assumed that the distribution of particle diameters could be described by a two-parameter gamma distribution, so the fraction of particles with diameters between x and $x+\mathrm{d}x$ is given by

$$P(x)=\frac{1}{\beta^{\alpha}\Gamma(\alpha)}\mathrm{e}^{-x/\beta}x^{\alpha-1}\,\mathrm{d}x=\gamma_{\alpha\beta}(x)\,\mathrm{d}x \tag{6.36}$$

The mean diameter of this particle distribution is $\alpha\beta$ and its variance is $\alpha\beta^2$. The size distribution of blowing snow particles has been studied at Byrd (Budd *et al.*, 1966) and at Halley (Dover, 1993) Stations by collecting particles on Formvar-coated slides and then measuring the resulting impressions. Both studies showed that the gamma distribution was a reasonable fit to the observations (Figure 6.19) and that the diameter corresponding to the peak of the distribution decreased with increasing height above the snow surface. This is to be expected since the larger particles, which have the highest fall velocities, will not, on average, remain in suspension long enough to reach any great height.

In the discussion above, we have assumed that a blowing snow particle remains unchanged from the time it leaves the saltation layer until it re-impacts with the ground. However, if the air in the blowing snow layer is not saturated with respect to ice then particles may sublime while in suspension. Smaller particles sublime more slowly than larger ones so sublimation may modify the particle size distribution by preferentially converting the larger particles to smaller ones. This may increase the tendency of the modal particle diameter to decrease with height (Figure 6.19), since particles at higher levels will, on average, have had longer to sublime than those closer to the ground. Evaporation of snow particles may make a significant contribution to the surface mass balance but

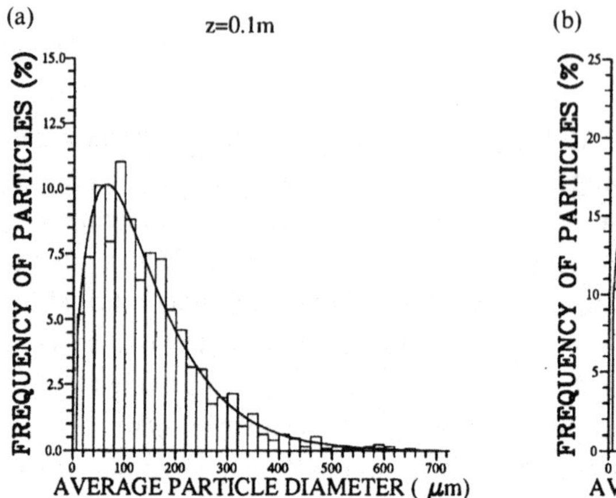

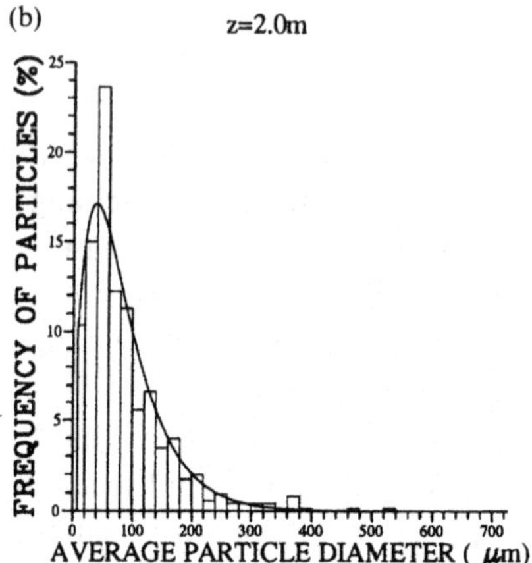

Figure 6.19 Blowing snow particle size distributions measured at heights of 0.1 m (left-hand diagram) and 2.0 m (right-hand diagram) above the snow surface at Halley Station. Gamma distribution functions (Eq. (6.36)) have been fitted to the observed distribution. From Dover (1993).

observations at Halley suggest that the air in the near-surface layer rapidly becomes saturated following the onset of blowing snow and hence that equilibrium sublimation rates are small.

At wind speeds just above the blowing snow threshold most of the horizontal transport of snow is carried by the saltation layer. However, as the wind speed increases, suspension transport rapidly dominates. Using some of the ideas expounded above, it is possible to develop expressions for the suspended snow transport but this has met with limited success because of simplifications inherent in the theory. A full numerical model of blowing snow transport, which includes the effects of sublimation, has been able to reproduce observed transport rates (Dover 1993). The results of a number of field studies suggest that the total drift transport in the blowing snow layer, Q, measured in g m^{-1} s^{-1}, can be related to windspeed, U, in m s^{-1}, at some reference level by an expression of the form

$$\ln(Q)=cU+d \tag{6.37}$$

Values for the constants c and d, determined in a number of Antarctic field studies, are given in Table 6.3. An alternative power law expression was fitted to measurements made at Mizuho by Takahashi (1985b):

$$Q=7.18\times10^{-4}U^{5.17} \tag{6.38}$$

This expression is for U measured in m s^{-1} at 1 m and, again, gives Q in units of g m^{-1} s^{-1}. These relationships between total drift transport and wind speed are

Table 6.3 *Values of the parameters c and d (see Eq. (6.37)) determined in a number of Antarctic field studies. These parameter values apply to wind speed at 3 m, measured in m s^{-1}, and give drift transport in g m^{-1} s^{-1}.*

Reference	Location	c	d
Budd *et al.* (1966)	Byrd Station	0.235	2.72
Kobayashi (1978)	Mizuho Station	0.397	−0.276 to −2.12
Dover (1993)	Halley Station	0.632	−3.69

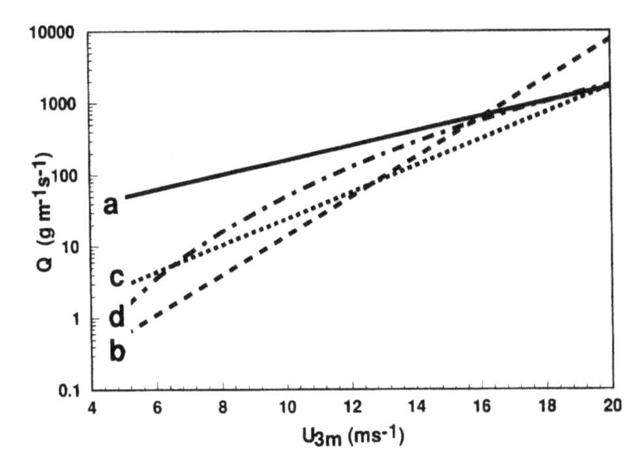

Figure 6.20 Total drift transport as a function of wind speed at 3 m. (a) Budd *et al.* (1966), (b) Dover (1993) – observations, (c) Dover (1993) – numerical model, (d) Takahashi (1985b).

shown in Figure 6.20, together with predictions from the numerical model of Dover (1993).

6.4.3 Blowing snow and the Antarctic climate

The studies described above clearly demonstrate the power of the wind to transport snow. For a steady wind speed of 10 m s^{-1}, the mean drift transport rate predicted by the models of Eq. (6.37) and Table 6.3 is about 655 tonnes per metre per year. Estimates have been made of the flux of blowing snow across the Antarctic coastline in relation to the overall mass balance of the ice sheets (Section 4.4). There are considerable uncertainties in these estimates, partly because of differences between the various drift transport models but also because all models predict a very strong dependence of drift flux on wind speed. It is thus necessary to know details of the wind speed distribution at coastal locations to make realistic calculations. A further difficulty is the assumption that sufficient surface snow is available to supply the drift. After a period of continuous ablation this may not be true. Recent estimates (Giovinetto *et al.*, 1992) suggest that up to 10% of the snowfall on the Antarctic continent may be removed by the flux of blowing snow across the coast. Although this is a significant fraction, it should be recalled that the uncertainties in the net accumulation are at least this large.

Blowing snow may also be of great importance on a regional scale because of its capacity to re-distribute precipitation. Since drift transport depends principally on wind speed, a divergent surface wind field will be associated with a divergent blowing snow mass flux and, hence, a net removal of snow from the region. Conversely, regions of surface wind convergence will tend to be associated with enhanced deposition of snow. Surface winds over much of Antarctica are strongly influenced by the surface slope so it is likely that the contribution of drift transport to net accumulation will also be governed by topography. For example, the convex coastal slopes are associated with an accelerating (and hence divergent) katabatic flow and are thus likely to be regions where snow is removed by drift transport. Takahashi *et al.* (1988) used a two-layer katabatic wind model (see Section 4.3) combined with the drift flux parameterisation given by Eq. (6.38) to estimate drift transport divergence at Mizuho Station and concluded that about 100 mm water equivalent per year was removed by this process. This is a very significant fraction of the annual precipitation at this location, estimated at 140–260 mm.

In extreme circumstances, drift flux divergence can remove all of the precipitation falling onto an area, resulting in a negative surface mass balance and the formation of a blue ice zone. A number of blue ice fields have been observed in the lee of the Yamato Mountains, near Mizuho Station. Application of a two-dimensional version of the model of Takahashi *et al.* (1988) to the local topography demonstrated the existence of strong drift flux divergence in this region. Blowing snow has also been implicated in the formation of other Antarctic blue ice fields (van den Broeke and Bintanja, 1995).

Blowing snow may also play a role in forcing katabatic winds. At −20°C, a drift snow loading of 0.01 kg m^{-3} will increase the air density by the same amount as would a 2.5°C isobaric cooling and such loadings may easily be reached in the lowest few metres at high wind speeds. Wendler *et al.* (1993) observed that katabatic forcing is insufficient to maintain katabatic winds of more than about 12 m s^{-1} in the interior of Adélie Land and suggested that the increase in air density caused by blowing snow, together with the cooling associated with sublimation of blowing snow, may provide a feedback mechanism by which stronger winds may be maintained. The latter effect may be quite important for air in katabatic flows, which undergoes adiabatic warming as it descends the slope and is thus generally unsaturated. Studies with a katabatic wind model that incorporated both effects (Gosink, 1989) indicate, however, that neither effect becomes significant until wind speeds exceed 28 m s^{-1}; hence blowing snow effects need only be considered in the strongest of katabatic flows.

6.5 Mesocyclones

Satellite imagery has been available since the early 1960s and benefited many areas of meteorology by virtue of its capability to give a synoptic view of large areas of the Earth. For routine analysis it can show the cloud associated with the major weather systems as well as smaller scale features such as squall lines, thunderstorms and mesoscale convective systems. However, in the polar regions, where *in situ* data were sparse, it revealed a completely new phenomenon that had not been resolved by the widely spaced surface observing network, but that could give severe weather and be a hazard over the oceans and in the coastal regions. This was the mesocyclone or polar low, which is a relatively short-lived, sub-synoptic-scale low-pressure system that imagery has revealed to be a feature of both polar regions. With their limited horizontal scale of less than 1000 km and lifecycle of normally under 24 h, mesocyclones are extremely difficult to forecast and our understanding of their development and occurrence is still well behind that of the synoptic-scale disturbances. In this section we will examine our present knowledge of mesocyclones in the high southern latitudes and consider their geographical occurrence and the mechanisms behind their formation and compare the systems observed with those found in the Arctic.

Mesocyclones over the high-latitude areas of the Northern Hemisphere (where they are often referred to as polar lows) have been studied since the 1960s but those occurring in the Southern Hemisphere were not investigated in detail until high-resolution satellite imagery became generally available (Lyons, 1983; Auer 1986; Sinclair, 1986). In the pre-satellite era the very sparse observing network in the Antarctic did not allow the resolution of sub-synoptic-scale weather systems and the rapid variations in surface meteorological parameters occasionally observed at the coastal stations as a result of the passage of these disturbances could rarely be interpreted correctly without data on the broader scale environment. As satellite imagery of the Antarctic has been studied more intensively, the high frequency with which mesocyclones occur has become apparent and a number of climatological and diagnostic studies have been carried out to investigate their relationship to the synoptic-scale circulation and to help in understanding the mechanisms behind their formation. Although a great deal of research has taken place during the last 30 years into Northern Hemisphere polar lows, the results of this work are only of limited value in investigating mesocyclones around the Antarctic since the topographic, oceanographic and meteorological conditions in the two polar regions are very different. However, with the completion of a number of detailed case studies of Antarctic mesocyclones and long-term climatological investigations, we are now beginning to gain some insight into the mechanisms behind their formation and their relationship to the larger-scale flow and the major climatological regimes of the Southern Hemisphere.

A recurring problem in the study of mesocyclones is that of nomenclature. The literature on Southern Hemisphere mesoscale vortices has used a bewildering variety of terms to describe the phenomena observed and the terms 'polar low', 'Antarctic coastal vortex', 'polar air cloud system' and 'comma cloud' have all been used to describe these disturbances. Here the term 'mesocyclone' is used as a generic description of the phenomenon and it is only when considering the different classes of vortex that additional terms are introduced.

6.5.1 General characteristics

Cyclonic disturbances in the atmosphere occur on a range of length scales from tens of metres to several thousand kilometres and embrace a wide range of phenomena. The mesocyclones discussed here, however, are the cold air vortices forming to the south of the main polar front and usually develop in outbreaks of polar air well removed from pre-existing frontal cloud bands. Since they are normally observed and tracked on satellite imagery, the term mesocyclone is generally used to refer to the vortices with a horizontal length scale of more than a few tens of kilometres. Small vortices linked to topographic features, such as von Kármán vortex streets, are excluded since they are not exclusively a polar phenomenon. The upper size limit of mesocyclones is usually taken as a diameter of 1000 km, although satellite imagery has shown that the vast majority of mesocyclones are less than 500 km in extent. This was evident in the study of Turner and Thomas (1994), who examined 162 summer-season vortices around the Antarctic Peninsula and found that 43% had a diameter of less than 200 km, 51% were in the range 200–500 km and only 6% were greater than 500 km (see Table 6.4). Larger vortices were almost exclusively found in the Bellingshausen Sea and at more northern latitudes. A similar result was found by Heinemann (1990) in his study of mesocyclones over the eastern Weddell Sea, where, of the 195 mesocyclones identified in satellite imagery, three-quarters had a diameter of less than 400 km and over half had a lifetime of less than 12 h. Examination of satellite imagery during other months in the extended summer period has confirmed that most mesocyclones off Halley Station have a diameter of several hundred kilometres and last only a few hours. Some vortices with the structure of mid-latitude depressions can be smaller than 1000 km, but these can usually be distinguished from true mesocyclones by their cloud pattern, which resembles that of a baroclinic wave with warm, cold and occluded fronts.

Studies based on satellite imagery have shown that mesocyclones can exist over a wide range of timescales from a few hours to, in exceptional cases, several days, depending on the particular forcing conditions. However, most are transitory features and last for less than one day. The vortex examined by Turner et al. (1993a) existed for over three days and remained quasi-stationary near the coast of the eastern Weddell Sea for much of this time, but the synoptic-scale

Table 6.4 *A breakdown of the mesocyclones found by Turner and Thomas (1994) in their analysis of satellite imagery of the area around the Antarctic Peninsula for the period December 1983 to February 1994.*

			Diameter (km)				
Vortex type	Description	Number	<200	200–500	>500	Spiral	Comma
Halley (H)	Over eastern Weddell Sea	17	7	10	0	6	11
Polar low (P)	To west of synoptic-scale low	35	19	14	2	8	27
Trough (T)	Associated with synoptic-scale trough	30	12	17	1	11	19
Small synoptic (S)	Having the form of a short baroclinic wave	21	2	14	5	8	13
Front (F)	Embedded in front	16	13	3	0	3	13
Ice edge (E)	Observed close to the ice edge	16	8	8	0	6	10
Over water (W)	Occurring over the ocean and well removed from synoptic-scale disturbances	12	9	3	0	6	6
Over ice (I)	Occurring over land or sea ice and well removed from synoptic-scale disturbances	15	0	14	1	5	10
Total		162	70	83	9	53	109

circulation and the baroclinic zone in which the system developed changed little over this period and were responsible for its longevity. Other cases have indicated that the more long-lived vortices appear to be associated with quasi-stationary flow on the synoptic scale such as large, slow-moving synoptic-scale low-pressure systems over the ocean. Many of these large lows can be much more complex than analysed on the operational charts and can contain a number of mesoscale vortices in their declining stages. In the Northern Hemisphere these are often referred to as a 'merry-go-round' type of polar low, although this type of development is quite rare in both polar regions. Some studies have examined only the longer-lived mesocyclones, but here we will impose no restriction in terms of lifetime and all mesoscale vortices not linked to topography and apparent on satellite imagery or detected by surface observing systems will be considered.

Northern Hemisphere polar lows are primarily an oceanic, winter-season phenomenon, very few systems having been observed to develop over the land or to occur during the summer months. This situation contrasts with that in the Southern Hemisphere, where, despite the lack of Antarctic-wide, multi-year studies of mesocyclone activity, it is clear from satellite imagery that Antarctic mesocyclones occur all year round (Heinemann, 1990; Turner and Thomas, 1994) and appear to have similar characteristics in all seasons. This must be a result of the differences in lower tropospheric and oceanographic conditions between the two polar regions and will be considered further in Section 6.5.5.

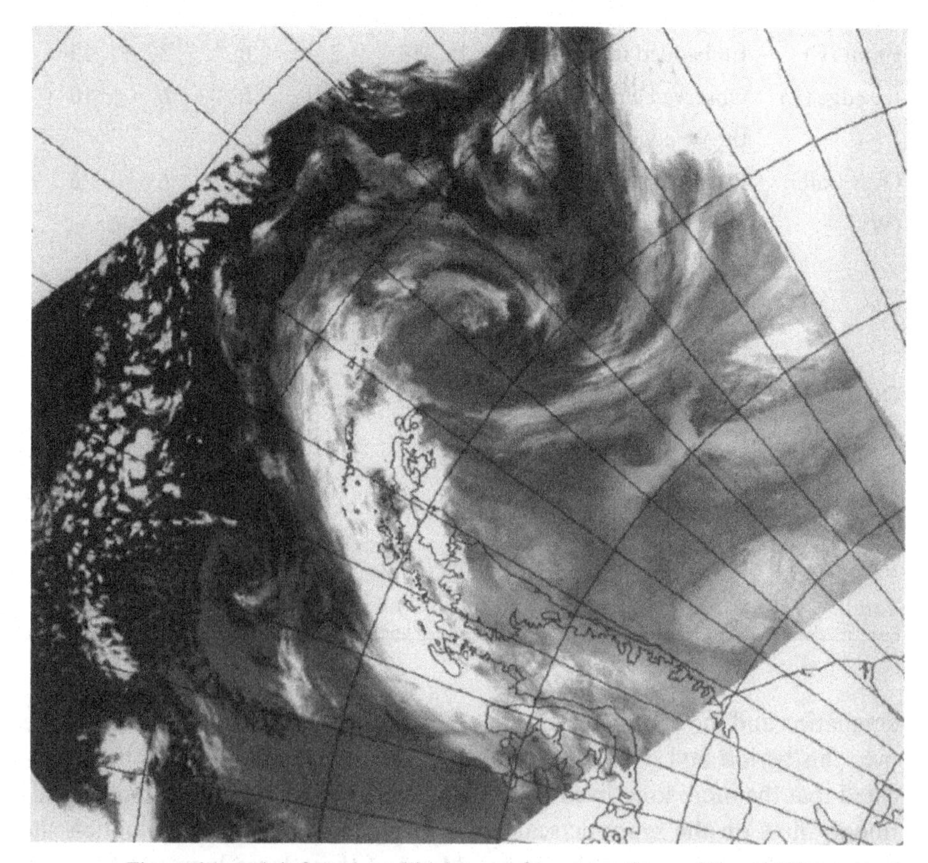

Figure 6.21 A infrared satellite image of a mesocyclone with spiraliform cloud signature over the Bellingshausen Sea at 10:51 GMT 14 September 1993.

The cloud signatures of mesocyclones observed on satellite imagery have often been considered to be of two types – spiraliform and comma-shaped. The former are characterised by a circular, symmetric pattern with rings of cloud encircling the centre of the low, as can be seen in Figure 6.21. The second group have the cloud organised into a comma shape with the head being near the centre of the low and an elongated tail extending towards regions of higher pressure. A typical mesoscale vortex with a comma-shaped cloud band is shown in Figure 6.22. Since mesocyclones are vorticity maxima they will all have a significant degree of rotation apparent in the cloud associated with the low. The difference between the spiraliform and comma-shaped appearance of the lows is caused by the degree of translation of the system as a whole, often as a result of the synoptic-scale flow in which the systems are embedded. The spiral type of cloud signature is found in association with mesocyclones developing in regions with little large-scale forcing and they therefore show a signature of pure rotation. The lows with comma-shaped cloud, on the other hand, are usually found in a moderate or strong, southerly synoptic flow and the elongated appearance of the cloud is

Figure 6.22 An infrared satellite image of a mesocyclone with comma-shaped cloud signature at 10:54 GMT 25 July 1993.

caused by the translation of the system as a whole. Infra-red satellite imagery suggests that comma-shaped clouds usually have cloud tops extending to middle and high levels, whereas the spiraliform systems tend to be confined to the lower troposphere. It should be pointed out that the term 'comma cloud' is used here to describe the cloud signatures associated with certain vortices and does not imply a particular type of vortex or formation mechanism. In particular, the term should not be confused with the class of vortex observed over the Pacific and Atlantic Oceans by Reed and Blier (1986) and called 'comma clouds' in their papers. The vortices that they studied were found at lower latitudes close to the polar front and are certainly cold-air, mesoscale vortices. However, their use of the term 'comma cloud' has caused some confusion in the literature and will be used here to describe the cloud signature only.

The surface pressure signatures of mesocyclones are hard to determine since so many vortices occur in remote regions where there are no instrumental measurements. The small, short-lived vortices over the eastern Weddell Sea observed by Heinemann (1990), which passed close to Belgrano Station on the Filchner–Ronne Ice Shelf, had only a weak pressure perturbation of 1–2 hPa. Slightly larger pressure signals of typically 3–5 hPa were reported by Bromwich (1991) for mesocyclones over the southwestern Ross Sea, where a good network of automatic weather stations was available to provide surface observations. On the other hand, the vortex examined by Turner *et al.* (1993a), which developed over Halley Station on the eastern side of the Weddell Sea, had a maximum pressure difference from the surrounding area of 18 hPa as determined from the surface observations, but this was an exceptional case with a lifetime of three days and the pressure anomaly is thought to have been greater than that measured for the majority of cases observed in the Antarctic. As would be expected with the small surface pressure perturbations associated with most mesocyclones, the winds around these systems are not particularly strong. Examination of satellite altimeter and scatterometer winds over the ice-free ocean suggests that speeds of the order of 10–15 m s^{-1} are quite common near the centre of mesocyclones, but few have winds in excess of gale force.

One characteristic of mesocyclones noted in several studies is that they often occur in outbreaks of several lows with a series of vortices forming and dissipating over a number of days. Such events have been observed over the eastern Weddell Sea (Heinemann, 1990) and the Bellingshausen Sea (Turner and Thomas, 1994), suggesting a strong link with the synoptic-scale circulation. A typical situation over the ocean is when a synoptic low becomes slow-moving and draws cold air equatorwards on its western flank. With the de-stabilizing effect of cold advection aloft and the input of heat from the ocean into the lowest layer of the atmosphere, the conditions are created for the serial development of mesocyclones. An example of several mesocyclones developing in close proximity in a southerly airstream over the Bellingshausen Sea is shown in Figure 6.23. Another situation

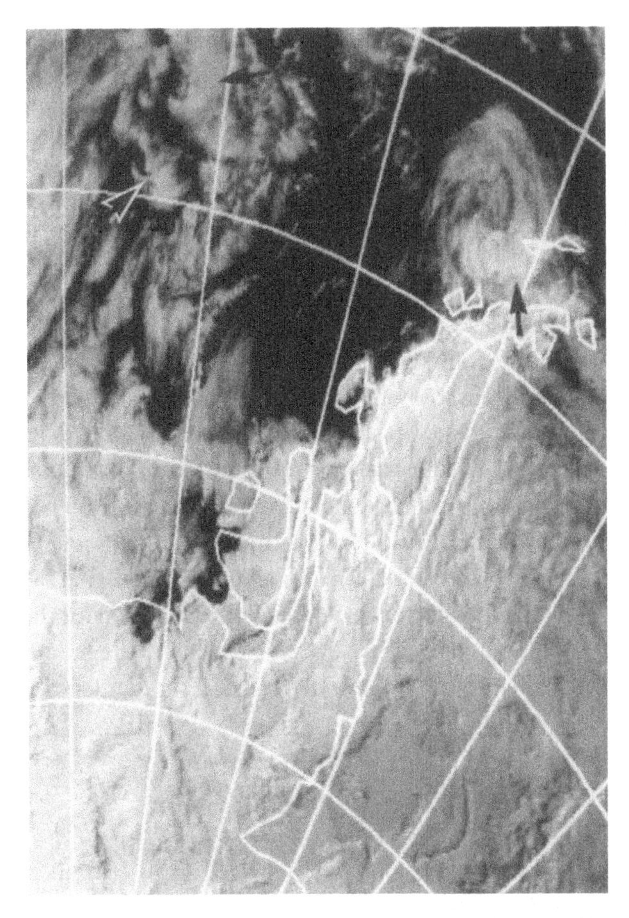

Figure 6.23 Three mesocyclones (indicated by arrows) close to the tip of the Antarctic Peninsula as seen in a NOAA infrared AVHRR image for 21:00 GMT 8 January 1984.

in which several mesocyclones can occur together is in the centre of a declining extra-tropical cyclone when the main low-pressure system is dissipating and several mesoscale lows develop within the general area of low pressure.

6.5.2 Climatology

Since the Northern Hemisphere polar low is a phenomenon of the winter season, it is not surprising that the first studies of Antarctic mesocyclones concentrated on the same season and the summer vortices were only considered at a later stage. The first Antarctic-wide climatological investigation of mesocyclones was carried out by Carleton and Carpenter (1989a; 1990), who examined seven years of winter season (June–September) Defense Meteorological Satellite Program (DMSP) imagery and investigated the occurrence of vortices with spiral and comma-shaped clouds and their relationship to the broad-scale circulation pattern. Subsequent investigations have been more limited in areal extent but have considered other seasons and incorporated more short-lived systems, which were excluded by Carleton and Carpenter. These have included investigations of vortices around the Antarctic Peninsula and Weddell Sea (Turner and Row, 1989; Carrasco and Bromwich 1993b; Turner and Thomas, 1994), over the eastern Weddell Sea (Heinemann, 1990) and in the Ross Sea sector (Bromwich, 1991).

Geographical distribution

The work of Carleton and Carpenter showed the extent to which mesocyclones are a feature of all oceanic areas of the high-latitude Southern Hemisphere and their study even considered vortices well into the mid-latitude areas. During their analysis of satellite imagery they observed vortices over all the ice-free and sea-ice-covered areas of the high-latitude Southern Hemisphere and identified some systems over the continent itself. Their study included all mesocyclones with a spiraliform cloud signature, but only those with a comma-shaped cloud pattern which had a lifetime of a least 24 h. This restriction was imposed in order that the short-lived, non-developing vortices were not included, but this criterion has caused difficulties when comparing their results with those from subsequent investigations. In their analysis the location of each vortex was noted and the results compiled into monthly maps showing each mesocyclone identified for the seven-winter period 1977–83. Their map of all vortices identified for the month of June is shown in Figure 6.24. Here the mesocyclones with comma (spiral)

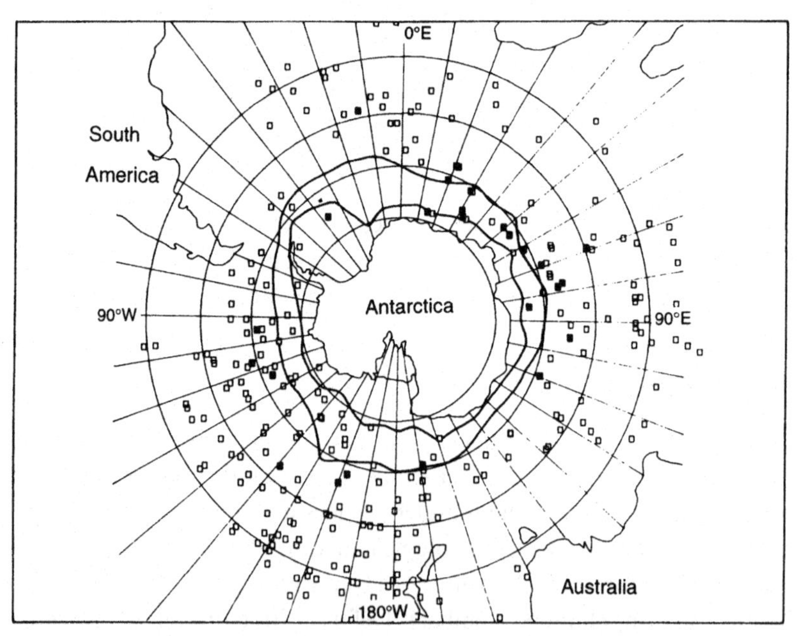

Figure 6.24 Polar low locations for the month of June during the period 1977–83. Commas (spirals) are shown by open (filled) squares. Maximum and minimum sea ice limits for the month are also shown. From Carleton and Carpenter (1989a). Copyright by A. Deepak Publishing.

cloud signatures are shown as open (filled) squares together with the maximum and minimum extents of the sea ice for the month. It is clear that mesocyclones with comma-shaped clouds are far more common than those with spiral cloud bands and the latter class comprised only 10% of the total. However, over the sea-ice-covered ocean and close to the edge of the sea ice the percentage of spiral systems was larger and was occasionally as great as 40% of the total. This was assumed to be because of the reduced synoptic-scale activity at these higher latitudes, which provides little translation of the vortices and allows them to develop a more circular symmetric cloud signature. A slightly higher ratio of spiraliform to comma cloud signatures was found by Turner and Thomas (1994) in their study of summer-season coastal vortices, in which one third of the total were spiraliform systems. This higher proportion was assumed to be a result of the reduced level of synoptic activity during the summer months and lower wind speeds over the ocean.

Subsequent studies of mesocyclones around the coastal region have shown that many more vortices are occurring than were recognised in the Carleton and Carpenter (1990) investigation. It can be seen in Figure 6.25, taken from the Heinemann study, that a maximum density of over three vortices per 250 km×250 km area and per 60 days was found over the ice-free ocean close

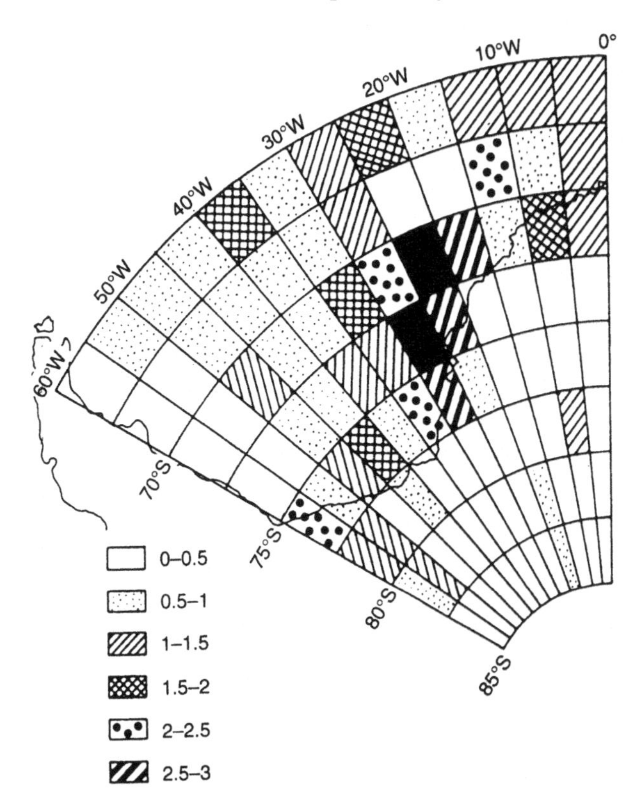

Figure 6.25 The frequency of mesocyclones over the eastern Weddell Sea during the summer months of 1983–88. Units are vortices per 250 km×250 km and per 60 days. From Heinemann (1990).

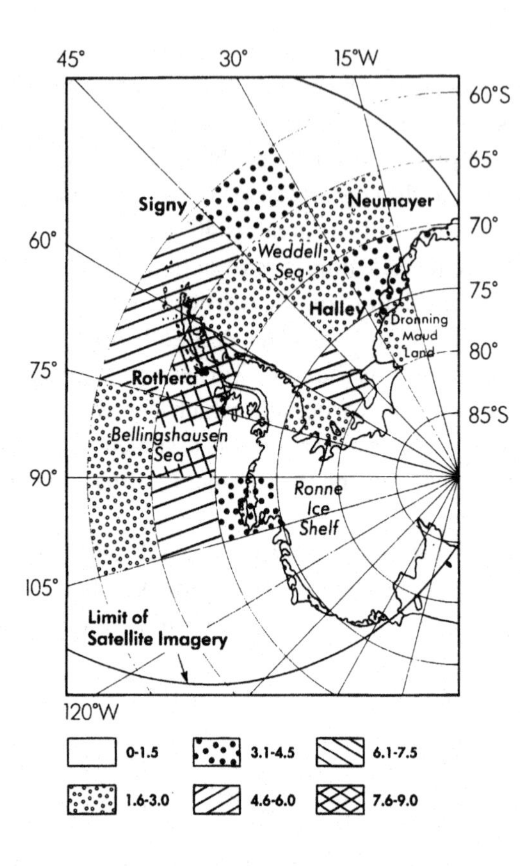

Figure 6.26 The monthly spatial frequency of mesocyclones around the Antarctic Peninsula during the period December 1983 to February 1984. Units are number of vortices per 15° longitude by 5° latitude per month normalised with respect to satellite coverage. From Turner and Thomas (1994), *International Journal of Climatology*. Copyright (1994) Royal Meteorological Society. Reprinted by permission of John Wiley & Sons Ltd.

to the Halley Research Station. Over the central part of the Weddell Sea, which is ice-covered even in summer, there was almost no mesocyclone activity, thus suggesting the importance of fluxes of heat and moisture from the ice-free ocean.

The large number of mesocyclones around the Antarctic was also apparent in the study of summertime mesoscale vortices over an area around the Antarctic Peninsula and the Weddell Sea by Turner and Thomas (1994). They identified a similar high density of vortices close to Halley Station, but also found other areas where mesocyclones occurred frequently. In summer, the Bellingshausen Sea to the west of the Antarctic Peninsula, is largely ice-free and, because synoptic-scale disturbances become slow-moving on the western side of the Peninsula, there are frequent outbreaks of cold, continental air northwards over this area in which many mesocyclones develop. They normalised the number of observed vortices according to the availability of imagery and produced a map of the mean number of systems per 5 degrees of latitude by 15 degrees of longitude and per month for their three-month study period. This is shown in Figure 6.26 and the vortex maximum of 7.6–9 mesocyclones per month over the Bellingshausen Sea is apparent in the latitude band 65° S to 70° S. The frequencies of vortex formation in Figures. 6.25 and 6.26 refer to differently sized latitude/longitude boxes and periods but when this is taken into account it is clear that the results of the two studies give similar orders of magnitude for the number of vortices over the eastern Weddell Sea, with a figure of 0.5–2.0 per month and per 250 km×250 km box being representative for the summer months.

Carleton and Carpenter (1989b) examined a number of aspects of the geographical distribution of mesocyclones during the 1977–83 winter months, including its links with the large-scale circulation. Most vortices were found to occur between 40° S and 60° S, with a peak at 45° S to 50° S and the greatest variability in the zonally averaged frequency of mesocyclones was also in this latitude band. Outside the 40° S to 60° S band the number of vortices observed decreased rapidly towards the tropics and Antarctic continent. Carleton and Carpenter (1989b) also found large meridional variations in the position of mesocyclone activity, which appear to be strongly linked to the locations of the long waves, and to wave number one in particular. A shift of this wave between the early and late part of the winter causes the area of maximum mesocyclone activity to transfer from the South Pacific to the Indian Ocean, resulting in these areas showing the greatest variability in mesocyclone occurrence. Since the position of the wave number one trough also affects the locations of the extra-tropical cyclones, there was a strong correlation between the positions of the mesocyclones and the frontal lows. The longitudes with fewest vortices were found to be around South America and the western South Atlantic, suggesting that few outbreaks of polar air occur in these regions.

During the whole of the Carleton and Carpenter study only three mesocyclones were observed over the Antarctic continent itself and there was a marked drop in the number of mesocyclones south of the edge of the continent in all the winter months examined. The lack of vortices over the ice may have been because of the inevitable use of only infra-red satellite imagery during this winter season study, which severely limited the identification of some types of vortex. During the winter the snow surface of the Antarctic is very cold and has a similar temperature to that of much of the low- and medium-level cloud associated with mesocyclones. With only infra-red imagery available, identification of vortices over the continent is very difficult and may have contributed to the limited number of systems identified. Subsequent studies have recognised that significant numbers of mesocyclones can form in some parts of the lower lying areas of the continent, although the higher areas of the plateau still appear to have few vortices at any time of the year. Bromwich (1989c) found frequent mesoscale cyclogenesis taking place in the region of the eastern Ross Ice Shelf and around Terra Nova Bay, with one or two new vortices forming each week. Some of these were detected on DMSP satellite imagery, but the majority had no cloud associated with them during their early stages of development, indicating the value of automatic weather station data in identifying some types of mesocyclone. A similar association of mesocyclones with low-lying areas of the Antarctic was found by Turner and Thomas (1994) who observed several vortices each month on satellite imagery of the Ronne Ice Shelf.

The location of maximum mesocyclone activity is also correlated with the extent of the sea ice and sea surface temperature anomalies on a regional basis. The connection with sea ice extent comes about as a result of the persistent southerly flow to the west of low-pressure centres, which both advects sea ice northwards and brings very cold Antarctic air masses to lower latitudes. Even though the capping effect of the sea ice reduces surface heat fluxes over a greater area, the advection of very cold air to lower latitudes can give very unstable conditions in the lower troposphere beyond the ice edge and outbreaks of mesocyclones. This will be particularly noticeable when the depressions are less mobile and the southerly flow can become established over several days. This reduction in static stability can also take place when sea surface temperatures in a particular area have a positive anomaly, giving increased heat fluxes into the lower layers of the atmosphere. This relationship was quantified by Carleton and Carpenter (1989a), who showed that there is a positive correlation between mesocyclone activity and positive SST anomalies during the winter months with the vortices being found downstream of the anomalies.

Links with the synoptic-scale circulation

The areal distribution of mesocyclones shows a high degree of variability, both within individual seasons and on an inter-annual basis. This is largely because of changes in the atmospheric circulation, including the location of the long waves and synoptic-scale weather systems, as well as other factors that affect the climate hemisphere-wide, such as sea surface temperature. The link with synoptic-scale activity has been examined for the winter season by Carleton and Carpenter (1990), who found a statistically significant relationship between the latitudes at which mesocyclones and frontal wave cyclones were occurring within individual seasons. Subsequent work for the same season (Carleton and Fitch, 1993) found that outbreaks of mesocyclones were associated with strong, horizontal gradients of 1000–500 hPa thickness and, at higher latitudes, with upper level cold pools. As will be discussed in Section 6.23, a strong horizontal gradient in lower tropospheric temperature is important for the growth of mesocyclones through baroclinic instability, as are upper cold pools that reduce the stability of the atmosphere.

Turner and Thomas (1994) also examined the relationship between mesocyclones and the major synoptic disturbances around the Antarctic Peninsula–Weddell Sea area. Their study related mesocyclone occurrence to the synoptic environment, the geographical location and the position of the sea ice. They assigned each vortex to one of eight classes, as shown in Table 6.4. The greatest number of vortices were found in the following synoptic situations.

(i) In the southerly airstream to the west of major, synoptic lows. Mesocyclones resembling these were some of the earliest to be identified in the Northern Hemisphere where they were known as polar air depressions (Meteorological Office, 1962).
(ii) In association with synoptic-scale troughs, often extending from the quasi-stationary, barotropic low-pressure systems which frequently occur in the Antarctic coastal region. Carleton and Carpenter (1989a) found that such a mesocyclone could evolve from a declining frontal cyclone.
(iii) Close to major frontal bands. This category may be indicative of developments on the fronts but no detailed case studies of this type of vortex have been carried out to date.
(iv) Forming in the centre of declining lows – the 'merry-go-round' type of development.

The work of Turner and Thomas also involved the development of a graphical representation to show the links between mesocyclone occurrence and the synoptic environment and the results for December 1983 are shown in Figure 6.27, in which the horizontal scale indicates the day of the month and the numbers of mesocyclones forming on each day are listed along the bottom. The passage of mobile, frontal low-pressure systems from west to east is represented by the diagonal lines whereas the 64 individual mesocyclones identified during the month are shown by the small letters which are the class abbreviations given in Table 6.4. The vertical location of the letters gives the approximate longitude at which each of the vortices was observed; the vertex of Figure 6.27 indicates locations from the Bellingshausen Sea at the top to the Weddell Sea just below the centre. Mesocyclones occurring on the continent itself are indicated on the bottom row.

During the month, four major frontal lows moved eastwards from the Bellingshausen Sea and each was accompanied by an outbreak of mesocyclones over the ocean area to the west of the Antarctic Peninsula. This was usually one or two days after the passage of the low across the Peninsula, by which time the southerly flow behind the low had brought cold air over the relatively warm, unfrozen ocean. Figure 6.27 also shows that the maximum mesocyclone activity during the period occurred over the Bellingshausen Sea with fewer systems being found over the Antarctic Peninsula and Weddell Sea. Of the 64 lows identified, only two were located over the continent itself, showing that mesocyclones are predominantly an oceanic phenomenon.

Another study of the links between mesocyclone formation and the synoptic environment was carried out by Heinemann (1990), who examined the relationship between summer-season mesocyclones over the eastern Weddell Sea and the atmospheric circulation as determined from the operational surface and 500 hPa analyses produced by ECMWF. He found that the more long-lived mesocyclones occurred in association with synoptic-scale lows located east-northeast of

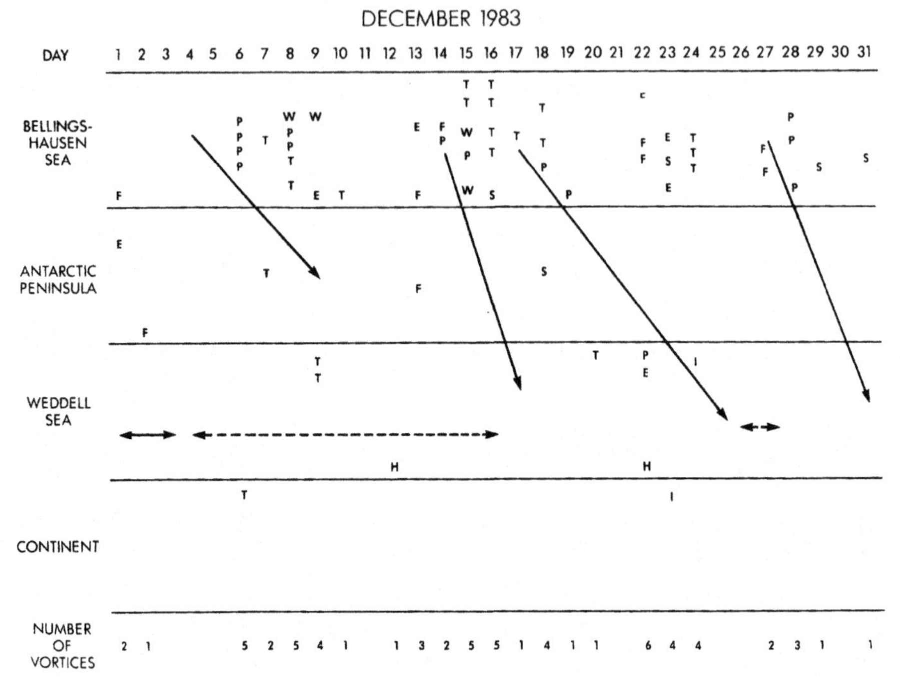

Figure 6.27 A schematic representation of the synoptic-scale atmospheric flow and mesoscale vortex activity for December 1983. The horizontal scale indicates days of the month, with the vertical axis showing locations around the Antarctic Peninsula. The solid lines indicate depressions moving eastwards and the broken lines show quasi-stationary anticyclones. The different classes of mesocyclone (see Turner and Thomas (1994)) are indicated as Halley (H), Polar Low (P), Trough (T), Small Synoptic (S), Front (F), Ice Edge (E), Over Water (W) and Over Ice (I). From Turner and Thomas (1994), *International Journal of Climatology*. Copyright (1994) Royal Meteorological Society. Reprinted by permission of John Wiley & Sons Ltd.

the Weddell Sea when the circulation re-enforced the low-level, off-continent easterly drainage flow in the region of Halley Station, bringing cold air over the relatively warm ice-free ocean. However, as with the Turner and Thomas (1994) study, the correlation between short-lived mesocyclones, which make up the bulk of the systems observed in this area, and the synoptic-scale flow was weak, suggesting that developments in the eastern Weddell Sea are the result of local forcing mechanisms.

Studies in the Northern Hemisphere have identified occasions when cold-air mesoscale vortices to the north of the polar front can merge with the main frontal band to form an 'instant occlusion'. In these situations the comma-shaped cloud band of the mesoscale system joins with the frontal wave to create a synoptic-scale disturbance with the appearance of a mature mid-latitude low-pressure system. Carleton and Carpenter (1989a) noted such developments taking place around the Antarctic, although they seem to be comparatively rare.

However, when they do occur, the resulting vigorous extra-tropical cyclones can have a far more significant circulation than either the mesocyclone or the original frontal wave would have had.

As observed in the Northern Hemisphere, mesocyclones are often associated with negative anomalies in the 500 hPa height field that usually arise from the presence of upper-level troughs. Turner and Thomas (1994) found that their 162 summer mesocyclones had a mean 500 hPa height anomaly of −5.3 dm with the largest anomalies for the various classes being for the small baroclinic waves (−10 dm), mesocyclones embedded in the synoptic lows (−10 dm), polar lows (−7 dm) and troughs (−9 dm). On the other hand, the vortices found near the edge of the sea ice and well removed from the major weather systems had only small height anomalies, indicating that they are only shallow, near-surface features with little connection to the upper-air conditions. Similarly, Heinemann (1990), found no link between mesocyclone activity over the eastern Weddell Sea and the circulation at the 500 hPa level.

Temporal variability

Since no Antarctic-wide, year-round studies of the occurrence of mesocyclones have been carried out to date it is not possible to consider the distribution of systems throughout the year with any degree of accuracy. One of the main problems is that, when only a few years of data for a particular season are examined, the variations in the synoptic-scale circulation dominate the changes observed between years and the effects of some of the hemisphere-wide cycles affecting the development of mesocyclones may be obscured. A further difficulty is that the studies of mesocyclones in the summer and winter seasons have used satellite imagery from different sources and the criteria for the inclusion of mesocyclones were not always the same so that the density of systems observed cannot be compared directly. Therefore the only season that we can examine for inter-annual variability is winter, for which we have the results from the seven-winter study of Carleton and Carpenter (1990). For the whole period they found that the greatest number of vortices occurred in July (the mean number observed was 43.6) and the minimum in September (mean 14.4), which they linked to the meridional migration of the circumpolar trough associated with the semi-annual pressure oscillation (van Loon (1967), see also Section 3.3). They also noted a decrease in the number of vortices in the Antarctic coastal region between June and September; they suggested that this may be linked to the increase in extent of the sea ice, which reduced the surface heat and moisture fluxes over large areas. This is in contrast to the case of frontal cyclones, which show a clear shift polewards as the winter progresses.

The other major inter-seasonal change in the mesocyclone activity is the shift westwards of the vortex maximum from the South Pacific to the southwest Indian Ocean during the June to September period. This can be seen in Figure

6.28, which shows the monthly mesocyclone variations by longitude. As discussed earlier, this has been linked to an amplification of the wave number one trough close to the longitude of western Australia, showing the sensitivity of the mesocyclone distribution to circulation changes on the hemispheric scale.

For the summer season there are no long-term studies of mesocyclone development and we have to rely on a number of single-year investigations to try to determine the months of greatest activity. In the area of the Antarctic Peninsula, December was found to be particularly active in the studies of Carrasco and Bromwich (1992), who examined August 1989 to February 1990, and of Turner and Thomas (1994), who considered December 1983 to February 1984. However, in the former study the highest spatial frequency was found in the Weddell Sea whereas in the latter most systems were observed in the Bellingshausen Sea. With such high variability in mesocyclone activity, longer term investigations are needed before we can determine how the number of systems changes throughout the summer season with any confidence.

Investigation of the inter-annual variability of atmospheric systems requires the availability of long time series of satellite images and the visual analysis of a great many individual scenes. Not surprisingly, there have been few studies that have attempted to process such long series of data over the Southern

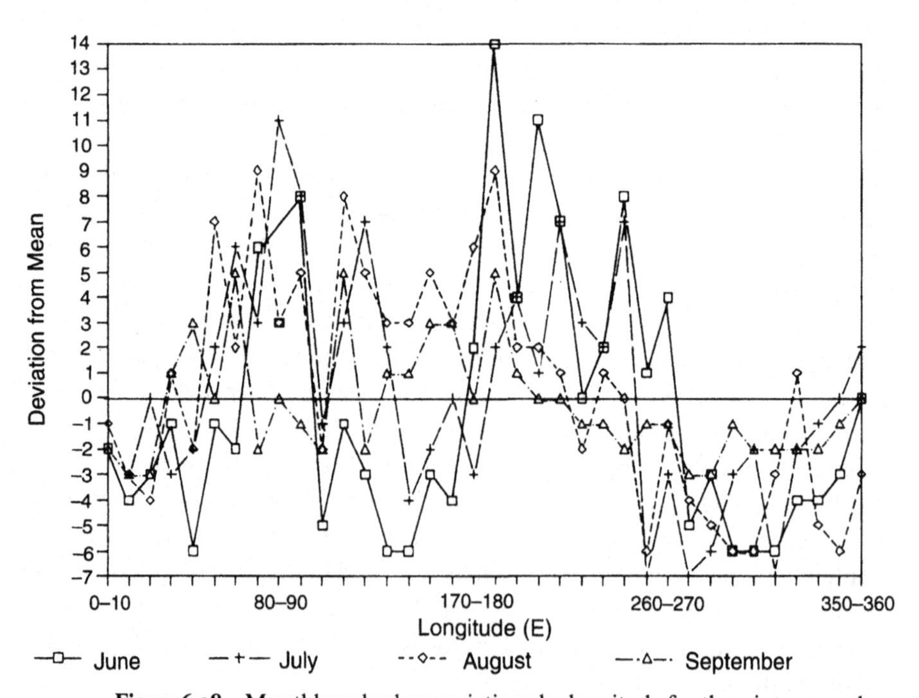

Figure 6.28 Monthly polar low variations by longitude for the winter months of 1977–83. Values in each 10° zone are shown as the departure from the hemispheric mean in that month. From Carleton and Carpenter (1989a). Copyright by A. Deepak Publishing.

Hemisphere, that by Carleton and Carpenter (1989a) for the winter season having been the only one to have examined a significant number of years for mesocyclone activity. Figure 6.29 has been prepared from their data and shows the number of mesocyclones per winter for 1977–83. They were found to vary in number from 203 in 1979 to 86 in 1977 with a mean of 137 and a standard deviation of 44. Such a large variability is consistent with the significant changes noted among years in the main tropospheric parameters, such as the height and wind fields, the locations of extra-tropical cyclones, sea ice extent and sea surface temperature. We know that the FGGE year of 1979 was a very anomalous period with strong westerlies and deep synoptic lows resulting in a strong meridional component to the circulation (Physick, 1981). This is assumed to be responsible for the large number of mesocyclones observed, because of the transport of cold polar air masses to lower latitudes. During this winter the mesocyclones were fairly evenly distributed around the Southern Hemisphere with no preferred sectors being apparent. Zonally, the vortices were found over the strong gradient of surface pressure anomaly just south of mid-latitudes and in the eastern hemisphere.

The number and location of mesocyclones in the Southern Hemisphere have been examined in association with El Niño–Southern Oscillation (ENSO) data,

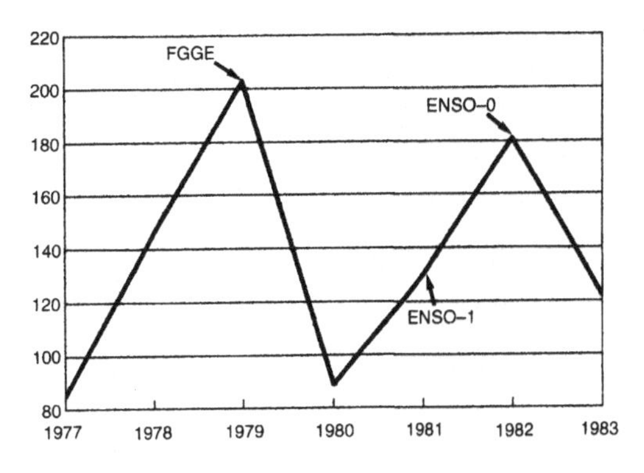

Figure 6.29 The number of mesocyclones observed during the winters of 1977–83 over the Southern Hemisphere. Adapted from Carleton and Carpenter (1989a).

ENSO being one of the major teleconnections linking the atmospheric and ocean circulations (see Section 7.2). The Carleton and Carpenter study encompassed the major ENSO 'warm' event of 1982–83 and allowed the comparison of mesocyclone activity throughout this and the preceding years. It was found that there were considerably more mesocyclones in 1982 than in the previous winter and that the location of maximum activity shifted from around 90° E to the area of New Zealand and the west Pacific. This would appear to have been the result of the combined effects of changes in the sea ice extent and the enhancement of the long-wave activity in the New Zealand sector.

Some intercomparisons of mesocyclone activity between two years have also been carried out and, although these cannot provide information on vortex activity in relation to the major climatic cycles of the Southern Hemisphere, they can be of value in highlighting the developments that take place in contrasting years. One such study was carried out by Carleton and Fitch (1993), who compared mesocyclone activity during the winters of 1988 and 1989 and found associations with the extent of the sea ice and the katabatic wind regime.

A further study was by Carrasco and Bromwich (1993b), who considered vortex formation in the Bellingshausen Sea over the summers of 1983–84 and 1989–90 and contrasted the systems found in the light of the synoptic environment. The 1989–90 summer had most mesocyclones over the western Weddell Sea whereas 1983–84 was characterised by a large number of mesocyclones on the other side of the Antarctic Peninsula, in the Bellingshausen Sea. The later period had a deeper, climatological low-pressure centre north of Marie Byrd Land, which resulted in stronger northwesterly winds bringing warm, mid-latitude air to the region west of the Peninsula – conditions that resulted in record-breaking high temperatures over the Antarctic Peninsula. This would have had the effect of stabilising the lower troposphere and inhibiting mesocyclone activity. The reason for the greater number of vortices to the east of the Peninsula during this year is less clear, but could have been a strengthening of the low-level thermal gradients over the Weddell Sea or the development of lows in the lee of the Peninsula because of topographic effects.

6.5.3 Development mechanisms and structure

The development of mesocyclones

As with Northern Hemisphere polar lows, it is thought that most Antarctic mesocyclones form through the processes of baroclinic instability and cyclonic vorticity advection. The baroclinicity is a result of moderate-to-strong low-level thermal gradients, while the vorticity advection takes place through upper-level short-wave troughs. Baroclinic zones can be found in several areas of the Antarctic, including the northern limit of the sea ice, at the land–sea boundary

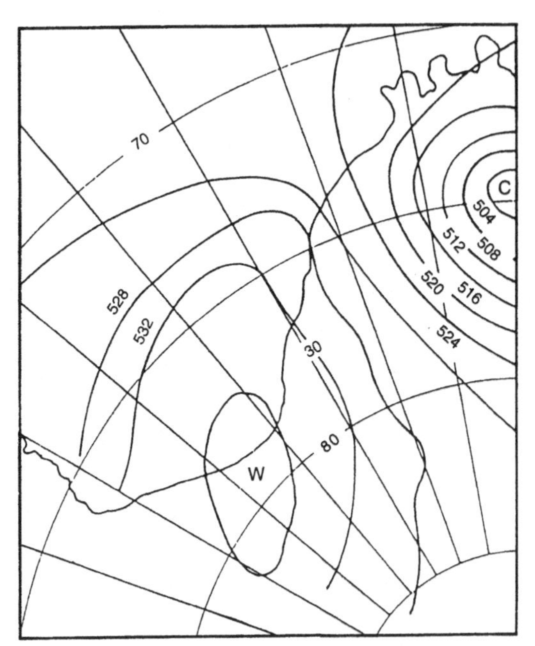

Figure 6.30 A map of the 1000–500 hPa thickness field (dm) over the Ronne Ice Shelf, Dronning Maud Land and the southern Weddell Sea at 12:00 GMT 3 January 1986 derived from satellite sounder data. From J. Turner *et al.*, *J. Geophys. Res.*, **98**, 1328, 1993, copyright by the American Geophysical Union.

in summer and where the katabatic winds descend to the low-lying coastal areas. The thermal gradients at the limit of the sea ice and near the coast are rarely strong, but the thermal contrast is enough to create weak frontal zones that can often be seen on satellite imagery as a band of cloud, which appear similar to the 'Arctic fronts' observed in the north. Synoptic forcing can also play a role in enhancing the thermal gradient, as was found by Heinemann (1990), who observed that mesocyclone activity over the eastern Weddell Sea often occurred in strong easterly, off-shore flow which brought cold continental air into close proximity to more moderate air over the ocean. In the case of the major mesocyclone over Halley Station investigated by Turner *et al.* (1993a), it was found that an upper-level cold trough had brought very cold air, extending through a deep layer of the troposphere, to the area inland of Halley Station while a relatively warm air mass had been advected southwards to just off the coast. As can be seen in Figure 6.30, this had created a strong thermal gradient in the 1000–500 hPa thickness field where values varied from 500 dm inland of Halley to 536 dm over the Ronne Ice Shelf. It was in this strong baroclinic zone that the vigorous mesocyclone developed. Another area where strong low-level baroclinicity occurs is around the southwestern Ross Sea, where very cold air emerges into the coastal region after having descended from the interior plateau down valley glaciers. Bromwich (1991), in his study of mesocyclones in this area, used automatic weather station data to investigate the vortices and confirmed the presence of the thermal gradients. The baroclinicity was assumed to be quite shallow, which would be consistent with the theoretical results of Mansfield (1974), who showed that the limited horizontal extent of polar lows in the Arctic was a result of the shallow depth of the baroclinic zones.

Climatological investigations and case studies have confirmed the correlation between the development of mesocyclones and aspects of the synoptic-scale circulation, such as strong gradients in the 1000–500 hPa thickness (Carleton and Fitch, 1993; Turner *et al.*, 1993a). In particular, many occur in association with upper-level troughs extending equatorwards from the polar vortex that transport Antarctic air to lower latitudes and provide cyclonic vorticity advection. These upper troughs are often to the west of surface synoptic-scale lows where there is also a negative anomaly of 500 hPa height and 1000–500 hPa thickness. Such cooling in the mid-troposphere, combined with heating at low-levels from the relatively warm ocean, de-stabilises the atmosphere and aids the development of vortices in the shallow baroclinic zones. The weak static stability in the area of many mesocyclone developments is apparent from the cumuliform clouds that are apparent on satellite imagery of systems in the early stages of development.

Certain areas experience many mesocyclone developments because of a combination of favourable synoptic conditions and ice-free ocean. This is the case in the eastern Weddell Sea, where there is a climatological trough extending down the coast of Dronning Maud Land from the synoptic-scale low which is often found north of SANAE Station, a prevailing easterly off-shore wind and ice-free conditions off the coast for most of the summer. A typical example of a mesocyclone in the eastern Weddell Sea is shown in Figure 6.31. The leading edge of the trough, which is often located over the ocean, provides a favourable environment for cyclogenesis since this region will aid vertical motion. One outbreak of mesocyclones in this area was investigated by Heinemann (1990), who showed that an upper-level cold pool was associated with the outbreak of a series of vortices by using model fields and satellite sounder data. In the case of the Turner *et al.* (1993a) system this also developed in an environment of forcing by an upper-level trough, but here the quasi-stationary nature of the synoptic situation for three days allowed the vortex to exist for much longer than do the majority of systems.

Fluxes of sensible and latent heat over the ice-free ocean are important in injecting heat and moisture into the lowest layers of the atmosphere, but play a smaller role in the development of mesocyclones than they do in the Northern Hemisphere. However, once a mesocyclone has formed, convection is important in the development of the low, as is demonstrated by the number of systems seen emerging from fields of cumulus cloud. Around the Antarctic, the air–sea temperature differences rarely exceed 15–20°C so that the very large heat fluxes found in the Arctic, which arise as a result of low air temperatures and relatively high sea surface temperatures in certain areas, do not occur. When the limited fluxes are coupled with the generally more zonal circulation of the Southern

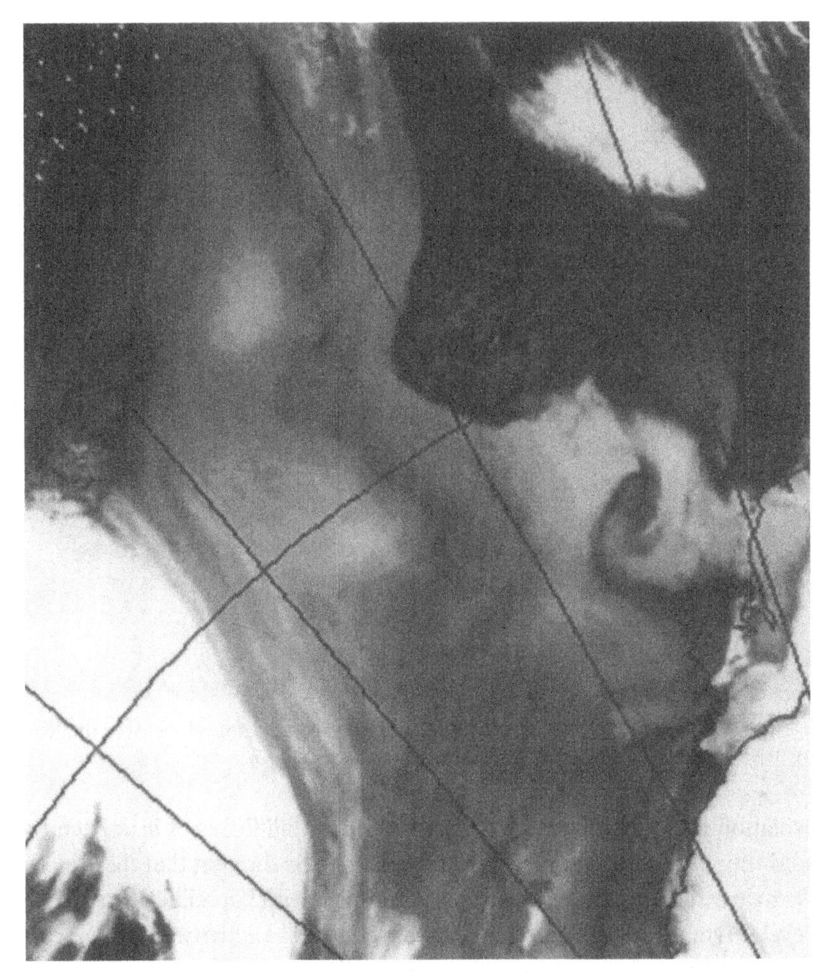

Figure 6.31 An infrared satellite image of a mesocyclone off Halley Station at 21:44 GMT 5 November 1993.

Hemisphere and the limited incursions of cold air equatorwards, then it can be understood why the deep convection of the northern polar lows does not occur in the Antarctic coastal region. However, it has been suggested that the mesocyclones with the greatest convective activity in the Southern Hemisphere may be located at relatively northerly latitudes, such as just to the south of New Zealand, where satellite imagery has shown many vigorous systems. As discussed in Section 3.2, the environmental lapse rates found around the Antarctic are remarkably stable so that the convective cloud associated with the mesocyclones tends to be trapped within the unstable boundary layer.

A number of mesocyclones have been observed to form on the low-lying ice shelves well removed from synoptic disturbances (Figure 6.32). Although the lack of data may mean that some of these developed in troughs or areas of low pressure that were not resolved by the observing network, many undoubtedly develop in

Figure 6.32 An infrared satellite image of a mesocyclone over the Ronne Ice Shelf at 21:09 GMT 4 May 1993.

isolation from larger scale systems. With the small fluxes of heat from the surface and the very stable nature of the atmosphere, it is thought that these mesocyclones form in response to dynamical forcing, coupled with baroclinic instability from low-level thermal gradients, rather than from convective activity. Low-level convergence can develop as a result of the katabatic winds flowing from the high plateau region and descending the valley glaciers to the coastal area or flat inland regions of the ice shelves (see Section 6.1.5). This air from the interior is also very cold and, although some adiabatic warming takes place as a result of descent, the air is still colder than the surrounding air when it reaches the level of the ice shelves. This causes baroclinic zones to be established close to areas of outflow from valley glaciers, resulting in the frequent development of mesocyclones. These areas of persistent katabatic winds and moderate baroclinicity need not just be on the ice shelves but can also be observed over the ocean where the winds extend out beyond the coast giving convergent flow and conditions suitable for cyclogenesis well removed from the steep topography that generated the strong winds.

Topography also plays a role in the development of some mesocyclones through the generation of cyclonic vorticity in the lee of mountain barriers. This results in the development of a trough of low pressure on the downwind side of the barrier and the formation of lee lows that can break away and move downstream. On satellite imagery many of these small vortices have the appearance of mesocyclones with the characteristic comma or spiraliform cloud signature and it is often difficult to tell whether the vortex formed as a result of the topographic forcing or through baroclinic or convective processes. Turner and Row (1989) observed a number of mesocyclones to the east of the Antarctic Peninsula in satellite imagery and this barrier, which has an average elevation of more than 2 km, has a major influence on cyclogenesis over the western Weddell Sea. It has also been suggested that the development of vortices over the eastern Weddell Sea is influenced by the high topography of Coats Land to the east. Here there is frequent easterly flow down from the plateau, resulting in vortex stretching and an increase in vorticity in the coastal region. The role that this plays in mesoscale cyclogenesis has yet to be fully investigated, but the frequent occurrence of mesocyclones over the ice-free ocean has been well documented.

Structure and lifecycle

The major problem in investigating the structure of mesocyclones has been the scarcity of data available for use in detailed case studies. Research into polar lows in the Northern Hemisphere has been able to draw on observations from aircraft campaigns, ship reports and some radiosonde ascents from isolated islands, whereas the very limited number of reporting stations in the Antarctic has meant that data from satellites and automatic weather stations have had to play a greater role than they have done in Arctic studies. Satellite imagery is capable of providing information on the number of mesocyclones and their variability, but it can only help in investigating the structure of the vortices via interpretation of the cloud signatures. In their satellite study of mesocyclones Carleton and Fitch (1993) examined the lifecycle of the vortices observed in infra-red satellite imagery and identified four stages of their development. These were

(i) incipient, when no clear centre of circulation was apparent and detection of the system may have been through enhanced convection;
(ii) developing, with increased organisation becoming evident with deepening and thickening of the stratiform cloud;
(iii) mature, having well-defined spiraliform or hooked cloud signatures; and
(iv) dissipating, characterised by a disorganised cloud signature.

Such a classification of the stages of mesocyclone development is useful when assembling a climatology of vortices and considering the number that attain a particular degree of activity, but it provides no information on the thermal and motion fields associated with the lows. However, other data are available that can fulfil this need. In particular, the TOVS data from the NOAA satellites (Section 2.6.2) are especially valuable since they allow the computation of the thicknesses and mean temperatures of layers of the troposphere – Turner *et al.* (1993a) made extensive use of TOVS data to investigate the structure of the Halley low discussed below. This case is not claimed to be typical of Antarctic mesocyclones

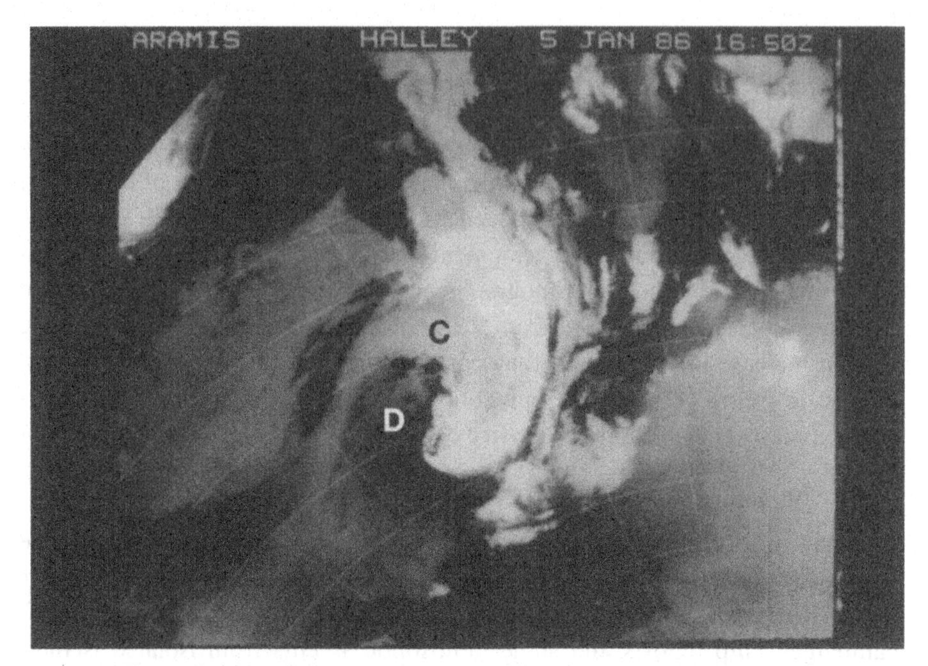

Figure 6.33 An infrared satellite image of a large mesocyclone off Halley Station at 16:50 GMT 5 January 1986.

since it lasted for three days and was larger than the majority of systems observed. However, it is one of the few cases for which extensive satellite, surface and upper air *in situ* data are available, and will serve to illustrate the complex structure of these vortices. An infra-red satellite image of the mesocyclone at its fully developed stage on 5 January 1986 is shown in Figure 6.33 and the formation and structure of the low are described in the following section.

A mesoscale vortex over Halley Station

The vortex formed early on 3 January 1986 when the eastern Weddell Sea was located under a strong thermal gradient throughout the lower troposphere, separating warm air that had been advected polewards down the eastern Weddell Sea and cold air descending from the Antarctic plateau. This thermal gradient was apparent in the 1000–500 hPa thickness field derived from the TOVS data which are shown in Figure 6.30. The three-hourly surface observations from Halley show that the low formed inland of the station, which registered a drop in surface pressure of 13 hPa in the 24 h prior to 12 GMT 3 January. However, the vortex soon crossed the coast, where it was first observed in the satellite imagery as a 300 km diameter spiral of cloud. As the continental air outbreak reached the coast a cold front developed, with the warm, maritime air ascending the leading edge of the tongue of cold air. This cold front is marked as 'C' in Figure 6.33. The Halley radiosonde ascent for 12 GMT 3 January showed the intrusion of cold air extending to 930 hPa with the warm, maritime air above.

The air from the interior of the continent was very dry throughout the troposphere and could be followed on imagery of coarse (40 km) resolution created from the 6.7 μm channel data available from the TOVS. These data can be used in a similar way to the water vapour imagery from the Meteosat and GOES satellites, although the resolution is much reduced (Turner and Ellrott, 1992). The water vapour imagery had a 'dark band' behind the cold front, which an isentropic analysis showed was associated with dry air that had descended from the upper troposphere. The dark band coincided with the cloud-free area behind the cold front seen on the infra-red imagery and labelled as 'D' in Figure 6.33.

During the following two days the comma of cloud associated with the cold front thickened and moved offshore to the southwest of the vortex centre. As the cold, continental air was advected over the ocean, some convective cloud developed, indicating instability; however, this never extended to higher than about 600 hPa. The cloud-free slot behind the front continued to be visible in the infra-red and water vapour imagery throughout the period and the hook of frontal cloud appeared very similar to that observed with many mesocyclones seen around the Antarctic. The radiosonde ascents from Halley and the infra-red imagery showed that the height of the frontal cloud was about 700 hPa, whereas the cloud at the centre of the low was found to extend to 450 hPa. The low had started to dissipate by 6 January as the blocking high over the Weddell Sea moved away and the trough over the interior declined, thus reducing the thermal gradient over the area.

The large amounts of data available for this case allowed the development of conceptual models for the mesocyclone at two stages of its development. These were for 00 GMT 4 January (Figure 6.34), when the cold front had just crossed over the coast, and 17 GMT 5 January (Figure 6.35), when the structure of the low was most developed. In both of these figures the dry slot seen in the water vapour imagery is indicated, together with the areas of ascent and descent determined from the isentropic analysis. At the earlier time there were two main regions of significant vertical motion: an area of descent just inland of Halley Station and low-level, slantwise ascent up the developing cold front. By 5 January the cold front was well out over the Weddell Sea with a broad area of descending air to the south. The ascent up the frontal surface at this time spiralled into the centre of the low with the cloud reaching about 500 hPa.

It is still unclear how much precipitation is associated with mesocyclones in general since most of the systems are located over the ocean well away from observing stations. In this case the low gave 24 h of precipitation in the form of snow as cloud at the centre of the vortex crossed Halley Station. However, other studies of mesocyclones in this area observed much thinner cloud associated with the vortices, suggesting that they usually give little precipitation. The results of one of the first investigations of mesocyclones in the Antarctic with passive microwave imagery (Carleton *et al.*, 1993) tend to confirm this impression of

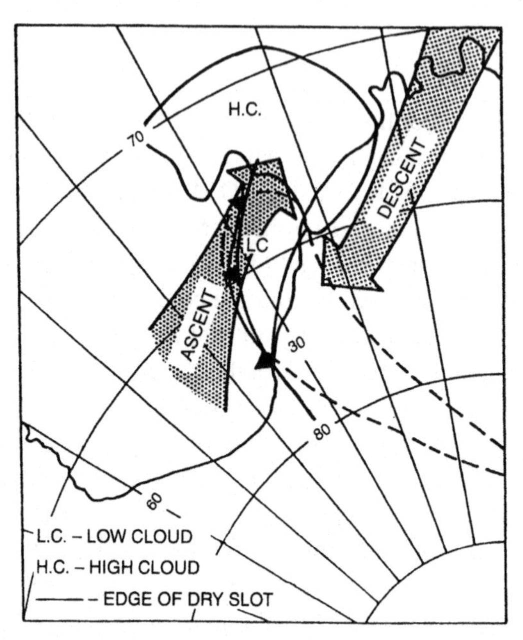

Figure 6.34 A conceptual model of the mesocyclone at 00:00 GMT 4 January 1986. From J. Turner *et al.*, *J. Geophys. Res.*, **98**, 1335, 1993, copyright by the American Geophysical Union.

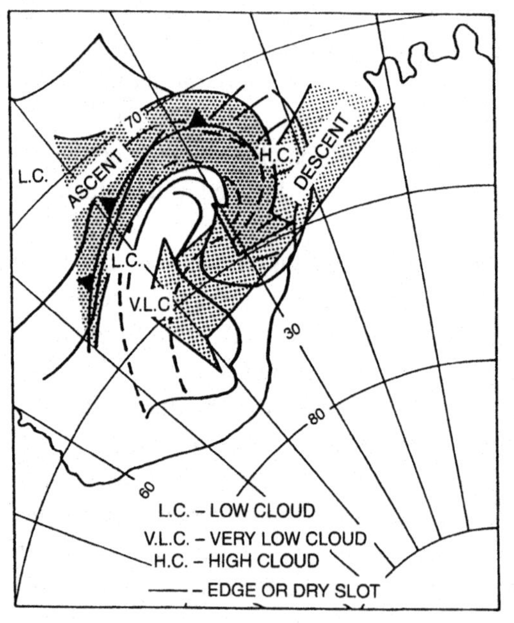

Figure 6.35 A conceptual model of the mesocyclone at 17:00 GMT 5 January 1986. From J. Turner *et al.*, *J. Geophys. Res.*, **98**, 1335, 1993, copyright by the American Geophysical Union.

relatively small amounts of liquid precipitation, although the algorithms for precipitation determination at high latitudes are still poorly tested.

6.5.4 **Modelling**

The current generation of global numerical weather prediction models, which are used to produce the routine forecasts over the Antarctic and surrounding sea areas, have a horizontal grid length in the range of 100–200 km, which is adequate to represent the synoptic-scale disturbances that are the main concern of the forecasting centres. Because the average mesocyclone has a diameter of less than 500 km, the models are not well suited to resolving these small disturbances, since there will only be three or four grid points across the typical vortex. There is a further problem in that the models are heavily reliant on satellite sounder data for defining the initial conditions of the integrations and these profiles, as transmitted over the Global Telecommunications System, typically have a horizontal resolution of several hundred kilometres and do not provide an adequate representation of the vortices. The sounder data also have poor vertical resolution, especially in the lowest layers of the atmosphere, which is a further difficulty in the analysis of the many shallow vortices observed. Turner and Thomas (1994) examined the capability of the UK Meteorological Office global, coarse-mesh model to represent the 162 summer-season mesocyclones that developed close to the Antarctic Peninsula. They found that a total of 23 mesocyclones were represented on the surface pressure analyses and, not surprisingly, all but one of these had a horizontal diameter of greater than 200 km. Most were the small synoptic-scale disturbances which were found mainly over the central and northern parts of the Bellingshausen Sea. The polar low class of mesocyclone, which usually forms on the western side of a depression, was very poorly handled and only one of the 35 systems identified on the satellite imagery appeared on the surface pressure analyses. This is because of the very limited number of observations over the ocean and the inability of the models to represent adequately the processes causing their development, such as strong surface heat fluxes and low-level baroclinic zones. None of the mesocyclones found over the continent appeared on the operational charts, even though a number had a diameter of more than 500 km.

Bromwich (1991) studied mesocyclone activity over the southwestern Ross Sea and examined the ability of the Australian hemispheric surface analyses to represent these systems. The success in analysing the mesocyclones varied from month to month and ranged from 0 to 60% with an average of 33%. This is comparable to the number of larger mesocyclones found on the UK Meteorological Office charts in the Turner and Thomas (1994) study described above. In their study of the mesocyclone over Halley Station, Turner *et al.* (1993a) found that the operational analyses produced by the UK Meteorological Office had a reasonable representation of this system with a low in the correct position close to the location of the cloud band in the satellite imagery. The analysis of the low by the data assimilation scheme was helped by the daily radiosonde ascents made at Halley Station and the three-hourly surface observations, which were also available in near real time for use in the analysis process.

High-resolution numerical models have only just begun to be used for the investigation of Antarctic mesocyclones and it will be some years before we can be confident of their ability to represent these mesoscale disturbances correctly. Some of the first experiments concerned with the use of high-resolution models were carried out by Münzenberg *et al.* (1992), who used the following two models to examine situations over the eastern Weddell Sea when mesocyclones were observed in satellite imagery.

(i) A high-resolution non-linear grid point model used in a quasi-linear mode. This model was used to examine vortices close to Halley Station using an instability analysis and, although this model had no parameterisation of physical processes, it proved valuable in gaining insight into the mechanisms behind the development of the vortices.

(ii) The Norwegian DNMI model, which was used in sensitivity studies to examine the role of various physical parameters, such as roughness length, topography and near-surface moisture. Although this model was run for cases in which mesocyclones were known to be in the area, no vortices were found to develop during the integrations. This result indicates the shortcomings of the present generation of models and the need for better parameterisation schemes for physical processes.

Studies have confirmed the importance of topography in cyclogenesis close to Halley Station, through vortex stretching as air descends from the high interior of the continent, and the need for accurate representation of moisture for the simulation of convection. The limited success in the simulations is consistent with studies from the Arctic, wherein it has proved very difficult to represent some types of polar low, especially where convection is thought to be important in the development. High-resolution models have the potential to be of value in mesocyclone research, although much work remains to be carried out before they can represent correctly the mesocyclones observed around the Antarctic.

6.5.5 Comparison with Northern Hemisphere polar lows

In the Northern Hemisphere, research into polar lows has concentrated on the most vigorous systems since these have the greatest impact on maritime operations and bring severe weather to coastal areas when they make landfall. Such systems can have winds of storm or hurricane force and can give extensive snowfall, which often brings disruption to areas on land. Understanding the dynamics of these systems is an important first step in being able to forecast their formation and development and this has driven a number of major research studies such as the Polar Lows Project (Lystad, 1986).

The most vigorous polar lows are characterised by deep convective cloud, indicating instability through a substantial layer of the atmosphere and abundant moisture. These systems occur in only a few areas of the Northern Hemisphere where very cold air with temperatures as low as −30°C flows off the ice and over the relatively warm water where sea surface temperatures can be as high as 10°C. Areas where such conditions occur include the Barents and

Norwegian Seas in the northeast Atlantic, the Gulf of Alaska in the North Pacific and the Sea of Japan. These large air–sea temperature differences result from the transport polewards of warm water to high latitudes, creating steep gradients of sea surface temperature and low-level baroclinic zones in the atmosphere. In the Southern Hemisphere, the location of the main landmasses promotes more zonal ocean currents (Section 3.5.2) and there is only limited meridional flow. Therefore the large air–sea temperature differences found in parts of the Arctic do not occur in the south. The only exception is within the polynyas which occur during the winter months when wind and ocean forcing open up ice-free areas close to the Antarctic continent (see Section 4.2.3). Then the air–sea temperature difference can be comparable to those found in areas such as the Norwegian Sea, although the polynyas are only of relatively limited horizontal extent.

The magnitudes of the sensible (H_S) and latent (H_L) heat fluxes over the polar oceans where mesocyclones develop, can be computed from the equations of Reed and Albright (1986):

$$H_S = -\rho C_H C_p V(T_A - T_S) \quad (6.39)$$

$$H_L = -\rho C_E L V(Q_A - Q_S) \quad (6.40)$$

where ρ is the density of air (1.275 kg m^{-3}), C_H and C_E the surface exchange coefficients (1×10^{-3}), C_p the specific heat of air (1.01×10^3 J kg^{-1} K^{-1}), L the latent heat of evaporation (2.47×10^6 J kg^{-1}), V the wind speed at 10 m and T_A (Q_A) and T_S (Q_S) the air temperature (specific humidity) at 10 m and at the sea surface respectively.

Tables 6.5 and 6.6, taken from Turner *et al.* (1993b), show sensible and latent heat fluxes for a number of cases of Northern Hemisphere polar lows and Antarctic mesocyclones. The fluxes are for the early stages of development of the systems and are significantly larger in the Northern Hemisphere cases when the winds have increased to gale or storm force. However, even with the relatively low wind speeds shown in the tables, the Arctic systems typically had sensible heat fluxes of 200–300 W m^{-2}, whereas, in the Antarctic coastal region, the figures were of the order of only 50 W m^{-2}. With sea surface temperatures south of the Antarctic Convergence rarely greater than 2–3°C during the Austral summer (Bottomley *et al.*, 1990) and air temperatures of no lower than about −10°C in most coastal areas, it can be seen that sensible heat fluxes do not approach the levels found in the Arctic. An exception is where the strong katabatic drainage flow brings very cold air down to the coast and out over the open water. Such a situation occurs in parts of the Ross Sea sector and flux estimates for a mesocyclone studied by Bromwich (1987) are given in Table 6.5. The latent

Table 6.5 *Environmental conditions and computed sensible heat fluxes for various Northern Hemisphere polar lows.*

Case	T_S (°C)	T_A (°C)	V (m s^{-1})	Track (km)	H_S (W m^{-2})	Region	Energy input (MJ m^{-2})	H_L (W m^{-2})
Rasmussen (1985)	6	−14	13	75	335	Bear Island	1.9	177
Harrold and Browning (1969)	3	−15	10	150	230	Iceland	3.4	105
Businger (1987)	5*	−8	13	200	220	Gulf of Alaska	3.4	150
Shapiro *et al.* (1987)	4	−7	15	n/a	210	Jan Mayen	n/a	146
Businger and Baik (1991)	3	−8	15	400	210	Bering Sea	5.6	151
Rabbe (1987)	5	−1	25	n/a	95	Norwegian Sea	n/a	n/a
Reed (1979)	7	0	17	n/a	155	Pacific Ocean	n/a	n/a

Notes:
* Climatological value
n/a, not available

heat fluxes in the Antarctic are particularly small compared with those found in the north, where strong, cold and dry winds blow over areas with relatively high sea surface temperatures. Under these conditions latent heat fluxes in the range of 140–180 W m^{-2} are found, compared with values of around 20 W m^{-2} over the eastern Weddell Sea.

The total energy added to the lowest layer of the atmosphere by sensible and latent heat fluxes will depend on the time that an air mass is over the open water and the length of its track over the ocean. Estimates of this are given in the tables for the cases in the two polar regions. Again, it can be seen that the energy added in the Arctic cases is almost an order of magnitude greater than that for the vortices around the Antarctic. The exception is with the Ross Sea case, for which the energy input starts to become comparable to that in some of the cases in the

Table 6.6 *Environmental conditions and computed sensible heat fluxes for various Southern Hemisphere mesocyclones.*

Case	T_S (°C)	T_A (°C)	V (m s^{-1})	Track (km)	H_S (W m^{-2})	Region	Energy input (MJ m^{-2})	H_L (W m^{-2})
Turner and Warren (1988b)	−1	−4	10	100	40	Halley Station	0.4	21
Heinemann (1990)	−1	−7	7	100	55	Halley Station	0.8	n/a
Bromwich (1987)	−1	−15	7.5	75	135	Ross Sea	1.3	n/a

north, although the length of the track over the ocean is quite limited. This is also the case during the winter months, when substantial surface fluxes can occur within polynyas, but the short track of the air over the open water places a limit on the energy that is added to the lowest layers.

The environmental lapse rate is one of the most important factors influencing the form of a mesoscale vortex and it determines the depth of the convection that can develop. If deep convective cloud is to form then the atmosphere must be conditionally unstable through a considerable layer; these conditions can come about through a combination of heating below and cooling aloft. Reed and Blier (1986) used a stability index to examine the lapse rate conditions that occurred during the development of polar lows over the Pacific Ocean. This was based on the difference between the temperature of surface air raised adiabatically to 500 hPa and the environmental temperature at that level. This quantity has large positive values under stable conditions and negative values when the atmosphere is unstable. Reed and Blier found that air masses over the Pacific containing polar lows (which they called comma clouds) had index values of less than 10°C. In northerly latitudes, where the more convectively active polar lows occur, the index will be much lower than this. Turner *et al.* (1993b) calculated the value of the stability index for Antarctic radiosonde ascents made during the whole of 1980 at Faraday Station on the Antarctic Peninsula and at Halley Station. For Halley Station they found that the index was almost always greater than 10°C and during the winter months was usually in the range 20–30°C. This is consistent with the very weak vortices noted on the satellite imagery. At Faraday Station, there was a smaller annual cycle in the index values and in all seasons the values frequently decreased to 5–7°C. However, throughout the year neither station ever had an ascent that gave a negative value indicating unstable conditions up to 500 hPa. The conditions observed arise through a combination of limited surface fluxes and more zonal atmospheric flow than is found in the Northern Hemisphere. This is consistent with the satellite imagery, which shows only shallow convection over the ocean. The index figures for the Antarctic do, however, suggest that, at latitudes around 60° S to 65° S the mesocyclones are similar to the 'comma cloud' type of polar low considered by Reed and Blier (1986). However, the figures agree with the observational data that systems with deep convection are not found in the Antarctic and that only under conditions of very strong meridional flow will moderate convection occur in the more northerly, oceanic parts of the Antarctic.

Although the above discussion has highlighted the differences between the mesocyclones observed in the two polar regions, there are some similarities. For example, the strong link between vortex activity and negative temperature and height anomalies at the 500 hPa level are a feature in the north and south (Businger, 1985; Turner and Thomas 1994). The tendency for comma-shaped cloud signatures to be more common than spiraliform vortices has also been observed. Heinemann (1990) also noted that the 195 vortices found in 232 days over the eastern Weddell Sea is comparable to the frequency of systems found in the Northern Hemisphere by Forbes and Lottes (1985).

Of the different types of Northern Hemisphere polar lows, the vortices described by Reed (1979) and called comma clouds by him appear to be most similar to the vortices found in the Antarctic. This type of system forms much closer to the polar front than do the more convective disturbances and is often found in the unstable air to the west of a synoptic depression. With its limited convective activity, association with major lows and short lifetime and horizontal scale, it resembles many of the mesocyclones found around the Antarctic in areas such as the Bellingshausen Sea.

The polar low of the Arctic is always considered a winter season phenomenon and it is true that the type of system with deep convection and active cumulonimbus cells is not found during the summer months. However, there are many less-well-developed mesoscale vortices in the high northern latitudes and these can be found throughout the year in cold, unstable northerly airstreams. These do not have the very strong winds of the convective systems and have not received as much attention in the literature. However, they have much in common with Antarctic mesocyclones.

6.5.6 Climatological consequences of mesocyclones

The importance of mesocyclones in the general circulation of the high southern latitudes is still uncertain and further research is required before their role can be assessed fully. As discussed in Section 4.3, mesocyclones are thought to play only a minor role in the vorticity budget of the Antarctic and most are shallow, short-lived systems which affect only a limited area. However, the increased wind speeds associated with the lows will give stronger surface fluxes of sensible and latent heat over the ocean and produce a cooling of the upper layers. This in turn will give greater downwards mixing of the less saline ice melt water, de-stabilising the water column. The effects of many mesocyclones could result in a significant effect on the physical and biological processes close to the northern limit of the ice edge and over the ice-free ocean. Since there are strong links between the occurrence of mesocyclones and the katabatic winds emerging from valleys in certain parts of the Antarctic, a further consequence of the lows will be to extend the influence of the katabatic winds to lower latitudes (Carleton and Fitch, 1993).

Perhaps the aspect of mesocyclones about which we know least is their associated precipitation. The first studies using passive microwave imagery to estimate precipitation in the Antarctic have shown that mesocyclones do appear to give some precipitation (Carleton *et al.*, 1993), although the algorithms used to identify precipitation over the ocean from SSM/I radiances are not well proven for high latitudes. However, even were the mesocyclones found to give significant amounts of snow or rain, the studies of the distribution of vortices show that most occur over the ocean so that they will have no impact on the mass balance of the Antarctic. Some more long-lived mesocyclones, such as that examined by Turner *et al.* (1993a), were found to have given significant amounts of snow over part of the Antarctic coastal region, but we know that these are rare events and hence probably have no major climatological effect.

One role that mesocyclones have been identified as playing in synoptic-scale developments is in the creation of 'instant occlusions'. As discussed in Section 6.5.2, the merging of mesocyclones with frontal waves can result in lows with the characteristics of mature depressions that will have more far-reaching consequences on the weather conditions than would the original mesocyclones. However, there have not been any detailed case studies of instant occlusions in the Antarctic and it is not clear how many such events take place. The climatological importance of these events is therefore still to be determined.

6.5.7 **Future research needs**

There are many questions remaining regarding mesocyclones around the Antarctic and further observational and modelling studies are required before a reasonably complete understanding of their formation, distribution and role in the circulation of the Southern Hemisphere can be obtained. The most significant handicap to research is still the lack of data with which to examine the structure of the vortices. Since most occur over the ocean, satellite data will be an important tool and the latest generation of satellite instruments will play an important role in this work. Sounding instruments such as the TOVS have been used to examine the thermal structure of mesocyclones and the next generation of sounders, such as the Advanced Microwave Sounder Unit, will be able to give improved thermal data. Passive microwave instruments have the ability to give data on the humidity, cloud liquid water and precipitation fields, and the retrieved information should improve during the next few years as processing algorithms are developed. These data should help answer the questions regarding the precipitation that falls from mesocyclones. The surface circulation can already be obtained by using surface wind vectors from the scatterometer on the ERS-1 satellite (Lachlan-Cope *et al.*, 1994) and future satellite missions will also fly improved versions of this instrument with a wider swathe. This will allow climatological data to be produced on the strength of mesocyclones.

Although the first flights by an instrumented aircraft through an Antarctic mesocyclone have taken place (Heinemann, personal communication), this powerful research tool has yet to be exploited fully around the Antarctic. In the Northern Hemisphere aircraft data from polar lows have given great insight into the development of these systems (Shapiro *et al.*, 1987) and in the Antarctic they would also allow the validation of the satellite data.

As discussed earlier, numerical models have only just begun to be used to investigate mesocyclones in the Antarctic. Since mesoscale models have had little success so far in simulating polar lows in the relatively data-rich Arctic, it will be some time before they will have the capability of giving realistic simulations of systems in the south. However, in the long term they have great potential to aid mesocyclone research and to provide answers to questions regarding mesocyclone formation and structure.

Chapter 7

Climate variability and change

In common with that of the northern polar regions, the Antarctic climate exhibits a much greater degree of inter-annual and inter-decadal variability than that which is observed at lower latitudes. Although the reasons for this are not fully understood, it is clear that the complex interactions between atmospheric circulation, the oceans and the cryosphere can give rise to a number of positive feedbacks that can result in enhanced climate variability. In the first section of this chapter we shall describe some aspects of the variability of the present-day Antarctic climate and examine possible mechanisms thought to be responsible for driving the observed variations. Some links have been established between climate variations in Antarctica and those occurring elsewhere. We discuss such 'teleconnections' in Section 7.2. Finally, in Section 7.3, we consider how the Antarctic climate might change as a result of global warming brought about by the release of 'greenhouse' gases into the atmosphere. The response of the Antarctic ice sheets to such warming and the consequences for global sea level changes have aroused much interest in recent years.

Our understanding of the nature of climate variability in Antarctica is necessarily limited because the available climate records are short and restricted in geographical coverage. Very few climatological stations were in operation before the International Geophysical Year of 1957/58 and, even after the IGY, most stations were established in coastal regions. Only two stations have operated on the East Antarctic plateau for longer than 20 years and, apart from the west coast of the Antarctic Peninsula, West Antarctica is almost devoid of long-term climatological observations. Deployment of automatic weather stations (Section 2.3) has improved the spatial coverage of the climatological observing network in recent years but few AWSs have been in operation for long enough to provide useful information on climate variability. Furthermore, problems with the accuracy and reliability of these records and the restricted range of parameters observed mean that records from a manned observing station are generally of greater value in long-term studies.

A number of proxy climatological indicators have been used to extend Antarctic climate records both spatially and temporally. Ice cores preserve a record of atmospheric conditions prevailing when the snow which formed the ice fell. Most notably, the ratio of oxygen isotopes in the ice depends largely on the temperature at which the water vapour condensed to form precipitation. The oxygen isotope 'thermometer' may be calibrated by comparing oxygen isotope ratios in the upper levels of an ice core with a temperature record from a nearby station. Isotope thermometry has been applied to deep ice cores, such as that obtained at Vostok station (Jouzel *et al.*, 1987), to create proxy temperature records covering hundreds of thousands of years. Such studies are outside the scope of this book. However, isotope temperature records from shallower cores can be used to extend our limited instrumental climate records back by 100–200 years. The most valuable cores for these studies are those obtained from regions of relatively high snow accumulation since this facilitates the dating of the annual layers revealed by the annual cycle in isotope ratios. The mass of ice between annual layers is a direct measure of the net accumulation during that year, from which a record of precipitation variations can be reconstructed.

A second technique involves the reconstruction of surface temperature histories from temperature profiles measured in boreholes in the ice. In order to invert the temperature profile, assumptions need to be made about the accumulation rate and the thermal properties of the ice must be known accurately. The technique has not yet been widely applied but has been shown to be of use in cases in which interpretation of the oxygen isotope record has proved difficult (Nicholls and Paren, 1993; Cuffey *et al.*, 1994).

7.1 Variations in the historical climate record

7.1.1 Temperature variations

The nature of the variability

Time series of annual mean surface air temperature at a selection of Antarctic stations are shown in Figure 7.1. These include the longest continuous temperature record from Antarctica (Orcadas Station, in the South Orkney Islands), a long record from the west coast of the Antarctic Peninsula (Faraday Station), two stations representative of the coast of East Antarctica (Halley and Mirny)

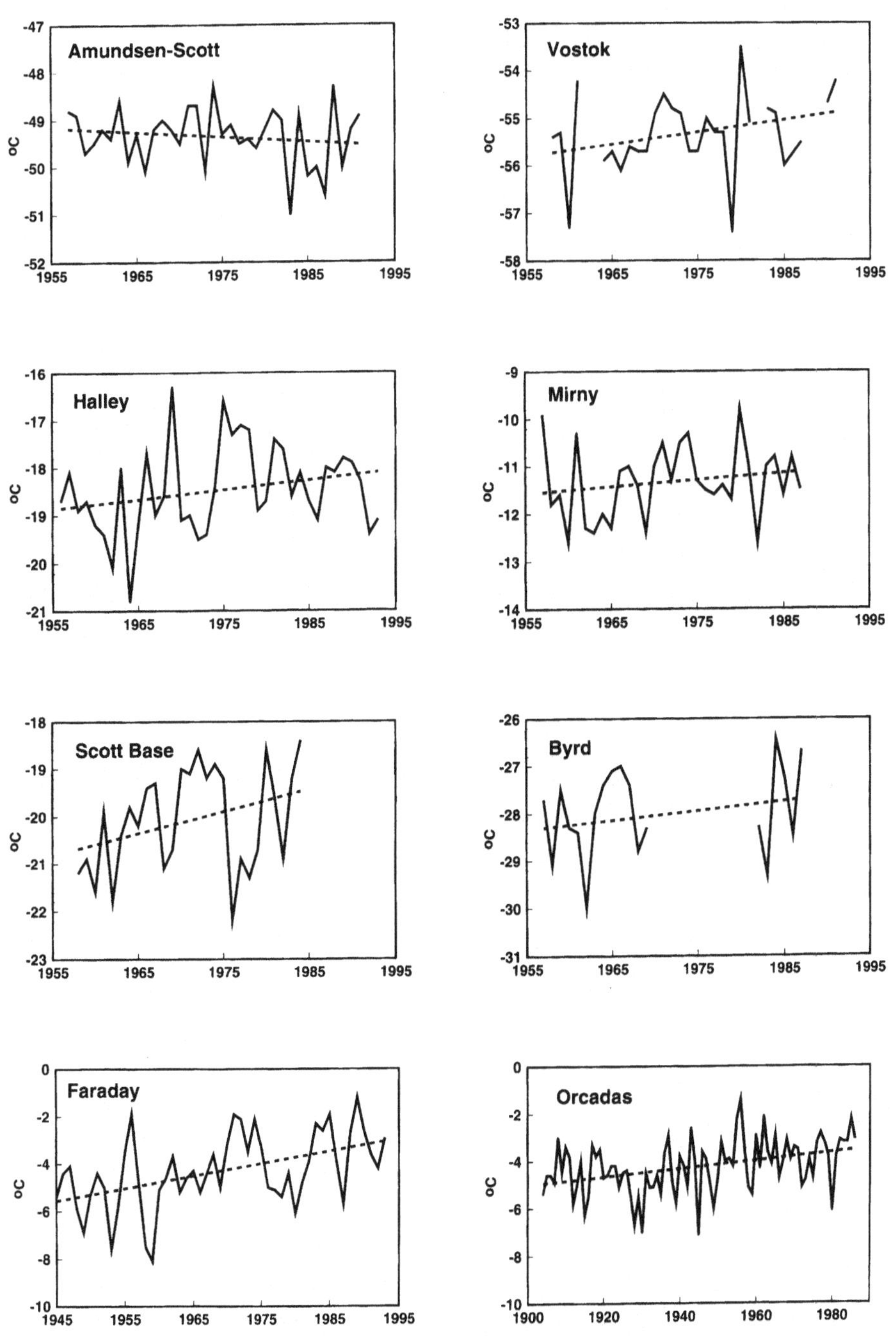

Figure 7.1 Time series of annual mean surface air temperature at a number of Antarctic stations. Data are from Jones and Limbert (1987), updated from recent reports in some cases. Data for Byrd for 1982–87 are from Stearns *et al.* (1993) for an automatic weather station.

Table 7.1 *Long-term means, standard deviations and least-squares linear trends for the annual mean temperature time series shown in Figure 7.1.*

Station	Period	Mean (°C)	Standard deviation (°C)	Trend (°C/year)
Orcadas	1904–86	−4.2	1.12	+0.017
Faraday	1945–93	−4.3	1.56	+0.052
Byrd	1957–69	−28.0	0.94	+0.020
	1982–87*			
Halley	1956–93	−18.5	0.94	+0.020
Mirny	1957–87	−11.3	0.76	+0.014
Scott Base	1958–84	−20.1	1.09	+0.046
Amundsen–Scott	1957–91	−49.3	0.61	−0.010
Vostok	1958–91	−55.3	0.83	+0.025

Notes:

* Data from an AWS for this period.

and two stations (Amundsen–Scott and Vostok) on the East Antarctic plateau. Although rather short and incomplete, the record from Byrd Station has been included as the only record representative of conditions on the West Antarctic plateau. The latter part of this record (1982–1987) was obtained from an automatic weather station.

Least-squares regression lines have been fitted to the records shown in Figure 7.1 in order to reveal long-term trends in temperature. The trends thus obtained are listed in Table 7.1, together with the long-term mean and standard deviation of temperature at each station. A number of features are apparent from these records.

(i) All records exhibit a high degree of inter-annual variability, with the greatest variations occurring along the west coast of the Antarctic Peninsula. The level of variability is considerably greater than that found at mid-latitudes (e.g. for London, the standard deviation of the annual mean temperature is 0.5°C) and comparable to that found in the Arctic (e.g. for Anchorage, Alaska, the standard deviation of annual mean temperature is 1.1°C). However, Arctic stations that exhibit this level of variability have a very continental climate, with a large annual temperature range (26°C for Anchorage). Faraday Station, with an annual temperature range of only 11°C, has a maritime climate and is exceptional in showing such a high level of variability.

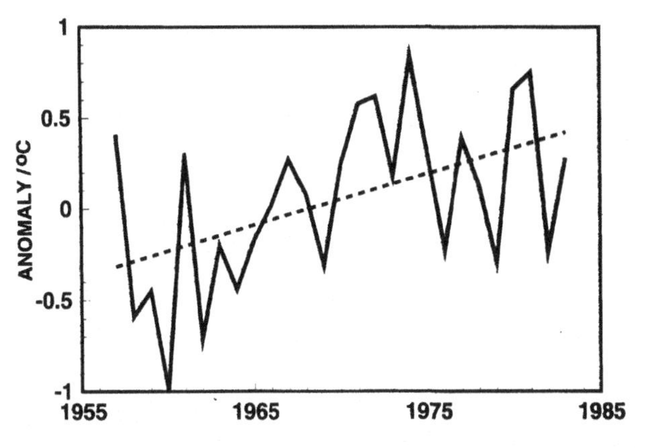

Figure 7.2 The areally-averaged Antarctic mean annual temperature anomaly, expressed as a deviation from the 1957–75 mean. Data from Raper *et al.* (1984).

(ii) Significant year-to-year persistence of temperature anomalies is seen in the Faraday Station record but is much less apparent in the records from East Antarctica.

(iii) The spatial pattern of the variations is complex, with little correlation being observed among the various records shown in Figure 7.1. In particular, the large temperature variations observed on the west coast of the Antarctic Peninsula are poorly correlated with temperatures in East Antarctica (King, 1994) and the Orcadas Station record shows little correlation with mean Antarctic temperatures (Raper et al., 1984). This means that it is not possible to infer the history of continental Antarctic temperatures from the longer series of observations available from the Antarctic Peninsula or South Orkney Islands.

(iv) Most records exhibit small warming trends. The exception is that of Amundsen–Scott Station, which shows a slight cooling over the period of the record. The warming on the west coast of the Antarctic Peninsula is greater than that found elsewhere in Antarctica but the shortness of the available records, together with the high level of inter-annual variability, makes the determination of trends difficult. Sansom (1989) found that the observed trends at Scott Base, Faraday, Mirny and Amundsen–Scott Stations over the period 1957–86 were not statistically significant if proper allowance was made for the high degree of serial autocorrelation (i.e. year-to-year persistence) in the time series. However, following recent record high temperatures in the Antarctic Peninsula (Morrison, 1990), the temperature trend at Faraday Station over the period 1945–90 now appears to be significant at the 99% confidence level (King, 1994; Stark, 1994). Raper *et al.* (1984) calculated seasonal and annual average temperatures for Antarctica by taking an areally weighted mean of all available station observations. The annual mean Antarctic temperature (Figure 7.2) shows a warming trend of 0.029°C per year over the period 1957–82, significant at the 95% confidence level. The greatest contribution to this

> trend comes from stations in the Antarctic Peninsula. However, even the smaller warming trends typical of East Antarctic coastal stations are large compared with the 0.0037°C per year warming of the Southern Hemisphere north of 60° S over the period 1957–82 (Jones *et al.*, 1986). Although climate model experiments (discussed in Section 7.3) do indicate that the polar regions may experience amplified warming in response to smaller global temperature increases, there is insufficient evidence at present to link the observed Antarctic temperature trends to the much smaller hemispheric and global trends.

Little is known about temperature variations at levels other than the surface. The sparseness of the radiosonde network, together with the problems of reconciling measurements made with different radiosonde types even within a single station record (see Chapter 2) make such studies difficult. Trenberth and Olson (1989) analysed upper tropospheric and lower stratospheric temperatures measured at McMurdo and Amundsen–Scott Stations between 1957 and 1986. After carefully making allowances for missing data, they concluded that there were no significant trends at any level at McMurdo. At Amundsen–Scott significant cooling trends were observed in October at the 100 hPa level and above. This cooling can be entirely attributed to the reduction in solar heating of the stratosphere resulting from increasingly depleted levels of stratospheric ozone since the early 1970s – the so-called 'Antarctic Ozone Hole' (Farman *et al.*, 1985). A small warming trend of about 0.2°C per decade has been observed in the lower Antarctic troposphere (850–300 hPa) between 1958 and 1987 (Angell, 1988).

Regional temperature variations may be brought about by two mechanisms. First, the surface energy balance may be altered by variations in (for example) cloudiness or albedo. Secondly, changes in atmospheric circulation may give rise to different patterns of advection of warm and cold air. Quantifying atmospheric circulation variations in Antarctica is very difficult because of the sparsity of the observing network. Jones and Wigley (1988) constructed a gridded sea-level pressure data set for the Antarctic region covering the period 1957–85. These data show slightly lower pressures in the most recent decade compared with the earlier part of the series, which is consistent with observed warming trends if these were caused by increased cyclonic activity around Antarctica. However, the spatial patterns of temperature and pressure change are not clearly correlated on a regional scale. On an inter-annual (rather than inter-decadal) time scale, Raper *et al.* (1984) found negative correlations between Antarctic temperatures and the strength of the circumpolar westerlies.

Interannual circulation changes can have a major effect on synoptic-scale weather systems because they can modulate the frequency of occurrence of disturbances and affect the position of storm tracks (Chapter 5). Many workers have therefore used the series of operational surface and upper-air analyses produced since the 1950s to examine the inter-annual variability of the atmospheric circulation over the Southern Ocean and Antarctic continent. These analyses have shown that the interannual variability of synoptic activity over the high-latitude areas of the Southern Hemisphere is large (Streten, 1980a) and therefore results in a high degree of variability in many of the meteorological parameters, such as precipitation and temperature, measured in the Antarctic coastal area. This large variability is also reflected in the latitudinal position of the circumpolar trough. However, the degree of variability varies around the continent and is largest in the northwest Ross Sea and to the northeast of the Weddell Sea (Streten, 1980b). Having surveyed the general pattern of observed temperature variations in Antarctica, we shall now focus on two regions which have received special attention because of the nature or magnitude of the observed variability.

Extreme variability in the Antarctic Peninsula

The extreme inter-annual variability and large trend seen in the Faraday Station record (Figure 6.1) are also present in records from the Marguerite Bay area some 300 km further south (King, 1994) and may extend as far south as Alexander Island (Harangozo *et al.*, 1994). However, the geographical extent of this region of extreme variations is clearly limited since the Faraday Station record correlates less well with that from the South Shetland Islands at the tip of the Antarctic Peninsula than it does with the Marguerite Bay record and correlations with continental Antarctic records are not statistically significant. The first eigenfunction of a principal component analysis of Antarctic temperatures carried out by Raper *et al.* (1984) shows large weightings over the western Antarctic Peninsula with little structure over the rest of the continent. This indicates the limited extent of the region of extreme variability and the high proportion (35.8%) of the total variance explained by this component highlights the large contribution of Peninsula temperature variations to the variability of the Antarctic climate as a whole.

The greatest contribution to inter-annual temperature variations comes from the winter months. Figure 7.3 shows the long-term means, standard deviations and trends of monthly mean temperatures at Faraday Station. Inter-annual temperature variations are relatively small between September and March but the standard deviations of monthly mean temperatures exceed 3°C between June and September. The warming trend is also more pronounced during the winter months, leading to a secular decrease in the annual temperature range and a change in the seasonal cycle of temperatures (Stark, 1994). There is a high degree of persistence of temperature anomalies from one year to the next. Sansom (1989) has shown that Faraday Station monthly temperature anomalies show statistically significant autocorrelation out to 19 months lag. Such a high degree of persistence suggests that temperatures in this region are being influenced by changes in ocean circulation and temperatures.

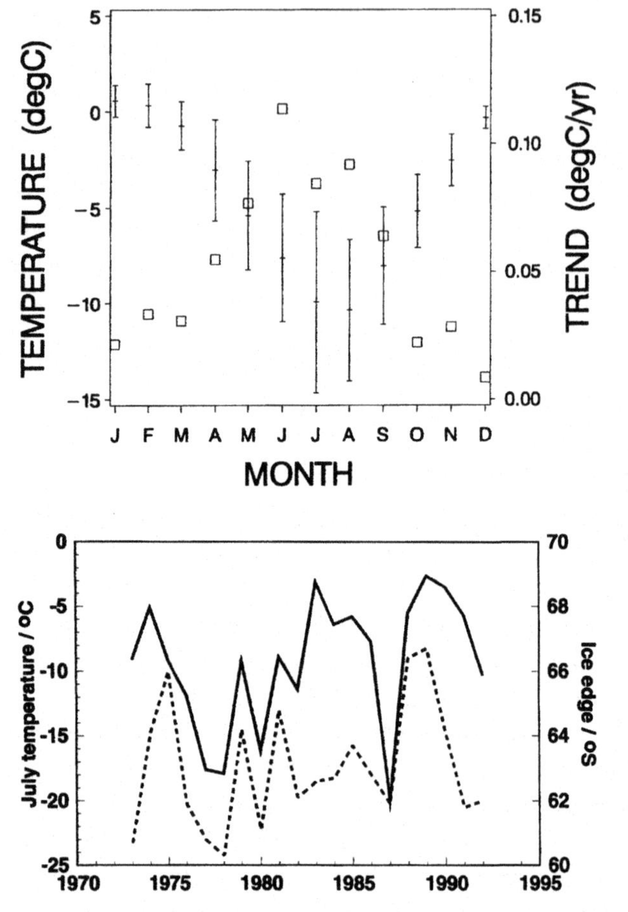

Figure 7.3 Monthly mean temperature statistics for Faraday Station, 1945–90. The points with error bars show the monthly mean temperature and its interannual standard deviation. The squares indicate the long-term temperature trend for each individual month. From King (1994), *International Journal of Climatology*. Copyright (1994) Royal Meteorological Society. Reprinted by permission of John Wiley & Sons Ltd.

Figure 7.4 Average July temperatures at Faraday Station (solid line) and the latitude of the northern edge of the pack ice at 70° W in July (broken line) for the period 1973–92. Anomalously cold temperatures at Faraday Station are associated with anomalously extensive pack ice (i.e. the ice edge being to the north of its mean position) and *vice versa.*

The west coast of the Antarctic Peninsula is the only region of Antarctica in which significant correlations have been found between temperatures and sea ice extent (Weatherly *et al.*, 1991; King, 1994). Winter temperatures are strongly anticorrelated with sea ice extent in the Bellingshausen Sea, with low winter temperatures associated with anomalously extensive sea ice and *vice versa* (Figure 7.4). Sea ice can influence temperatures in two ways; first by reducing the heat flux from ocean to atmosphere and secondly by reducing the fraction of solar radiation absorbed at the ocean surface. Since lower temperatures will promote the growth of sea ice there is potential for a feedback mechanism and it would seem likely that the extreme climatic sensitivity of the region results, at least in part, from the effects of this feedback. Unlike the coast of East Antarctica, the west Antarctic Peninsula coast lies close to the ice edge throughout much of the winter and may thus be strongly affected by small changes in the position of the ice edge.

Routine satellite measurements of sea ice extent are only available from 1973 onwards (see Section 3.5). The winter ice extent in the Bellingshausen Sea shows considerable inter-annual variability over this period and the record is not long enough to establish whether there is a long-term trend in ice extent associated with the observed temperature increase (Gloersen and Campbell, 1988). However, if relationships between ice extent and winter temperatures derived from the short satellite record held good over a longer period then a retreat of the winter ice edge by 1° over 40 years would explain the observed temperature trend. Given the large inter-annual variability in ice extent, such a small trend would be very hard to detect.

West Antarctic Peninsula temperatures are influenced by factors other than sea ice extent. A stronger northerly component to the atmospheric circulation in the vicinity of the Peninsula results in increased warm air advection and is associated with above-average winter temperatures in the region (King, 1994). Schwerdtfeger (1976) indicated that there might be an association between a strengthening of the circumpolar westerlies and rising Peninsula temperatures during the 1970s, but this relationship does not seem to hold over a longer time period (King, 1994). Above-average winter temperatures in the Peninsula are also associated with increased cloudiness (King, 1994). Although increased cloud cover could be associated with enhanced warm advection, it will also tend to increase temperatures directly by increasing the downwards component of long-wave radiation at the surface. Cloud cover will also reduce insolation, but this has a smaller effect on the surface energy balance during the winter when surface albedo is high and insolation is, in any case, small. Increased cloudiness thus tends to warm the surface during the polar winter (Arking (1991); see also Section 3.1).

Several environmental changes have occurred in the west Antarctic Peninsula that are consistent with the warming trends seen in station temperature records. Extensive ablation has been observed on low-lying glaciers and small ice caps (Splettstoesser, 1992) and the Wordie Ice Shelf in Marguerite Bay has decreased in area from 2000 km^2 in 1966 to 700 km^2 in 1989 (Doake and Vaughan, 1991). Reduced snow and ice cover and a milder climate have allowed increased colonisation by plants at some sites (Fowbert and Lewis Smith, 1994). It has also been suggested (Fraser *et al.*, 1992) that changes in the relative abundance of some penguin species are consistent with a long-term decline in summer sea ice extent to the west of the Peninsula. All of these observations are consistent with the regional nature of the temperature changes observed in station records and suggest that the warming trend has continued for some time. Later in this section we shall examine additional evidence for the prolonged nature of this trend.

Variations on the East Antarctic plateau

The temperature record from Amundsen–Scott Station is unique amongst those shown in Figure 7.1 in that it exhibits a cooling trend, albeit a rather small one.

The only other record from the East Antarctic Plateau, from Vostok, shows a warming trend similar to that found at many coastal stations. Both stations are situated in regions of very uniform topography and are likely to be representative of a large surrounding area. It thus seems likely that, although much of the East Antarctic plateau may have warmed slightly since 1957, a considerable area has also cooled over this period.

The most marked changes at Amundsen–Scott have occurred since 1976. Between 1976 and 1985 there was a significant decrease in insolation during January and February associated with an increase in cloud cover during these months (Dutton *et al.*, 1991). Insolation values made a partial recovery between 1986 and 1989. Concurrent with these changes, temperatures during January, February and March increased, but April and May temperatures decreased rapidly. The warming during the late summer months appears to be caused by decreased long-wave cooling associated with the increased cloud cover. However, no significant cloud cover changes appear to have occurred during April and May, suggesting a dynamical origin for the winter cooling.

Neff (1992) has speculated that the observed trends could be associated with changes in tropospheric and stratospheric circulation resulting from the spring depletion of stratospheric ozone which also started to become significant from the mid-1970s onward (Farman *et al.*, 1985). The distribution of mid-tropospheric wind directions found at Amundsen–Scott is strongly bimodal, with peaks at 135° and 315°. The former direction is approximately parallel to the 3000 m contour, showing that winds from this sector are strongly controlled by the underlying topography whereas winds from the latter direction are blowing from the Weddell Sea sector and are responsible for the majority of the transport of moisture to the South Pole (Hogan *et al.*, 1982). Between 1980 and 1990, the frequency of winds from 315° during January and February increased relative to the frequency of winds from 135°, thus explaining the observed increase in cloudiness. Neff has argued that springtime ozone depletion leads to a reduction in the stability of the stratosphere, allowing the stratospheric vortex and upper tropospheric circulation to remain strongly coupled to the underlying topography for longer into the spring warming season. The breakdown of the vortex and transport of warmer air into both stratosphere and troposphere may then be delayed and the warming and increased cloudiness seen in late summer result from the topographically bound circulation finally breaking down at this late stage, allowing warm, moist air to reach the plateau. It is not clear how this is related to the cooling trends observed in April and May, but the two phenomena appear to be connected. The processes linking temperature changes to variations in tropospheric and stratospheric circulation are clearly complex and further study is required to understand these mechanisms fully.

Extending the historical temperature record

It is of great interest to know whether the period covered by the instrumental records (basically the second half of the twentieth century) is typical of the recent past or whether it is, in some respects, anomalous. Although few permanent stations were established before 1945, a fair number of scientific expeditions have visited Antarctica from the late nineteenth century onwards. Given the high level of inter-annual variability seen in Antarctica, it is difficult to compare the short records available from these expeditions with the later measurements from permanent stations. Jones (1990) assembled a catalogue of observations made by early twentieth century expeditions. Two areas have been visited sufficiently frequently for these data to provide a useful extended climate record. The meteorological records of expeditions to the Antarctic Peninsula area indicate that this region was between 0.3 and 3.0°C cooler during the early years of the twentieth century than it was during the reference period 1957–86, suggesting that the region has been warming throughout the twentieth century. The other area which was frequently visited by early expeditions is McMurdo Sound in the Ross Sea, where the current McMurdo Station and Scott Base are situated. Expedition records from this area indicate that early twentieth century temperatures showed a degree of variability similar to that of the recent record and were not significantly warmer or cooler than those measured during 1957–86.

If we wish to reconstruct the temperature record before the early twentieth century or extend it to areas not reached by the early expeditions, we must resort to proxy techniques, such as oxygen isotope thermometry. The temperature record deduced from oxygen isotope ratios in shallow ice cores from the Antarctic Peninsula area is complex, with a cooling trend indicated from the mid nineteenth century to 1950 and warming thereafter (Peel, 1992b). This conflicts with the steady post-1880 warming trend seen in Southern Hemisphere instrumental records (Jones *et al.*, 1986). However, Peel points out that the isotopic signal observed in the snow depends not only on condensation temperatures at the core site but also on the source region of the water which formed the ice. Peninsula ice cores reveal a strong isotopically 'warm' anomaly between 1973 and 1977. This does not correspond to a warm temperature anomaly in the station records but it is the period during which a large polynya was observed to form in the northeastern Weddell Sea (Section 3.5). It is likely that the Weddell Polynya provided an additional source of moisture for precipitation falling on the east side of the Antarctic Peninsula and the relatively low sea surface temperatures in the polynya (compared with the open ocean to the north of the ice edge, the 'normal' water source for precipitation) could have caused the observed isotopic anomaly. Bearing this in mind, it is possible to re-interpret the isotopically warm conditions of the mid nineteenth century as a period during which there was a greater fraction of open water within the pack ice as a result of a rather different prevailing atmospheric circulation pattern. The increased source of water vapour from relatively cold seas would then give rise to an isotopically 'warm' signal in precipitation, even with Peninsula temperatures somewhat cooler than at present, thus reconciling the isotope record with the observed Southern Hemisphere temperature trend. In order to separate the effects of changes in temperature of the source and deposition regions unambiguously, it is necessary to examine changes in additional isotope ratios, such as that between hydrogen and deuterium.

Isotopic records from ice cores obtained from other regions of Antarctica indicate a complex pattern of temperature variations over the past 500 years (Mosley-Thompson, 1992). Anomalies and trends are not well correlated between sites, suggesting that marked regional climatic variations have occurred in the past.

7.1.2 Precipitation variations

Since the saturation vapour pressure of water, and hence the moisture-holding capacity of air, increases rapidly and non-linearly with increasing temperature, one might expect to see a strong correlation between variations in temperature and in precipitation. The existence of such a correlation in a spatial sense was demonstrated by Robin (1977), who showed that the spatial distribution of accumulation over the Antarctic ice sheets closely followed the geographical variation of saturation vapour pressure immediately above the surface inversion. A number of other studies (Muszynski and Birchfield, 1985; Giovinetto *et al.*, 1990; Fortuin and Oerlemans, 1990) have shown that there is a strong correlation between the spatial distributions of accumulation and surface air temperature. On the basis of these results, it has often been assumed that this correlation also holds good for temporal variations. However, we shall see that this is not always the case.

The difficulties associated with measuring precipitation directly at Antarctic stations have been discussed in Chapter 2. Because of these problems, most of our information on temporal variations in precipitation comes from analysis of annual accumulation layers in ice cores. However, as demonstrated in Section 3.4.2, analysis of the frequency with which precipitation is reported at a station can provide a useful and robust proxy measurement, particularly if one is interested in relative variations rather than absolute values. Over the period 1956–93 the frequency with which precipitation has been reported in synoptic observations at Faraday has increased by about 20%. As seen earlier, there has also been a significant warming over this period but there is little correlation between year-to-year variations in precipitation frequency and variations in temperature. This suggests that factors other than temperature variations contribute to the observed variability in precipitation on the west coast of the Antarctic Peninsula.

High precipitation appears to be associated with low mean annual surface pressure in the Bellingshausen Sea, indicating that increased or more vigorous cyclonic activity in this region may lead to enhanced precipitation (Turner and Colwell, 1995).

Analysis of ice cores from the Antarctic Peninsula (Peel, 1992b) confirms the regional nature of this trend. At three sites, the average increase in accumulation since 1950 is 1.16% per year, or 15.2% °C^{-1} given the observed warming trend. This is significantly greater than the 10% °C^{-1} rate of change of saturation vapour pressure at −19°C (the average temperature of the core sites). Thus, temperature changes alone do not seem sufficient to explain the observed trends and it is likely that concurrent circulation changes have occurred. Studies of the atmospheric transport of water vapour to Antarctica, discussed in Section 4.4, have shown that circulation variations are probably the primary source of interannual variability in precipitation (Connolley and King, 1993).

Recent increases in accumulation rate have also been observed in East Antarctica. Isaksson and Karlén (1994) have shown that accumulation variations determined from ice cores in Dronning Maud Land correlate quite closely with the Halley temperature record over the period 1956–89. At Dome C, accumulation since 1965 has been 30% higher than that between 1955 and 1965 (Petit *et al.*, 1982). In Wilkes Land, accumulation rates between 1975 and 1985 were some 20% higher than the mean for 1930–85 (Morgan *et al.*, 1991). We have already seen that temperature trends in East Antarctica are quite small and the correspondingly small changes in moisture-holding capacity of the air cannot explain such large increases in precipitation. Once again, circulation variations would appear to be important, although the nature of such changes remains to be quantified.

7.2 Interactions with the tropical and mid-latitude circulation

Although the Antarctic continent is the most physically remote and inaccessible region on Earth, its climate is nevertheless inextricably linked with the atmospheric and oceanic conditions at lower latitudes. This comes about through its role as one of the two main heat sinks in the general circulation of the atmosphere and therefore results in it playing an important part in the global climate system. The general circulation of the atmosphere is driven by the large equator-to-pole temperature difference and the atmosphere and ocean respond to this temperature gradient by the transport of heat polewards. In the Southern Hemisphere much of this is carried by the transient eddies (extra-tropical cyclones) because the upper-air long waves (Rossby waves) here are usually weaker and more barotropic (i.e. the fields of geopotential and temperature are approximately in phase) than those found in the Northern Hemisphere. This means that, around the Antarctic, they transport little heat polewards (van Loon (1979); see also Section 4.2). The role and magnitudes of the heat and momentum fluxes are dealt with in Chapter 4 so here we will examine how the landmass of the Antarctic interacts with the Rossby waves and how the synoptic and mesoscale conditions over Antarctica are affected by the mid-latitude circulation. The effects of atmospheric circulation variability outside the Antarctic on the high-latitude climatic conditions will also be considered, together with teleconnections that have been identified in recent years. These are statistically significant temporal correlations between meteorological parameters at widely separated points (Mo and White, 1985) and often indicate links between the climatic conditions of the tropical and mid-latitude/polar regions. The interactions between the mid-latitude and tropical atmospheric/oceanic systems will be discussed from the point of view of high-frequency fluctuations associated with the synoptic depressions and inter-annual timescales, with particular attention being paid to the teleconnections between the climate systems of El Niño/Southern Oscillation (ENSO) and the Antarctic.

7.2.1 Synoptic and mesoscale interactions

The synoptic-scale low-pressure systems found over the Southern Ocean and in the Antarctic coastal region consist of two main types of depression. First, those that have developed in the cyclogenetic regions of the mid-latitudes and moved polewards towards the Antarctic continent. Secondly, systems that have formed at higher latitudes, often within the circumpolar trough between 60° S and 70° S (Jones and Simmonds, 1993). When extra-tropical cyclones are near the Antarctic coast they are usually in their mature or declining phase and are often slow-moving with disorganised low cloud. However, they can still give high winds in the coastal region as a result of a pronounced pressure gradient between the low-pressure centres and the area of high pressure that is usually present in the interior of the continent. The latitude of origin and tracks followed by the depressions are determined primarily by the upper level flow, but can be affected by other factors, such as the sea surface temperatures, the extent of the sea ice and the distribution and form of other depressions that have developed near to the continent. The extent of the sea ice around the continent, for example, and the track and location of the synoptic-scale depressions are closely connected, but the exact quantitative nature of the links has yet to be fully resolved. However, it is known that the extent of the sea ice can be forced by the near-surface wind (Enomoto and Ohmura, 1990) with strong meridional winds resulting in rapid ice advance or retreat. Cold, southerly flow on the western side of a depression can also result in cooling at the ice surface with the freezing over of leads which limits the heat flux into the atmosphere. It should be noted, however, that surface cooling can also occur under anticyclonic conditions when the absence of cloud allows the surface to lose heat rapidly through long-wave radiation. The position of the northern limit of the sea ice and the sea ice concentration over weekly to inter-annual timescales are therefore often good indicators of depression activity and the depression tracks over the period. The location and concentration of sea ice may affect the subsequent development and tracks followed by synoptic-scale depressions, although the exact mechanisms behind this link are not clear at present. It has been suggested that areas of reduced sea ice concentration may have an effect on the formation of new cyclonic disturbances through increased sensible heat transfer into the lower troposphere from the ice-free ocean. This was proposed as the main reason behind the markedly different circulation regimes found over the Amundsen–Bellingshausen Sea during the winters of 1973 and 1974 in the study by Ackley and Keliher (1976). They found that, during the first part of the 1973 winter, anticyclonic conditions gave northerly flow and resulted in negative ice anomalies with large fluxes of heat into the atmosphere compared with those in 1974, so that during the second half of the 1973 winter there were many depressions over the area that Ackley and Keliher suggested may have been the result of the increased surface fluxes.

Apart from synoptic-scale systems, mesocyclones are also frequently observed to develop around the coast of the Antarctic, especially in regions of increased baroclinicity, e.g. areas in which long waves are amplified with equatorial advection of cold air. In addition, strong thermal gradients can also be found where the cold air from the interior emerges from the steep valley glaciers as a strong katabatic wind. At most coastal sites such winds do not blow strongly all the time but vary in strength over periods of several days or longer. This may be explained in terms of the fact that the katabatic winds can be enhanced by the synoptic-scale flow, such as when a large depression is to the east of a valley with a resultant southerly geostrophic flow over the region (Streten, 1968). Such conditions can also advect the sea ice away from the coast and create large areas of open water, even during the winter months when the air temperature is well below zero. This results in large fluxes of heat and moisture into the atmosphere, which de-stabilise the lower troposphere and favour mesocyclogenesis. The effects of the strong offshore flow can also be felt further north through the strengthening of the baroclinicity near the ice edge (Kottmeier and Hartig, 1990) giving greater mesocyclogenesis there and possibly leading to an intensification of the original synoptic-scale low. Such feedbacks are clearly complex and can also involve cloud–radiation interactions in the marginal ice zone (Cahalan and Chiu, 1986).

Recent studies have shown that the Antarctic continent does not simply play a passive role in the meridional transport of heat and moisture, but rather is involved in a complex interaction with the long waves. Although the large-amplitude long waves are most pronounced at mid-latitudes, they extend into the

Antarctic region and are affected by the high topography of the continent. When the long waves have a large amplitude there is a greater exchange of air between the Antarctic and mid-latitudes with warm, moist air penetrating into the interior. Such situations appear to be responsible for many of the major precipitation events over the plateau and the rapid rises of surface temperature that have been observed at the South Pole (see Section 5.3.4).

A number of indices have been developed to quantify changes in the long-wave structure that take place across the Southern Hemisphere. One of the most studied is the Trans-Polar Index (TPI) which reflects changes on the hemispheric scale via wave number one. The TPI is the surface pressure anomaly difference between Stanley, Falkland Islands and Hobart, Tasmania (Pittock, 1984). Positive values reflect a displacement of the wave number one towards South America whereas negative values indicate displacement towards Australia. During the winter, the surface pressure anomalies of the Antarctic Peninsula and Australasian regions tend to be out of phase (Carleton, 1992), indicating the extent to which the climate of the Antarctic and lower latitude areas are linked through hemisphere-wide teleconnections. Variations in the TPI on the inter-annual time scale have been linked to sea ice conditions in the Scotia Sea (Rogers and van Loon, 1982) and to the cyclone tracks in the Drake Passage area (Mayes, 1985).

The hemispheric long waves also play a role in the maintenance of the 'coreless' winter, a phenomenon that results in little temperature decrease over the interior of the continent after the rapid drop through the autumn and early winter. This is apparent over the entire interior of the continent and comes about as the long waves move in response to seasonal changes in the heating rates of the land and ocean areas of the mid-latitude regions and higher latitude areas of the Southern Hemisphere (van Loon (1967); see Section 3.2).

Mechoso (1980) suggested that, as the propagating Rossby waves come up against the sloping ice surface of the continent, strong baroclinicity may be generated in the coastal region. This in turn would imply greater cyclogenesis around the continent, which has indeed been detected in numerical analyses (Jones and Simmonds, 1993) and in satellite imagery (Turner and Thomas, 1994). Further evidence of the influence of the continent on the mid-latitude long waves comes from the numerical studies of James (1988). Using a relatively simple barotropic model he reproduced many of the features of the Southern Hemisphere winter season tropospheric flow and, in particular, the split jet close to New Zealand. This feature has been observed in routine numerical analyses, as can be seen in the map of zonal anomaly of the mean streamfunction ψ^* (the wind blows along the isopleths of the streamfunction) at 250 hPa taken from James (1988) and shown in Figure 7.5. The data for this figure were computed from the ECMWF analyses for the period 1979–84 and clearly show the split in the mid-latitude tropospheric jet south of Australia and the region of climatological blocking over New Zealand. The presence of high-pressure systems in this

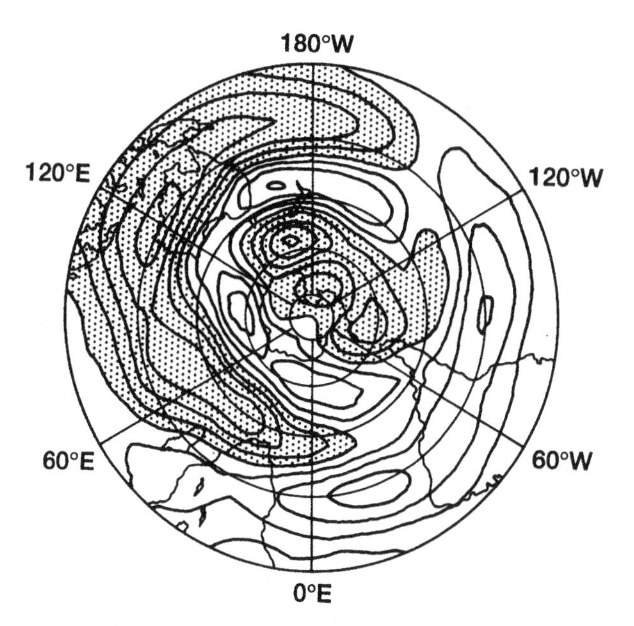

Figure 7.5 The zonal anomaly of the mean streamfunction ψ^* at 250 hPa for the Southern Hemisphere, 1979–84, based on ECMWF analyses. The chart shows the split jet over New Zealand. The contour interval is $2.5 \times 10^6\, m^2\, s^{-1}$ and negative values are stippled. After James (1988).

region results in small zonally averaged winds over New Zealand and, as will be discussed later, is felt to be important in the teleconnections in the region of the Pacific Ocean.

James attributed the split jet to the topographic forcing by the Antarctic and the asymmetry of the continent, where the highest orography is displaced by some 10° of latitude from the pole. The split jet develops in association with an upper air blocking pattern close to New Zealand with a ridge extending south near 160° E towards the Antarctic. Within this regime there is also a trough extending over the Tasman Sea, which is one of the major cyclogenetic areas in the southern hemisphere. Trenberth (1980) had previously identified a link between the asymmetric topography of the Antarctic and the amplitude and phase of the 500 hPa long waves in the Southern Hemisphere. This work related the amplitude and phase of the long waves to the strength of the blocking ridge close to New Zealand and its associated downstream trough over the Tasman Sea. This trough is thought to have important regional and hemispheric consequences. For example, it has been noted that when it contains a cut-off low there is usually extensive precipitation and a temperature anomaly over eastern Australia. This may affect the phase of the southern oscillation via the north Australian–Indonesian low, as discussed below.

Numerical studies have also suggested that the vorticity transport resulting from the interaction of mid-latitude cyclones with the Antarctic circulation plays a central role in maintaining the katabatic circulation over the continent (James, 1989). This topic is discussed at length in Section 4.3.

The circumpolar trough, which rings the Antarctic between 60° S and 70° S, is the most pronounced feature of the Southern Hemisphere mean surface pressure field. It is present because of the large number of depressions found in this zone which have either moved south from mid-latitudes or developed in the Antarctic coastal zone. A characteristic feature of the surface pressure values measured at stations in the coastal zone is a semi-annual oscillation (see Figure 3.17). This comes about because of changes in the position and intensity of the circumpolar trough over the year. Monthly mean sea level pressure fields show that the trough is furthest south and most pronounced in the intermediate seasons and moves north and weakens in the summer and winter. The cycle takes this form because of the phase difference between the seasonal cycle of surface pressure values over the Antarctic and the sub-Antarctic latitudes and results in changes in a number of aspects of the high-latitude climate, including the strength of the westerlies and precipitation measured at the coastal stations, which is greatest at times when the trough is further south and more intense (see Figure 3.30). Across the latitude band of 40° S to 60° S, the semi-annual oscillation results in the zonal westerlies being strongest in the intermediate months of February/March and September/October and weakest in the summer and winter periods of December/January and July/August. Over the 30° S to 50° S zone the winds show a variability that is 180° out of phase with respect to the above. This is illustrated in Figure 7.6, which shows the pressure difference between 40° S to 60° S and 30° S to 50° S over the Pacific during the period 1972–77. Changes in the circulation characteristics of the mid-latitude region, e.g. cyclogenesis areas, are consequently reflected in the position and depth of the circumpolar trough and in aspects of the climate of the coastal region of the Antarctic.

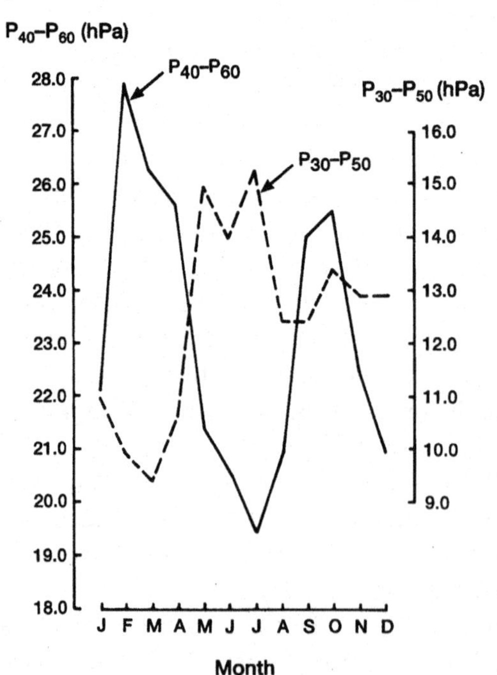

Figure 7.6 The annual cycle of pressure gradients between 40° S and 60° S, and between 30° S and 50° S (1972–77) averaged over the Pacific longitudes 110° E to 70° W. From Carleton (1992), after Streten and Zillman (1984).

7.2.2 Teleconnections with the El Niño–Southern Oscillation phenomenon

Background to the ENSO phenomenon

ENSO is the most pronounced inter-annual climatic variation found on Earth and it has a profound effect on the atmospheric and oceanic conditions across the Pacific Basin and in other parts of the world. It is a climate cycle involving the atmospheric and oceanic circulations of the Pacific, but many climatic variations beyond the tropics have been attributed to ENSO, including drought in India, floods in the southwestern USA and, more indirectly, drought in the Sahel. Attention has recently focused on the ENSO signal that can be detected in and around the Antarctic continent and on the role played by the continent in this complex atmospheric/oceanic interaction. However, the study of ENSO in the Antarctic is made difficult by the relatively short data record, compared with those for the tropics and mid-latitude areas. The problem is compounded by the poor observational network in many parts of the Southern Hemisphere, especially in the South Pacific. Nevertheless, the first studies of the effects of ENSO on the Antarctic climate have been carried out and are summarised in this section.

The Southern Oscillation (SO) was first identified in the 1920s and involves fluctuation of the atmospheric mass across the tropical Pacific with periods of 'warm', 'cold' and intermediate conditions in the near-surface layer of the eastern Pacific Ocean. During the more frequent 'cold' conditions, the equatorial region is characterised by strong easterly trade winds and a gradient in surface pressure across the Pacific with high pressure in the east, near South America, and low pressure around Indonesia. At upper levels there is a westerly return flow and this flow pattern is known as the Walker circulation. The low pressure over the western Pacific results in active convective systems and extensive, heavy precipitation over many parts of Australasia.

The Walker circulation also has major oceanographic consequences, with a westwards-moving ocean current (the South Equatorial Current) near the surface and an eastwards-moving return current (the Equatorial Undercurrent) at lower levels. In the 'cold' phase of the cycle there is extensive upwelling of cold, nutrient-rich water off the western coast of South America, which is of great importance to local fisheries. During the less frequent 'warm' periods, a radical change in the circulation takes place which is known as El Niño. This is characterised by a reduction in the strength of the trade winds and a decrease in the east–west surface pressure gradient across the Pacific. This results in weaker convergence over the western Pacific and less convective activity and precipitation. During 'warm' ENSO events there is also an amplification of the seasonal cycle in the trade winds and the trough in the westerlies of the South Pacific Ocean. In the ocean, the currents are also reduced in strength and there is less upwelling off South America. The typical atmospheric and oceanographic conditions associated with the 'warm' and 'cold' phases of the ENSO are summarised in Figure 7.7, taken from Bigg (1990). El Niño events usually occur at intervals of several years, but can occasionally be separated by periods of up to a decade. Details of the tropical evolution of the ENSO can be found in Philander and Rasmussen (1985) and Rasmussen and Carpenter (1982).

An important index used in the study of the ENSO is the Southern Oscillation Index (SOI) which is computed from the twice-normalised Tahiti minus Darwin mean sea level pressure difference (Parker, 1983). (Twice-normalised here means that the pressure at *each* station is expressed as the difference from its long-term mean and then normalised with respect to its standard deviation before calculating the difference.) During 'cold' periods the SOI is positive whereas negative values are recorded during the 'warm' events. The values of the SOI since 1950 are shown in Figure 7.8. In recent decades there have been several 'warm' ENSO events when the SOI was large and negative. The most pronounced was the 1982–83 event, which was accompanied by major climatic fluctuations around the world. In examining the Antarctic ENSO signal we will employ the usual system of defining the years of the cycle and take the year of a 'warm' event as $year_0$ and the previous year as $year_{-1}$ etc.

Observed teleconnections

ENSO teleconnections across the Southern Ocean and in the Antarctic coastal region have been investigated by a number of workers, with some of them finding evidence of systematic circulation changes between the different phases of the ENSO cycle. Mo and White (1985) examined teleconnections across the whole Southern Hemisphere using Australian mean sea level pressure and 500 hPa height analyses for the period 1972–80 and found that monthly mean, zonally averaged anomalies of these two quantities showed two patterns of variability.

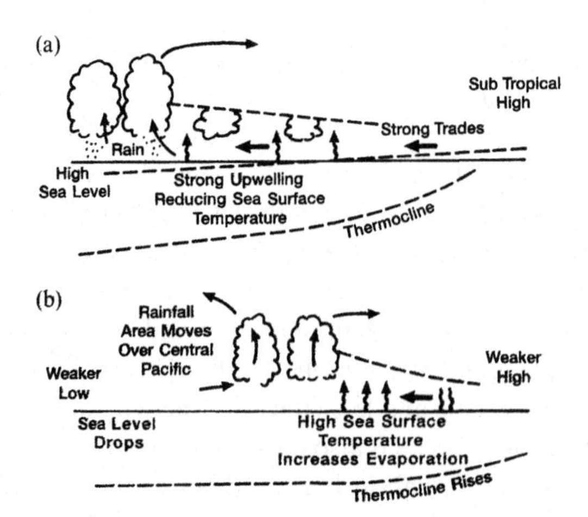

Figure 7.7 Schematic equatorial cross-sections of atmospheric and oceanic features in (a) the normal, Walker circulation and (b) the 'warm', El Niño circulation of the Southern Oscillation. After Bigg (1990).

First, an out-of-phase relationship between high and low latitudes both in the surface pressure and in 500 hPa data that they described as a 'seesaw'. Secondly, mid-latitude 500 hPa height anomalies that were negatively correlated with anomalies in the sub-tropics and polar regions. Both these patterns are weaker during the summer season and this was attributed to the weaker atmospheric circulation at that time of year. These patterns were established in parallel with strengthening and weakening of the westerlies in alternating zonal bands (Rogers and van Loon, 1982) so that stronger trade winds also occur with stronger westerlies north of 45° S (Trenberth, 1981). The Mo and White study found a high degree of teleconnectivity in the summer mean surface pressure field as far south as 70° S, but the authors cautioned over the validity of this result because of doubts concerning the quality of the numerical model data from the Antarctic continent.

Other studies have shown a connection between ENSO and the temporal evolution of mean sea level pressure anomalies over the extratropical Southern Hemisphere. In particular, Krishnamurti *et al.* (1986) found regions near to the Antarctic where up to 30% of the pressure variance was on ENSO timescales (30–50 months) and identified a propagation of pressure anomalies between the South and North Polar regions. They also traced the pressure reversals of El Niño 'warm' years 1965 and 1969 back to the Antarctic and identified a propagation equatorwards of sea level pressure anomalies across the high-latitude area of the Southern Hemisphere from the start of ENSO 'warm' events. van Loon and Shea (1985) also found that significant reversals occurred in the signs of the seasonal mean sea level pressure anomalies across the Southern Hemisphere over the winters preceding an ENSO event (year$_{-1}$) and the winter of the event (year$_0$). Although their data did not extend south of 50° S, there was a suggestion of an ENSO precursor in the northern part of the Weddell Sea. Carleton (1988) examined this proposal in detail and found evidence of enhanced cyclonicity with lower surface pressure in this area in the spring season

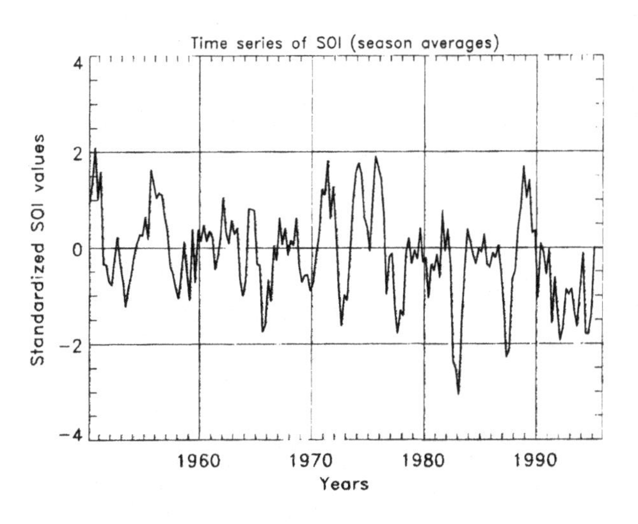

Figure 7.8 The Southern Oscillation Index since 1950. The units are standard deviations from the climatological monthly mean sea level pressure differences between Tahiti and Darwin.

preceding a 'warm' event. This contrasts with the anomalously anticyclonic conditions and generally northerly flow over the region in the spring, one year before a 'warm' event. These results were obtained by compositing monthly mean sea level pressure charts for ENSO $year_0$ and $year_{-1}$ for nine 'warm' events that occurred since 1950. The results were statistically significant for most of the Weddell Sea area and provided convincing evidence for a precursive ENSO signal in the Antarctic coastal region.

Work on ENSO teleconnections also suggested a link with the Weddell Sea sea ice, with reduced sea ice concentrations in the early summer of 'warm' years compared with the same period in $year_{-1}$. This was felt to be a result of the enhanced southerly flow during the 'warm' year giving divergence of the pack ice. This signal could also be seen in the sea ice duration data for the South Orkney Islands, where the ice was found to clear later during 'warm' years (mean date 9 December) compared with during $year_{-1}$ (mean date 12 November). The long series of surface meteorological data from these islands also suggested that the pressure and wind reversal in this region had become more marked during the ENSO events since the early 1950s, with a possible displacement westwards of the anticyclonic–cyclonic anomalies. There is also some evidence that the late summer sea ice variations for 'warm' and 'cold' SO events have become more pronounced since 1950 in the Weddell Sea.

With this evidence of statistically significant links between the SO and surface pressure anomalies around the Antarctic, the synoptic-scale activity around the continent can be expected to show distinct changes during ENSO cycles with subsequent effects on elements measured at the coastal stations, such as precipitation. Streten (1975) noted major changes in depression activity over the South Pacific in the autumn, winter and spring prior to the 'warm' event of 1972–73. At that time, the number of cloud vortices was observed to increase by about 20% compared with that during the more normal years 1971–72. Another aspect of the atmospheric circulation over the Southern Ocean and its variation during the ENSO cycle was considered by Carleton and Whalley (1988), who investigated the winter-season hemispheric flux polewards of sensible heat during the years 1973–77. This work was carried out using an analysis of synoptic-scale cloud signatures in satellite imagery to determine the number and location of systems and data on eddy heat flux produced in earlier work. They found a connection between the efficiency of the depressions in transporting eddy sensible heat polewards and the phase of the Southern Oscillation. It was shown that the depressions were more efficient in transporting heat when the SOI was low, this being the case for all stages of the lifecycle of the vortices. These results are consistent with the out-of-phase relationship of the zonal flow between the tropical, mid-latitude and polar regions of the Southern Hemisphere so that the efficiency of the meridional eddy heat transport by the depressions in the winter season increases when the westerlies south of about 45° S are weaker than those to the north of

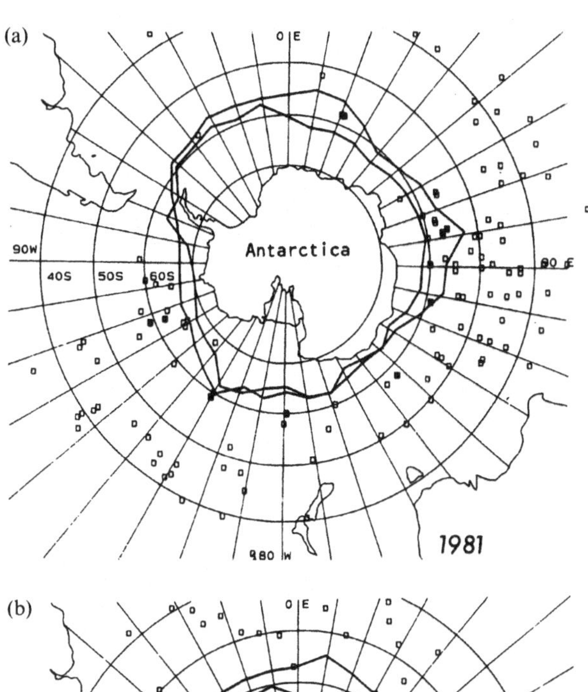

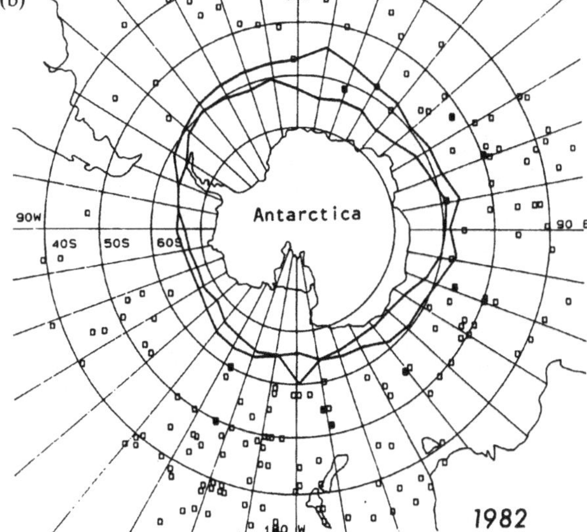

Figure 7.9 Mesocyclone locations and maximum/minimum sea ice extent for the winters of (a) 1981 and (b) 1982. The lows with comma-shaped (spiraliform) cloud signatures are shown by open (filled) squares. From Carleton and Carpenter (1989a).

this latitude and the annual cycle of the circumpolar trough and the trades is enhanced (Carleton and Whalley, 1988).

As the synoptic-scale activity varies with the SOI, so mesocyclones, that often form in outbreaks of cold, Antarctic air moving northwards, could also be expected to change. The relationship of the occurrence of mesocyclones and ENSO was investigated by Carleton and Carpenter (1989a) as part of their study into mesocyclone activity during the seven winters of 1977–83. This period included the 'warm' ENSO event of 1982–83 and they examined vortex development during the winters of 1981 (year_{-1}) and 1982 (year_0). As can be seen in Figure 7.9, they found many more vortices in 1982 than they had in the preceding

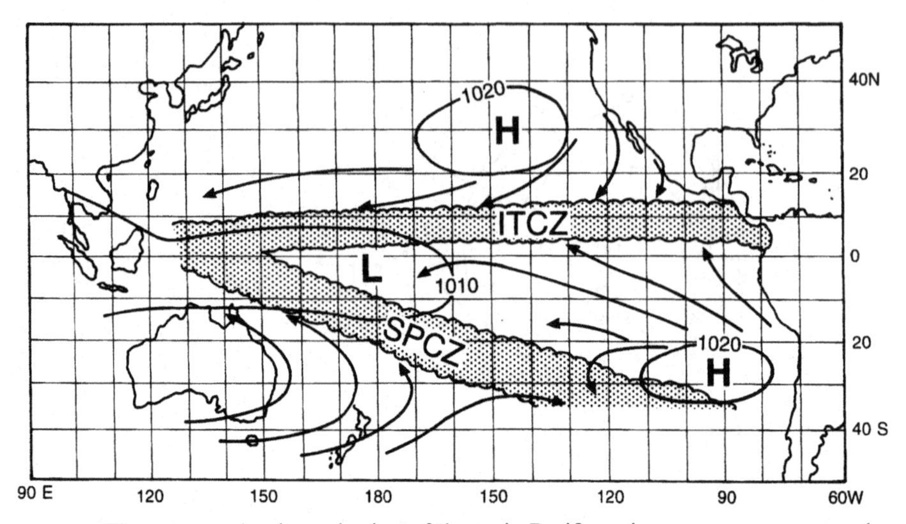

Figure 7.10 A schematic view of the main Pacific region convergence zones, the Intertropical Convergence Zone (ITCZ) and the South Pacific Convergence Zone (SPCZ). Also shown are the annual mean sea level pressure contours and surface wind streamlines. After Vincent (1994).

year and a major shift in longitude of the region of maximum vortex activity. Figure 7.9 shows that the location of most vortices moved eastwards from around 90° E in 1981 to the area around New Zealand in 1982. The M1 index of meridional pressure anomaly for this sector (the pressure difference between 147.5° E and 176° W near 43° S, Trenberth (1976)) shows a marked change between the two years with values of +11 hPa and −8.7 hPa in 1981 and 1982 respectively. Thus the change in mesocyclone occurrence between the two years is linked with a change from a reduced to a more amplified cycle of the climatological trough in the Tasman Sea. During 'warm' ENSO events in general, large numbers of mesocyclones were observed southeast of Australia and around New Zealand in outbreaks of cold, polar air (Carleton and Carpenter, 1990). However, in the year before a 'warm' event, when the amplitude of the trough is suppressed, there are few vortices because of the limited number of cold air outbreaks. Carleton cautions, however, that systematic adjustments of the M1 with SOI alterations are not apparent in all parts of the observational record. The longitude of maximum mesocyclone formation has also been linked to the TPI (Carleton and Carpenter, 1990).

Other precursors to ENSO events in the high southern latitudes were found by Trenberth (1976) and Trenberth and Shea (1987). Trenberth (1976) used sea level pressure anomalies to identify areas where systematic leads to SO 'warm' events occurred and found indications three to four seasons in advance of ENSO events over New Zealand and a link to changes in the South Pacific Convergence Zone (see Figure 7.10 for the location of the SPCZ). In a case study of the 1982/83 ENSO event, Mo *et al.* (1987) showed that the Antarctic was involved in the establishment of blocking in the New Zealand region. They found that, at this time, the block was close to its climatological position, with a ridge extending south of New Zealand to the coast of the Antarctic and a deep cut-off low over the Tasman Sea. With the upstream and downstream troughs being located near 160° W and 90° E, this system thus extended for a considerable distance around the coast of the Antarctic and resembled the flow pattern that Trenberth (1980) felt might link the ENSO with the New Zealand block. The dynamical evolution of the New Zealand block was further investigated by Mo *et al.* (1987) using a general circulation model. Using data for June 1982 and altering the sea surface temperatures, orography and other factors affecting the thermal forcing in the model, they concluded that the block was strongly influenced by the land–sea heat contrast near Australia, but that cold air outbreaks from Antarctica were of most importance. Smith and Stearns (1993a), using results from their analysis of temperature behaviour across the continent, have suggested that the cold air outbreaks from the Antarctic could be related to the similar temperature anomalies noted on the Ross and Amery ice shelves that contrasted with conditions along the Wilkes Land coast. As suggested by Parish and Bromwich (1987), the topography of Antarctica concentrates the drainage flow from the interior plateau within specific regions so that the cold air outflow which plays a part in the establishment of the New Zealand block may be a result of the topography of the Antarctic. Smith and Stearns (1993a) created a model of the conditions along the Antarctic coast between the Ross and Amery ice shelves at different stages of the SOI by using their temperature and pressure anomaly data and a knowledge of the climatological upper air conditions. They found that the surface anomalies would indeed amplify the 500 hPa trough–ridge–trough pattern in this sector of the Antarctic prior to the SOI minimum and so suggested that the topographically induced temperature differences along the coast could be responsible for maintaining the New Zealand block.

The SPCZ is thought to be an important component in ENSO and has recently been linked to climate variations in the Antarctic. This feature is a large, persistent convective cloud band that has been studied extensively since it was first recognised when satellite imagery became available in the 1960s. It stretches from New Guinea east-southeastwards to about 30° S, 120° W and has a more zonal orientation on its western side (Vincent, 1994). It is a region in which many mid-latitude depressions develop, move polewards and decline and it is maintained by tropical–extra-tropical interactions. The SPCZ is located at the boundary of the two main centres of action in the SO that are located over Indonesia and the tropical South Pacific. Its location is closely tied to the underlying sea surface temperatures and shows a displacement north and east during El Niño events. This was the case during the 1972/73 'warm' event that was investigated by Streten (1975). This movement comes about because of longitudinal shifts in the long waves, which is consistent with the findings of van Loon (1984), who reported changes in the annual cycle of the climatological westerly trough and trade winds between the winters of $years_0$ of 'warm' ENSO events, when an enhanced cycle is found, and the suppressed nature of the cycle in earlier years. When surface pressure is low around New Zealand, the main depression track in the area is displaced northwards, with cold fronts moving over Australia and declining in the SPCZ so the observed negative surface pressure anomaly identified one or two years ahead of the ENSO 'warm' events results in changes in the storm tracks near New Zealand.

An Antarctic ENSO signal was reported by Savage *et al.* (1988) using data from South Pole Station and the automatic weather stations deployed across the continent since 1980. They found a correlation between the 'warm' phase of ENSO and surface temperatures at the South Pole, which manifested itself as cold temperatures in the year following a minimum in the SOI. For example, after the low SOI values observed in 1982 the mean annual temperature at Pole Station in 1983 was −51°C or 3.3 standard deviations below the 1957–86 mean. In the period since the station was established in 1957, similar links between the SOI in $year_0$ of a 'warm' event and air temperatures at the pole in $year_{+1}$ were also found in 1958/59, 1969/70 and 1972/73. However, the coupling of 1964–66 was not well resolved and none was found in 1976. Overall, the relationship between the SOI and temperature at the pole is significant at the 90% confidence level, as can be seen in the graph showing the SOI and pole temperatures which is reproduced in Figure 7.11 from Savage *et al.* (1988). Savage *et al.* also considered the effect of the 1982/83 ENSO 'warm' event on the anomalies of temperature and surface wind at manned stations and AWS sites in the Ross Sea–Ross Ice Shelf sector of the continent and found a coherent pattern in these quantities. The analysis suggested that anomalously cold conditions in this area a year after an ENSO event were associated with an enhanced surface layer inversion and a stronger katabatic drainage flow from the interior. Savage *et al.* concluded that this anomalous temperature and wind regime was maintained by a large-scale flow that was dynamically linked to the phenomenon of ENSO.

The nature of the Antarctic circulation anomalies associated with the SO was further investigated by Smith and Stearns (1993a; 1993b) who examined the horizontal pressure and temperature patterns over Antarctica in relation to the phase of the cycle. They created composite pressure and temperature fields both for 'warm' and for 'cold' ENSO years from 24 manned stations that had a meteorological record spanning about 30 years. They found that there was a distinct change in sign of the pressure anomalies surrounding the minima in the SOI, with the values changing from positive before the minimum to negative afterwards (Figure 7.12). The largest anomalies were near Scott Base and Mawson Station with the smallest extending across the continent from Casey Station to Amundsen-Scott Station and the Antarctic Peninsula. This is

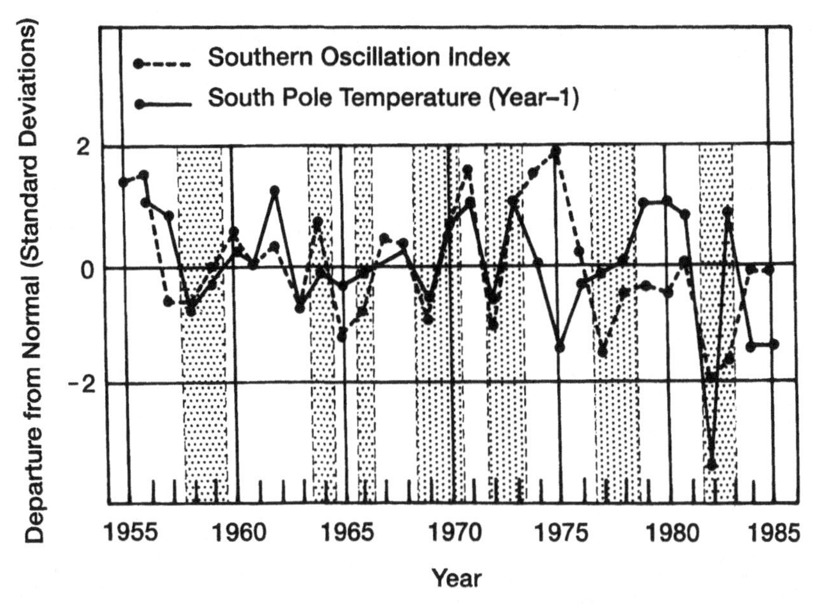

Figure 7.11 Normalised time series of the annual values of the SOI and of the annual mean temperatures from the following year (lag +1) at Amundsen–Scott (South Pole) Station. Shading shows the seven ENSO 'warm' episodes that have occurred since 1955. After Savage *et al.* (1988).

consistent with the changes found in the pressure anomalies from $year_{-1}$ to $year_0$ by Carleton (1988) and van Loon and Shea (1985; 1987). The temperature anomalies, which are also shown in Figure 7.12, also indicate a change in sign across the minimum in the SOI for some parts of the continent.

Although an ENSO signal can be found in the Antarctic sea level pressure fields and sea ice extent data, some studies have suggested that ENSO may, at least in part, be a response to forcing *by* the Antarctic. For example, Chiu (1983) not only found a link between the SOI in March/April and the area of sea ice in the subsequent July–December period over the eight years 1973–80, but also a lag of the SOI with the area of sea ice in the latter part of the year. However, care must be taken when examining these relationships because both the sea ice and the SOI data have large temporal autocorrelations. Nevertheless, Carleton (1989) extended this work by accounting for the autocorrelations and examining regional associations. He found that some sea ice–SOI links were significant in particular areas, such as the regions of extensive ice production in the Ross and Weddell Seas. Here, the sea ice changes lag behind the SOI but precede changes in the eccentricity of the circumpolar vortex and, particularly, in wave number one. This long wave can be monitored by use of the TPI discussed earlier and links have been suggested between this and ENSO. Carleton (1989) examined the period 1973–82 in detail and found that an increased ice extent in late winter was associated with a displacement of wave number one towards

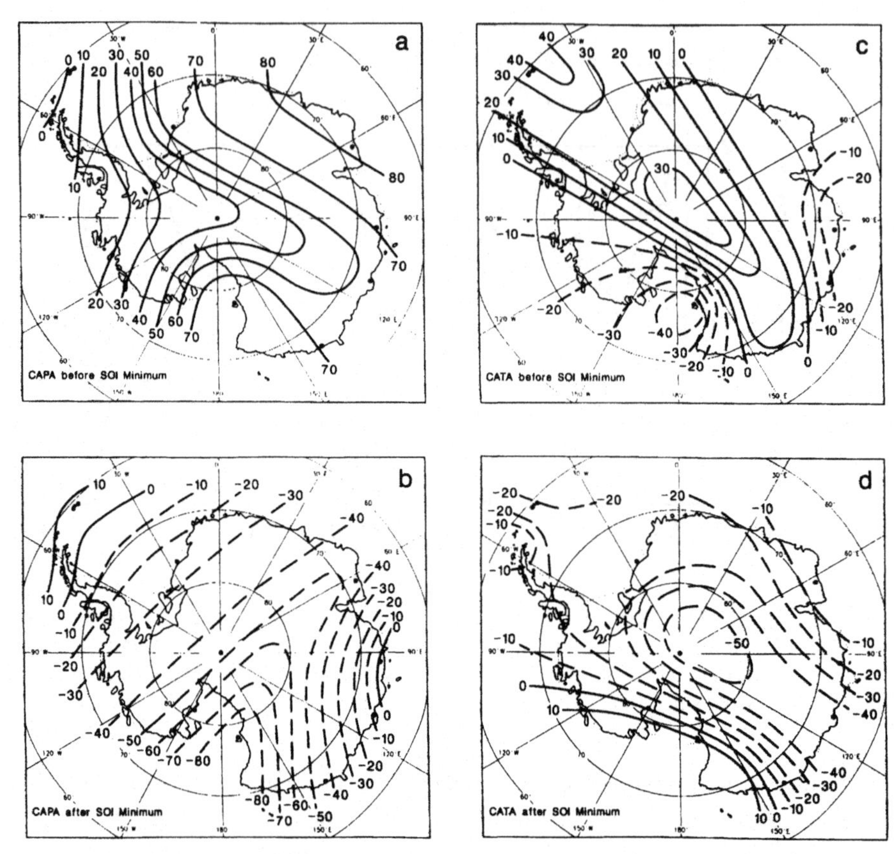

Figure 7.12 Composite maps for the SOI minimum. (a) The Composite Annual Pressure Anomaly (CAPA) before minimum. (b) The CAPA after minimum. (c) The Composite Annual Temperature Anomaly (CATA) before minimum. (d) The CATA after minimum. Pressures are contoured every 10 hPa and temperatures every 10°C. Negative anomalies are shown by dashed contours. From Smith and Stearns, In: *Antarctic Meteorology and Climatology: Studies Based on Automatic Weather Stations*, Antarctic Research Series, Volume 61, D. H. Bromwich and C. R. Stearns, editors, American Geophysical Union, Washington, pp. 155, copyright by the American Geophysical Union

Australia (TPI negative) and an amplification of the Tasman Sea trough during the August/September period. These changes take place about five months before the Walker Circulation adjusts as indicated by the SOI. Model studies involving coupled atmosphere–ocean models also suggest a connection between sea ice and ENSO, although there are still considerable differences among the current models being used. However, work by Mitchell and Hills (1986) and Simmonds and Dix (1986) suggested a reasonably strong and significant response of the tropical pressure/height field to prescribed anomalies of the Antarctic winter sea ice extent. A further consequence of the reduction in the extent of ice in the models is to reduce the strength of the westerlies.

A conceptual model for the teleconnection between the Antarctic and the tropics has been proposed by Budong *et al.* (discussed by Carleton (1992)). They assumed that changes in the heat loss from the Southern Ocean come about because of changes in the sea ice resulting from variations in depression tracks. The sea ice variations in turn give rise to changes in the frequency of extra-tropical cyclones and the meridional extension of the subtropical high in the South Pacific. This area of high pressure is a function of the Hadley circulation and is closely associated with the sea surface temperature zonal anomalies of the equatorial Pacific. In this model the Antarctic heat flux anomalies are linked to the tropical Pacific via the Peru Current. The link between the Antarctic Ocean and the tropical regions therefore takes place via the eastern sides of the three Southern Ocean gyres which have periods of between three and five years (Budd, 1991).

7.2.3 Future research needs

Much more work needs to be carried out on the low-frequency links between the Antarctic and lower latitudes using both observational data and coupled atmosphere–ocean–sea ice models. The relatively short record of meteorological observations from the Antarctic means that other data, such as ice cores, will have to be used to investigate ENSO events prior to 1957, when a reasonable network of observations was lacking. Many questions have arisen from the work carried out to date, with perhaps the most important being the need to understand the dynamical means by which the tropical signal of ENSO affects the Antarctic. More specifically, the role of the South Pacific Convergence Zone and the split jet phenomenon near New Zealand need to be better understood and this can probably be best accomplished through theoretical and numerical modelling studies.

7.3 Future climate predictions – Antarctica in a 'greenhouse' climate

The Antarctic ice sheets contain enough water to raise global sea levels by 65 m if completely melted. Although a catastrophic collapse of the major part of the ice cap is unlikely on a timescale of decades or even centuries, the potential contribution to a rise in sea level from even a relatively modest reduction in ice volume is large enough to have caused concern and has generated interest in how the Antarctic regional climate may respond to global changes. Of particular interest are changes in precipitation and evaporation over the continent resulting from a general warming of the atmosphere in response to increased levels of carbon dioxide and other 'greenhouse' gases.

In Chapter 4 we showed that general circulation models (GCMs) are now capable of making reasonable simulations of the present Antarctic climate when sea ice cover and sea surface temperatures are prescribed using climatological observations. However, in order to make useful climate *forecasts*, it is necessary to couple an atmospheric GCM to an ocean and sea ice model so that ice cover and sea surface temperatures can also be determined prognostically. Attempts to run such fully coupled models have encountered problems of 'climate drift', particularly in the polar regions. Some coupled models appear to exhibit a steady decline in Antarctic sea ice extent, even in runs in which the atmospheric radiative forcing is held at its present-day value (Cattle *et al.*, 1992). These problems will undoubtedly be resolved as sea ice and ocean models are improved and better parameterisations are developed for fluxes at the atmosphere–sea ice boundary. At the time of writing, however, most estimates of changes in Antarctic surface accumulation have been based on experiments with atmosphere-only GCMs in which sea ice cover and/or sea surface temperatures are varied by some specified amount which is meant to reflect the effect of warmer global temperatures.

The simplest model experiments for studying enhanced 'greenhouse' warming are those in which a step change (most commonly a doubling) is made in the concentration of carbon dioxide and the model is then run until a new equilibrium climate state is achieved. Such experiments have generally been carried out using an atmospheric GCM coupled to a simple mixed-layer ocean model and a crude thermodynamic sea ice model (e.g. Mitchell *et al.*, 1990). Although the increases in globally averaged surface temperatures predicted by different models vary in the range 1.5–4.5°C, the regional patterns of predicted warming are reassuringly similar. All models indicate that the largest equilibrium warming occurs in the polar regions in winter as a result of the strong feedbacks associated with the removal of sea ice. A warming of up to 10°C in winter is indicated around Antarctica.

Equilibrium response experiments are relatively cheap to run and simple to interpret but they may give a false impression of the response of the climate to a gradual increase in greenhouse gas concentrations. The atmosphere may respond relatively rapidly to changes in radiative forcing but the large thermal inertia of the oceans will strongly modify the transient response of the coupled system. Experiments with an atmospheric GCM coupled to a full ocean circulation model (Stouffer *et al.*, 1989) in which atmospheric carbon dioxide concentrations were increased by 1% per year show a marked hemispheric asymmetry in the warming pattern. After 70 years of model integration (by which time carbon dioxide concentrations have doubled), the largest warming still occurs in the Arctic sea ice zone but is reduced to 5°C, compared with 7°C for the equilibrium response experiment. However, in the Antarctic sea ice zone, where an equilibrium response experiment indicated a 12°C warming, the transient response is a warming of only 2°C. The contrasts between the two polar regions appear to be due to different oceanic circulation regimes in the North Atlantic and Southern Ocean. Under present climatic conditions, surface waters sink in both regions, transporting heat from the atmosphere and upper ocean layers to the deep ocean. In the North Atlantic, this circulation is largely driven by thermohaline convection. As the upper ocean warms, convection becomes less vigorous and less heat is transported into the deep ocean, leading to a feedback mechanism that allows the surface waters to warm rapidly. In the Southern Ocean, in contrast, the circulations which transport heat to the deep ocean are more strongly controlled by wind stress, so the deep ocean can continue to act as a heat buffer as the surface waters warm. As discussed above, the current generation of coupled models has a number of weaknesses but the indication is that the atmosphere and oceans surrounding Antarctica are unlikely to warm by more than 1–2°C over the next 50 years.

Given a prediction of warming around Antarctica, there are a number of ways in which one can attempt to estimate the corresponding changes in accumulation over the Antarctic ice sheets. The simplest assumption is that the change in precipitation will be proportional to the change in saturation vapour pressure. As discussed in Section 7.1.2, this argument is based on observations of the *spatial* distributions of precipitation and temperature in the present-day climate and there is strong evidence that circulation changes may be as important as temperature variations are in determining precipitation trends. Fortuin (1992) has carried out experiments with an axisymmetric model that represented the mean meridional circulation over Antarctica and had coastal temperatures imposed as a boundary condition. With a 1.2°C increase in coastal temperatures specified, the model predicted an annual increase in accumulation of 117.2 km^3 water equivalent (about 7% of the present-day accumulation). This is significantly smaller than the 146.0 km^3 increase predicted from the change in saturation vapour pressure, suggesting that the weaker circulation in the warmer climate somewhat offsets the increase in humidity. However, Fortuin's model did not include the transport of water vapour by weather systems, which are known to play an important role in the water vapour budget (Section 4.4). Changes in this component of the circulation can only be assessed by conducting GCM experiments.

A number of experiments have been conducted with GCMs to study the sensitivity of the Antarctic regional climate to prescribed changes in Southern Ocean sea surface temperatures and sea ice extent or concentration. A decrease in winter sea ice extent or reduction in ice concentration within the pack ice zone seems an inevitable consequence of a warmer climate – an increase of 10 W m^{-2} in the average heat flux from the upper ocean to the sea ice could eliminate most of the Antarctic pack ice (Budd, 1991). Mitchell and Senior (1989) carried out an experiment in which all sea ice north of 67.5° S was removed and replaced by ocean at a temperature of −1.8°C. Not unexpectedly, this produced a strong local warming in the region from which ice was removed, but temperatures over Antarctica only increased by 1–2°C. Sea level pressure fell by nearly 4 hPa at 65° S as a result both of the thermodynamic forcing and of the reduction in surface roughness associated with replacing the sea ice by open ocean. A small increase in the strength of the circumpolar westerlies accompanied these changes. An analogous experiment, in which the fraction of open water within the pack ice was increased in steps from zero to 100%, was carried out by Simmonds and Budd (1991). The fall in pressure over the sea ice zone seen in the Mitchell and Senior experiment was confirmed in this study, but the strength of the Southern Hemisphere westerlies did not change monotonically with increasing open water fraction but rather reached a maximum with 80% open water. This appears to be largely due to complex changes in the baroclinicity of the Weddell Sea region that affect the zonally averaged flow. In a fully coupled model, changes in the strength of the westerlies will force changes in Southern Ocean circulation, leading to the possibility of feedbacks that are not represented in atmosphere-only models.

Precipitation changes in the Simmonds and Budd experiment are also discussed in Budd and Simmonds (1991). Precipitation over Antarctica was observed to increase monotonically with increasing open water fraction. The 20% open water fraction experiment was taken as the control run because it approximates to present conditions in the sea ice zone. Increasing the open water fraction to 100% caused precipitation to increase by 23% of its control value and evaporation to decrease by 7%, giving a 48% increase in net accumulation. The greatest increases occurred in the high-precipitation zone on the coastal slopes of Antarctica. A further experiment was carried out in which global sea surface temperatures were set to their predicted equilibrium values following a doubling of atmospheric CO_2 concentration. In this experiment, precipitation increased by 47% and, although evaporation also increased by 20%, net accumulation still increased by 68%. Although experiments such as these should not be regarded as climate predictions, they provide some guidance on the likely sensitivity of the Antarctic climate to changes on a hemispheric or global scale.

What are the consequences of these climate changes for the Antarctic ice sheets? In Section 4.4 we saw that the mass balance of the ice sheets is probably nearly neutral in the present climate, with a close balance between accumulation, iceberg calving and basal melting of ice shelves. The increase in accumulation associated with a warmer climate will contribute to a positive mass balance until the ice sheets respond dynamically and calving rates also increase to balance the greater input of water. In contrast to temperate glaciers, the Antarctic ice sheets do not have an extensive ablation zone. Even if coastal temperatures rose by several degrees, surface melting would still be an insignificant contribution to the total mass balance, thus the short-term consequence of globally warmer temperatures is likely to be an increase in the mass of the Antarctic ice sheets and a consequent negative contribution to sea-level rise. Given the uncertainties in global climate predictions and in the response of the Antarctic regional climate, it is very difficult to make quantitative estimates of this contribution. Current best estimates (Warrick and Oerlemans, 1990) suggest that growth of the Antarctic ice sheets may contribute up to 6 mm in global sea level fall by 2030 if 'greenhouse' gas emissions continue to grow at their current rate.

In the longer term, the situation is more complex. As the oceans around Antarctica start to warm and the extent of the sea ice zone is reduced, warmer water may penetrate under the major ice shelves, increasing basal melt rates. Budd *et al.* (1987) estimated that a 4°C rise in ocean temperatures could increase basal melt rates by 10 m per year, leading to a removal of the ice shelves in as short a period as 50 years. Although this has no direct consequences for sea level rise, since floating ice displaces its own weight of water, the removal of the fringing ice shelves is likely to induce changes in the grounded ice sheets. Of particular concern is the West Antarctic Ice Sheet, which is grounded below sea level over much of its bed and is thus potentially unstable. Model studies by Budd *et al.* (1987) showed that melting of the West Antarctic Ice Sheet could contribute 1 m to global sea level rise after 500 years and 4 m after 5000 years, by which time the ice sheet would have almost completely disappeared. The East Antarctic Ice Sheet is believed to be more stable and will respond more slowly to changes in climatic forcing.

There are considerable uncertainties involved in all of these predictions. Global and regional climate models need to be improved in order to provide predictions with a greater level of confidence. Such improvements will come about as a result of further observational studies aimed at improving our understanding of the workings of the Antarctic climate system. In order to obtain better estimates of the changes in ice sheet volume associated with climate changes, it will be necessary to couple ice sheet models to existing atmosphere–ocean GCMs, which will be a formidable task. However, even the current tentative predictions indicate potentially large changes in sea level associated with climate change in Antarctica. The social and economic consequences of these changes are such that research into the Antarctic climate system and its role in global climate change must remain a high priority for environmental scientists.

Appendix A

A chronological list of stations that have made multi-year meteorological observations in the Antarctic and on the sub-Antarctic Islands

WMO No.	No. on Fig. 1.1	u/a stn[3]	Name	Operating country	Latitude[1]	Longitude[1]	Elevation (m)	Period covered by meteorological record
			Cape Adare	UK	71.3° S	170.2° E	6	1899–1900, 1911
			Hut Point	UK	77.9° S	166.7° E	12	1902–04
			Snow Hill	Sweden	64.4° S	57.0° W	13	1902–03
88968	105		Orcadas	Argentina	60.7° S	44.7° W	6	1903–
88903			Grytviken[4]	UK	54.3° S	36.5° W	3	1905—[5]
			Husvik	UK	54.2° S	36.7° W	?	1909–52, 1954–55, 1957–59
			Stromness	UK	54.1° S	36.7° W	?	1910–53
			Cape Evans	UK	77.6° S	166.4° E	20	1911–12, 1915–17
			Framheim	Norway	78.6° S	163.6° W	40	1911–12

WMO No.	No. on Fig. 1.1	u/a stn[3]	Name	Operating country	Latitude[1]	Longitude[1]	Elevation (m)	Period covered by meteorological record
			Godthul	UK	54.3° S	36.3° W	?	1911–15, 1922–29
			Leith Harbour	UK	54.1° S	36.7° W	?	1911, 1918–61
			Ocean Harbour/New Fortuna Bay	UK	54.3° S	36.3° W	?	1911–20
			Cape Denison	Australia	67.1° S	142.7° E	6	1912–13
			Prince Olaf Harbour	UK	54.1° S	37.1° W	?	1916, 1919–31
		✓	Little America	USA	78.6° S	163.9° W	40	1929–30, 1934–35, 1940–41, 1956–58
			Ushuaia	Argentina	54.8° S	68.3° W	7	1931–
			Campbell Island	New Zealand	52.0° S	169.0° E		1941–
88938			Deception Island	UK/ Argentina	63.0° S	60.7° W	8	1944–67
88949			Port Lockroy	UK	64.5° S	63.3° W		1944–54[5]
88963	102		Esperanza	Argentina	63.4° S	57.0° W	13	1945–
88940			Hope Bay	UK	63.4° S	57.0° W		1945–48, 1952–63
89066	84		San Martin	Argentina	68.1° S	67.1° W	4	1946–
88961			Stonington Island	UK	68.2° S	67.0° W		1945–49, 1958–59, 1960–75
88952/ 89063	87	✓	Argentine Islands/Faraday	UK	65.4° S	64.4° W	11	1947–
			Melchior	Argentina	64.3° S	63.0° W	8	1947–60
89042	104		Signy Island	UK	60.7° S	45.6° W	6	1947–
88934			Admiralty Bay	UK	62.1° S	58.4° W	9	1948–60
			Macquarie Island	Australia	54.5° S	158.9° E	30	1949–

WMO No.	No. on Fig. 1.1	u/a stn[3]	Name	Operating country	Latitude[1]	Longitude[1]	Elevation (m)	Period covered by meteorological record
		✓	Maudheim	Norway/ UK/ Sweden	71.0° S	10.9° W	38	1950–52
			Port-Martin	France	66.8° S	141.4° E	14	1950–52
			Almirante Brown	Argentina	64.9° S	62.9° W	7	1951–59, 1968–
			View Point	UK	63.5° S	57.4° W	?	1953–60
89564	10	✓	Mawson	Australia	67.6° S	62.9° E	16	1954–
			General Belgrano I	Argentina	78.0° S	38.8° W	50	1955–79
88959			Horseshoe Island	UK	67.8° S	67.3° W	9	1955–68
89642	28	✓	Dumont d'Urville	France	66.7° S	140.0° E	43	1956–
89022	108	✓	Halley	UK	75.5° S	26.4° W	39	1956–
89664	52	✓	McMurdo	USA	77.9° S	166.7° E	24	1956–
			Oazis	USSR	66.3° S	100.7° E	28	1956–58
			Pionerskaya	USSR	69.7° S	95.5° E	2740	1956–58
			Shackleton	UK	78.0° S	37.2° W	58	1956–57
			Byrd	USA	80.0° S	120.0° W	1530	1957–70 AWS since 1980
89611	17	✓	Casey/Wilkes	Australia	66.3° S	110.5° E	41	1957–
89571	14	✓	Davis	Australia	68.6° S	78.0° E	13	1957–64, 1969–
			Ellsworth	USA	77.7° S	41.1° W	42	1957–62
			Hallett	New Zealand/USA	72.3° S	170.3° E	5	1957–64
89592	15	✓	Mirny	USSR, now Russia	66.5° S	93.0° E	30	1957–
			Norway Station	Norway	70.5° S	2.5° W	55	1957–61
			Scott Base	New Zealand	77.9° S	166.7° E	14	1957–
89009	65	✓	South Pole. Later Amundsen–Scott	USA	90° S		2800	1957–

WMO No.	No. on Fig. 1.1	u/a stn[3]	Name	Operating country	Latitude[1]	Longitude[1]	Elevation (m)	Period covered by meteorological record
89532	5	✓	Syowa	Japan	69.0° S	39.6° E	21	1957, 1959–61, 1966–
			Baudouin	Belgium	70.4° S	24.3° E	37	1958–60, 1964–66
89057	92		Capitán Arturo Prat	Chile	62.5° S	59.7° W	5	1958–
89059	101		General Bernardo O'Higgins	Chile	63.3° S	57.9° W	10	1958–
89606	22	✓	Vostok	USSR, later Russia	78.5° S	106.9° E	3488	1958–61, 1963–93
89056	93		Presidente Frei Montalva/ Teniente Rodolfo Marsh	Chile	62.4° S	58.9° W	10	1960–
89512	–	✓	Novolaza-revskaja	USSR, now Russia	70.8° S	11.8° E	99	1961–
			Eights	USA	75.2° S	77.2° W	420	1962–65
89001		✓	SANAE	South Africa	70.3° S	2.4° W	52	1962–93
88958			Adelaide Island	UK	67.8° S	67.9° W	26	1962–77
88970	90		Teniente Matienzo	Argentina	65.0° S	60.1° W	32	1962–72, 1975
89542	7	✓	Molodeznaja	USSR, now Russia	67.7° S	45.9° E	40	1963–
89061	86		Palmer	USA	64.8° S	64.1° W	8	1965–
			Plateau	USA	79.2° S	40.5° E	3624	1965–69
			Petrel	Argentina	63.5° S	56.3° W	18	1967–1976, 1977
89050	96	✓	Bellings-hausen	USSR, now Russia	62.2° S	58.9° W	16	1968–
88962/ 89065	82		Fossil Bluff[2]	UK	71.3° S	68.3° W	55	1961, 1962, 1968–75
			Vanda	New Zealand	77.5° S	161.6° E	94	1969–70, 1974
89055	103	✓	Marambio	Argentina	64.2° S	56.7° W	198	1970–

89657		✓	Leningrad-skaja	USSR, now Russia	69.5° S	159.4° E	295	1971–
			Mizuho	Japan	70.7° S	40.3° E	2230	1976–78
89062	83		Rothera	UK	67.5° S	68.1° W	16	1974, 1976–
			Siple	USA	75.9° S	84.2° W	1050	1978–82
			Corbeta Uruguay	Argentina	59.5° S	27.3° W	14	1979–80
			Primavera	Argentina	64.2° S	61.0° W	50	1979–
			Arctowski	Poland	62.1	58.5	?	1980–91 or 92
89967	107		General Belgrano II	Argentina	77.9° S	34.6° W	32	1980–
			Russkaya	USSR	74.8° S	136.9° W	100	1980–?
89002	111	✓	Von Neumayer	Federal Republic of Germany	70.7° S	8.4° W	50	1981–
89058	94		Great Wall	People's Republic of China	62.2° S	59.0° W	10	1985–
			Artigas	Uruguay	62.2° S	58.9° W		1987–
89524	4		Asuka	Japan	71.5° S	24.1° E	931	1987–
			Commandte Ferraz	Brazil	62.1° S	58.4° W	?	1987–
			Dakshin Gangotri	India	70.1° S	12.0° E	?	1987–91
89662	43		Terra Nova Bay[2]	Italy	74.7° S	164.1° E	80	?
89674	54		Williams Field	USA	77.9° S	167.0° E	8	?
89064	91		Juan Carlos[2]	Spain	62.7° S	60.4° W	?	?
89053	95		Jubany	Argentina	62.2° S	58.6° W	4	1988–
89251	97		King Sejong	South Korea	62.2° S	58.7° W	10	1988–
89054	98		Dinamet	Uruguay	62.2° S	58.8° W	10	?
89573	13		Zhongshan	People's Republic of China	69.4° S	76.4° E	18	1989–
89514	2		Maitri	India	70.8° S	11.7° E	117	1992–

This appendix was compiled from information in Schwerdtfeger (1984), the list of Antarctic expeditions produced by Headland (1989), the list of UK Antarctic meteorological records (British Antarctic Survey, 1995), the planning documents for the SCAR FROST project and the annual reports produced by SCAR listing stations operating in the Antarctic. All currently operating stations are included, as well as all closed stations that have previously operated for at least two winters.

Notes:

1 Some of the stations have occupied slightly different locations over the years. Here we give the current position.

2 Currently a summer-only station.

3 This refers to stations that have had an operational radiosonde programme during some period of their existence.

4 Few data since 1982.

5 With some gaps in the data record.

Appendix B

A chronological list of automatic weather stations that have been deployed in the Antarctic and on the sub-Antarctic Islands

Name	No. on Fig. 1.1	AWS/WMO Numbers[1]	Operating country	Latitude	Longitude	Elevation (m)	Period of operation
Asgard		8908	USA	77.6° S	160.1° E	1750	1980–82
Byrd	77	89324	USA	80.0° S	119.4° W	1530	1980–88, 1990–
D10	27	89832	USA	66.7° S	139.8° E	243	1980–
D17			USA	66.7° S	139.7° E	438	1980
Dome C II	24	89828	USA	75.1° S	123.4° E	3250	1980–82, 1984–
Ferrell	55	89872	USA	77.9° S	170.8° E	45	1980–
Manning		8905	USA	78.7° S	166.9° E	66	1980–85
Marble Point	49	89866	USA	77.4° S	163.7° E	84	1980–
D57		8916	USA	68.1° S	137.5° E	2105	1981–87
Jimmy		8911	USA	77.9° S	166.8° E	200	1981–82, 1987–90
Laurie		8910	USA	77.5° S	170.1° E	23	1981–85

Name	No. on Fig. 1.1	AWS/WMO Numbers[1]	Operating country	Latitude	Longitude	Elevation (m)	Period of operation
Meeley		8915	USA	78.3° S	170.2° E	49	1981–85
D47	26	89834	USA	67.4° S	138.7° E	1560	1982–90, 1992–
GEO3 (Phillpot)	9	89762	Australia	68.7° S	61.1° E	1830	1982–
Siple		89284	USA	75.9° S	84.0° W	1054	1982–92
Whitlock	47	89865	USA	76.2° S	168.4° E	275	1982–
D80	25	89836	USA	70.0° S	134.9° E	2500	1983–
Nancy		8908	USA	77.9° S	168.2° E	25	1983
Windless Bight		8918	USA	77.7° S	167.7° E	40	1983–85
172.5 West		8923	USA	78.3° S	172.5° W	42	1984
Arrival Heights		8909	USA	77.9° S	166.7° E	202	1984
Byrd Glacier		8921	USA	80.0° S	165.0° E	75	1984
Fogle		8909	USA	77.8° S	166.7° E	202	1984–85
GC41 (Radok)	20	89813	Australia	71.6° S	111.3° E	2761	1984–
GC46 (P Schwerdt-feger)	21	89805	Australia	74.1° S	109.8° E	3096	1984–
Inexpressible Island		8922	USA	74.9° S	163.6° E	80	1984
Manuela	42	89864	USA	74.9° S	163.7° E	80	1984–
Marilyn	56	89869	USA	80.0° S	165.1° E	75	1984–
Martha I		8923	USA	78.3° S	72.5° W	20	1984–86
Tiffany		8911	USA	78.0° S	168.2° E	25	1984–85
A028 (Loewe)	19	1170	Australia	68.4° S	112.2° E	1622	1985–90
Gill	59	89376	USA	80.0° S	178.6° W	55	1985–
Larsen Ice Shelf	99	89262	USA	66.9° S	60.9° W	17	1985–
Schwerdt-feger	57	89868	USA	79.9° S	170.0° E	60	1985–

Name	No. on Fig. 1.1	AWS/WMO Numbers[1]	Operating country	Latitude	Longitude	Elevation (m)	Period of operation
Allison		8900	USA	89.8° S	60.0° W	2835	1986–87
Bowers		8909	USA	85.2° S	163.4° E	2090	1986
Butler Island	100	89266	USA	72.2° S	60.2° W	91	1986–
Clean Air	64	89208	USA	90.0° S	0.0° W	2835	1986–
Elaine	58	89873	USA	83.1° S	174.2° E	60	1986–
GF08	17	89803	Australia	68.5° S	102.1° E	2123	1986–
Law Dome Summit	23	89811	Australia	66.7° S	112.7° E	1366	1986–
Lettau	60	89377	USA	82.5° S	174.4° W	55	1986–
Patrick		8905/ 8901	USA	89.9° S	45.0° E	2835	1986–87
Uranus-Glacier	81	89264	USA	71.4° S	68.9° W	780	1986–
Buckle Island		8928	USA	66.9° S	163.2° E	520	1987–88
Dolleman Island		8917	USA	70.6° S	61.0° W	396	1987–88
Martha II	61	89374	USA	78.4° S	173.4° W	18	1987–92
Scott Island	33	89371	USA	67.4° S	180.0° W	30	1987–
Lynn	36	89860	USA	74.2° S	160.4° E	1772	1988–
Shristi	39	89862	USA	74.7° S	161.6° E	1200	1988–92
Sushila		8921	USA	74.4° S	161.3° E	1430	1988–91
Cape Adams			USA	75.0° S	62.5° W	25	1989–92
Mt Erebus		8911	USA	77.5° S	167.1° E	3700	1989–90
Pat		8931	USA	74.8° S	163.1° E	30	1989–90
Pegasus			USA	78.0° S	166.6° E	50	1989
Racer Rock	89	89261	USA	64.1° S	61.6° W	17	1989–
Sandra	37	89861	USA	74.5° S	160.5° E	1525	1989–94
Cape Denison	30	8907	USA	67.0° S	142.7° E	31	1990–
LGB10	11	89758	Australia	71.3° S	59.2° E	2620	1990–
Pegasus North	53	89667	USA	77.9° S	166.5° E	10	1990–

Name	No. on Fig. 1.1	AWS/WMO Numbers[1]	Operating country	Latitude	Longitude	Elevation (m)	Period of operation
Port Martin	29	89643	USA	66.8° S	141.4° E	39	1990–
AGO-A77			USA	77.5° S	23.7° W	1545	1991–
Casey Airstrip	18	89810	Australia	66.3° S	110.8° E	390	1991–
LGB20	8	89757	Australia	73.8° S	55.7° E	2741	1991–
Linda	55b	89769	USA	78.5° S	168.4° E	50	1991–
Minna Bluff	55a	89768	USA	78.6° S	166.7° E	920	1991–
Pegasus South	51	8937	USA	78.0° S	166.6° E	10	1991–
Young Island	32	89660	USA	66.2° S	162.3° E	30	1991–
Bonapart Point	88	89269	USA	64.8° S	64.1° W	8	1992–
Mount Howe	68	89349	USA	87.3° S	149.5° W	2400	1992–93
Mount Siple	79	89327	USA	73.2° S	127.1° W	230	1992–
Possession Island	46	89879	USA	71.9° S	171.2° E	30	1992–
Willie Field	50	8901	USA	77.9° S	167.0° E	40	1992–
AGO-A81		89705	USA	81.5° S	3.7° E	2410	1993–94, 1996–
Henry	66	89108	USA	89.0° S	1.0° W	2755	1993–
Kelly		8921	USA	89.0° S	179.6° W	2950	1993–94
LGB35	12	89568	Australia	76.0° S	65.0° E	2342	1993–
Lindsay		8986	USA	89.0° S	89.9° W	2815	1993–94
Nico	62	89799	USA	89.0° S	89.7° E	2935	1993–
Penguin Point	31	89847	USA	67.6° S	146.2° E	30	1993–
AGO-A80			USA	80.7° S	20.4° W	1200	1994–
Brianna	71	21362	USA	83.9° S	134.1° W	549	1994–
Cape Webb	31a	8933	USA	67.9° S	146.8° E	37	1994–
Doug		21359	USA	82.3° S	113.2° W	1433	1994–
Elizabeth	73	89332	USA	82.6° S	137.1° W	549	1994–

Name	No. on Fig. 1.1	AWS/WMO Numbers[1]	Operating country	Latitude	Longitude	Elevation (m)	Period of operation
Erin	70	21363	USA	84.9° S	128.8° W	1006	1994–
Harry		21355	USA	83.0° S	121.4° W	945	1994–
LGB46	12a	89577	Australia	75.9° S	71.5° E	2352	1994–
LGB59	12b	89774	Australia	73.5° S	76.78° E	2537	1994–
Recovery Glacier	112	8932	USA	80.8° S	22.3° W	1220	1994–
Santa Claus I	85	8910	USA	65.00° S	65.7° W	25	1994–
Sky Hi	80	8917	USA	75.0° S	70.8° W	1395	1994–
Sutton		8939	USA	67.1° S	141.4° E	871	1994–
Theresa	72	21358	USA	84.6° S	115.8° W	1463	1994–
J C	69	21357	USA	85.1° S	135.52° W	549	1994–
Dome F	6	8982	USA	77.3° S	39.7° E	3810	1995–
Limbert		8925	USA	75.4° S	59.9° W	40	1995–
Relay Station	6a	89744	USA	74.0° S	43.1° E	3353	1995–
AGO–A84		8932	USA	84.4° S	23.9° W	2103	1996–
D-57		21360	USA	68.2° S	137.5° E	2105	1996–
Cape King	44	7351	Italy	73.6° S	166.6° E	163	?
Cape Philips	45	7379	Italy	73.1° S	169.6° E	550	?
Cape Ross	48	89666	Italy	76.7° S	163.0° E	143	?
Drescher	109	89214	Federal Republic of Germany	72.87° S	19.03° W	20	?
Enigma Lake		7354	Italy	74.7° S	164.0° E	210	?
Filchner	106	89258	Federal Republic of Germany	77.08° S	51.22° W	20	?
Hi Priestley Glacier	35	7355	Italy	73.6° S	160.7° E	1980	?
Nansen Ice Sheet	40	7350	Italy	74.8° S	163.3° E	50	?

Name	No. on Fig. 1.1	AWS/WMO Numbers[1]	Operating country	Latitude	Longitude	Elevation (m)	Period of operation
Nordenski-old	110	89014	Finland	73.05° S	13.38° W	?	?
Priestley Glacier	38	7352	Italy	74.3° S	163.2° E	640	?
Terra Nova Bay	43	89662	Italy	74.7° S	164.1° E	88	?
Tourmaline Plateau	34	7356	Italy	74.1° S	163.4° E	1700	?

This appendix was compiled from information in the annual publications on the USA's AWSs (e.g. Savage *et al.* (1985) onwards) and the planning documents for the SCAR FROST project.

Note:

[1] The four digit AWS numbers are the identifiers used by SERVICE ARGOS. Over the years, different AWSs have used the same identifier and occasionally one AWS has had different numbers during its period of operation. Some AWSs have five digit WMO numbers if the data have been put onto the Global Telecommunications System.

References

Ackley, S. F. and T. E. Keliher (1976), Antarctic sea ice dynamics and its possible climatic effects, *AIDJEX Bulletin* **33**, 53–76.

Allan, T. D. and T. H. Guymer (1984), SEASAT measurements of wind and waves on selected passes over JASIN, *Int. J. Remote Sensing* **5**, 379–408.

Allison, I. (1989), Pack-ice drift off East Antarctica and some implications, *Ann. Glaciol.* **12**, 1–8.

Allison, I., R. E. Brandt and S. G. Warren (1993b), East Antarctic sea ice – albedo, thickness distribution, and snow cover, *J. Geophys. Res.* **98**, 12417–29.

Allison, I., G. Wendler and U. Radok (1993a), Climatology of the East Antarctic Ice Sheet (100° E to 140° E) derived from automatic weather stations, *J. Geophys. Res. – Atmos.* **98**, 8815–23.

Alvarez, J. A. (1958), An anomalous July over the southern parts of South America, *Notos* **7**, 3–5.

Alvarez, J. A. and B. J. Lieske (1960), The Little America blizzard of May 1957. In: *Proceedings of the Symposium on Antarctic Meteorology, Melbourne, Australia*, Australian Bureau of Meteorology, Melbourne, pp. 115–27.

Ambach, W. (1974), The influence of cloudiness on the net radiation balance of a snow surface with high albedo, *J. Glaciol.* **13**, 73–84.

Anderson, P. S. (1993), Evidence for an Antarctic winter coastal polynya, *Antarctic. Sci.* **5**, 221–6.

Anderson, P. S. (1994), A method for rescaling humidity sensors at temperatures well below freezing, *J. Atmos. Ocean. Tech.* **11**, 1388–91.

Anderson, P. S., S. D. Mobbs, J. C. King, I. McConnell and J. M. Rees (1992), A microbarograph for internal gravity wave studies in Antarctica, *Antarctic Sci.* **4**, 241–8.

Andreas, E. L. (1987), A theory for the scalar roughness and the scalar transfer coefficient over snow and sea ice, *Bound.-Layer Meteorol.* **38**, 159–84.

Andreas, E. L. and A. P. Makshtas (1985), Energy exchange over antarctic sea-ice in the spring, *J. Geophys. Res.* **90**, 7199–212.

Angell, J. K. (1988), Variations and trends in tropospheric and stratospheric global temperatures, 1957–87, *J. Climate* **1**, 1296–313.

Argentini, S., G. Mastrantonio, G. Fiocco and R. Occone (1992), Complexity of the wind field as observed by a sodar system and by automatic weather stations on the Nansen Ice Sheet, Antarctica, during summer 1988–89: two case studies, *Tellus* B **44**, 422–9.

Arking, A. (1991), The radiative effect of clouds and their impact on climate, *Bull. Amer. Meteor. Soc.* **72**, 795–813.

Arya, S. P. (1988), *Introduction to Micrometeorology*, Academic Press, San Diego, 307 pp.

Astapenko, P. D. (1964), *Atmospheric Processes in the High Latitudes of the Southern Hemisphere. Section II of the IGY Programme (Meteorology) No 3. Israel Program for Scientific Translations, Jerusalem*, Oldbourne Press, London, 286 pp.

Auer, A. H. (1986), An observational study of polar air depressions in the Australian region. In: *Preprint Volume Second International Conference on Southern Hemispheric Meteorology, 1–5 December 1986*, American Meteorological Society, Boston, pp. 46–9.

Baines, P. G. and K. Fraedrich (1989), Topographic effects on the mean tropospheric flow patterns around Antarctica, *J. Atmos. Sci.* **46**, 3401–15.

Ball, F. K. (1956), The theory of strong katabatic winds, *Australian J. Phys.* **9**, 373–86.

Ball, F. K. (1960), Winds on the ice slopes of Antarctica. In: *Antarctic Meteorology*, Pergamon Press, London, pp. 9–16.

Bamber, J. L. and A. R. Harris (1994), The atmospheric correction for satellite infrared radiometer data in polar regions, *Geophys. Res. Lett.* **21**, 2111–4.

Barkstrom, B. R. and G. L. Smith (1984), The Earth Radiation Budget Experiment (ERBE), *Bull. Amer. Met. Soc.* **65**, 1170–85.

Bengtsson, L. (1989), Numerical weather prediction at the Southern Hemisphere. In: *Proceedings of the Third International Conference on Southern Hemisphere Meteorology and Oceanography, Buenos Aires, Argentina, 13–17 November 1989*, AMS, Boston, pp. 1–3.

Bigg, G. R. (1990), El Niño and the Southern Oscillation, *Weather* **45**, 2–8.

Bintanja, R. and M. R. van den Broeke (1995), Momentum and scalar transfer coefficients over aerodynamically smooth Antarctic surfaces, *Bound.-Layer Meteorol.* **74**, 89–111.

Bottomley, M., C. K. Folland, J. Hsiung, R. E. Newell and D. E. Parker (1990), *Global Ocean Surface Temperature Atlas*, Meteorological Office, Bracknell.

British Antarctic Survey (1995), Handlist of meteorology records, *British Antarctic Survey Archives*, Cambridge.

Bromwich, D. H. (1979) Precipitation and accumulation estimates for East Antarctica, derived from rawinsonde information. Ph.D. Thesis, University of Wisconsin-Madison, 142 pp.

Bromwich, D. H. (1987), A case study of mesoscale cyclogenesis over the southwestern Ross Sea, *Antarctic. J. of the US* **22(5)**, 254–6.

Bromwich, D. H. (1988), Snowfall in high southern latitudes, *Rev. Geophys.* **26**, 149–68.

Bromwich, D. H. (1989a), An extraordinary katabatic wind regime at Terra Nova Bay, Antarctica, *Mon. Wea. Rev.* **117**, 688–95.

Bromwich, D. H. (1989b), Satellite analysis of antarctic katabatic wind behaviour, *Bull. Amer. Meteor. Soc.* **70**, 738–49.

Bromwich, D. H. (1989c), Subsynoptic-scale cyclone developments in the Ross Sea sector of the Antarctic. In: *Polar and Arctic Lows*, P. F. Twitchell, E. Rasmussen and K. L. Davidson, editors, A. Deepak, Hampton, Virginia, pp. 331–45.

Bromwich, D. H. (1990), Estimates of Antarctic precipitation, *Nature*. **343**, 627–9.

Bromwich, D. H. (1991), Mesoscale cyclogenesis over the southwestern Ross Sea linked to strong katabatic winds, *Mon. Wea. Rev.* **119**, 1736–52.

Bromwich, D. H. (1992), A satellite case study of a katabatic surge along the Transantarctic Mountains, *Int. J. Remote Sensing* **13**, 55–66.

Bromwich, D. H., J. F. Carrasco, Z. Liu and R.-Y. Tzeng (1993), Hemispheric atmospheric variations and oceanographic impacts associated with katabatic surges across the Ross Ice Shelf, Antarctica, *J. Geophys. Res.* **98**, 13045–62.

Bromwich, D. H., J. F. Carrasco and C. R. Stearns (1992), Satellite observations of katabatic wind propagation for great distances across the Ross Ice Shelf, *Mon. Wea. Rev.* **120**, 1940–9.

Bromwich, D. H., Y. Du and T. H. Parish (1994), Numerical simulation of winter katabatic winds from West Antarctica crossing the Siple Coast and the Ross Ice Shelf, *Mon. Wea. Rev.* **122**, 1417–35.

Bromwich, D. H. and D. D. Kurtz (1984), Katabatic wind forcing of the Terra Nova Bay polynya, *J. Geophys. Res.* **89**, 3561–72.

Bromwich, D. H. and Z. Liu (1995), An observational study of the springtime Siple Coast confluence zone. In: *Preprints: Fourth Conference on Polar Meteorology and Oceanography*, American Meteorological Society, Boston, pp. 272–7.

Bromwich, D. H., T. R. Parish and C. A. Zorman (1990), The confluence zone of the intense katabatic winds at Terra Nova Bay, Antarctica, as derived from airborne sastrugi surveys and mesoscale numerical modelling, *J. Geophys. Res.* **95**, 5495–509.

Bromwich, D. H. and C. J. Weaver (1983), Latitudinal displacement from main moisture sources controls $\delta^{18}O$ of snow in coastal Antarctica, *Nature* **301**, 145–7.

Brost, R. A. and J. C. Wyngaard (1978), A model study of the stably stratified planetary boundary layer, *J. Atmos. Sci.* **35**, 1427–40.

Brown, R. A. and L. Zing (1994), Estimating central pressures of oceanic midlatitude cyclones, *J. Appl. Met.* **33**, 1088–95.

Budd, W. F. (1966), The drifting of nonuniform snow particles. In: *Antarctic Research Series, Volume 9: Studies in Antarctic Meteorology*, M. J. Rubin, editor, American Geophysical Union, Washington, pp. 59–70.

Budd, W. F. (1991), Antarctica and global change, *Climatic Change*. **18**, 271–99.

Budd, W. F., W. J. R. Dingle and U. Radok (1966), The Byrd Snow Drift Project: Outline and basic results. In: *Antarctic Research Series, Volume 9: Studies in Antarctic Meteorology*, M. J. Rubin, editor, Washington, American Geophysical Union, pp. 71–134.

Budd, W. F., B. J. McInnes, D. Jenssen and I. N. Smith (1987), Modelling the response of the West Antarctic Ice Sheet to a climatic warming. In: *Dynamics of the West Antarctic Ice Sheet*, C. J. van der Veen and J. Oerlemans, editors, Reidel, Dordrecht, pp. 321–58.

Budd, W. F. and I. Simmonds (1991), The impact of global warming on the Antarctic mass balance and global sea level. In: *International Conference on the Role of the Polar Regions in Global Change*, Vol. II, G. Weller, C. L. Wilson and B. A. B. Severin, editors, University of Alaska, Fairbanks, pp. 489–94.

Budd, W. F. and I. N. Smith (1985), The state of balance of the Antarctic Ice Sheet, an updated assessment 1984. In: *Glaciers, Ice Sheets and Sea Level: Effects of a CO_2-induced Climatic Change*, National Academy Press, Washington, pp. 172–7.

Businger, J. A., J. C. Wyngaard, Y. Izumi and E. F. Bradley (1971), Flux-profile relationships in the atmospheric surface layer, *J. Atmos. Sci.* **28**, 181–9.

Businger, S. (1985), The synoptic climatology of polar-low outbreaks, *Tellus* A **37**, 419–32.

Businger, S. (1987), The synoptic climatology of polar-low outbreaks over the Gulf of Alaska and the Bering Sea, *Tellus* A **39**, 307–25.

Businger, S. and J.-J. Baik (1991), An Arctic hurricane over the Bering Sea, *Mon. Wea. Rev.* **119**, 2293–322.

Cahalan, R. F. and L. S. Chiu (1986), Large-scale short-period sea ice atmosphere interaction, *J. Geophys. Res.* **91**, 10709–17.

Carleton, A. M. (1979), A synoptic climatology of satellite observed extratropical cyclone activity for the Southern Hemisphere winter, *Arch. Met. Geophys. Biokl. B.* **27**, 265–79.

Carleton, A. M. (1987), Satellite-derived attributes of cloud vortex systems and their application to climate studies, *Rem. Sens. Environ.* **22**, 271–96.

Carleton, A. M. (1988), Sea ice–atmosphere signal of the Southern Oscillation in the Weddell Sea, Antarctica, *J. Clim.* **1**, 379–88.

Carleton, A. M. (1989), Antarctic sea–ice relationships with indices of the atmospheric circulation of the Southern Hemisphere, *Climate Dynamics* **3**, 207–20.

Carleton, A. M. (1992), Synoptic interactions between Antarctica and lower latitudes, *Aust. Met. Mag.* **40**, 129–41.

Carleton, A. M. and D. A. Carpenter (1989a), Satellite climatology of 'polar air' vortices for the Southern Hemisphere winter. In: *Polar and Arctic Lows*, P. F. Twitchell, E. A. Rasmussen and K. L. Davidson, editors, A. Deepak, Hampton, pp. 401–13.

Carleton, A. M. and D. A. Carpenter (1989b), Intermediate-scale sea ice–atmosphere interactions over high southern latitudes in winter, *Geo. Journal* **18**, 87–101.

Carleton, A. M. and D. A. Carpenter (1990), Satellite climatology of 'polar lows' and broad-scale climatic associations for the southern hemisphere, *Int. J. Climatol.* **10**, 219–46.

Carleton, A. M. and D. Whalley (1988), Eddy transport of sensible heat and the life history of synoptic systems: a statistical analysis for the Southern Hemisphere winter, *Met. Atmos. Phys.* **38**, 140–52.

Carleton, A. M. and M. Fitch (1993), Synoptic aspects of Antarctic mesocyclones, *J. Geophys. Res.* **98**, 12997–3018.

Carleton, A. M., L. A. McMurdie, H. Zhao, K. B. Katsaros, N. Mognard and C. Claud (1993), Satellite microwave sensing of Antarctic ocean mesocyclones. In: *Proceedings of the Fourth International Conference on Southern Hemisphere Meteorology and Oceanography. March 29–April 2, 1993, Hobart, Australia*, AMS, Boston, pp. 497–8.

Carlson, T. N. (1991), *Mid-latitude Weather Systems*, Harper Collins Academic, London, 507 pp.

Carrasco, J. F. and D. H. Bromwich (1992), Mesoscale cyclogenesis over the southeastern Pacific Ocean, *Ant. J. of the US.* **27(5)**, 289–91.

Carrasco, J. F. and D. H. Bromwich (1993a), Mesoscale cyclogenesis dynamics over the southwestern Ross Sea, Antarctica, *J. Geophys. Res.* **98 D7**, 12973–95.

Carrasco, J. F. and D. H. Bromwich (1993b), Interannual variation of mesoscale cyclones near the Antarctic Peninsula. In: *Proceedings of the Fourth International Conference on Southern Hemisphere Meteorology and Oceanography. March 29–April 2, 1993, Hobart, Australia*, AMS, Boston, pp. 499–500.

Carroll, J. J. (1982), Long-term means and short-term variability of the surface energy balance components at the South Pole, *J. Geophys. Res.* **87**, 4277–86.

Carsey, F. D. (1980), Microwave observations of the Weddell Polynya, *Mon. Wea. Rev.* **108**, 2032–44.

Carter, D. J. T. (1993), Global wave height climatologies from ERS-1 altimeter data, and comparisons with those from GEOSAT. In: *Proceedings First ERS-1 Symposium – Space at the Service of our Environment, Cannes, France, 4–6 November 1992*, ESA, Noordvijk, pp. 489–92.

Cattle, H., J. M. Murphy and C. A. Senior (1992), The response of the Antarctic climate in general circulation model experiments with transiently increasing carbon dioxide concentrations, *Phil. Trans. R. Soc. Lond.* ***B.*** **338**, 209–18.

Cavalieri, D. J., P. Gloersen and W. J. Campbell (1984), Determination of sea ice parameters with the Nimbus-7 SMMR, *J. Geophys. Res.* **89**, 5355–69.

Cavalieri, D. J. and S. Martin (1985), A passive microwave study of polynyas along the Antarctic Wilkes Land coast. In: *Oceanology of the Antarctic Continental Shelf (Antarctic Research Series, Volume 43)*, S. S. Jacobs, editor, American Geophysical Union, Washington, pp. 227–52.

Cavalieri, D. J. and C. L. Parkinson (1981), Large-scale variations in observed Antarctic sea ice extent and associated atmospheric circulation, *Mon. Wea. Rev.* **109**, 2323–36.

Cavalieri, D. J. and H. J. Zwally (1985), Satellite observations of sea ice, *Adv. Space Res.* **5**, 247–55.

Cerni, T. A. and T. R. Parish (1984), A radiative model of the stable nocturnal boundary layer with application to the polar night, *J. Clim. Appl. Met.* **23**, 1563–72.

Chang, T. C., P. Gloersen, T. Schmugge, T. T. Wilheit and H. J. Zwally (1976), Microwave emission from snow and glacier ice, *J. Glaciol.* **16**, 23–39.

Chiu, L. S. (1983), Antarctic sea ice variations 1973–1980. In: *Variations in the Global Water Budget*, A. Street-Perrot, M. Beran and R. Ratcliffe, editors, Reidel Publishing, Dordrecht, pp. 301–11.

Ciais, P., J. C. W. White, J. Jouzel and J. R. Petit (1995), The origin of present-day Antarctic precipitation from surface snow deuterium excess data, *J. Geophys. Res.* **100**, 18917–27.

Claud, C., N. M. Mognard, K. B. Katsaros, A. Chedin and N. A. Scott (1993), Satellite observations of a polar low over the Norwegian Sea by special sensor microwave imager, Geosat, and TIROS-N operational vertical sounder, *J. Geophys. Res.* **98**, 14487–506.

Comiso, J. C. (1994), Surface temperatures in the polar regions from Nimbus 7 temperature humidity infrared radiometer, *J. Geophys. Res.* **99**, 5181–200.

Connolley, W. M. and H. Cattle (1994), The Antarctic climate of the UKMO Unified Model, *Antarctic Sci.* **6**, 115–22.

Connolley, W. M. and J. C. King (1993), Atmospheric water vapour transport to Antarctica inferred from radiosonde data, *Quart. J. Roy. Met. Soc.* **119**, 325–42.

Court, A. (1949), *Meteorological data for Little America III*, Monthly Weather Review Supplement No. 48, American Meteorological Society, Boston.

Cuffey, K. M., R. B. Alley, P. M. Grootes, J. M. Bolzan and S. Anandakrishnan (1994), Calibration of the δ^{18}O isotopic palaeothermometer for central Greenland, using borehole temperatures, *J. Glaciol.* **40**, 341–9.

Culf, A. D. and J. F. R. McIlveen (1993), Acoustic observations of the peripheral Antarctic boundary layer. In: *Waves and Turbulence in Stably Stratified Flows*, S. D. Mobbs and J. C. King, editors, Oxford University Press, Oxford, pp. 139–54.

d'Aguanno, J. (1986), Use of AVHRR data for studying katabatic winds in Antarctica, *Int. J. Remote Sensing* 7, 703–13.

Dalrymple, P. C., H. Lettau and S. Wollaston (1966), South Pole Micrometeorology Program. In: *Studies in Antarctic Meteorology; Antarctic Research Series*, Vol. 9, M. J. Rubin, editor, American Geophysical Union, Washington, pp. 13–57.

Dalu, G. A., M. Baldi, M. D. Moran, C. Nardone and L. Sbano (1993), Climatic atmospheric outflow at the rim of the Antarctic continent, *J. Geophys. Res.* **98**, 12955–60.

del Guasta, M., M. Morandi, L. Stefanutti, J. Brechet and J. Piquad (1993), One year of cloud lidar data from Dumont d'Urville (Antarctica). 1. General overview of geometrical and optical properties, *J. Geophys. Res.* **98**, 18575–87.

Doake, C. S. M. and D. G. Vaughan (1991), Rapid disintegration of the Wordie Ice Shelf in reponse to atmospheric warming, *Nature* **350**, 328–30.

Dolgina, I. M. (1962), *Soviet Antarctic Expeditions: Third Continental Expedition 1958–9. Meteorological Observations II*, Arctic and Antarctic Scientific Research Institute, Leningrad, 477 pp.

Dolgina, I. M., M. A. Marshunova and L. S. Petrova (1976), *Reference Book of the Climate of Antarctica.* Volume 1. Radiation, Arctic and Antarctic Scientific Research Institute, Leningrad, 213 pp (In Russian).

Dover, S. E. (1993) Numerical modelling of blowing snow. Ph.D. Thesis, University of Leeds, 237 pp.

Dozier, J. and S. G. Warren (1982), Effect of viewing angle on the infrared brightness temperature of snow, *Water Resources Research* **18**, 1424–34.

Drewry, D. J., editor (1983), *Antarctica: Glaciological and Geophysical Folio*, Scott Polar Research Institute, Cambridge.

Dutton, E. G., R. S. Stone, D. W. Nelson and B. G. Mendonca (1991), Recent interannual variations in solar radiation, cloudiness and surface temperature at the South Pole, *J. Climate* **4**, 848–58.

Ebert, E. E. (1988), A pattern recognition technique for distinguishing surface and cloud types in the polar regions, *J. Clim. Appl. Meteorol.* **26**, 1412–27.

Ebert, E. E. (1989), Analysis of polar clouds from satellite imagery using pattern recognition with a statistical cloud analysis scheme, *J. Appl. Met.* **28**, 382–99.

Ebert, E. E. (1992), Pattern recognition analysis of polar clouds during summer and winter, *Int. J. Remote Sensing* **13**, 97–109.

Edwards, N. R. (1992) Numerical modelling of turbulence in the stably stratified atmosphere. Ph.D. Thesis, University of Leeds.

Egger, J. (1985), Slope winds and the axisymmetric circulation over Antarctica, *J. Atmos. Sci.* **42**, 1859–67.

Egger, J. (1992), Topographic wave modification and the angular momentum balance of the Antarctic troposphere, *J. Atmos. Sci.* **49**, 327–34.

Egger, J., C. Wamser and C. Kottmeier (1993), Internal atmospheric gravity waves near the coast of Antarctica, *Bound.-Layer Meteorol.* **66**, 1–17.

Eicken, H., M. A. Lange, H. W. Hubberten and P. Wadhams (1994), Characteristics and distribution patterns of snow and meteoric ice in the Weddell Sea and their contribution to the mass balance of sea ice, *Annales Geophysicae* **12**, 80–93.

Elliot, W. P. and D. J. Gaffen (1991), On the utility of radiosonde humidity archives for climate studies, *Bull. Amer. Meteor. Soc.* **72**, 1507–20.

Enomoto, H. and A. Ohmura (1990), The influence of atmospheric half-yearly cycle on the sea ice extent in the Antarctic, *J. Geophys. Res.* **95**, 9497–511.

Farman, J. C., B. G. Gardiner and J. D. Shanklin (1985), Large losses of total ozone in Antarctica reveal seasonal ClO_x/NO_x interaction, *Nature* **315**, 207–10.

Fitt, R. N., F. Whitby and J. Brown (1979), The impact of FGGE data on prognosis in the Australian region. In: *Preprints of Australia–New Zealand GARP Symposium*, ANMRC, Melbourne, pp. 1–8.

Fogg, D. E. (1992), *A History of Antarctic Science*, Cambridge University Press, Cambridge, 484 pp.

Foldvik, A. and T. Gammelsrød (1988), Notes on Southern Ocean hydrography, sea-ice and bottom water formation, *Palaeogeography. Palaeoclimatology Palaeoecology* **67**, 3–17.

Forbes, G. S. and W. D. Lottes (1985), Classification of mesoscale vortices in polar air streams and the influence of the large-scale environment on their evolutions, *Tellus* A **37**, 132–55.

Fortuin, J. P. F. (1992) The surface mass balance and temperature of Antarctica. Ph.D. Thesis, University of Utrecht, 106 pp.

Fortuin, J. P. F. and J. Oerlemans (1990), Parameterisation of the annual surface temperature and mass balance of Antarctica, *Ann. Glaciol.* **14**, 78–84.

Foster, T. D. and E. C. Carmack (1976), Frontal zone mixing and Antarctic Bottom Water formation in the southern Weddell Sea, *Deep Sea Research* **23**, 301–17.

Fowbert, J. A. and R. I. Lewis Smith (1994), Rapid population increases in native vascular plants in the Argentine Islands, Antarctic Peninsula, *Arctic Alpine Res.* **26**, 290–6.

Fraser, W. R., W. Z. Trivelpiece, D. G. Ainley and S. G. Trivelpiece (1992), Increases in Antarctic penguin populations: reduced competition with whales or a loss of sea ice due to environmental warming?, *Polar Biol.* **11**, 525–31.

Fuji, Y. (1979), Sublimation and condensation at the ice sheet surface of Mizuho station, Antarctica, *Antarctic Record* **67**, 51–63.

Gallée, H. and G. Schayes (1992), Dynamical aspects of katabatic wind evolution in the Antarctic coastal zone, *Bound.-Layer Meteorol.* **59**, 141–61.

Gardiner, B. G. and J. D. Shanklin (1989), *Measurements of Solar and Terrestrial Radiation at Faraday and Halley*, British Antarctic Survey, Cambridge, 45 pp.

Garrett, J. F. (1980), Availability of the FGGE drifting buoy system data set, *Deep Sea Research* A **27**, 1083–6.

Garratt, J. R. (1992), *The Atmospheric Boundary Layer*, Cambridge University Press, Cambridge, 316 pp.

Gates, E. M. and W. C. Thompson (1985), Simulated rime icing of some wind speed sensors, *J. Atmos. Ocean. Tech.* **3**, 273–82.

Genthon, C. (1994), Antarctic climate modelling with general circulation models of the atmosphere, *J. Geophys. Res.* **99**, 12953–61.

Giovinetto, M. B. (1964), The drainage systems of Antarctica. In: *Antarctic Research Series*, Volume 2: *Antarctic Snow and Ice Studies*, M. Mellor, editor, American Geophysical Union, Washington, pp. 127–55.

Giovinetto, M. B. and C. R. Bentley (1985), Surface balance in ice drainage systems of Antarctica, *Ant. J. of the U.S.* **20**(4), 6–13.

Giovinetto, M. B., D. H. Bromwich and G. Wendler (1992), Atmospheric net transport of water vapour and latent heat across 70° S, *J. Geophys. Res.* **97**, 917–30.

Giovinetto, M. B., N. M. Waters and C. R. Bentley (1990), Dependence of Antarctic surface mass balance on temperature, elevation and distance to open water, *J. Geophys. Res.* **95**, 3517–31.

Glazman, R. E. and A. Greysukh (1993), Satellite altimeter measurements of surface wind, *J. Geophys. Res.* **98**, 2475–83.

Gloersen, P. and W. J. Campbell (1988), Variations in the Arctic, Antarctic and global sea ice covers during 1978–1987 as observed with the Nimbus 7 scanning multichannel microwave radiometer, *J. Geophys. Res.* **93**, 3564–72.

Gloersen, P., T. C. Chang, T. T. Wicheit and W. Nordberg (1974), Microwave maps of the polar ice of the earth, *Bull. Amer. Met. Soc.* **55**, 1442–8.

Gloersen, P., W. J. Campbell, D. J. Cavalieri, J. C. Comiso, C. L. Parkinson and H. J. Zwally (1992), *Arctic and Antarctic Sea Ice, 1978–1987*, NASA, Washington, DC, 290 pp.

Gloersen, P., W. Nordberg, T. J. Schmugge and T. T. Wilheit (1973), Microwave signatures of first-year and multi-year sea ice, *J. Geophys. Res.* **78**, 3564–72.

Gosink, J. P. (1989), The extension of a density current model of katabatic winds to include the effects of blowing snow and sublimation, *Bound.-Layer Meteorol.* **49**, 367–94.

Grenfell, T. C. (1983), A theoretical model of the optical properties of sea ice in the visible and near infrared, *J. Geophys. Res.* **88**, 9723–35.

Guymer, L. B. (1978), *Operational Applications of Satellite Imagery to Synoptic Analysis in the Southern Hemisphere*, Australian Bureau of Meteorology, Melbourne, Technical report 29.

Harangozo, S. A. (1994), Interannual atmospheric circulation–sea ice extent relationships in the Southern Ocean: An analysis for the west Antarctic Peninsula region. In: *Preprints, Sixth Conference on Climate Variations. Nashville, Tennessee, January 23–28, 1994*, American Meteorological Society, Boston, pp. 364–7.

Harangozo, S. A., S. R. Colwell and J. C. King (1994), Interannual and long-term air temperature variability in the Southern Antarctic Peninsula from a reconstructed record for eastern

Alexander Island. In: *Preprints, 6th Conference on Climate Variations*, American Meteorological Society, Boston.

Harrold, T. W. and K. A. Browning (1969), The polar low as a baroclinic disturbance, *Quart. J. Roy. Met. Soc.* **95**, 710–23.

Hart, T., P. Steinle, P. Riley, R. Seaman, W. Bourke and J. Le Marshall (1993), Increasing the effectiveness of satellite temperature soundings in the Australian Bureau of Meteorology's global analysis and prediction system. In: *Preprints, Fourth International Conference on Southern Hemisphere Meteorology and Oceanography. March 29–April 2, 1993. Hobart, Australia*, American Meteorological Society, Boston, pp. 30–1.

Headland, R. K. (1989), *Chronological List of Antarctic Expeditions and Related Historical Events*, Cambridge University Press, Cambridge, 730 pp.

Heinemann, G. (1989), On the roughness length Z_0 of the snow surface of the Filchner–Ronne Ice Shelf, *Polarforschung*. **59**, 17–24 (In German).

Heinemann, G. (1990), Mesoscale vortices in the Weddell Sea region (Antarctica), *Mon. Wea. Rev.* **118**, 779–93.

Herman, G. F. and W. T. Johnson (1980), Arctic and Antarctic climatology of a GLAS general circulation model, *Mon. Wea. Rev.* **108**, 1974–91.

Hogan, A. W. (1975), Summer ice crystal precipitation at the south pole, *J. Appl. Met.* **14**, 246–9.

Hogan, A., S. Barnard, J. Samson and W. Winters (1982), The transport of heat, water vapour, and particulate matter to the South Polar Plateau, *J. Geophys. Res.* **87**, 4287–92.

Holton, J. R. (1979), *Introduction to Dynamic Meteorology*, Academic Press, New York.

Hoskins, B. J. and P. Berrisford (1988), A potential vorticity perspective of the storm of 15–16 October 1987, *Weather*, 122–9.

Houghton, J. T. (1984), *Remote Sensing of Atmospheres*, Cambridge University Press, Cambridge.

Howarth, D. A. (1983), Seasonal variations in the vertically integrated water vapor transport fields over the southern hemisphere, *Mon. Wea. Rev.* **111**, 1259–72.

Hsiung, J., and R. E. Newell, 1983: The principal nonseasonal modes of variation of global sea surface temperature. *J. Phys. Ocean.* **13**, 1957–67.

Hughes, N. A. (1984), Global cloud climatologies: a historical review, *J. Clim. Appl. Meteorol.* **23**, 724–51.

Ingleby, N. B. and R. A. Bromley (1991), A diagnostic study of the impact of SEASAT scatterometer winds on numerical weather prediction, *Mon. Wea. Rev.* **119**, 84–103.

Ingram, W. J., C. A. Wilson and J. F. B. Mitchell (1989), Modelling climate change: an assessment of sea-ice and surface albedo feedbacks, *J. Geophys. Res.* **94**, 8609–22.

Inoue, J. (1989a), Surface drag over the snow surface of the Antarctic plateau. 1. Factors controlling surface drag over the katabatic wind region, *J. Geophys. Res.* **94**, 2207–17.

Inoue, J. (1989b), Surface drag over the snow surface of the Antarctic plateau. 2. Seasonal change of surface drag in the katabatic wind region, *J. Geophys. Res.* **94**, 2219–24.

Isaksson, E. and W. Karlén (1994), High resolution climatic information from short firn cores, western Dronning Maud Land, *Climatic Change*. **26**, 421–34.

Jacka, T. H., 1983, *A Computer Data Base for Antarctic Sea Ice Extent*. ANARE Research Notes, volume 13. Antarctic Division, Hobart, 54 pp.

Jacobs, S. S. (1993), A recent sea-ice retreat west of the Antarctic Peninsula, *Geophys. Res. Lett.* **20**, 1171–4.

Jacobs, S. S., H. H. Hellmer, C. S. M. Doake, A. Jenkins and R. M. Frolich (1992), Melting of ice shelves and the mass balance of Antarctica, *J. Glaciol.* **38**, 375–87.

James, I. N. (1988), On the forcing of planetary scale Rossby waves by Antarctica, *Quart. J. Roy. Met. Soc.* **114**, 619–37.

James, I. N. (1989), The Antarctic drainage flow: implications for hemispheric flow on the southern hemisphere, *Antarctic. Sci.* **1**, 279–90.

Jeffries, M. O., R. A. Shaw, K. Morris, A. L. Veazey and H. R. Rouse (1994), Crystal structure, stable isotopes ($\delta^{18}O$), and development of sea ice in the Ross, Amundsen and Bellingshausen seas, Antarctica, *J. Geophys. Res.* **99**, 985–95.

Jenkins, A. and C. S. M. Doake (1991), Ice–ocean interaction on Ronne Ice Shelf, Antarctica, *J. Geophys. Res.* **96**, 791–813.

Jones, D. A. and I. Simmonds (1993), A climatology of Southern Hemisphere extratropical cyclones, *Climate Dynamics* **9**, 131–45.

Jones, D. A. and I. Simmonds (1994), A climatology of Southern Hemisphere anticyclones, *Climate Dynamics* **10**, 333–48.

Jones, P. D. (1990), Antarctic temperatures over the present century – A study of the early expedition record, *J. Climate* **3**, 1193–203.

Jones, P. D. (1991), Southern Hemisphere sea-level pressure data: an analysis and reconstructions back to 1951 and 1911, *Int. J. Climatol.* **11**, 585–607.

Jones, P. D. and D. W. S. Limbert (1987), *A Data Bank of Antarctic Surface Temperature and Pressure Data. Office of Energy Research, Report no. TR038*, USA Department of Energy, Washington, 52 pp. (Available from: National Technical Information Service, USA Department of Commerce, Springfield, VA 22161)

Jones, P. D., S. C. Raper and T. M. L. Wigley (1986), Southern Hemisphere surface air temperature variations: 1851–1984, *J. Clim. Appl. Met.* **25**, 1213–30.

Jones, P. D. and T. M. L. Wigley (1988), Antarctic gridded sea level pressure data: an analysis and reconstruction back to 1957, *J. Climate* **1**, 1199–220.

Jouzel, J. and L. Merlivat (1984), Deuterium and O-18 in precipitation – modeling the isotopic effects during snow formation, *J. Geophys. Res.* **89**, 1749–57.

Jouzel, J., C. Lorius and V. N. Petit (1987), Vostok ice core: a continuous isotope temperature record over the last climatic cycle (160,000 years), *Nature*. **329**, 403–8.

Juckes, M. N., I. N. James and M. Blackburn (1994), The influence of Antarctica on the momentum budget of the southern extratropics, *Quart. J. Roy. Met. Soc.* **120**, 1017–44.

Keller, L. M., G. A. Weidner, and C. R. Stearns, 1990, Antarctic automatic weather station data for the calender year 1989: Department of Meteorology, University of Wisconsin–Madison, 354 pp.

Kikuchi, T. (1988), A case study of a wave-like cloud and gravity wave in the lower troposphere in Mizuho Plateau, Antarctica, *Bound.-Layer Meteorol.* **43**, 403–9.

Kikuchi, T. and Y. Ageta (1989), A preliminary estimate of inertia effects in a bulk model of katabatic wind. In: *Proceedings of the Symposium on Polar Meteorology and Glaciology, No. 2*, National Institute of Polar Research, Tokyo, pp. 61–9.

King, J. C. (1989), Low-level wind profiles at an antarctic coastal station, *Antarctic Sci.* **1**, 169–78.

King, J. C. (1990), Some measurements of turbulence over an antarctic ice shelf, *Quart. J. Roy. Met. Soc.* **116**, 379–400.

King, J. C. (1993a), Control of near-surface winds over an Antarctic ice shelf, *J. Geophys. Res.* **98**, 12949–54.

King, J. C. (1993b), Contrasts between the Antarctic stable boundary layer and the mid-latitude nocturnal boundary layer. In: *Waves and Turbulence in Stably Stratified Flows*, S. D. Mobbs and J. C. King, editors, Oxford University Press, Oxford, pp. 105–20.

King, J. C. (1994), Recent climate variability in the vicinity of the Antarctic Peninsula, *Int. J. Climatol.* **14**, 357–69.

King, J. C. and P. S. Anderson (1988), Installation and performance of the STABLE instrumentation at Halley, *BAS Bull.* **79**, 65–77.

King, J. C. and P. S. Anderson (1994), Heat and water vapour fluxes and scalar roughness lengths over an Antarctic ice shelf, *Bound.-Layer Meteorol.* **69**, 101–21.

King, J. C., P. S. Anderson, M. C. Smith and S. D. Mobbs (1995), Surface energy and water balance over an Antarctic ice shelf in winter. In: *Preprints, Fourth Conference on Polar Meteorology and Oceanography*, American Meteorological Society, Boston, pp. 79–81.

King, J. C., S. D. Mobbs, M. S. Darby and J. M. Rees (1987), Observations of an internal gravity wave in the lower troposphere at Halley, Antarctica, *Bound.-Layer Meteorol.* **39**, 1–13.

Kobayashi, D. (1978), Snow transport by katabatic winds in Mizuho Camp area, East Antarctica, *J. Meteorol. Soc. Japan*. **56**, 130–9.

Kobayashi, S. (1985), Annual precipitation estimated by blowing snow observations at Mizuho Station, East Antarctica, 1980. In: *Proceedings of the Seventh Symposium on Polar Meteorology and Glaciology*, S. Kawaguchi, editor, National Institute of Polar Research, Tokyo, pp. 117–22.

Kobayashi, S., N. Ishikawa, T. Ohata and S. Kawaguchi (1983), Observations of an atmospheric boundary layer at Mizuho Station using an acoustic sounder. In: *Proceedings of the 5th NIPR Symposium on Polar Meteorology and Glaciology*, K. Kusunoki, editor, National Institute of Polar Research, Tokyo, pp. 37–49.

Kodama, Y. and G. Wendler (1986), Wind and temperature regima along the slope of Adélie Land, Antarctica, *J. Geophys. Res.* **91**, 6735–41.

Kondo, J. and H. Yamazawa (1986), Measurement of snow surface emissivity, *Bound.-Layer Meteorol.* **34**, 415–6.

König, G. (1985), Roughness length of an Antarctic ice shelf, *Polarforschung* **55**, 27–32.

König-Langlo, G. and E. Augstein (1994), Parameterization of the downward long-wave radiation at the Earth's surface in polar regions, *Meteorol. Zeitschrift*. **3**, 343–7.

Kottmeier, C. (1986), The influence of baroclinicity and stability on the wind and temperature conditions at the Georg von Neumayer antarctic station, *Tellus* A **38**, 263–76.

Kottmeier, C. and D. Engelbart (1992), Generation and atmospheric heat exchange of coastal polynyas in the Weddell Sea, *Bound.-Layer Meteorol.* **60**, 207–34.

Kottmeier, C. and R. Hartig (1990), Winter observations of the atmosphere over Antarctic sea ice, *J. Geophys. Res.* **95**, 16551–60.

Kottmeier, C., J. Olf, W. Frieden and R. Roth (1992), Wind forcing and ice motion in the Weddell Sea region, *J. Geophys. Res.* **97**, 20373–83.

Krishnamurti, T. N., S.-H. Chu and W. Iglesias (1986), On the sea level pressure of the Southern Oscillation, *Arch. für Met., Geophys. und Biokl.* A **34**, 385–425.

Kuhn, M. (1969), *Preliminary Report on Meteorological Studies at Plateau Station, Antarctica*, Meteorology Department, University of Melbourne, Melbourne, 20 pp.

Kuhn, M., L. S. Kundla and L. A. Stroschein (1977a), The radiation budget at Plateau Station, Antarctica. In: *Meteorological Studies at Plateau Station*, Antarctic Research Series, Vol 25, J. A. Businger, editor, American Geophysical Union, Washington, pp. 41–73.

Kuhn, M., H. H. Lettau and A. J. Riordan (1977b), Stability-related wind spiralling in the lowest 32 meters, American Geophysical Union *Antarctic Research Series*, Vol 25, J. A. Businger, editor, American Geophysical Union Washington, D.C., pp. 93–111.

Kuhn, P. M., L. P. Stearns and J. R. Stremikis (1967), *Atmospheric Infrared Radiation over the Antarctic (ESSA Technical Report IER 55-IAS 2)*, USA Department of Commerce, Boulder, Colorado, pp. 16+appendices.

Kukla, G. and P. Robinson (1980), Annual cycle of surface albedo, *Mon. Wea. Rev.* **108**, 56–68.

Kurtz, D. D. and D. H. Bromwich (1983), Satellite observed behaviour of the Terra Nova Bay polynya, *J. Geophys. Res.* **88**, 9717–22.

Kurtz, D. D. and D. H. Bromwich (1985), A recurring, atmospherically forced polynya in Terra Nova Bay. In: *Oceanology of the Antarctic Continental Shelf (Antarctic Research Series, Volume 43)*, S. S. Jacobs, editor, American Geophysical Union, Washington, pp. 177–201.

Lachlan-Cope, T. A. (1992), The use of a simultaneous physical retrieval scheme for satellite derived atmospheric temperatures: Weddell Sea, Antarctica, *Int. J. Remote Sensing* **13**, 141–54.

Lachlan-Cope, T. A., J. Turner, J. P. Thomas and E. A. Rasmussen (1994), High latitude mesoscale atmospheric circulation features observed with the ERS-1 scatterometer. In: *Proc 2nd ERS-1 Symposium, Hamburg, Germany, 11–14 October 1993*, ESA, Noordwijk, pp. 809–14.

Lamb, H. H. (1959), The southern westerlies: a preliminary survey, main characteristics and apparent associations, *Quart. J. Roy. Met. Soc.* **85**, 1–23.

Lamb, H. H. (1982), The climate environment of the Arctic Ocean. In: *The Arctic Ocean*, L. Rey, editor, John Wiley and Sons, New York, pp. 135–61.

Lamb, H. H. and G. P. Britton (1955), General atmospheric circulation and weather variations in the Antarctic, *Geog. J.* **121**, 334–49.

Lange, M., S. F. Ackley, G. S. Dieckmann, H. Eicken and P. Wadhams (1989), Development of sea ice in the Weddell Sea, Antarctica, *Ann. Glaciol.* **12**, 92–6.

Lange, M. A. and H. Eicken (1991), The sea ice thickness distribution in the northwestern Weddell Sea, *J. Geophys. Res.* **96**, 4821–37.

Large, W. G. and H. van Loon (1989), Large-scale low frequency variability of the 1979 FGGE surface buoy drifts and winds over the Southern Hemisphere, *J. Phys. Ocean.* **19**, 216–32.

Launiainen, J. and T. Vihma (1994), On the surface heat fluxes in the Weddell Sea. In: *The Polar Oceans and Their Role in Shaping the Global Environment*, O. M. Johannessen, R. D. Muench and J. E. Overland, editors, American Geophysical Union, Washington, DC, pp. 399–419.

Le Marshall, J. F. and G. A. M. Kelly (1981), A January and July climatology of the Southern Hemisphere based on daily numerical analyses 1973–77, *Aust. Met. Mag.* **29**, 115–23.

Lied, N. T. (1964), Stationary hydraulic jumps in a katabatic flow near Davis, Antarctica, 1961, *Aust. Met. Mag.* **47**, 40–51.

Liljequist, G. H. (1957), *Energy Exchange of an Antarctic Snow Field. Norwegian–British–Swedish Antarctic Expedition 1949–52*, Vol. II, Part 1, Norsk Polarinstitutt, Oslo, 298 pp.

Limbert, D. W. S. (1963), The snow accumulation budget at Halley Bay in 1959, and associated meteorological factors, *BAS Bull.* **2**, 73–92.

Limbert, D. W. S., S. J. Morrison, C. B. Sear, P. Wadhams and M. A. Rowe (1989), Pack-ice motion in the Weddell sea in relation to weather systems and determination of a Weddell Sea sea-ice budget, *Ann. Glaciol.* **12**, 104–12.

Liu, Z. and D. H. Bromwich (1993), Acoustic remote sensing of planetary boundary layer dynamics near Ross Island, Antarctica, *J. Appl. Met.* **32**, 1867–82.

Loewe, F. (1962), On the mass economy of the interior of the Antarctic ice cap, *J. Geophys. Res.* **67**, 5171–7.

Loewe, F. (1970), The transport of snow on ice sheets by the wind. In: *Studies on Drifting Snow. Publ. 13*, Meteorology Department, University of Melbourne, Melbourne.

Lutz, H. J., W. L. Smith and E. Raschke (1990), A note on the improvement of TIROS operational vertical sounder temperature retrievals above the Antarctic snow and ice fields, *J. Geophys. Res.* **95**, 11747–54.

Lutz, H.-J. and W. L. Smith (1988), TOVS over polar regions. In: *The Technical Proceedings of the 4th International TOVS Study Conference, Igls, Austria*, W. P. Menzel, editor, Cooperative Institute for Meteorological Satellite Studies, University of Wisconsin, Madison, pp. 168–81.

Lyons, S. W. (1983), Characteristics of intense Antarctic depressions near the Drake Passage. In: *Preprints, First International Conference Southern Hemisphere Meteorology, 31 July–6 August 1983*, AMS, Boston, pp. 238–40.

Lystad, M., editor (1986), *Polar Lows in the Norwegian, Greenland and Barents Seas. Final Report, Polar Lows Project*, Norwegian Meteorological Institute, Oslo, Norway, 196 pp.

Mahrt, L. and S. Larsen (1982), Small scale drainage front, *Tellus* **34**, 579–87.

Mahrt, L. and W. Schwerdtfeger (1970), Ekman spirals for exponential thermal wind, *Bound.-Layer Meteorol.* **1**, 137–45.

Manabe, S. and D. Hahn (1981), Simulation of atmospheric variability, *Mon. Wea. Rev.* **109**, 2260–86.

Manins, P. C. and B. L. Sawford (1979), A model of katabatic winds, *J. Atmos. Sci.* **36**, 619–30.

Mansfield, D. A. (1974), Polar lows: the development of baroclinic disturbances in cold air outbreaks, *Quart. J. Roy. Met. Soc.* **100**, 541–54.

Martin, P. J. and D. A. Peel (1978), The spatial distribution of 10 m temperatures in the Antarctic Peninsula, *J. Glaciol.* **20**, 311–7.

Massom, R. (1991), *Satellite Remote Sensing of Polar Regions*, Lewis Publishers, Florida, 307 pp.

Masuda, K. (1990), Atmospheric heat and water budgets of polar regions: analysis of FGGE data. In: *Proceedings of the Third NIPR Symposium on Polar Meteorology and Glaciology*, National Institute of Polar Research, Tokyo, pp. 79–88.

Masuda, K., T. Takashima and Y. Takayama (1988), Emissivity of pure and sea waters for the model sea surface in the infrared window regions, *Rem. Sens. Environ.* **24**, 313–29.

Mather, K. B. and G. S. Miller (1967), *Notes on Topographic Factors Affecting the Surface Wind in Antarctica, with Special Reference to Katabatic winds; and Bibliography*, University of Alaska, Fairbanks, 63 pp. (Technical Report, Grant no. GA–900)

Mayes, P. R. (1985), Secular variations in cyclone frequencies near the Drake Passage, southwest Atlantic, *J. Geophys. Res.* **90**, 5829–39.

Maykut, G. A. (1978), Energy exchange over young sea ice in the central Arctic, *J. Geophys. Res.* **83**, 3646–58.

McClain, E. P., W. G. Pichel, C. C. Walton, Z. Ahmad and J. Sutton (1983), Multi-channel improvements to satellite derived global sea surface temperatures, *Adv. Space Res.* **2**, 43–7.

Mechoso, C. R. (1980), The atmospheric circulation around Antarctica: linear stability and finite amplitude interactions with migrating cyclones, *J. Atmos. Sci.* **37**, 2209–33.

Meinardus, W. and L. Mecking (1911), Tägliche synoptische Wetterkarten der höheren südlichen Breiten von Oktober 1901 bis März

1904. In: *Deutsche Südpolar-Expedition 1901–1903, Meteorologischer Atlas*, Berlin.

Meteorological Office (1962), *A Course in Elementary Meteorology*, HMSO, London, Met O 707, 189 pp.

Meteorological Office (1982), *Observer's Handbook*, HMSO, London, 220 pp.

Meteorological Office (1995), *Marine Observer's Handbook*, HMSO, London, 227 pp.

Mitchell, J. F. B. and C. A. Senior (1989), The antarctic winter; simulations with climatological and reduced sea-ice extents, *Quart. J. Roy. Met. Soc.* **115**, 225–46.

Mitchell, J. F. B. and T. S. Hills (1986), Sea-ice and the antarctic winter circulation: a numerical experiment, *Quart. J. Roy. Met. Soc.* **112**, 953–69.

Mitchell, J. F. B., S. Manabe, V. Meleshko and T. Tokioka (1990), Equilibrium climate change – and its implications for the future. In: *Climate Change. The IPCC Scientific Assessment*, J. T. Houghton, G. J. Jenkins and J. J. Ephraums, editors, Cambridge University Press, Cambridge, pp. 131–72.

Mo, K. C., J. Pfaendtner and E. Kalnay (1987), A GCM study on the maintenance of the June 1982 blocking in the southern hemisphere, *J. Atmos. Sci.* **44**, 1123–42.

Mo, K. C. and G. H. White (1985), Teleconnections in the Southern Hemisphere, *Mon. Wea. Rev.* **113**, 22–37.

Mobbs, S. D. and J. M. Rees (1989), Studies of atmospheric internal gravity waves at Halley Station, Antarctica, using radiosondes, *Antarctic Sci.* **1**, 65–75.

Morgan, V. I., I. D. Goodwin, D. M. Etheridge and C. W. Wookey (1991), Evidence from Antarctic ice cores for recent increases in snow accumulation, *Nature*. **354**, 58–60.

Morley, B. M., E. E. Uthe and W. Viezee (1989), Airborne lidar observations of clouds in the Antarctic troposphere, *Geophys. Res. Lett.* **16**, 491–4.

Morris, E. M. and D. Vaughan (1994), Snow surface temperatures in West Antarctica, *Antarctic Sci.* **6**, 529–35.

Morrison, S. J. (1990), Warmest year on record in the Antarctic Peninsula?, *Weather*. **45**, 231–2.

Mosley-Thompson, E. (1992), Paleoenvironmental conditions in Antarctica since A.D. 1500: ice core evidence. In: *Climate since A.D. 1500*, R. S. Bradley and P. D. Jones, editors, Routledge, London, pp. 572–91.

Motoi, T., N. Ono and M. Wakatsuchi (1987), A mechanism for the formation of the Weddell Polynya in 1974, *J. Phys. Ocean.* **17**, 2241–7.

Münzenberg, A., R. Engels and H.-D. Schilling (1992), Numerical studies of the formation of Antarctic mesoscale vortices in the Weddell Sea region. In: *Applications of New Forms of Satellite Data in Polar Low Research. Proceedings of the Polar Low Workshop Hvanneyri, Iceland 23–26 June 1992*, J. Turner and E. Rasmussen, editors, EGS Polar Low Working Group, Cambridge, pp. 33–9.

Murphy, B. F. and I. Simmonds (1993), An analysis of strong wind events simulated in a GCM near Casey in the Antarctic, *Mon. Wea. Rev.* **121**, 522–34.

Murray, R. J. and I. Simmonds (1991a), A numerical scheme for tracking cyclone centres from digital data. Part I: development and operation of the scheme, *Aust. Met. Mag.* **39**, 155–66.

Murray, R. J. and I. Simmonds (1991b), A numerical scheme for tracking cyclone centres from digital data. Part II: application to January and July general circulation model simulations, *Aust. Met. Mag.* **39**, 167–80.

Muszynski, I. and B. E. Birchfield (1985), The dependence of Antarctic accumulation rates on surface temperature and elevation, *Tellus* A **37**, 204–8.

Nakamura, N. and A. H. Oort (1988), Atmospheric heat budgets of the polar regions, *J. Geophys. Res.* **93**, 9510–24.

Naval Oceanography Command Detachment (1985) *Sea Ice Climatic Atlas*: Asheville, NC, Naval Oceanography Command Detachment, v. 1. Antarctica, NAVAIR 50-IC-540.

Neal, A. B. (1972), Cyclones and anticyclones in November 1969 and June 1970, *Aust. Met. Mag.* **20**, 217–30.

Neff, W. D. (1980) An observational and numerical study of the atmospheric boundary layer overlying the East Antarctic ice sheet. Ph.D. Thesis, University of Colorado, 272 pp.

Neff, W. D. (1992), On the influence of stratospheric stability on lower tropospheric circulations over the South Pole. In: *Preprints, Third Conference on Polar Meteorology and Oceanography*, American Meteorological Society, Boston, pp. 115–20.

Newton, C. W. and E. O. Holopainen, eds (1990), *Extratropical Cyclones: The Erik Palmén Memorial Volume*. American Meteorological Society, Boston, 262 pp.

Nicholls, K. W. and J. G. Paren (1993), Extending the Antarctic meteorological record using ice-sheet temperature profiles, *J. Climate*. **6**, 141–50.

Nowlin, W. D. J. and J. M. Klinck (1986), The physics of the Antarctic Circumpolar Current, *Rev. Geophys.* **24**, 469–91.

O'Connor, W. P. and D. H. Bromwich (1988), Surface airflow around Windless Bight, Ross Island, Antarctica, *Quart. J. Roy. Met. Soc.* **114**, 917–38.

O'Connor, W. P., D. H. Bromwich and J. F. Carrasco (1994), Cyclonically forced barrier winds along the Transantarctic Mountains near Ross Island, *Mon. Wea. Rev.* **122**, 137–50.

Offiler, D. (1990), Wind fields and surface fluxes. In: *Microwave Remote Sensing for Oceanographic and Marine Weather Forecast Models*, R. A. Vaughan, editor, Kluwer Academic Publishers, pp. 355–74.

Ohata, T., N. Ishikawa, S. Kobayashi and S. Kawaguchi (1985a), Heat balance at the snow surface in a katabatic wind zone, East Antarctica, *Ann. Glaciol.* **6**, 174–7.

Ohata, T., S. Kobayashi, N. Ishikawa and S. Kawaguchi (1985b), Structure of the katabatic winds at Mizuho Station, East Antarctica, *J. Geophys. Res.* **90**, 10651–58.

Orlanski, I. (1975), A rational subdivision of scales for atmospheric processes, *Bull. Amer. Met. Soc.* **56**, 527–30.

Overland, J. E. (1984), Scale analysis of marine winds in straits and along mountainous coasts, *Mon. Wea. Rev.* **112**, 2532–6.

Parish, T. R. (1977) A low-level jet of cold air near the tip of the Antarctic Peninsula. An example of inertial winds. M.S. Thesis, University of Wisconsin-Madison, 47 pp.

Parish, T. R. (1980) Surface winds in East Antarctica. Ph.D. Thesis, University of Wisconsin-Madison, 121 pp.

Parish, T. R. (1981), The katabatic winds of Cape Denison and Port Martin, *Polar Rec.* **20**, 525–32.

Parish, T. R. (1982), Surface airflow over East Antarctica, *Mon. Wea. Rev.* **110**, 84–90.

Parish, T. R. (1983), The influence of the Antarctic Peninsula on the wind field over the western Weddell Sea, *J. Geophys. Res.* **88**, 2684–92.

Parish, T. R. (1984), A numerical study of strong katabatic winds over Antarctica, *Mon. Wea. Rev.* **112**, 545–54.

Parish, T. R. (1992), On the role of Antarctic katabatic winds in forcing large-scale tropospheric motions, *J. Atmos. Sci.* **49**, 1374–85.

Parish, T. R. and D. H. Bromwich (1986), The inversion wind pattern over West Antarctica, *Mon. Wea. Rev.* **114**, 849–60.

Parish, T. R. and D. H. Bromwich (1987), The surface wind field over the Antarctic ice sheets, *Nature* **328**, 51–4.

Parish, T. R. and D. H. Bromwich (1989), Instrumented aircraft observations of the katabatic wind regime near Terra Nova Bay, *Mon. Wea. Rev.* **117**, 1570–85.

Parish, T. R. and D. H. Bromwich (1991), Continental-scale simulation of the Antarctic katabatic wind regime, *J. Climate*. **4**, 135–46.

Parish, T. R. and G. Wendler (1991), The katabatic wind regime at Adélie Land, Antarctica, *Int. J. Climatol.* **11**, 97–107.

Parish, T. R. and K. T. Waight (1987), The forcing of antarctic katabatic winds, *Mon. Wea. Rev.* **115**, 2214–26.

Parish, T. R., D. H. Bromwich and R.-Y. Tzeng (1994), On the role of the Antarctic continent in forcing large-scale circulations in high southern latitudes, *J. Atmos. Sci.* **51**, 3566–79.

Parish, T. R., P. Pettré and G. Wendler (1992), On the interaction between the katabatic wind regime and large-scale tropospheric forcing near Adélie Land, Antarctica. In: *Preprints, Third Conference on Polar Meteorology and Oceanography*, American Meteorological Society, Boston, pp. (J6)141–4.

Parker, D. E. (1983), Documentation of a Southern Oscillation index, *Met. Mag.* **112**, 184–8.

Parkinson, C. L. (1992), Interannual variability of monthly southern ocean sea ice distributions, *J. Geophys. Res.* **97**, 5349–63.

Parkinson, C. L. and D. J. Cavalieri (1982), Interannual sea-ice variations and sea-ice/atmosphere interactions in the Southern Ocean, *Ann. Glaciol.* **3**, 249–54.

Peel, D. A. (1992a), Spatial temperature and accumulation rate variations in the Antarctic Peninsula. In: *The Contribution of Antarctic Peninsula Ice to Sea Level Rise*, E. M. Morris, editor, British Antarctic Survey, Cambridge, pp. 11–15.

Peel, D. A. (1992b), Ice core evidence from the Antarctic Peninsula region. In: *Climate since A.D. 1500*, R. S. Bradley and P. D. Jones, editors, Routledge, London, pp. 549–71.

Pendlebury, S. and G. Reader (1993), Case study of numerical prediction of an extreme wind event in Antarctica. In: *Preprints, Fourth International Conference on Southern Hemisphere Meteorology and Oceanography. March 29–April 2, 1993. Hobart, Australia*, American Meteorological Society, Boston, pp. 124–5.

Périard, C. and P. Pettré (1993), Some aspects of the climatology of Dumont d'Urville, Adélie Land, Antarctica, *Int. J. Climatol.* **13**, 313–27.

Petit, J. R., J. Jouzel, M. Pourchet and L. Merlivat (1982), A detailed study of snow accumulation and stable isotope content in Dome C (Antarctica), *J. Geophys. Res.* **87**, 4301–8.

Pettré, P. and J.-C. André (1991), Surface-pressure change through Loewe's phenomena and katabatic flow jumps: study of two cases in Adélie Land, Antarctica, *J. Atmos. Sci.* **48**, 557–71.

Pettré, P., C. Payan and T. R. Parish (1993), Interaction of katabatic flow with local thermal effects in a coastal region of Adélie Land, East Antarctica, *J. Geophys. Res.* **98**, 10429–40.

Pettré, P., M. F. Renaud, M. Déqué, S. Planton and J. C. André (1990), Study of the influence of katabatic flows on the antarctic circulation using GCM simulations, *Meteorol. Atmos. Phys.* **43**, 187–95.

Petty, G. W. (1990) On the response of the Special Sensor Microwave/Imager to the marine environment. Implications for atmospheric parameters retrievals. Ph.D. Dissertation, University of Washington, 291 pp.

Petty, G. W. and K. B. Katsaros (1990), Precipitation observed over the South China Sea by the Nimbus-7 scanning multichannel microwave radiometer during winter MONEX, *J. Appl. Met.* **29**, 273–87.

Philander, S. G. and E. M. Rasmussen (1985), The Southern Oscillation and El Niño, *Adv. in Geophys.* A **28**, 197–215.

Phillpot, H. R. (1968), *A Study of the Synoptic Climatology of the Antarctic. Technical Report 12*, International Antarctic Meteorology Research Centre, Melbourne, 139 pp.

Phillpot, H. R. (1991), The derivation of 500 hPa height from automatic weather station surface observations in the Antarctic continental interior, *Aust. Met. Mag.* **39**, 79–86.

Phillpot, H. R. and J. W. Zillman (1970), The surface temperature inversion over the Antarctic continent, *J. Geophys. Res.* **75**, 4161–9.

Physick, W. L. (1981), Winter depression tracks and climatological jet streams in the southern hemisphere during the FGGE year, *Quart. J. Roy. Met. Soc.* **107**, 883–98.

Pittock, A. B. (1984), On the reality, stability and usefulness of Southern Hemisphere teleconnections, *Aust. Met. Mag.* **32**, 75–82.

Rabbe, A. (1987), A polar low over the Norwegian Sea, 29 February–1 March 1984, *Tellus* A **39**, 326–33.

Radok, U. and R. C. Lile (1977), A year of snow accumulation at Plateau Station. In: *Meteorological Studies at Plateau Station, Antarctica. Antarctic Research Series*, Vol. 25, J. A. Businger, editor, American Geophysical Union, Washington, DC, pp. 17–26.

Radok, U., D. Jenssen and B. McInnes (1987), *On the Surging Potential of Polar Ice Streams*, USA Department of Commerce, Springfield, Report DOE/ER/60197-H1, 62 pp.

Radok, U., N. Streten, and G. E. Weller (1975), Atmosphere and ice, *Oceanus* **18**, 16–27.

Raper, S. C. B., T. M. L. Wigley, P. R. Mayes, P. D. Jones and M. J. Salinger (1984), Variations in surface air temperatures: part 3. The Antarctic, 1957–1982, *Mon. Wea. Rev.* **112**, 1341–53.

Raschke, E., editor (1987), *Report of the International Satellite Cloud Climatology Project (ISCCP) Workshop on Cloud Algorithms in the Polar Regions*, World Climate Programme, Geneva, 12 pp.

Rasmussen, E. (1985), A case study of a polar low development over the Barents Sea, *Tellus* A **37**, 407–18.

Rasmussen, E. A., T. S. Pedersen, L. T. Pedersen and J. Turner (1992), Polar lows and arctic instability lows in the Bear Island region, *Tellus* A **44**, 133–54.

Rasmussen, E. M. and T. H. Carpenter (1982), Variations in tropical sea surface temperature and surface wind fields associated with the Southern Oscillation/El Niño, *Mon. Wea. Rev.* **110**, 354–84.

Reed, R. J. (1979), Cyclogenesis in polar airstreams, *Mon. Wea. Rev.* **107**, 38–52.

Reed, R. J. and M. D. Albright (1986), A case study of explosive cyclogenesis in the eastern Pacific, *Mon. Wea. Rev.* **114**, 2297–319.

Reed, R. J. and W. Blier (1986), A case study of a comma cloud development in the Eastern Pacific, *Mon. Wea. Rev.* **114**, 1681–95.

Rees, J. M. and Rottman,J. W. (1994), Analysis of solitary disturbances over an Antarctic ice shelf, *Bound.-Layer Meteorol.* **69**, 285–310.

Rees, J. M. and S. D. Mobbs (1988), Studies of internal gravity waves at Halley Base, Antarctica using wind observations, *Quart. J. Roy. Met. Soc.* **114**, 939–66.

Rees, J. M., I. McConnel, P. S. Anderson and J. C. King (1994), Observations of internal gravity waves over an Antarctic ice shelf using a microbarograph array. In: *Stably Stratified Flows. Flow and Dispersion over Topography*, I. P. Castro and N. J. Rockliff, editors, Oxford University Press, Oxford, pp. 61–79.

Rees, W. G. (1993), Infrared emissivity of Arctic winter snow, *Int. J. Remote Sensing*. **14**, 3069–73.

Reynolds, J. M. (1981), The distribution of mean annual temperatures in the Antarctic Peninsula, *BAS Bull.* **54**, 123–33.

Robin, G de Q. (1977), Ice cores and climatic change, *Phil. Trans. R. Soc. Lond.* Series B **280**, 143–68.

Rockey, C. C. and D. A. Braaten (1995), Characterization of polar cyclonic activity and relationship to observed snowfall events at McMurdo Station, Antarctica. In: *Preprints, Fourth Conference on Polar Meteorology and Oceanography, January 15–20 1995, Dallas, Texas*, AMS, Boston, pp. 244–5.

Rogers, J. C. and H. van Loon (1982), Spatial variability of sea level pressure and 500 mb height anomalies over the Southern Hemisphere, *Mon. Wea. Rev.* **110**, 1375–92.

Rossow, W. B. (1987), Application of ISCCP cloud algorithm to satellite observations of the polar regions. In: *Report of the International Satellite Cloud Climatology Project (ISCCP) Workshop on Cloud Algorithms in the Polar Regions*, E. Raschke, editor, World Climate Programme, Geneva, pp. Appendix c.7 1–3.

Rossow, W. B. (1993), Clouds. In: *Atlas of Satellite Observations Related to Global Change*, R. J. Gurney, J. L. Foster and C. L. Parkinson, Cambridge University Press, Cambridge, pp. 141–63.

Rotman, S. R., A. D. Fisher and D. H. Staelin (1982), Inversion for physical characteristics of snow using passive radiometric observations, *J. Glaciol.* **28**, 179–85.

Rubin, M. J. and M. B. Giovinetto (1962), Snow accumulation in central West Antarctica as related to atmospheric and topographic factors, *J. Geophys. Res.* **67**, 5163–70.

Rusin, N. P. (1961), *Meteorological and Radiational Regime of Antarctica*, Israel Program for Scientific Translations, Jerusalem, 355 pp.

Sansom, J. (1989), Antarctic surface temperature time series, *J. Climate* **2**, 1164–72.

Sato, N., K. Kikuchi, S. C. Barnard and A. W. Hogan (1981), Some characteristic properties of ice crystal precipitation in the summer season at South Pole Station, Antarctica, *J. Met. Soc. Jap.* **59**, 772–80.

Satow, K. (1985), Variability of surface mass balance in the Mizuho Plateau, Antarctica. In: *Proceedings of the Seventh Symposium on Polar Meteorology and Glaciology*, S. Kawaguchi, editor, National Institute of Polar Research, Tokyo, pp. 132–40.

Saunders, R. W. (1986), An automated scheme for the removal of cloud contamination from AVHRR radiances over Western Europe, *Int. J. Remote Sensing* **7**, 867–86.

Savage, M. L., C. R. Stearns and D. Fleming (1985), *Antarctic Automatic Weather Station Data for the Calendar Year 1980*, Department of Meteorology, University of Wisconsin, Madison, 72 pp.

Savage, M. L., C. R. Stearns and G. A. Weidner (1988), The Southern Oscillation signal in Antarctica. In: *Preprint Volume Second Conference on Polar Meteorology and Oceanography, Madison, Wisconsin, March 29–31 1988*, AMS, Boston, pp. 141–4.

Schiffer, R. A. and W. B. Rossow (1983), The International Satellite Cloud Climatology Project (ISCCP), *Bull. Amer. Met. Soc.* **64**, 779–84.

Schmitt, W. (1957), Synoptic meteorology of the Antarctic. In: *Meteorology of the Antarctic*, M. P. Van Rooy, editor, Weather Bureau, Department of Transport, Pretoria, pp. 209–31.

Schnack-Schiel, S. (1987), *The Winter Expedition of R. V. Polarstern to the Antarctic (ANT B/1–3). Berichte zur Polarforschung*, Vol. 39, Alfred-Wegener-Institut, Bremerhaven, 259 pp.

Schwalb, A. (1978), *NOAA Technical Memorandum*, NESS 95: *The TIROS-N/NOAA A-G Satellite Series*, NOAA, Washington, DC.

Schwalb, A. (1982), *NOAA Technical Memorandum*, NESS 116: *Modified Version of the TIROS-N/NOAA A-G Satellite Series (NOAA E-J) – Advanced TIROS-N (ATN)*, NOAA, Washington, DC, 23 pp.

Schwerdtfeger, W. (1970), The climate of the Antarctic. In: *Climates of the Polar Regions*, S. Orvig, editor, Elsevier, New York, pp. 253–355.

Schwerdtfeger, W. (1976), Changes of temperature field and ice conditions in the area of the Antarctic Peninsula, *Mon. Wea. Rev.* **104**, 1441–3.

Schwerdtfeger, W. (1979), Meteorological aspects of the drift of ice from the Weddell Sea towards the mid-latitude westerlies, *J. Geophys. Res.* **84**, 6321–8.

Schwerdtfeger, W. (1984), *Weather and Climate of the Antarctic*, Elsevier, Amsterdam, 261 pp.

Schwerdtfeger, W., L. M. de la Canal and J. Scholten (1959), *Meteorologia descriptiva del sector Antartico Sudamericano*, Instituto Antartico Argentino, Buenos Aires, 425 pp.

Scorer, R. S. (1986), *Cloud Investigation by Satellite*, Ellis Horwood, Chichester, 316 pp.

Shapiro, M. A., L. S. Fedor and T. Hampel (1987), Research aircraft measurements of a polar low over the Norwegian Sea, *Tellus* A **39**, 272–306.

Shearman, R. J., 1983: The Meteorological Office Main Marine Data Bank. *Marine Observer* **53**, 208–17.

Simmonds, I. (1990), Improvements in general circulation model performance in simulating Antarctic climate, *Antarctic Sci.* **2**, 287–300.

Simmonds, I. and M. Dix (1986), The circulation changes induced by the removal of Antarctic sea ice in a July general circulation model. In: *Preprints Volume, Second International Conference on Southern Hemisphere Meteorology, Wellington, New Zealand December 1–5 1986*, AMS, Boston, pp. 107–10.

Simmonds, I. and R. Law (1995), Associations between Antarctic katabatic flow and the upper level winter vortex, *Int. J. Climatol.* **15**, 403–21.

Simmonds, I. and W. F. Budd (1991), Sensitivity of the southern hemisphere circulation to leads in the Antarctic pack ice, *Quart. J. Roy. Met. Soc.* **117**, 1003–24.

Sinclair, M. R. (1981), Record-high temperatures in the Antarctic – a synoptic case study, *Mon. Wea. Rev.* **109**, 2234–42.

Sinclair, M. R. (1986), Investigation of polar air stream cyclogenesis in the New Zealand area using quasi-Lagrangian diagnostics. In: *Preprints, Second International Conference on Southern Hemisphere Meteorology, 1–5 December 1986*, American Meteorological Society, Boston, pp. 50–1.

Sinclair, M. R. (1993), Cyclone tracks for the Southern Hemisphere derived from geostrophic relative vorticity. In: *Preprints, Fourth International Conference on Southern Hemisphere Meteorology and Oceanography. March 29–April 2, 1993. Hobart, Australia*, American Meteorological Society, Boston, pp. 72–3.

Sinclair, M. R. (1994), An objective cyclone climatology for the Southern Hemisphere, *Mon. Wea. Rev.* **122**, 2239–56.

Sinclair, M. R. (1995), A climatology of cyclogenesis for the Southern Hemisphere, *Mon. Wea. Rev.* **123**, 1601–19.

Slutz, R. J., S. J. Lubker, J. D. Hiscox, S. D. Woodruff, R. L. Jenne, D. H. Joseph, P. M. Steurer, and J. D. Elms, 1985, Comprehensive Ocean–Atmosphere Data Set (COADS), Release 1. Cooperative Institute for Research in Environmental Sciences (CIRES), Environmental Research Laboratories (ERL), National Center for Atmospheric Research (NCAR) and National Climatic Data Center (NCDC): Boulder, CIRES, 267 pp.

Smiley, V. N., B. M. Whitcomb, B. M. Morley and J. A. Warburton (1980), Lidar determination of atmospheric ice crystal layers at south pole during clear-sky precipitation, *J. Appl. Met.* **19**, 1074–90.

Smith, S. R. and C. R. Stearns (1993a), Antarctic climate anomalies surrounding the minimum in the southern oscillation index. In: *Antarctic Meteorology and Climatology: Studies Based on Automatic Weather Stations*, Antarctic Research Series, Volume 61, D. H. Bromwich and C. R. Stearns, editors, American Geophysical Union, Washington, pp. 149–74.

Smith, S. R. and C. R. Stearns (1993b), Antarctic pressure and temperature anomalies surrounding the minimum in the Southern Oscillation index, *J. Geophys. Res.* **98**, 13071–83.

Smith, S., R. D. Muench and C. Pease (1990), Polynyas and leads: an overview of physical processes and environment, *J. Geophys. Res.* **95**, 9461–79.

Smith, W. L., H. M. Woolf, C. M. Hayden and A. J. Schreiner (1985), The simultaneous retrieval export package. In: *The Technical Proceedings of the Second International TOVS Study Conference*, W. P. Menzel, editor, Cooperative Institute for Meteorological Satellite Studies, University of Wisconsin, Madison, pp. 224–53.

Solopov, A. V. (1969), *Oases in Antarctica*, Israel Program for Scientific Translations, Jerusalem, 146 pp.

Sorbjan, Z., Y. Kodama and G. Wendler (1986), Observational study of the atmospheric boundary layer over Antarctica, *J. Clim. Appl. Met.* **25**, 641–51.

Splettstoesser, J. (1992), Antarctic global warming?, *Nature* **355**, 503.

Stark, P. S. (1994), Climatic warming in the central Antarctic Peninsula area, *Weather* **49**, 215–20.

Stearns, C. R. and G. A. Weidner (1993), Sensible and latent heat flux estimates in Antarctica. In: *Antarctic Meteorology and Climatology: Studies Based on Automatic Weather Stations, Antarctic Research Series, Volume 61*, D. H. Bromwich and C. R. Stearns, editors, American Geophysical Union, Washington, pp. 109–38.

Stearns, C. R. and G. Wendler (1988), Research results from Antarctic automatic weather stations, *Rev. Geophys.* **26**, 45–61.

Stearns, C. R., L. M. Keller, G. A. Weidner and M. Sievers (1993), Monthly mean climatic data for Antarctic automatic weather stations. In: *Antarctic Research Series*, Volume 61: *Antarctic Meteorology and Climatology: Studies Based on Automatic Weather Stations*, D. H. Bromwich and C. R. Stearns, editors, American Geophysical Union, Washington, pp. 1–21.

Stone, R. S. (1993), Properties of austral winter clouds derived from radiometric profiles at the South Pole, *J. Geophys. Res.* **98**, 12961–71.

Stouffer, R. J., S. Manabe and K. Bryan (1989), Interhemispheric asymmetry in climate response to a gradual increase in atmospheric CO_2, *Nature* **342**, 660–2.

Streten, N. A. (1968), Some features of mean annual windspeed data for coastal East Antarctica, *Polar Rec.* **14**, 315–22.

Streten, N. A. (1975), Satellite derived influences to some characteristics of the South Pacific atmospheric circulation associated with the Niño event of 1972–73, *Mon. Wea. Rev.* **103**, 989–95.

Streten, N. A. (1980a), Some synoptic indices of the Southern Hemisphere mean sea level circulation 1972–77, *Mon. Wea. Rev.* **108**, 18–36.

Streten, N. A. (1980b), Antarctic meteorology: the Australian contribution past, present and future, *Aust. Met. Mag.* **28**, 105–40.

Streten, N. A. and D. J. Pike (1980), Characteristics of the broadscale antarctic sea ice extent and the associated atmospheric circulation 1972–1977, *Arch. für Met., Geophys. und Biokl.* A **29**, 279–99.

Streten, N. A. and A. J. Troup (1973), A synoptic climatology of satellite observed cloud vortices over the Southern Hemisphere, *Quart. J. Roy. Met. Soc.* **99**, 56–72.

Streten, N. A. and J. W. Zillman (1984), Climates of the South Pacific. In: *Climates of the Oceans*, H. van Loon, editor, Elsevier, Amsterdam.

Swithinbank, C. (1973), Higher resolution satellite pictures, *Polar Rec.* **16**, 739–42.

Tabler, R. D. (1980), Self-similarity of wind profiles in blowing snow allows outdoor modeling, *J. Glaciol.* **6**, 71–5.

Takahashi, S. (1985a), Estimation of precipitation from drifting snow observations at Mizuho Station in 1982. In: *Memoires of the National Institute of Polar Research*, Special Issue 39, National Institute of Polar Research, Tokyo, pp. 123–31.

Takahashi, S. (1985b), Characteristics of drifting snow at Mizuho Station, Antarctica, *Ann. Glaciol.* **6**, 71–5.

Takahashi, S., R. Naruse, M. Nakawo and S. Mae (1988), A bare ice field in East Queen Maud Land, Antarctica, caused by horizontal divergence of drifting snow, *Ann. Glaciol.* **11**, 156–60.

Taljaard, J. J. (1965), Cyclogenesis, cyclones and anticyclones in the Southern Hemisphere during the period June to December 1958, *Notos* **14**, 73–84.

Taljaard, J. J. (1967), Development, distribution and movement of cyclones and anticyclones in the Southern Hemisphere during the IGY, *J. Appl. Met.* **6**, 973–87.

Taljaard, J. J., H. van Loon, H. L. Crutcher, and R. L. Jenne, 1969, *Temperatures, Dewpoints and Heights at Selected Pressure Levels, Climate of the Upper Air: Part 1. Southern Hemisphere*: Washington, D.C., Chief of Naval Operations, Volume 1 NAVAIR 50-1C-55, 135 pp.

Tanaka, S., T. Yoshino, T. Yamanouchi and S. Kawaguchi (1982), On the satellite remote measurements of vertical temperature profile of the Antarctic atmosphere. In: *Proceedings of the Fourth Symposium on Polar Meteorology and Glaciology*, K. Kusunoki, editor, National Institute of Polar Research, Tokyo, pp. 94–100.

Tchernia, P. (1980), *Descriptive Regional Oceanography*, Pergamon Press, Oxford, 253 pp.

Thompson, L. G. and E. Mosley-Thompson (1982), Spatial distribution of microparticles within snow-fall, *Ann. Glaciol.* **3**, 300.

Tournadre, J. and R. Ezraty (1990), Local climatology of wind and sea state by means of satellite radar altimeter measurements, *J. Geophys. Res.* **95**, 18255–68.

Trenberth, K. E. (1976), Fluctuations and trends in indices of the Southern Hemisphere circulation, *Quart. J. Roy. Met. Soc.* **102**, 65–75.

Trenberth, K. E. (1980), Planetary waves at 500 mb in the southern hemisphere, *Mon. Wea. Rev.* **108**, 1378–89.

Trenberth, K. E. (1981), Interannual variability of the Southern Hemisphere 500 mb flow: regional characteristics, *Mon. Wea. Rev.* **109**, 127–36.

Trenberth, K. E. and D. J. Shea (1987), On the evolution of the Southern Oscillation, *Mon. Wea. Rev.* **115**, 3078–96.

Trenberth, K. E. and J. G. Olson (1988), An evaluation and intercomparison of global analyses from NMC and ECMWF, *Bull. Amer. Meteor. Soc.* **69**, 1047–57.

Trenberth, K. E. and J. G. Olson (1989), Temperature records at the South Pole and McMurdo Sound, *J. Climate*. **2**, 1196–206.

Tsang, L. and J. A. Kong (1977), Theory for thermal microwave emission from a bounded medium containing spherical scatterers, *J. Applied Physics*. **48**, 3593–99.

Turner, J. and S. R. Colwell (1995), Temporal variability of precipitation over the western Antarctic Peninsula. In: *Preprints, Fourth Conference on Polar Meteorology and Oceanography*, American Meteorological Society, Boston, pp. 113–6.

Turner, J. and H. Ellrott (1992), High latitude moisture structure determined from HIRS water vapour imagery, *Int. J. Remote Sensing*. **13**, 81–95.

Turner, J., J. R. Eyre, D. Jerrett and E. McCallum (1985), The HERMES system, *Met. Mag.* **114**, 161–73.

Turner, J., T. A. Lachlan-Cope, D. E. Warren and C. N. Duncan (1993a), A mesoscale vortex over Halley Station, Antarctica, *Mon. Wea. Rev.* **121**, 1317–36.

Turner, J., T. A. Lachlan-Cope and J. P. Thomas (1993b), A comparison of Arctic and Antarctic mesoscale vortices, *J. Geophys. Res.* **98**, 13019–34.

Turner, J., T. A. Lachlan-Cope, J. P. Thomas and S. Colwell (1995), The synoptic origins of precipitation over the Antarctic Peninsula, *Antarctic. Sci.* **7**, 327–37.

Turner, J. and M. Row (1989), Mesoscale vortices in the British Antarctic Territory. In: *Polar and Arctic Lows*, P. F. Twitchell, E. Rasmussen and K. L. Davidson, editors, A Deepak, Hampton, Virginia, pp. 347–56.

Turner, J. and J. P. Thomas (1994), Summer-season mesoscale cyclones in the Bellingshausen–Weddell region of the Antarctic and links with the synoptic-scale environment, *Int. J. Climatol.* **14**, 871–94.

Turner, J. and D. E. Warren (1988a), Cloud track winds in the polar regions from sequences of AVHRR images, *Int. J. Remote Sensing*. **4**, 695–703.

Turner, J. and D. E. Warren (1988b), The structure of sub-synoptic-scale vortices in polar airstreams from AVHRR and TOVS data. In: *Proceedings of the Second Conference on Polar Meteorology and Oceanography March 29–31 1988 Madison, WI*, C. Stearns, editor, American Meteorological Society, Boston, MA.

Tzeng, R.-Y., D. H. Bromwich and T. R. Parish (1993), Present-day Antarctic climatology of the NCAR Community Climate Model Version 1, *J. Climate* **6**, 205–26.

Tzeng, R.-Y., D. H. Bromwich, T. R. Parish and B. Chen (1994), NCAR CCM2 simulation of the modern Antarctic climate, *J. Geophys. Res.* **99**, 23131–48.

UNESCO (1978), *World Water Balance and Water Resources of the Earth*, UNESCO Press, Paris.

Valtat, B. (1959), Loewe's Phenomenon. In: *Antarctic Meteorology*, Pergamon, London, pp. 67–70.

van den Broeke, M. and R. Bintanja (1995), The interaction of katabatic winds and the formation of blue ice areas in East Antarctica, *J. Glaciol.* **41**, 395–407.

van Loon, H. (1960), Features of the atmospheric circulation in the South Pacific Ocean during the whaling seasons 1955–1956 and 1956–1957. In: *Antarctic Meteorology*, Pergamon, London, pp. 274–80.

van Loon, H. (1967), The half-yearly oscillations in middle and high southern latitudes and the coreless winter, *J. Atmos. Sci.* **24**, 472–86.

van Loon, H. (1972), Temperature, pressure and wind in the Southern Hemisphere. In: *Meteorology of the Southern Hemisphere. Meteorological Monographs, Volume 18 no. 35*, C. W. Newton, editor, American Meteorological Society, Boston, pp. 25–100.

van Loon, H. (1979), The association between latitudinal temperature gradient and eddy transport. Part I: transport of sensible heat in winter, *Mon. Wea. Rev.* **107**, 525–34.

van Loon, H. (1984), The Southern Oscillation, part III: associations with the trades and with the trough in the westerlies of the South Pacific Ocean, *Mon. Wea. Rev.* **112**, 947–54.

van Loon, H. and J. C. Rogers (1981), Remarks on the circulation over the southern hemisphere in FGGE and on its relation to the phases of the Southern Oscillation, *Mon. Wea. Rev.* **109**, 2255–59.

van Loon, H. and D. J. Shea (1985), The Southern Oscillation part IV: the precursors south of 15° S to the extremes of the Oscillation, *Mon. Wea. Rev.* **113**, 2063–74.

van Loon, H. and D. J. Shea (1987), The Southern Oscillation, VI, anomalies of sea level pressure on the southern hemisphere and of Pacific sea surface temperature during the development of a warm event, *Mon. Wea. Rev.* **115**, 370–79.

Viehoff, T. (1991), Sea ice observations in the Weddell Sea. In: *Proceedings of the Fifth AVHRR Data Users Meeting*, EUMETSAT, Darmstadt, pp. 483–9.

Vincent, D. G. (1994), The South Pacific Convergence Zone (SPCZ): a review, *Mon. Wea. Rev.* **122**, 1949–70.

Voskresenskii, A. I. and K. I. Chukanin (1986), Main features of atmospheric circulation over Antarctica. In: *Climate of Antarctica*, Russian translations series, I. M. Dolgin, editor, A. A. Balkema, Rotterdam, pp. 178–85.

Wadhams, P. (1994), Sea ice thickness changes and their relation to climate. In: *The Polar Oceans and Their Role in Shaping the Global Environment*, O. M. Johannessen, R. D. Muench and J. E. Overland, editors, American Geophysical Union, Washington, DC, pp. 337–61.

Wadhams, P., C. B. Sear, D. R. Crane, M. A. Rowe, S. J. Morrison and D. W. S. Limbert (1989), Basin-scale ice motion and deformation in the Weddell Sea during winter, *Ann. Glaciol.* **12**, 178–86.

Wadhams, P., M. A. Lange and S. F. Ackley (1987), The ice thickness distribution across the Atlantic sector of the Antarctic Ocean in midwinter, *J. Geophys. Res.* **92**, 14535–52.

Wallace, J. M. and P. V. Hobbs (1977), *Atmospheric Science. An Introductory Survey*, Academic Press, New York, 467 pp.

Walsh, K. and J. McGregor (1993), Simulations of Antarctic climate using a limited area model. In: *Preprints, Fourth International Conference on Southern Hemisphere Meteorology and Oceanography*, Hobart, pp. 463–4.

Warren, G. R. (1994), *Sea Surface Temperature Analysis. NMC Operations Bulletin No. 25*, Bureau of Meteorology, Melbourne.

Warren, S. G. (1982), Optical properties of snow, *Rev. Geophys. Space Phys.* **20**, 67–89.

Warren, S. G., C. J. Hahn, J. London, R. M. Chervin and R. L. Jenne (1986), *Global Distribution of Total Cloud and Cloud Type Amounts Over Land*, Vol. NCAR Technical Note TN-273+STR/DOE Technical Report ER/60085-H1, NCAR, Boulder, 29 pp.

Warren, S. G., C. J. Hahn, J. London, R. M. Chervin and R. L. Jenne (1988), *Global Distribution of Total Cloud and Cloud Type Amounts Over the Ocean*, Vol. NCAR Technical Note TN-317+STR/DOE Technical Report ER/0406, NCAR, Boulder, 42 pp.

Warrick, R. and J. Oerlemans (1990), Sea Level Rise. In: *Climate Change. The IPCC Scientific Assessment*, J. T. Houghton, G. J. Jenkins and J. J. Ephraums, editors, Cambridge University Press, Cambridge, pp. 257–81.

Wattam, S. and J. Turner (1996), Weather forecasting for aviation and marine operations in the Antarctic Peninsula region, *Meteorol. Appl.* **2**, 323–32.

Weatherly, J. W., J. E. Walsh and H. J. Zwally (1991), Antarctic sea ice variations and seasonal air temperature relationships, *J. Geophys. Res.* **96**, 15119–30.

Webb, D. J., P. D. Killworth, A. C. Coward, and S. R. Thompson, (1991), *The FRAM Atlas of the Southern Ocean:* Natural Environment Research Council, Swindon.

Weller, G. (1968a), *The Heat Budget and Heat Transfer Processes in Antarctic Plateau Ice and Sea Ice. Australian National Antarctic Research Expedition (ANARE), Scientific Report 4A, No. 102*, Australian National Antarctic Research Expedition, Melbourne, 155 pp.

Weller, G. (1968b), Heat energy transfer through a four-layer system: air–snow–sea ice–water, *J. Geophys. Res.* **73**, 1209–20.

Weller, G. (1980), Spatial and temporal variations in the South Polar surface energy balance, *Mon. Wea. Rev.* **108**, 2006–14.

Wendler, G., J. C. André, P. Pettré, J. Gosink and T. Parish (1993), Katabatic winds in Adélie coast. In: *Antarctic Research Series, Volume 61: Antarctic Meteorology and Climatology: Studies Based on Automatic Weather Stations*, D. H. Bromwich and C. R. Stearns, editors, American Geophysical Union, Washington, pp. 23–46.

Wendler, G., N. Ishikawa and Y. Kodama (1988), The heat balance of the icy slope of Adélie Land, Eastern Antarctica, *J. Appl. Met.* **27**, 52–65.

White, F. D. J. and R. A. Bryson (1967), The radiative factor in the mean meridional circulation of the Antarctic atmosphere during the polar night. In: *WMO Technical Note*, No. 87: *Polar Meteorology. Proceedings of the WMO/SCAR/ICPM Symposium on Polar Meteorology, Geneva, 5–9 September 1966*, World Meteorological Organization, Geneva, pp. 199–224.

Woodruff, S. D., R. J. Slutz, R. L. Jenne, and P. M. Steurer (1987), A comprehensive ocean–atmosphere data set. *Bull. Amer. Met. Soc.* **68**, 1239–50.

Worby, A. P. and I. Allison (1991), Ocean–atmosphere energy exchange over thin, variable concentration Antarctic pack ice, *Ann. Glaciol.* **15**, 184–90.

World Meteorological Organization (1992), *WMO Manual on the Global Data Processing System, Volume II, Number 485*, World Meteorological Organization, Geneva.

Xu, J.-S., H. von Storch and H. van Loon (1990), The performance of four spectral GCMs in the Southern Hemisphere: the January and July climatologies and the semiannual wave, *J. Climate* **3**, 53–70.

Yamanouchi, T. and T. P. Charlock (1994), Radiative effects of clouds, ice sheet and sea ice in the Antarctic. In: *Snow and Ice Covers: Interactions with the Atmosphere and Ecosystems (Proceedings of Yokohama Symposia J2 and J5, July 1993). IAHS Publication no. 223*, H. G. Jones, T. D. Davies, A. Ohmura and E. M. Morris, editors, IAHS, Wallingford, pp. 29–34.

Yamanouchi, T., K. Suzuki and S. Kawaguchi (1987), Detection of clouds in Antarctica from infrared multispectral data of AVHRR, *J. Met. Soc. Jap.* **65**, 949–62.

Yasunari, T. and S. Kodama (1993), Intraseasonal variability of katabatic wind over East Antarctica and planetary flow regime in the Southern Hemisphere, *J. Geophys. Res.* **98**, 13063–70.

Young, M. V., G. A. Monk and K. A. Browning (1987), Interpretation of satellite imagery of a rapidly deepening cyclone, *Quart. J. Roy. Met. Soc.* **113**, 1089–115.

Zibordi, G. and G. P. Meloni (1991), Classification of Antarctic surfaces using AVHRR data: a multispectral approach, *Antarctic. Sci.* **3**, 333–8.

Zillman, J. W. (1967), The surface radiation balance in high southern latitudes. In: *Polar Meteorology. Proceedings of the WMO/SCAR Symposium on Polar Meteorology, Geneva, 5–9 September 1966*, Vol. WMO Technical Note No. 87, WMO, Geneva, pp. 142–71.

Zillman, J. W. (1972), Solar radiation and sea-air interaction south of Australia. In: *Antarctic Research Monograph No. 19*, American Geophysical Union, Washington, pp. 11–40.

Zwally, H. J. (1977), Microwave emissivity and accumulation rate of polar firn, *J. Glaciol.* **18**, 195–215.

Zwally, H. J. and P. Gloersen (1977), Passive microwave images of the polar regions and research applications, *Polar Rec.* **18**, 431–50.

Zwally, H. J., J. C. Comiso, C. L. Parkinson, W. J. Campbell, F. D. Carsey, and P. Gloersen (1983a), *Antarctic Sea Ice, 1973–1976*. Report NASA SP-459: Washington, DC, NASA, 206 pp.

Zwally, H. J., C. L. Parkinson and J. C. Comiso (1983b), Variability of Antarctic sea ice and changes in carbon dioxide, *Science* **220**, 1005–12.

Index